Sustainable Development Policy Directory

Sustainable Development Policy Directory

W. Alan Strong
School of Built Environment
University of Ulster
UK

and

Lesley A. Hemphill
School of the Built Environment
University of Ulster
UK

Blackwell
Publishing

Blackwell Publishing Ltd

Editorial offices:
Blackwell Publishing Ltd, 9600 Garsington Road, Oxford OX4 2DQ, UK
Tel: +44 (0)1865 776868
Blackwell Publishing Inc., 350 Main Street, Malden, MA 02148-5020, USA
Tel: +1 781 388 8250
Blackwell Publishing Asia Pty Ltd, 550 Swanston Street, Carlton, Victoria 3053, Australia
Tel: +61 (0)3 8359 1011

First published 2006 by Blackwell Publishing Ltd

ISBN-13: 978-1-4051-2150-7
ISBN-10: 1-4051-2150-5

Library of Congress Cataloging-in-Publication Data
Strong, W. Alan.
Sustainable development policy directory / W. Alan Strong, Lesley A. Hemphill.–1st ed.
p. cm.
Includes bibliographical references and index.
ISBN-13: 978-1-4051-2150-7 (alk. paper)
ISBN-10: 1-4051-2150-5 (alk. paper)
1. Sustainable development–Directories. 2. Environmental policy–Directories.
I. Hemphill, Lesley A. II. Title.

HC79.E5S773 2005
338.9′27′025–dc22

2005013102

A catalogue record for this title is available from the British Library

Set in 10/12.5pt Stone Sans
by Graphicraft Limited, Hong Kong
Printed and bound in India
by Replika Press Pvt, Ltd, Kundli

The publisher's policy is to use permanent paper from mills that operate a sustainable forestry policy, and which has been manufactured from pulp processed using acid-free and elementary chlorine-free practices. Furthermore, the publisher ensures that the text paper and cover board used have met acceptable environmental accreditation standards.

For further information on Blackwell Publishing, visit our website:
www.thatconstructionsite.com

Contents

CONTENTS

Foreword

One of the hardest things to do is to implement policy. That was the lesson of the current Government's first term. The path to on-the-ground realisation of the policy-maker's intentions is strewn with booby traps. The institutions or mechanisms for implementation may be bureaucratically paralysed, or may deflect a good intention so much that it has an unintended negative effect, or they may be absent altogether. People active on the policy's journey into reality may be short of the necessary knowledge and skills, or time and resources. The policy itself may be at fault, crafted in isolation in Whitehall or Brussels and horribly short of realism about how and where it is to be implemented.

This Directory means that a whole range of built environment professionals need never be caught out by the perversities of policy – in its formulation or on its journey into the real life of you and me. To have in one place the means of not only burrowing down into any one policy, but also of exploring the interconnections is a precious resource for any professional – 'sustainability literate' or not.

No less importantly, I am sure that even the shortest dip into this Directory will bring more understanding of both the challenges of sustainable development and the importance of its role in meeting challenges. If we are to meet our economic, social and environmental goals *at the same time*, rather than in damaging trade-off, as at present, then we will have to make radical improvements to the efficiency with which we use energy and raw materials, and learn to think about very different areas of policy in a 'joined up' way.

At every turning point of history – and we are certainly at one of those – engineers and scientists have been active and influential. Today, nobody who delivers our built environment can be excused from taking responsibility for the future. Armed with this Directory their contribution can only be made more effective.

Sara Parkin OBE
Forum for the Future

Preface

Sustainable development has gained international recognition as a global challenge and the one which can unify aspirations in pursuit of improved quality of life. Kofi Annan (United Nations, Director General) encapsulated this global drive at the World Conference on Sustainable Development in Johannesburg 2002 in stating, 'Sustainable development is the foremost development challenge facing the international community'. Sustainable development remains a complex subject area, requiring an in-depth knowledge of current acceptable practice, innovative approaches and a strong commitment at global and local levels.

The built environment provides the vehicle for communities and industry to develop and represents the product of urban planning and architectural processes within a defined spatial organisation in meeting the basic needs of shelter, food, safety and employment. Sustainable development provides the structure by which the built environment develops across the world and requires a range of innovative political and social attributes, as well as a raft of strategic and legislative instruments to guide and inform this process. It is anticipated that, as information becomes more precise on measuring the real consequences of man's actions on the environment and the planet as a whole, the means of tackling this damage will become more sophisticated and exact. However until such a point is reached the value of current and historic data and policy-based information cannot be overestimated.

The rationale for this book emerged from the quest to develop an enhanced understanding and knowledge of key current and historical sustainability legislative drivers exerting influence on the built environment as a whole. Discussions with the University of Ulster Visiting Professor in Sustainability Development, Dr Jim McQuaid, resulted in the embryonic idea of developing a policy-based directory of current and historic sustainability related policies. It could be used to help inform the readership on sustainability policy and action within the built environment across four distinct jurisdictions, including international, European, the United Kingdom and Northern Ireland. This Directory should help various built environment disciplines to broaden their knowledge of the impact of sustainability on their own sector and interaction with others.

The publication has been developed by the Sustainable Development Group (SDG) within the School of the Built Environment, University of Ulster, to meet a niche in the built environment market for a reference document on sustainable development related policy documents. The SDG forms part of the Royal Academy of Engineers 'Visiting Professor in Engineering Design for Sustainable Development' Scheme, which incorporates a network of 26 UK universities. This network provides the opportunity for academics to share and exchange knowledge on sustainable development for teaching and learning purposes in a number of engineering disciplines.

This Directory is designed as a desktop reference guide for local authorities, built environment professionals, environmental consultants, professional bodies, educationalists and students. It provides an integrated picture of sustainable development across thirteen built environment themed chapters which should help readers concerned with sustainability policy to broaden their knowledge on the key issues. The Directory is divided into chapters covering key built environment themes of biodiversity, climate change, construction, energy, environment, planning, pollution, social issues, sustainable development policy and practice, transport, urban development, waste management and water. Each chapter covers key elements and highlights the sustainability challenges for that theme, as well as listing the relevant sustainability-related policy documents alongside a brief summary and a URL web-based link for the four separate jurisdictions. Therefore the reader can obtain information on specific policy documents and follow the web-link via an internet browser to access a copy of the actual document. It is acknowledged that webpage addresses change but enough detail is provided for any references to the policy or document to be found through the common search engines.

The authors wish to acknowledge and thank a range of parties for their advice, encouragement and steadfast support throughout the compilation of this Directory, including Professor Alistair Adair (Head of School), staff and colleagues in the School of the Built Environment at the University of Ulster, the Royal Academy of Engineering, the University of Ulster Sustainability Visiting Panel, and especially the authors' respective families and friends.

W. Alan Strong BSc, C Eng, C Env, FICE, FCIWEM
Senior Lecturer in Environmental Engineering
Leader, Sustainable Development Group
University of Ulster

Dr Lesley A. Hemphill BSc, PhD
Project Officer, Sustainable Development Group
University of Ulster

1 Contextual Background

1.1 Sustainable Development Context

This chapter charts the progress of the sustainable development concept from its early origins in environmental policy through to its established position on the world's political stage. Reference is made to key publications, conferences and government strategies each aimed at translating the sustainability aspiration into a tangible, realisable goal.

1.1.1 Origins of Sustainable Development

Sustainable development is a relatively new concept, first developed in the 1980s to stress the need for the simultaneous achievement of development and environmental goals. However the origins of the concept can be traced back to the increased concern shown for the environment in the 1960s and 1970s. By this stage crisis thinking on the world's environmental problems became commonplace as doubts were raised about the potential of the world's resource base to sustain the established patterns of human activity. It is within this preservation and limitation context that early ideas about sustainable development emerged. Around this time three landmark publications brought into question the assumption that there were 'no practical limits' to economic growth, these were *Limits to Growth* (Meadows *et al.*, 1972), *A Blueprint for Survival* (The Ecologist, 1972) and *Small is Beautiful* (Schumacher, 1974). These publications brought into focus the familiar conflict of the environment versus economic growth, implying that mixing both agendas involved a trade-off in the performance of one or the other.

This theme was taken up at the United Nations (UN) Conference on the Human Environment held in Stockholm in 1972, which indicated that economic development, without proper regard to environmental constraints, was both wasteful and unsustainable. Whilst it is recognised that all growing economies face some deterioration in environmental quality, the modern sustainable development debate has tended to shift the focus away from growth versus the environment to a complementary approach. The UN Conference at Stockholm in 1972 and subsequent conferences in Bucharest (1974) and Vancouver (1976) helped paint a clearer picture of the type of response required from national governments to overcome environmental and development problems. These conferences helped to crystallise more than two decades of conservationist thinking and resulted in the publication of the *World Conservation Strategy* in 1980 (IUCN, 1980).

The *World Conservation Strategy* (WCS) signified the formal introduction of the term 'sustainable development', although it stressed sustainability in ecological terms with less concern for economic development. It was based upon three main objectives:

- to maintain essential ecological processes & life support systems;
- to preserve genetic diversity; and
- to ensure the sustainable use by us and our children of species and ecosystems.

However, of greater importance was the fact that the WCS indicated for the first time that development was a means to achieve conservation, instead of being viewed as an obstruction to it. Furthermore the strategy offered an 'intellectual framework and practical guidance' to government policy-makers to stimulate a focused approach to the management of resources and guidance on how this could be carried out (IUCN, 1980).

1.1.2 The WCED 'Brundtland' Report 1987

The World Commission on Environment and Development (WCED), an independent body set up in 1983 by the UN, produced a report entitled *Our Common Future* which re-examined the critical environmental and development problems in an attempt to formulate realistic proposals to solve them (WCED, 1987). The 'Brundtland' report, as it became known, authored by Dr. Gro Harlem Brundtland (Director General of the World Health Organisation), explicitly introduced the term 'sustainable development' as a conceptual bridge to bring together environmentalists and industrialists. A key part of the Brundtland report was the identification of six elements considered important in a successful sustainable development policy:

- Reviving economic growth
- Changing the quality of growth
- Meeting essential needs for jobs, food, energy, water and sanitation
- Ensuring a sustainable level of population
- Reorienting technology and managing risk
- Merging environment and economics in decision-making processes.

This report gave political support to sustainable development, resulting in many countries beginning to promote it as a desirable policy. It also had a significant effect in defining what is meant by sustainable development, providing the most instantly recognisable interpretation of the term – the Brundtland definition:

> *'development that meets the needs of the present without compromising the ability of future generations to meet their own needs.' (WCED, 1987)*

The report explained that this definition contains two key concepts. Firstly that of 'needs', in particular the essential needs of the world's poor, and secondly the idea of 'limitations' imposed by the state of technology and social organisation on the environment's ability to meet present and future needs (WCED, 1987).

1.1.3 Earth Summit I

The findings of the Brundtland report and the WCS increased the pressure for a new summit to follow up Stockholm, leading to the UN planning a Conference on Environment and Development to spell out how to achieve sustainable development. The Conference became known as the 'Earth Summit' and was held in Rio de Janeiro in 1992, drawing representatives from 178 countries (UNCED, 1992). A total of five important Agreements were reached at the conference, which together formed a framework for further action. These were:

- *The Rio Declaration on Environment and Development* – 27 principles guiding action on environment and development
- *Agenda 21* – a comprehensive action programme to help achieve a more sustainable pattern of development
- *A Statement of Principles* – to guide the management, conservation and sustainable development of all types of forest
- *A UN Framework Convention on Climate Change* – an agreement between countries for action to reduce the risks of global warming by limiting the emission of greenhouse gases
- *A Convention on Biological Diversity* – an agreement on how to protect the diversity of species and habitats in the world.

Agenda 21 proved to be one of the most significant outcomes of the Earth Summit, as it set out an action plan for achieving sustainable development in the twenty-first century. This document consisted of 40 chapters setting out policy areas dealing with social and economic issues, e.g. combating poverty, the protection of human health and the promotion of sustainable human settlements, the conservation and management of resources to allow for future development, the meeting of agricultural needs without destroying land, the protection of ecosystems, and the safeguarding of oceans and halting the spread of deserts (UNCED, 1992). The meeting and subsequent documents agreed at Rio were important because it produced an agreement in principle that nations would attempt to promote sustainable development. However as *Agenda 21* did not have any legal status to require governments and other groups to abide by it, its overall impact was somewhat diluted.

1.1.4 Local Agenda 21

Contained within *Agenda 21* was the introduction of what became known as 'Local Agenda 21' (LA21), which provided an important mechanism for identifying the actions needed to pursue sustainable development at the local level. *Agenda 21* stated:

> *'by 1996 [later extended to 2000] most local authorities in each country should have undertaken a consultative process with their populations and achieved a consensus on a Local Agenda 21 for the community.' (UNCED, 1992)*

This process of LA21 has come to mean two things, firstly a consultative process leading to wider involvement and empowerment, and secondly, implementing *Agenda 21* at a local level in such a way as to involve the community and other interested parties. In the UK, local authorities have been actively involved in developing their own LA21 policies as a response to that suggested in *Agenda 21*. This response has varied from an indicator-based approach highlighting the state of the local authority area and operations to detailed plans indicating the necessary targets and benchmarks required to meet sustainable development standards.

1.1.5 UK Response to Sustainable Development

An important recommendation of the Earth Summit was that individual countries should prepare strategies and action plans to help implement sustainable development, committing the world's nations to finding solutions to the problems associated with sustainability. The UK government has produced a number of reports in a direct response to this suggestion. The first of these UK sustainable development strategies was produced in 1994, building on material in the 1990 White Paper, *This Common Inheritance* (DOE, 1990). This first sustainable development strategy (DOE, 1994) was split into four key parts:

- considering the principles of sustainable development;
- reviewing the state of the UK environment by identifying trends and likely pressure points;
- reviewing the major sectors of the economy and identifying where the objectives of sustainability may require new policy responses; and
- finally discussing how action by central and local government, by business and industry, by voluntary bodies and by individuals could best help achieve sustainable development.

This strategy provided a very useful overview of how the UK could put sustainable development into practice whilst recognising the need for a change in attitudes as well as policy responses if the principles of sustainable development were to become more than aspirations.

In 1999 the UK Government produced a new sustainable development strategy, *A Better Quality of Life: A Strategy for Sustainable Development in the UK*, (DETR, 1999) following an extensive consultation phase (*Opportunities for Change* (DETR, 1998)). This document has conceptualised sustainable development and its progress within the UK. The new strategy was the first to make reference to the now commonly accepted four objectives of sustainable development:

- Social progress which recognises the needs of everyone
- Effective protection of the environment
- Prudent use of natural resources
- Maintenance of high and stable levels of economic growth and employment.

The strategy sets out the UK government's broad framework for making sustainable development a reality, identifying the following factors as being paramount in achieving this:

- Better decision-making – putting sustainable development at the heart of every government department's work and promoting the use of integrated impact assessment and appraisal tools in support of sustainability.
- Taxes, regulations and other policy instruments – e.g. fuel duty, levy on business use of energy, car tax reductions for small engine sizes, increased landfill tax charges.
- Information and involvement – improved awareness of sustainable development through the 'Are you doing your bit?' campaign and recommendations to the national curriculum review on sustainable development education from the Sustainable Development Education Panel.
- Research and advice – encouragement and support of research and development and the spread of best practice and networking.
- Advisory bodies – establishment of the Sustainable Development Commission to undertake an advisory role to the government on all matters relating to sustainable development.

As well as stating the government's priorities the strategy identified indicators as a key mechanism by which the UK could monitor and review progress. This culminated in the presentation of a list of 15 'headline' indicators that could be used as a national barometer of Sustainable Development (Table 1.1).

1.1.6 Earth Summit II

The year 2002 marked the 10-year anniversary of the Earth Summit held in Rio de Janeiro and therefore provided a milestone date to review progress towards achieving sustainable development. The United Nations decided that this could be best achieved by means of a follow-up Summit and hence 'Earth Summit II' or the World Summit on Sustainable Development (WSSD) was convened. Given the magnitude of problems associated with developing countries, Johannesburg, South Africa, was chosen as the ideal location to discuss and formulate new ideas on how to move the sustainable development agenda forward. The summit gained international support with 191 nations represented at Johannesburg.

The WSSD produced two key outputs:

- *The Johannesburg Declaration on Sustainable Development* – reaffirming the commitment of the world's nations to achieving sustainability.
- *A Plan of Implementation* – building on the achievements made since Earth Summit I and expediting the realisation of the remaining goals.

A wide range of concrete commitments and targets for action to achieve more effective implementation of sustainable development objectives were agreed to and reaffirmed by governments (detailed in the Johannesburg Plan of Implementation). Examples of some of the key themes and commitments discussed are presented below:

- *Poverty eradication* – halve, by the year 2015, the proportion of the world's people whose income is less than $1 a day and the proportion of people who suffer from hunger.
- *Water and sanitation* – halve, by the year 2015, the proportion of people who do not have access to basic sanitation.
- *Sustainable production and consumption* – encourage and promote the development of a 10-year framework of programmes to accelerate the shift towards sustainable consumption and production.
- *Energy* – programmes to diversify the energy supply, improve access to energy, remove market distortions, and establish domestic energy efficiency programmes.
- *Chemicals* – aim, by 2020, to use and produce chemicals in ways that do not lead to significant adverse effects on human health and the environment.
- *Management of the natural resource base* – programmes focusing on water, oceans and fisheries, the atmosphere, biodiversity, and forests.
- *Corporate responsibility* – actively promote corporate responsibility and accountability, including the full development and effective implementation of intergovernmental agreements and measures, international initiatives and public–private partnerships, and appropriate national regulations.
- *Health* – enhance health education with the objective of achieving improved health literacy on a global basis by 2010.

Table 1.1 UK Headline Indicators (*A Better Quality of Life*, 1999).

SD Pillar	Paragraph Reference	No.	Indicator
Maintaining high & stable levels of economic growth & employment	'Our economy must continue to grow' (3.12)	H1	Total output of the economy (GDP & GDP per head)
	'Investment (in modern plant machinery, as well as research & development) is vital to our future prosperity' (3.14)	H2	Total & social investment as a percentage of GDP
	'Maintain high & stable levels of employment so everyone can share greater job opportunities' (box after 1.8)	H3	Proportion of people of working age who are in work
Social progress which recognises the needs of everyone	'Tackling poverty & social exclusion' (7.17)	H4	Indicators of success in tackling poverty & social exclusion
	'Equip people with the skills to fulfil their potential' (3.16)	H5	Qualifications at age 19
	'Improve health of population overall' (3.17)	H6	Expected years of healthy life
	'Reduce the proportion of unfit housing stock' (3.18)	H7	Homes judged unfit to live in
	'Reduce both crime & fear of crime' (3.19)	H8	Level of crime
Effective protection of the environment	'Continue to reduce our emissions of greenhouse gases now, & plan for greater reductions in the longer term' (3.20)	H9	Emissions of greenhouse gases
	'Reduce air pollution & ensure air quality continues to improve through the longer term' (3.21)	H10	Days when air pollution is moderate or higher
	'Improve choice in transport; improve access to education, jobs, leisure & services; & reduce the need to travel' (7.29)	H11	Road traffic
	'Improving river quality' (8.30)	H12	Rivers of good or fair quality
	'Reverse the long-term decline in populations of farmland & woodland birds' (3.24)	H13	Populations of wild birds
	'Re-using previously developed land, in order to protect the countryside & encourage urban regeneration' (3.25)	H14	New homes built on previously developed land
Prudent use of natural resources	'Move away from disposal of waste, towards waste reduction, reuse, recycling & recovery' (box after 6.11)	H15	Waste arisings & management

- *Sustainable development of small-island developing states* – undertake initiatives by 2004 aimed at implementing the Global Programme of Action for the Protection of the Marine Environment from Land-based Activities to reduce, prevent and control waste and pollution, and their health-related impacts.
- *Sustainable Development for Africa* – improve sustainable agricultural productivity and food security in accordance with the UN Millennium Development Goals, in particular to halve by 2015 the proportion of people who suffer from hunger.
- *Institutional framework for sustainable development* – adopt new measures to strengthen institutional arrangements for sustainable development at international, regional and national levels.

Overall the Johannesburg Summit was considered to be successful in reaffirming sustainable development as a central element of the international agenda and giving new impetus to global action to fight poverty and protect the environment.

1.2 The Built Environment

The built environment represents the 'product' of urban planning and architectural processes, and of various construction activities that take place in a defined spatial organisation. These processes hope to meet the needs of communities, industries and governments, and require a knowledge and understanding of scientific and engineering principles. Buildings dominate the built environment and are supported by an infrastructure and utilities framework.

The adequacy of the built environment can be judged by its suitability in meeting people's basic needs such as shelter, food and employment. However, as nations develop, the provision is more readily assessed against a higher regime of criteria such as comfort, access, wealth and communications. In this regard the built environment impinges upon the natural environment to meet these needs of society, and is therefore at the boundary between environmental effects and human requirements.

1.2.1 The Built Environment Professions

Various professions become crucial to plan, control, regulate and implement the necessary structures needed within any built environment. These built environment professionals face the challenge of meeting the needs of people and the growing economy while attempting to maintain the nature, character and posterity of the natural environment. Table 1.2 highlights the key professions which collectively make up the built environment.

If a development was examined from start to finish it would primarily involve each of these professions at different stages of the building life-span as follows:

- *Stage 1: Design, Planning and Strategic Organisation* – Architect; Town Planner; Transport Planner; Civil and Environmental Engineer; Environmental Scientist; Quantity Surveyor.
- *Stage 2: Construction* – Builder; Civil Engineer; Environmental Engineer.
- *Stage 3: Post-Construction and Operations* – Chartered Surveyor; Housing Manager; Energy Manager; Environmental Health Officer.

No one profession is considered more important than the rest as it is the collective collaboration that helps contribute to a successful and ultimately sustainable built environment. Indeed the challenge for all built environment professionals is to contribute to inter-disciplinary teamwork to maximise the delivery of sustainable development.

1.2.2 The Built Environment and Sustainable Development

The present and future state of the built environment is the key to achieving the aspiration of sustainability in that it represents the best and worst opportunities for societies to survive and prosper in a stable and protected environment. Legislation can place restrictions on developments, international and national strategies can offer a basis for good practice, whilst professionals can apply scientific and engineering knowledge and expertise, all to provide an appropriate 'quality of life' for society. However there needs to be a comprehensive understanding of the collective issues for the built environment to properly align with the long-term delivery of sustainable development. The scope of the

Table 1.2 Built Environment Professions.

Architect	Builder	Chartered Surveyor
Civil Engineer	**BUILT ENVIRONMENT**	Energy Manager
Environmental Health Officer		Environmental Engineer
Environmental Scientist		Housing Manager
Quantity Surveyor	Town Planner	Transport Planner

Table 1.3 Scope of the Built Environment.

Triple Bottom Line	Built Environment Scope
Environment	Energy & Materials; Urban Form; Land Use & Pollution; Open Space; Climate Change; Biodiversity
Economics	Property Market; Labour Market; Household Income; Investment & Equity; Production & Distribution
Society	Security; Identity; Accessibility; Basic Needs (Water, Food, Shelter, Clothing)

built environment straddles a number of issues and disciplines that fit against the triple bottom line of sustainable development (see Table 1.3).

The built environment involves a whole life provision through planning, design, construction and maintenance of buildings and their support infrastructure. However, achieving and maintaining a sustainable built environment requires significant changes in behaviour patterns. The level at which materials, energy and space are used needs reduction so that buildings, settlements and transportation make fewer demands on resources and create less waste and pollution. The lack of a cross-disciplinary approach and integrated planning, in which there has been scant attention to the impact and needs of other professional activities, has resulted in unsustainable built environment practices:

- Land uses are dispersed and segregated, increasing need for travel.
- Transport infrastructure gives priority to the car over public transport.
- Tolerance of unsightly buildings and failure to recognise potential re-development opportunities leads to increased demand on 'green land'.
- Bad habits in domestic waste recycling leads to over-use of landfill sites and to potential pollution.
- Income groups are physically segregated making social cohesion more difficult.
- Inadequate energy management and auditing causes increased CO_2 emissions and global warming.
- Population growth adds further burden on energy, water, financial and habitat stocks.

Only through a concerted effort to influence the hearts and minds of society can these unsustainable built environment practices be reversed. This Directory considers sustainable development from a built environment perspective to enable due consideration to be made of how the diverse built environment disciplines can individually and collectively influence the achievement of sustainability. The importance placed on the built environment and developed areas to satisfy and fulfil the needs of both individuals and businesses will continue to increase apace. This emphasises the need to follow sustainable practices to ensure that the results of development pressures contribute positively to accomplishing a sustainable built environment.

1.2.3 Corporate and Individual Sustainability Responsibilities

Sustainability has wide-reaching implications for all sectors of society, including individuals and businesses, in addition to those for national and local governments. Indeed, if sustainability is to be successfully implemented, then the contribution made by individuals and businesses cannot be overstated. There is often a perceived conflict between the goals of the corporate society and individuals given the ever-increasing globalisation of world markets in which individual needs get misplaced by the quest of large corporations for increased productivity and profitability. A balance needs to be struck between corporate thinking and individual aspirations to secure sustainability.

Figure 1.1 highlights the differing agendas of corporate and individual societies viewed against the four UK sustainable development objectives. The figure shows that corporate sustainability is primarily focused on economic issues such as whole life costing and profit maximisation. In contrast individual sustainability is much more socially biased, concerned with issues such as health and quality of life. The key to achieving a balanced approach to sustainability is to ensure that both sectors recognise the important linkages across all four objectives, balancing primary sectoral aims against overall goals. Predominance in favour of any one objective would result in unsustainable practice.

CORPORATE MATTERS \ INDIVIDUAL MATTERS	A **Environmental protection**	B **Social progress**	C **Prudent use of natural resources**	D **Economic stability**
A **Environmental protection**		Public health	Environmental stewardship	Eco-purchasing
B **Social progress**	Corporate social responsibility		Behavioural change	Quality of life
C **Prudent use of natural resources**	Resource management	Sustainable employment		Value for money
D **Economic stability**	Whole life costing	Labour & community relations	Environmental management systems	

BALANCE

Figure 1.1 Balancing Corporate and Individual Sustainability Responsibilities.

1.3 Chapter Structure and Style

This Directory is structured around 14 chapters. Chapter 1 provides a brief introduction to the sustainable development topic area including details on its origin, importance to the built environment and information to help readers navigate their way through the book. Chapters 2–14 introduce individual built environment themes in alphabetical order, these being biodiversity, climate change, construction, energy, environment (general), planning, pollution, social issues, sustainable development policy and practice, transport, urban development, waste management and water. Each of these themed chapters is structured into two main parts. Part one provides a contextual overview of the theme, its key elements and information on the sustainability challenges facing each theme, whilst part two highlights the actual policy tables with an accompanying index to help find specific documents quickly.

Although each chapter is a stand-alone account of key sustainable development related documents there is a clear overlap in some of the issues addressed. For example climate change has close links to pollution, energy, transport, planning, water, construction, waste management, urban development, social issues and sustainable development policy and practice (see Figure 1.2). As a result of these inter-relationships some of the documents listed under climate change will also appear in these other chapters. Figure 1.2 encapsulates some of the main inter-relationships between the chapters and how these themes interact within the context of the built environment. It is important to emphasise that the goal of sustainability requires that themes should not be investigated in isolation and only with an integrated built environment based approach can society fully contribute to sustainable development.

1.4 Directory Explanatory Notes

This Directory provides a guide to the key policy documents produced in recent years on sustainable development. These key policy documents are structured into 13 themed chapters to enable the reader to go directly to the specific area of interest. In order to help aid the understanding of the reader the following points of clarification have been included:

- *Policy language* – This Directory presents policy documents from four separate levels of jurisdiction notably International, European, United Kingdom and Northern Ireland. Each chapter presents a series of policy tables which run sequentially from the highest (International) to the lowest jurisdiction (Northern Ireland). All themes have policy documentation from each level of the hierarchy

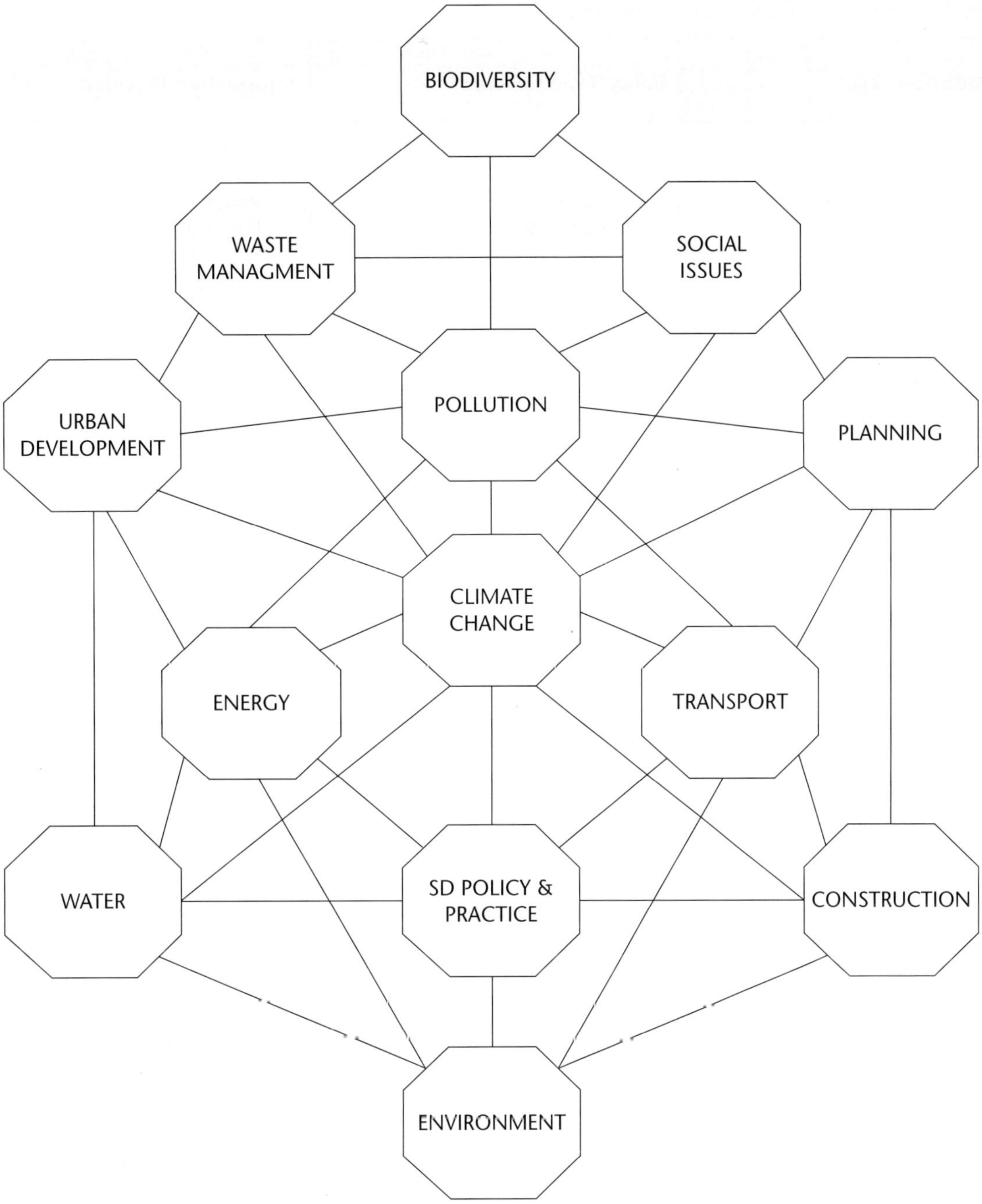

Figure 1.2 Inter-Relationships between Directory Chapters.

with the exception of construction and planning which do not have any international coverage. Within each jurisdiction there are different types of document overviewed with the breakdown of document type and information provided in Table 1.4.

- *Hierarchical nature of tables* – The four jurisdictions highlighted allow the policy tables to be presented in hierarchical tiers of importance. It therefore follows that policy documents found in the International jurisdiction will have dealt with global issues and actions necessary to contribute to sustainability on a global basis. Similarly, documents presented in the Northern Ireland jurisdiction only have implications for Northern Ireland citizens. In this regard it also follows that documents of global significance such as the Earth Summit, or documents at the level of European Directives will have an influence on policy responses at all the subsequent tiers of jurisdiction and should result in initiatives or policy documents aimed at meeting the objectives set in the higher level documents.

Table 1.4 Overview of Directory Structure.

Jurisdiction Level	Policy Type Covered	Information Provided
INTERNATIONAL	■ Conferences ■ Conventions ■ Protocols ■ Agreements	■ Policy type ■ Year ■ Venue ■ Purpose ■ Key outcome(s) ■ Website
EUROPEAN	■ Directives ■ Decisions ■ Communication ■ Recommendations ■ Miscellaneous	■ Policy type ■ Programme title ■ Policy objective ■ Community measure and date ■ Document reference ■ Website
UNITED KINGDOM	■ Strategies & reports ■ Consultation papers ■ Guidance Notes	■ Policy type ■ Document title and date ■ Objective ■ Contents ■ Website
NORTHERN IRELAND	■ Strategies & reports ■ Consultation papers ■ Guidance Notes	■ Policy type ■ Document title and date ■ Objective ■ Contents ■ Website

- *Chronological order* – Each of the policy tables is presented in chronological order to aid navigation. However it is important to note that some older landmark documents have also been included as well as the key documents published up to the end of December 2004. In some cases older documents have been superseded by more recent publications so it is important to sequentially search through the theme index to ensure the most up-to-date document is being examined.
- *United Kingdom and Northern Ireland jurisdictions* – Both these jurisdictions have been included as examples of how the International and European level sustainable development policy related documents filter down through to the local level. Broadly speaking the UK will follow the policy lead taken at European level with the devolved NI administration following the UK approach after an anticipated time-lag.
- *Legal status of documents* – It is important to realise that the policy documents presented in this Directory can have differing legal status. Some of the documents, once ratified, are mandatory and require those countries to put in place legally binding instruments to comply with their content. Some of the other documents represent guidance notes or good practice which provide governments with policy direction but do not require legal compliance.

1.5 Webpage Caveat

This Directory is designed to provide a brief overview on the purpose of the policy documents contained in the subsequent tables. The URL web-links provide a direct route to the actual policy documents and the detail contained in them. A concerted effort has been made by the authors to ensure that the webpage references included were active up to the end of December 2004. However despite these best efforts some webpage addresses change and documents can be removed or archived making the URL link inactive. If entering the URL address does not turn up the desired text, it may mean one of the following:

- the host webpage has been changed;
- the document has been superseded by a new version;

- the document has been archived; or
- the document is no longer in force.

In order to establish which of the above reasons apply it is possible to enter the full name of the document including the year it was published into one of the web-based search engines to indicate if a new webpage is revealed.

References

DETR (1998) *Opportunities for Change – A Consultation Document.* London, HMSO.
DETR (1999) *Sustainable Development Strategy for the UK: A Better Quality of Life.* London, HMSO.
DOE (1990) *This Common Inheritance.* London, HMSO.
DOE (1994) *Sustainable Development: The UK Strategy.* London, HMSO.
IUCN (1980) *World Conservation Strategy.* Gland, IUCN.
Meadows, D.H., Meadows, D.L., Randers, J., Behrens, W.W. (1972) *Limits to Growth.* New York, New American Library.
Schumacher, E.F. (1974) *Small is Beautiful: A Study of Economics as if People Mattered.* London, Abacus.
The Ecologist (1972) *Blueprint for Survival.* Harmondsworth, Penguin Books Ltd.
UNCED (1992) *United Nations Council for Economic Development – Earth Summit 1992.* London, Regency Press.
WCED (1987) *Our Common Future.* Oxford, Oxford University Press.

2 Biodiversity

2.1 Biodiversity and the Built Environment

Biodiversity is the abbreviated term for biological diversity, which describes the variety and variability of genes, species, populations and natural ecosystems. It refers to the number of species that can be found in nature, but there is no definitive species count. Predictions of the purported number of species lie at approximately 5 to 30 million, of which only 1.4 million have been properly categorised and defined. Against this uncertainty of species number and description is evidence of an ongoing loss of biodiversity, arising from the built environment and due primarily to growth of human settlements and exploitation of natural resources through habitat destruction, excessive harvesting, surface pollution and the inappropriate introduction of foreign plants and animals.

The United Nations Convention on Biological Diversity in 1992 was a landmark document as it set the international objectives for conservation of biological diversity, for sustainable use of its components and for the fair and equitable sharing of the benefits arising out of the utilization of genetic resources. The European Council Decision 93/626/EEC dealt with the conclusion of this Rio de Janeiro Convention on Biological Diversity by anticipating, preventing and attacking the causes of significant reduction or loss of biological diversity at source. This is because of its intrinsic value and its ecological, genetic, social, economic, scientific, educational, cultural, recreational and aesthetic values. The Decision concludes by promoting international, regional and global co-operation among states, intergovernmental organisations and the non-governmental sector.

The balance between biodiversity preservation and growth through the built environment remains crucial, requiring monitoring and reporting. The UK Biodiversity Action Plan in 1996 established a planned approach with clear biological targets for 391 species and 45 habitats, with monitoring and reporting to assess progress, evaluate priorities and address emerging issues. *Working with the Grain of Nature: A Biodiversity Strategy for England in 2002* built upon the foundations already laid, but recognised that biodiversity must be mainstreamed into all activities – working through policies on agriculture, water, fisheries, woodland management and urban areas in order to balance the impact of building activities. The strategy also set out a programme for the subsequent five years for the main policy sectors, so as to make the changes necessary to conserve, enhance and 'work with the grain' of nature and ecosystems rather than against them.

2.2 Key Biodiversity Elements

Biological diversity is classified into three levels of activity – genetic diversity, species diversity and ecosystem diversity. Preservation at these levels has ramifications for areas beyond environmental monitoring and impact to economic and social factors.

Genetic diversification is the sum of genetic information contained in the genes of individual plants, animals and micro-organisms. This is the domain of botanists, ecologists, environmentalists and scientists. Studies of genetic diversity engage with soils, oceans, wildlife, forests, woodlands and water in pursuit of a full understanding of the life of flora and fauna. *Species diversity* follows on from the genetic information and refers to the number of different living organisms on earth, while *ecosystem diversity*, also called habitat diversity, relates to the variety of habitats within an ecosystem.

The greatest species diversity is located in rainforests, coral reefs, geologically ancient lakes and coastal wetlands. Rainforests contain up to half of the total species count while covering only 7% of the earth's land surface. As only a limited amount of these rainforests have been properly explored, the actual count on the species residing there is likely to be significantly higher. These biological resources are a measure of the earth's natural environment or capital.

Species engage in many complex interactions with living organisms, mainly in soils, over a variety of timescales thereby subscribing to the production of organisms for food, pharmaceuticals and fibres, and to the health of the ecosystem. In addition, species are recognised as an indicator of economic or business capital development, being seen as a limit to increased prosperity. Historically, economic development has been limited by the availability of things such as labour, energy and finance, but the key resource issue is currently switching to this decline in living and biodiversity systems.

Biodiversity is critically important in sustaining food production, especially in light of the increasing rate of growth in world population. In addition, various animal and plant species are the source of medicines to cure a range of known diseases and could hold the key for medical research and development into diseases presently considered incurable. Rainforests play an important role in the planetary recycling of carbon, nitrogen and oxygen. They help to regulate the greenhouse effect by absorbing carbon dioxide out of the air and returning oxygen to the atmosphere, thereby influencing the growth capacity of the ecosystem. Finally, the planet is an interwoven ecosystem, showing that the existence of one species is important to that of another, so that the demise of one species may eventually lead to the loss of many others, leading to accelerated loss of important genetic information.

2.3 Sustainability Challenges for Biodiversity

Biodiversity is classically and primarily included in the environmental sector of study but contributes to all four objectives of Sustainable Development by its role in both natural and economic capitals. Areas of biological diversity link with each of the Sustainable Development Objectives below:

- *Effective protection of the environment* is achieved through biological and ecosystem support structures, and appropriate countryside stewardship. Protection of endangered species, wildlife and wildbird populations are all linked with built environment activities, whereas desertification also contributes to species depletion. Re-use of brown-field sites, pesticide control and minimisation, and land conservation can facilitate support for native species and natural habitats.
- *Social progress which recognises the needs of everyone* involves human health and quality of life, and these can be greatly enhanced by the development of pharmaceutical drugs, sourced from sound biodiversity management. Sensitive and balanced land-use planning along with the use of many aspects of biodiversity in landscaped developments make valuable inputs to good physical health and mental well-being, assisting social and human experience.
- *Prudent use of natural resources* is linked to the cause and effect of catastrophic events of all types. Inappropriate natural resource usage can lead to fish, bird and mammal extinctions, hedgerow depletion and land contamination. Humanity's role in stewardship for and as custodians of the earth is inconsistently practised and requires local product resourcing, maintenance of diverse ecology, waste minimisation and environment-friendly building specifications.
- *Maintenance of high and stable levels of economic growth and employment* arises through good crop and food productivity, controlled use of genetically modified food, strong agriculture and textile yields, innovative processes, appropriate land use planning and commitment to greening on the part of governments, and a whole life costing approach to all product and process systems. However the threat of increased globalisation of product and consumption markets undermines the proximity principle of sustainable development.

Several issues and challenges exist for environmentalists, economists, decision-makers and strategists in protecting and preserving the fundamental elements of biodiversity and directly allied areas, and this sector is being seen with equal importance to the other sustainable development themes. The following issues arise from literature, web pages, strategy reviews and learned opinion, and are given to set the Theme and Policy reference sections below in context.

- *Protection of species* and minimisation of species extinction is essential for maintenance of food chains and ecosystems, and species security for future generations.
- *Trade in endangered species*, including animals under threat of extinction such as elephants, rhinos, tigers, whales and tortoises, and the experimentation on or killing of animals for skins, horns, shells and organs, is strictly illegal and, whilst it may realise lucrative financial gains for poachers, it cannot be tolerated if irreversible biological diversity damage is to be avoided.
- *Robust international biodiversity plans* are needed to conserve natural habitats, whilst fully recognising the millions of years formation period, and the unique nature of some wildlife and habitats.
- *Desertification, land degradation and vegetation loss* can lead to human famine, social deprivation and reduced quality of life, with the critical removal of energy-yielding green cells. Woodland area maintenance can mitigate the effects of deforestation, but could also cause an increase in carbon dioxide and global warming, as a result of a loss in hedgerows. This also has further implications for wildlife in terms of loss of habitat and intensified rural development that depletes soil nutrients.
- *Use of genetically modified organisms and foods* has huge bio-safety implications and imposes a level of financial uncertainty that requires a whole life costing approach.
- *Use of man-made chemicals*, such as pesticides, which can infiltrate local waterways through surface run-off causing eutrophication and thereby affecting life's natural balance.
- *Biodiversity and natural heritage management* and maintenance are complemented through the designation of Areas of Natural Beauty, Areas of Special Scientific Interest and support for National Parks.
- *Development and enhancement of stronger international links* with key biodiversity organisations such as World Wildlife Fund, Friends of the Earth, the Royal Society for Protection of Birds and the Royal Society for the Prevention of Cruelty to Animals are required on the larger scale to give status and vision to the global biodiversity issues.

2.4 Biodiversity Theme Index

Jurisdiction	Tier	Title	Date	Page No.
INTERNATIONAL	Conferences	1st Meeting of The Conference of the Parties to the Convention on Biological Diversity (COP-1)	1994	19
		2nd Meeting of The Conference of the Parties to the Convention on Biological Diversity (COP-2)	1995	19
		3rd Meeting of The Conference of the Parties to the Convention on Biological Diversity (COP-3)	1996	19
		4th Meeting of The Conference of the Parties to the Convention on Biological Diversity (COP-4)	1998	19
		5th Meeting of The Conference of the Parties to the Convention on Biological Diversity (COP-5)	2000	20
		6th Meeting of The Conference of the Parties to the Convention on Biological Diversity (COP-6)	2002	20
		7th Meeting of The Conference of the Parties to the Convention on Biological Diversity (COP-7)	2004	20
	Conventions	Convention on Wetlands of International Importance (Ramsar)	1971	21
		Convention on International Trade in Endangered Species of Wild Fauna & Flora (CITES)	1973	21

Jurisdiction	Tier	Title	Date	Page No.
INTERNATIONAL	Conventions	Convention on the Conservation of Migratory Species (Bonn)	1979	22
		Convention on the Conservation of European Wildlife & Natural Habitats (Bern)	1979	22
		UN Convention on Biological Diversity (UNCED)	1992	23
		Convention to Combat Desertification	1994	23
	Protocols	Protocol on Environmental Protection to the Antarctic Treaty (Madrid Protocol)	1991	23
		Cartagena Protocol on Bio-safety	2000	24
	Agreements	Agreement on the Conservation of Bats in Europe (EUROBATS)	1991	24
		Agreement on the Conservation of Small Cetaceans in the Baltic & North Sea (ASCOBANS)	1992	25
		Agreement on the Conservation of African-Eurasian Migratory Water Birds (AEWA)	1995	25
		Agreement on Straddling & Highly Migratory Fish Stocks	1995	26
EUROPEAN	Council Directives	Conservation of Wild Birds (Birds Directive) *(CD 79/409/EEC Amended by CD 81/854/EEC; Com Dir 85/411/EEC; CD 86/122/EEC; CD 90/656/EEC; Com Dir 91/244/EEC; CD 94/24/EC; Com Dir 97/49/EC)*	1979 1981 1985 1986 1990 1991 1994 1997	27
		Importation of skins of certain seal pups & products derived therefrom *(CD 83/129/EEC Amended by CD 89/370/EEC)*	1983 1989	27
		Protection of Animals used for Experimental Purposes *(CD 86/609/EEC)*	1986	27
		Contained use of Genetically Modified Micro-organisms *(CD 90/219/EEC Amended by Com Dir 94/51/EC; Com Dir 98/81/EC)*	1990 1994 1998	28
		The Conservation of Natural Habitats & of Wild Fauna & Flora (Habitats Directive) *(CD 92/43/EEC Amended by CD 97/62/EC)*	1992 1997	28
		The Keeping of Wild Animals in Zoos *(CD 1999/22/EC)*	1999	28
		The Deliberate Release into the Environment of Genetically Modified Organisms *(Dir 2001/18/EC)*	2001	28
	Council Decisions	Conservation of Antarctic Marine Living Resources (Canberra Convention) *(C Dec 81/691/EEC)*	1981	29
		Bern Convention on the Conservation of the European Wildlife & Natural Habitats *(C Dec 82/72/EEC)*	1982	29
		Bonn Convention on the Conservation of Migratory Species of Wild Animals *(C Dec 82/461/EEC)*	1982	29
		The Rio de Janeiro Convention on Biological Diversity *(C Dec 93/626/EEC)*	1993	29
		International Dolphin Conservation Programme *(C Dec 1999/337/EC)*	1999	29

Jurisdiction	Tier	Title	Date	Page No.
EUROPEAN	Council Communication	Community Biodiversity Strategy *(COM (98) 42)*	1998	30
		Forests & Development: the EC Approach *(COM (1999) 554)*	1999	30
		Biodiversity Action Plan for the Conservation of Natural Resources *(COM (2001) 162)*	2001	30
		Biodiversity Action Plan for Agriculture *(COM (2001) 162)*	2001	31
		Biodiversity Action Plan for Fisheries *(COM (2001) 162)*	2001	31
		Biodiversity Action Plan for Economic & Development Co-operation *(COM (2001) 162)*	2001	31
		Towards a Thematic Strategy on the Sustainable use of Pesticides *(COM (2002) 349)*	2002	31
	Miscellaneous	Cetacean Protection *(C Reg No. 348/81)*	1981	32
		Pan-European Biological & Landscape Diversity Strategy (PEBLDS)	1996	32
		Endangered Species of Wild Fauna & Flora *(C Reg No. 338/97 Amended by C Reg No. 938/97; C Reg No. 2307/97; C Reg No. 2214/98; C Reg No. 1476/99; C Reg No. 2724/2000; C Reg No. 1579/2001; C Reg No. 2476/2001)*	1997 1998 1999 2000 2001	33
		Novel Foods & Novel Food Ingredients (GMO) *(Reg (EC) No. 258/97)*	1997	33
		Conservation & Sustainable Management of Forests in Developing Countries *(Reg No. 2492/2000)*	2000	33
		Transboundary Movement of Genetically Modified Organisms (GMOs) *(COM (2002) 85)*	2002	34
		Interpretation Manual of European Union Habitats	2003	34
		Natura 2000 and Forests 'Challenges & Opportunities' – Interpretation Guide	2003	35
		The Implementation of Directive 92/43/EEC on the Conservation of Natural Habitats & of Wild Fauna & Flora *(COM (2003) 845)*	2003	36
		LIFE-Nature: Communicating with Stakeholders & the General Public – *Best Practice Examples for Natura 2000*	2004	36
UK	Strategies & Reports	Setting the Agenda for Biodiversity Research – Monitoring & Indicators	2000	37
		Our Countryside: The Future – A Fair Deal for Rural England (Rural White Paper)	2000	37
		Countryside Survey 2000 – Accounting for Nature: Assessing Habitats in the UK Countryside	2000	38
		What is Sustainability – Biodiversity, the Environment & SD: A Summary Guide for Companies	2000	38
		Implementing EU Biodiversity Strategy: A UK Case Study	2000	39
		Sustaining the Variety of Life: 5 years of the UK Biodiversity Action Plan	2001	39
		1st Report by the UK under Article 17 on the Implementation of Directive 92/43/EEC on the Conservation of Natural Habitats & of Wild Fauna & Flora	2001	40

Jurisdiction	Tier	Title	Date	Page No.
UK	Strategies & Reports	Literature Review of the Implications of Climate Change for Species, Habitats & the wider UK Countryside	2001	40
		Science in Action for Biodiversity: A Report of the Biodiversity Working Group 1998–2001	2001	41
		Rural White Paper Implementation Plan	2002	41
		UK Indicators for Sustainable Forestry	2002	41
		Working with the Grain of Nature: A Biodiversity Strategy for England	2002	42
		Directive on the Conservation of Wild Birds: 79/409/EEC – 7th Report by the UK under Article 12: Implementation of National Provisions 1999–2001	2002	42
		UK Biodiversity Action Plan – tracking progress – results of 2002 reporting	2003	43
		Countryside Information for Policy – The Lessons from CS 2000	2003	43
		Measuring the progress of the Biodiversity Strategy for England: Baseline Assessment	2003	43
		Review of Marine Nature Conservation – Working Group Report to Government	2004	44
		The Darwin Initiative – 7th Annual Report: Protecting Biological Diversity Around the World	2004	44
		Rural Strategy 2004	2004	44
	Guidance Notes	Roots & Routes: Guidelines on Highways Works & Trees	1999	45
		Tree Preservation Orders: A Guide to the Law & Good Practice	2000	45
		Managing Coastal Activities: A Guide for Local Authorities	2004	46
NORTHERN IRELAND	Strategies & Reports	NI Biodiversity Strategy Proposals	1999	47
		Biodiversity in NI – Recommendations to Govt for a Biodiversity Strategy	2000	47
		The NI Countryside Survey 2000 – Summary Report on Broad Habitats	2000	47
		Partners in Protection – A Review of the Procedures for Notifying, Protection & Managing ASSIs in NI	2001	48
		Habitat Change in the NI Countryside – Technical Report to the NI Countryside Survey 2000	2002	48
		The NI Biodiversity Strategy	2002	49
		The Habitats Regulations – A Guide for Competent Authorities	2002	49
		Recent Changes in the Distribution of Red Squirrels in Northern Ireland	2003	49
		NI Habitat Action Plan – Blanket Bog Final Draft	2003	50
		Northern Ireland Habitat Action Plan – Lowland Heathland Final Draft	2003	50
		Northern Ireland Habitat Action Plan – Lowland Raised Bog Final Draft	2003	50
		Northern Ireland Habitat Action Plan – Maerl Beds Final Draft	2003	50

Jurisdiction	Tier	Title	Date	Page No.
NORTHERN IRELAND	Strategies & Reports	Northern Ireland Habitat Action Plan – Montane Heath Final Draft	2003	51
		Northern Ireland Habitat Action Plan – Mudflats Final Draft	2003	51
		Northern Ireland Habitat Action Plan – Saline Lagoons Final Draft	2003	51
		Northern Ireland Habitat Action Plan – Seagrass Beds Final Draft	2003	51
		Northern Ireland Habitat Action Plan – Sheltered Muddy Gravels Final Draft	2003	51
		Northern Ireland Habitat Action Plan – Species-Rich Hedgerows Final Draft	2003	51
		Northern Ireland Habitat Action Plan – Tidal Rapids Final Draft	2003	52
		Northern Ireland Habitat Action Plan – Upland Heath Final Draft	2003	52
		Causeway Coast AONB Summary Management Plan	2003	52
		Natural Heritage Strategic Plan 2003	2003	52
		A Framework for Biodiversity Publicity & Promotion in Northern Ireland	2004	53
		Northern Ireland Hare Survey 2004	2004	53
	Consultation Papers	ASSIs in NI – Consultation on Proposals for their Future Management & Protection	2003	53
		Giant's Causeway & Causeway Coast World Heritage Site – Management Plan Public Consultation Draft	2004	54
		National Parks & other Protected Landscape Areas – A Discussion of Options for Establishing National Parks & Managing Other Outstanding Landscapes in Northern Ireland	2004	54
	Guidance Notes	Guidelines for the Selection of Biological ASSIs in Northern Ireland	1999	55
		Nature in Your Neighbourhood: A Guide to Improving the Natural Environment & Biodiversity Where You Live	2004	55

2.5 Biodiversity Policy Tables

Table 2.5.1 International Level Biodiversity Policy.

Title	Venue	Purpose	Key Outcome(s)	Website
1st Meeting of The Conference of the Parties to the Convention on Biological Diversity (COP-1)	Nassau, Bahamas (1994)	At its first meeting, the Conference of the Parties decided on a medium-term programme of work for the period 1995–1997. Implementation of this programme has laid the groundwork for the long-term implementation of the Convention itself. In particular it has seen the development of a number of thematic work programmes, and identified a series of key cross-cutting issues relevant to all work programmes.	COP-1 produced 33 Documents and 13 Decisions. The main themes addressed included: ▪ Guidance to the financial mechanism; ▪ Medium-term programme of work.	http://www.biodiv.org/doc/decisions/COP-01-dec-en.pdf http://www.biodiv.org/doc/meetings/cop/cop-01/official/cop-01-17-en.pdf
2nd Meeting of The Conference of the Parties to the Convention on Biological Diversity (COP-2)	Jakarta, Indonesia (1995)	The second meeting of the Conference of the Parties investigated further implementation measures and explored in more depth the Global Environment Facility.	COP-2 produced 39 documents and 23 Decisions. The main themes addressed included: ▪ Marine & coastal biological diversity; ▪ Access to genetic resources; ▪ Conservation & sustainable use of biological diversity; ▪ Bio-safety.	http://www.biodiv.org/doc/decisions/COP-02-dec-en.pdf http://www.biodiv.org/doc/meetings/cop/cop-02/official/cop-02-19-en.pdf
3rd Meeting of The Conference of the Parties to the Convention on Biological Diversity (COP-3)	Buenos Aires, Argentina (1996)	The third meeting of the Convention provided an opportunity for an analysis of its achievements, but also for far-reaching self-criticism and for the identification of the even bolder actions that were required.	COP-3 produced 112 documents and 27 Decisions. The main themes addressed included: ▪ Agricultural biodiversity; ▪ Financial resources and mechanism; ▪ Identification, monitoring and assessment; ▪ Intellectual property rights.	http://www.biodiv.org/doc/decisions/COP-03-dec-en.pdf http://www.biodiv.org/doc/meetings/cop/cop-03/official/cop-03-38-en.pdf
4th Meeting of The Conference of the Parties to the Convention on Biological Diversity (COP-4)	Bratislava, Slovak Republic (1998)	The fourth meeting of the Conference of the Parties established a programme of work to cover the period from then until the seventh meeting and, more importantly, established a process to review the operations of the Convention and set out a longer term programme of work.	COP-4 produced 71 documents and 19 Decisions. The main themes addressed included: ▪ Inland water ecosystems; ▪ Review of the operations of the Convention; ▪ Article 8(j) and related issues (traditional knowledge); ▪ Benefit sharing.	http://www.biodiv.org/doc/decisions/COP-04-dec-en.pdf http://www.biodiv.org/doc/meetings/cop/cop-04/official/cop-04-27-en.pdf

Continued

Table 2.5.1 International Level Biodiversity Policy. *Continued*

Title	Venue	Purpose	Key Outcome(s)	Website
5th Meeting of The Conference of the Parties to the Convention on Biological Diversity (COP-5)	Nairobi, Kenya (2000)	This meeting saw the adoption of a number of reports including reports on the status of the biosafety protocol; the global environment facility; and access & benefit sharing. The meeting also focused on reviewing the implementation of the programme of work including the key mechanisms for implementation.	COP-5 produced 81 documents and 29 Decisions. The main themes addressed included: ▪ Dryland, Mediterranean, arid, semi-arid, grassland and savannah ecosystems; ▪ Sustainable use, including tourism; ▪ Access to genetic resources.	http://www.biodiv.org/doc/decisions/COP-05-dec-en.pdf http://www.biodiv.org/doc/meetings/cop/cop-05/official/cop-05-23-en.pdf
6th Meeting of The Conference of the Parties to the Convention on Biological Diversity (COP-6)	The Hague, Netherlands (2002)	The 6th Conference of the Parties sent a clear signal to the World Summit on Sustainable Development that conservation & the sustainable use of biodiversity are essential for achieving sus. dev. It was necessary to move from policy development & dialogue to action, & there was a shift in emphasis from the conservation of biodiversity to its sustainable use. The Conference discussed & agreed draft guidelines on access to & benefit-sharing of genetic resources, since an agreement on that subject was crucial to the process of sus. dev. Another important theme was forests, because they were rich in biodiversity & provided a wide range of services & products to many groups of people.	COP-6 produced 114 documents and 32 Decisions. The main themes addressed included: ▪ Forest ecosystems; ▪ Alien species; ▪ Benefit-sharing; ▪ Strategic plan 2002–2010.	http://www.biodiv.org/doc/decisions/COP-06-dec-en.pdf http://www.biodiv.org/doc/meetings/cop/cop-06/official/cop-06-20-en.pdf
7th Meeting of The Conference of the Parties to the Convention on Biological Diversity (COP-7)	Kuala Lumpur, Malaysia (2004)	At the 7th meeting the Conference of the Parties was invited to consider a framework for setting goals, targets & indicators & their application to the elaborated programmes of work. The framework should be viewed as being flexible, enabling the setting of national targets & indicators, which would be critical in the pursuit of the 2010 target.	COP-7 produced 91 documents and 36 Decisions. The main themes addressed included: ▪ Mountain ecosystems; ▪ Protected areas; ▪ Transfer of technology and technology cooperation.	http://www.biodiv.org/doc/decisions/COP-07-dec-en.pdf http://www.biodiv.org/doc/meetings/cop/cop-07/official/cop-07-21-part1-en.pdf

Convention on Wetlands of International Importance (Ramsar)	Ramsar, Iran (1971)	An intergovernmental treaty which provides the framework for national action and international co-operation for the conservation and wise use of wetlands and their resources.	The Conference adopted eleven Recommendations which are submitted to the Governments and other Bodies concerned for consideration and appropriate action. These Recommendations are appended as Annex II to this Final Act and relate to: conservation of the Wadden Sea, north western Europe; conservation of Thjorsarver, Iceland; conservation of North Bull Island, Ireland; conservation of Lakes Ab-i-Istada and Dasht-e-Nawar, Afghanistan; conservation of the Medway Estuary, England; oil pollution, pesticides; promotion of wetlands research; promotion of hunting research and education; African wetlands; wetlands and the Man and the Biosphere Programme.	http://www.ramsar.org/
Convention on International Trade in Endangered Species of Wild Fauna & Flora (CITES)	Washington, D.C. USA (1973)	An international agreement between Governments, with the aim of ensuring that international trade in specimens of wild animals and plants does not threaten their survival.	Convention introduced certain controls on the international trade in specimens of selected species. These require that all import, export, re-export and introduction from the sea of species covered by the Convention has to be authorised through a licensing system. The species covered by CITES are listed in three appendices, according to the degree of protection they need. ▪ Appendix I includes species threatened with extinction. Trade in specimens of these species is permitted only in exceptional circumstances. ▪ Appendix II includes species not necessarily threatened with extinction, but in which trade must be controlled in order to avoid utilisation incompatible with their survival. ▪ Appendix III contains species that are protected in at least one country, which has asked other CITES Parties for assistance in controlling the trade.	http://www.cites.org/eng/disc/text.shtml#texttop

Continued

Table 2.5.1 International Level Biodiversity Policy. *Continued*

Title	Venue	Purpose	Key Outcome(s)	Website
Convention on the Conservation of Migratory Species (Bonn Convention)	Bonn, Germany (1979)	The Parties: ▪ should promote, co-operate in and support research relating to migratory species; ▪ shall endeavour to provide immediate protection for migratory species included in Appendix I; and ▪ shall endeavour to conclude Agreements covering the conservation and management of migratory species included in Appendix II.	Appendix I lists the migratory species which are endangered. Parties that are Range States of a migratory species listed in Appendix I shall endeavour to: ▪ Conserve and, where feasible and appropriate, restore those habitats of the species which are of importance in removing the species from danger of extinction; ▪ Prevent, remove, compensate for or minimise, as appropriate the adverse effects of activities or obstacles that seriously impede or prevent the migration of the species; ▪ Prevent, reduce or control factors that are endangering or are likely to further endanger the species, including strictly controlling the introduction of, or controlling or eliminating, already introduced exotic species. Appendix II lists migratory species which have an unfavourable conservation status & which require international agreements for their conservation and management, as well as those which have a conservation status which would significantly benefit from the international co-operation that could be achieved by an international agreement.	http://www.cms.int/documents/index.htm
Convention on the Conservation of European Wildlife & Natural Habitats (Bern Convention)	Bern, Switzerland (1979)	To conserve wild flora and fauna and their natural habitats. To promote co-operation between states. To give particular emphasis to endangered and vulnerable species, including endangered and vulnerable migratory species.	The Contracting Parties have undertaken, to protect the habitats of wild flora & fauna & to give special attention to the conservation of wild flora & fauna species listed in Appendices I & II as well as to the protection of the fauna species listed in Appendix III. The Appendices to the convention concern the following: ▪ Appendix I – strictly protected flora species ▪ Appendix II – strictly protected fauna species ▪ Appendix III – protected fauna species ▪ Appendix IV – prohibited means & methods of killing, capture & exploitation.	http://conventions.coe.int/treaty/en/Treaties/Html/104.htm

			The Contracting Parties undertake to submit reports on their conservation policies every four years & to report to the standing committee every two years on any exceptions made under Article 9 of the Convention.	
UN Convention on Biological Diversity (UNCED)	Rio de Janeiro, Brazil (1992)	The objectives of this Convention, to be pursued in accordance with its relevant provisions, are the conservation of biological diversity, the sustainable use of its components & the fair & equitable sharing of the benefits arising out of the utilization of genetic resources. This includes by appropriate access to genetic resources & appropriate transfer of relevant technologies, taking into account all rights over those resources & to those technologies; & by appropriate funding.	The Convention sets out commitments for maintaining the world's ecological underpinnings as we go about the business of economic development. The Convention establishes three main goals: the conservation of biological diversity, the sustainable use of its components, and the fair and equitable sharing of the benefits from the use of genetic resources.	http://www.biodiv.org/convention/articles.asp
Convention to Combat Desertification	Paris, France (1994)	The objective of this Convention is to combat desertification & mitigate the effects of drought in countries experiencing serious drought and/or desertification, particularly in Africa, through effective action at all levels, supported by international co-operation & partnership arrangements, in the framework of an integrated approach which is consistent with Agenda 21, with a view to contributing to the achievement of sustainable development in affected areas.	The Contents of the Convention Document are: ■ Introduction on convention purpose ■ General provisions; Action programmes, scientific & technical co-operation & supporting measures; Institutions; Procedures; Final provisions ■ Annex I – Regional implementation annex for Africa ■ Annex II – Regional implementation annex for Asia ■ Annex III – Regional implementation annex for Latin America & the Caribbean ■ Annex IV – Regional implementation annex for the Northern Mediterranean ■ Annex V – Regional implementation annex for Central & Eastern Europe.	http://www.unccd.int/convention/text/convention.php
Protocol on Environmental Protection to the Antarctic Treaty (Madrid Protocol)	Madrid, Spain (1991)	The Parties commit themselves to the comprehensive protection of the Antarctic environment and dependent and associated ecosystems and hereby designate Antarctica as a natural reserve, devoted to peace and science.	Activities in the Antarctic Treaty area shall be planned & conducted so as to limit the adverse impacts on the Antarctic environment & dependent/associated ecosystems. The establishment of the Committee for Environmental Protection to provide advice & formulate recommendations to the parties in connection with the implementation of this Protocol.	http://www-old.aad.gov.au/information/treaty/protocoltxt.asp

Continued

Table 2.5.1 International Level Biodiversity Policy. *Continued*

Title	Venue	Purpose	Key Outcome(s)	Website
			A set of assessment procedures (Annex 1) for the prior assessment of the impacts of certain activities on the Antarctic environment (EIA) and other Annexes covering the following: ▪ Annex II – Conservation of Antarctic fauna & flora ▪ Annex III – Waste Disposal & Waste Management ▪ Annex IV – Prevention of Marine pollution ▪ Annex V – Area Protection & Management.	
Cartagena Protocol on Bio-safety	Cartagena, Columbia (2000)	In accordance with the precautionary approach contained in Principle 15 of the Rio Declaration on Environment and Development, the objective of this Protocol is to contribute to ensuring an adequate level of protection in the field of the safe transfer, handling and use of living genetically modified organisms (GMOs) resulting from modern biotechnology that may have adverse effects on the conservation and sustainable use of biological diversity, taking also into account risks to human health, and specifically focusing on transboundary movements.	The protocol sets out a comprehensive regulatory system of rules & procedures for ensuring the safe transfer, handling & use of GMOs, subject to transboundary movement.	http://www.biodiv.org/biosafety/default.aspx
Agreement on the Conservation of Bats in Europe (EUROBATS)	Bonn, Germany (1991)	Aimed to address threats to 37 bat species in Europe arising from habitat degradation, disturbance of roosting sites and harmful pesticides.	Parties to the Agreement agree to work through legislation, education, conservation measures and international co-operation towards the conservation of bats in Europe. The Parties fundamental obligations are to: ▪ prohibit the deliberate capture, keeping or killing of bats except under permit from its competent authority ▪ identify sites within its jurisdiction that are important to the conservation of bats and protect these sites from damage or disturbance ▪ promote public awareness	http://www.eurobats.org

			■ promote research programmes relating to conservation and management of bats ■ consider the potential effects of pesticides on bats, when assessing pesticides for use, and endeavour to replace timber treatment chemicals which are highly toxic to bats with safer alternatives.	
Agreement on the Conservation of Small Cetaceans in the Baltic & North Sea (ASCOBANS)	New York, USA (1992)	The aim of the Agreement is to promote close cooperation amongst Parties with a view to achieving and maintaining a favourable conservation status for small cetaceans.	The Parties undertake to cooperate closely in order to achieve & maintain a favourable conservation status for small cetaceans. A Conservation and Management Plan forming part of the Agreement obliges Parties to engage in habitat conservation & management, surveys & research, pollution mitigation & public information.	http://www.ascobans.org/index0101.html
Agreement on the Conservation of African-Eurasian Migratory Water Birds (AEWA)	The Hague, The Netherlands (1995)	The Agreement provides for co-ordinated and concerted actions to be taken by the Range States throughout the migration systems of the water birds to which it applies.	The AEWA covers 172 species of birds ecologically dependent on wetlands for at least part of their annual cycle, including many species of pelicans, storks, flamingos, ducks, waders, terns, gulls and geese. Parties to the Agreement are called upon to engage in a wide range of conservation actions which are described in a comprehensive Action Plan (2000 – onward). This detailed plan addresses such key issues as: species and habitat conservation, management of human activities, research and monitoring, education and information, and implementation. Parties shall take co-ordinated measures to maintain migratory water bird species in a favourable conservation status or to restore them to such a status. To this end they shall apply within the limits of their national jurisdiction the measures prescribed in Article III, together with the specific actions determined in the Action Plan provided for in Article IV, of this Agreement.	http://www.unep-aewa.org/eng/agree/agree_full.htm

Continued

Table 2.5.1 International Level Biodiversity Policy. *Continued*

Title	Venue	Purpose	Key Outcome(s)	Website
Agreement on Straddling & Highly Migratory Fish Stocks	New York, USA (1995)	The objective of this Agreement is to ensure the long-term conservation and sustainable use of straddling fish stocks and highly migratory fish stocks through effective implementation of the relevant provisions of the Convention.	The Agreement elaborates on the fundamental principle, established in the Convention that States should co-operate to ensure conservation & promote the objective of the optimum utilisation of fisheries resources within & beyond the exclusive econ. zone. It promotes good order in the oceans through the effective management & conservation of high seas resources by among other things: ■ Establishing, detailed minimum international standards for the conservation & management of straddling fish stocks & highly migratory fish stocks; ■ Ensuring that measures taken for the conservation & management of those stocks in areas under national jurisdiction & in the adjacent high seas are compatible & coherent; ■ Ensuring that there are effective mechanisms for compliance & enforcement of those measures on the high seas; and ■ Recognising the special requirements of developing States in relation to conservation & management as well as the development & participation in fisheries for the two types of stocks mentioned above.	http://www.un.org/Depts/los/convention_agreements/convention_overview_fish_stocks.htm

Table 2.5.2 European Level Biodiversity Policy.

Policy type	Programme Title	Policy Objective	Community Measure	Document Reference	Website
Council Directives	Conservation of Wild Birds	The long-term protection & conservation of all bird species naturally living in the wild within the European territory of the Member States (except Greenland).	▪ CD 79/409/EEC on the conservation of wild birds (02/04/79) ▪ Amended by CD 81/854/EEC (19/10/81) ▪ Comm Dir 85/411/EEC (25/07/85) ▪ CD 86/122/EEC (08/04/86) ▪ CD 90/656/EEC (04/12/90) ▪ Comm Dir 91/244/EEC (06/03/91) ▪ CD 94/24/EC (08/06/94) ▪ EU Council Dec amending the instruments on the accession of new Member States to the EU (01/01/95) ▪ Comm Dir 97/49/EC (29/07/97)	▪ Off Jnl L103, 25/04/79 ▪ Off Jnl L319, 07/11/81 ▪ Off Jnl L233, 30/08/85 ▪ Off Jnl L100, 16/04/86 ▪ Off Jnl L353, 17/12/00 ▪ Off Jnl L115, 08/05/91 ▪ Off Jnl L164, 30/06/94 ▪ Off Jnl L1, 01/01/95 ▪ Off Jnl L223, 13/08/97	http://europa.eu.int/scadplus/leg/en/lvb/l28046.htm
	Importation of Certain Seal Pups & Derived Products	To protect certain seal species by introducing Community legislation designed to protect seal pups against non-traditional hunting.	▪ CD 83/129/EEC concerning the importation into the Member States of skins of certain seal pups & derived products (28/03/83) ▪ Amended by CD 89/370/EEC (08/06/89)	▪ CD 83/129/EEC ▪ Off Jnl L91, 09/04/83 ▪ Off Jnl L163, 14/06/89	http://europa.eu.int/scadplus/leg/en/lvb/l28039.htm
	Protection of Animals used for Experimental Purposes	To protect animals used for experimental or scientific purposes by ensuring that such animals are adequately cared for & that no unnecessary pain or suffering is inflicted.	▪ CD 86/609/EEC on the approximation of the laws, regulations & administrative provisions of the Member States regarding the protection of animals used for experimental & other scientific purposes (24/11/86)	▪ CD 86/609/EEC ▪ Off Jnl L358, 18/12/86	http://europa.eu.int/scadplus/leg/en/lvb/l28104.htm

Continued

Table 2.5.2 European Level Biodiversity Policy. *Continued*

Policy type	Programme Title	Policy Objective	Community Measure	Document Reference	Website
Council Directives	Contained use of Genetically Modified Micro-organisms	To lay down common measures for the contained use of genetically modified micro-organisms for the purposes of protecting human health & the environment.	▪ CD 90/219/EEC on the contained use of genetically modified micro-organisms (23/04/90) ▪ Amended by Comm Dir 94/51/EC (07/11/94) ▪ Comm Dir 98/81/EC (26/10/98)	▪ COM (2001) 263 Final ▪ Off Jnl L117, 08/05/90 ▪ Off Jnl L297, 18/11/94 ▪ Off Jnl L330, 05/12/98	http://europa.eu.int/scadplus/leg/en/lvb/l21157.htm
	Natural Habitats (Natura 2000)	To maintain biodiversity by conserving natural habitats & wild flora & fauna in the European territory of the Member States.	▪ CD 92/43/EEC on the conservation of natural habitats & of wild fauna & flora (21/05/92) ▪ Amended by CD 97/62/EC (27/10/97)	▪ CD 92/43/EEC ▪ Off Jnl L206, 22/07/92 ▪ Off Jnl L305, 08/11/97	http://europa.eu.int/scadplus/leg/en/lvb/l28076.htm
	The Keeping of Wild Animals in Zoos	To adopt minimum standards for housing & caring for animals in zoos & to reinforce the role of zoos in conserving biodiversity.	▪ CD 1999/22/EC on the keeping of wild animals in zoos (29/03/99)	▪ CD 1999/22/EC ▪ Off Jnl L94, 09/04/99	http://europa.eu.int/scadplus/leg/en/lvb/l28069.htm
	The Deliberate Release into the Environment of Genetically Modified Organisms	In accordance with the precautionary principle, the objective of this Directive is to approximate the laws, regulations & administrative provisions of the Member States & to protect human health & the environment when: ▪ carrying out the deliberate release into the environment of genetically modified organisms for any other purposes than placing on the market within the Community, ▪ placing on the market genetically modified organisms as or in products within the Community.	▪ Directive 2001/18/EC of the European Parliament & the Council on the deliberate release into the environment of genetically modified organisms (12/03/01) ▪ Repealing CD 90/220/EEC	▪ Off Jnl L106/1 17/04/01	http://europa.eu.int/eur-lex/pri/en/oj/dat/2001/l_106/l_10620010417en00010038.pdf

Council Decisions	Conservation of Antarctic Marine Living Resources (Canberra Convention)	To protect the ecological balance in the Antarctic Ocean by means of international regulations on the management & conservation of its marine resources.	Council Decision 81/691/EEC on the conclusion of the Convention on the conservation of Antarctic marine living resources (04/09/81)	Off Jnl L252, 05/09/81	http://europa.eu.int/scadplus/leg/en/lvb/l28103.htm
	Bern Convention on the Conservation of the European Wildlife & Natural Habitats	To ensure the conservation of European wildlife & natural habitats by means of co-operation between States.	Council Decision 82/72/EEC concerning the conclusion of the Bern Convention on the conservation of European wildlife & natural habitats (03/12/81)	Off Jnl L38, 10/02/82	http://europa.eu.int/scadplus/leg/en/lvb/l28050.htm
	Bonn Convention on the Conservation of Migratory Species of Wild Animals	To develop international co-operation with a view to the conservation of migratory species of wild animals.	Council Decision 82/461/EEC on the conclusion of the Bonn Convention on the conservation of migratory species of wild animals (24/06/82)	Off Jnl L210, 19/07/82	http://europa.eu.int/scadplus/leg/en/lvb/l28051.htm
	The Rio de Janeiro Convention on Biological Diversity	Anticipating, preventing & attacking the causes of significant reduction or loss of biological diversity at source because of its intrinsic value & because of its ecological, genetic, social, economic, scientific, educational, cultural, recreational & aesthetic value. Promoting international, regional & global co-operation among states & inter-governmental organisations & the non-governmental sector.	Council Decision 93/626/EEC concerning the conclusion of the Convention on Biological Diversity (25/10/93)	Off Jnl L309, 13/12/93	http://europa.eu.int/scadplus/leg/en/lvb/l28102.htm
	International Dolphin Conservation Programme	To help reduce incidental dolphin mortality during tuna fishing.	Council Decision 1999/337/EC on the signature by the EC to the agreement on the international dolphin conservation programme (26/04/99)	Off Jnl L132, 27/05/99	http://europa.eu.int/scadplus/leg/en/lvb/l28083.htm

Continued

Table 2.5.2 European Level Biodiversity Policy. *Continued*

Policy type	Programme Title	Policy Objective	Community Measure	Document Reference	Website
Council Communications	Community Biodiversity Strategy	The Community Biodiversity Strategy set out in this Communication provides the framework for developing Community policies & instruments in order to comply with the CBD. The 1st Report from the EC to the Conference of the Parties of the CBD provides a summary assessment of the importance & status of biodiversity in the EU. It also provides a comprehensive overview of on-going & planned Community initiatives & instruments relevant to achieve the objectives of the CBD. This report therefore constitutes an important background document for the strategy.	Communication of the European Commission to the Council and the European Parliament on a European Community Biodiversity Strategy (04/02/98)	COM (98) 42	http://europa.eu.int/comm/environment/docum/9842en.pdf
	Forests & Development: the EC Approach	To define the objectives of the EC with regard to co-operation on the development of forestry, to identify areas where dialogue & assistance are required & set out the action planned to realise the objectives taking into account the experience gained in recent years.	Communication from the Commission to the Council & European Parliament on the EC approach to forests & development (04/11/99)	COM (1999) 554 Final	http://europa.eu.int/scadplus/leg/en/lvb/l28105.htm
	Biodiversity Action Plan for the Conservation of Natural Resources	Improving or maintaining the status of wild flora & fauna & their ecosystems & habitats. Building on & complementing existing Community environmental legislation & initiatives & making maximum use of these instruments in order to put the objectives set out in the Community biodiversity strategy into practice.	Commission Communication to the Council & European Parliament, on a biodiversity action plan for the conservation of natural resources (Vol. II) (27/03/01)	COM (2001) 162 Final	http://europa.eu.int/scadplus/leg/en/lvb/l28023.htm

Biodiversity Action Plan for Agriculture	Establishing an action plan to improve or maintain biodiversity status & prevent further biodiversity loss due to agricultural activities.	▪ Commission Communication to the Council & European Parliament on a biodiversity action plan for agriculture (Vol. III) (27/03/01)	▪ COM (2001) 162 Final	http://europa.eu.int/scadplus/leg/en/lvb/l28024.htm
Biodiversity Action Plan for Fisheries	Establishing an action plan to improve or maintain biodiversity status & preventing biodiversity loss due to fisheries & aqua-culture activities.	▪ Commission Communication to the Council & European Parliament on a biodiversity action plan for fisheries (Vol. IV) (27/03/01)	▪ COM (2001) 162 Final	http://europa.eu.int/scadplus/leg/en/lvb/l28025.htm
Biodiversity Action Plan for Economic & Development Co-operation	To reverse the loss of biodiversity & environmental resources in developing countries as well as reducing poverty.	▪ Communication from the Commission to the Council & European Parliament, on a biodiversity action plan for economic & development co-operation (Vol. V) (27/03/01)	▪ COM (2001) 162 Final	http://europa.eu.int/scadplus/leg/en/lvb/l28037.htm
Towards a Thematic Strategy on the Sustainable use of Pesticides	This Communication represents a major step in the preparation of the Thematic Strategy on the Sustainable Use of Pesticides. The main objectives of the thematic strategy are: ▪ to minimise the hazards & risks to health and environment from the use of pesticides; ▪ to improve controls on the use and distribution of pesticides; ▪ to reduce the levels of harmful active substances, in particular by replacing the most dangerous by safer (including non-chemical) alternatives; ▪ to encourage the use of low-input or pesticide-free crop farming; ▪ to establish a transparent system for reporting & monitoring progress including the development of appropriate indicators.	▪ Communication from the Commission to the Council, the European Parliament & the Economic & Social Committee – Towards a Thematic Strategy on the Sustainable use of Pesticides (01/07/02)	▪ COM (2002) 349	http://europa.eu.int/eur-lex/en/com/pdf/2002/com2002_0349en01.pdf

Continued

Table 2.5.2 European Level Biodiversity Policy. *Continued*

Policy type	Programme Title	Policy Objective	Community Measure	Document Reference	Website
Miscellaneous	Cetacean Protection	To adopt measures restricting imports into the Community of cetacean products.	▪ Council Reg (EEC) No. 348/81 on common roles for imports of whale & other cetacean products (20/01/81)	▪ Off Jnl L39, 12/08/81	http://europa.eu.int/scadplus/leg/en/lvb/l28038.htm
	Pan-European Biological & Landscape Diversity Strategy (PEBLDS)	The Pan-European Biological & Landscape Diversity Strategy is a European response to support implementation of the Convention on Biological Diversity. The Strategy was proposed in the Maastricht Declaration *Conserving Europe's Natural Heritage*, & builds on the Bern Convention, the European Conservation Strategy, the Dobrís & Lucerne Ministerial Conferences, UNCED, & other existing initiatives & programmes. The Strategy aims to strengthen the application of the Bern Convention in relation to the Convention on Biological Diversity, following the Monaco Declaration. The Strategy introduces a co-ordinating & unifying framework for strengthening & building on existing initiatives. Furthermore, the Strategy seeks to more effectively integrate ecological considerations into all relevant socio-economic sectors, & will increase public participation in, & awareness & acceptance of, conservation interests.	▪ Pan-European Strategy (1996)	▪ Nature & Environment No. 74 (Council of Europe)	http://www.strategyguide.org/fulltext.html

Endangered Species of Wild Fauna & Flora	To protect endangered species of wild fauna & flora & to guarantee their conservation by regulating trade therein.	▪ Council Regulation (EC) No. 338/97 on the protection of species of wild fauna & flora by regulating trade therein (09/12/96) ▪ Amended by: – Comm Reg (EC) No. 938/97 – Comm Reg (EC) No. 2307/97 – Comm Reg (EC) No. 2214/98 – Comm Reg (EC) No. 1476/1999 – Comm Reg (EC) No. 2724/2000 – Comm Reg (EC) No. 1579/2001 – Comm Reg (EC) No. 2476/2001	▪ Off Jnl L61, 03/03/97 ▪ Off Jnl L140, 30/05/97 ▪ Off Jnl L325, 27/11/97 ▪ Off Jnl L279, 16/10/98 ▪ Off Jnl L171, 07/07/99 ▪ Off Jnl L320, 18/12/00 ▪ Off Jnl L209, 02/08/01 ▪ Off Jnl L334, 18/12/01	http://europa.eu.int/scadplus/leg/en/lvb/l11023.htm
Novel Foods & Novel Food Ingredients (GMO)	To authorise the placing of novel foods & novel food ingredients on the market within the Community while taking account of requirements regarding public health, the environment & consumer information.	▪ Reg (EC) No. 258/97 of the European Parliament & the Council concerning novel foods & novel food ingredients (27/01/97)	▪ Off Jnl L43, 14/02/97	http://europa.eu.int/scadplus/leg/en/lvb/l21119.htm
Conservation & Sustainable Management of Forests in Developing Countries	To contribute to the conservation & sustainable management of forests in developing countries, in order that the latter may meet the economic, social & environmental demands placed upon forests.	▪ Reg (EC) No. 2492/2000 of the European Parliament & the Council on measures to promote the conservation & sustainable management of tropical forests & other forests in developing countries (07/11/00)	▪ Off Jnl L288, 15/11/00	http://europa.eu.int/scadplus/leg/en/lvb/l28087.htm

Continued

Table 2.5.2 European Level Biodiversity Policy. *Continued*

Policy type	Programme Title	Policy Objective	Community Measure	Document Reference	Website
Miscellaneous	Transboundary Movement of Genetically Modified Organisms (GMOs)	To implement the provisions of the Cartagena protocol on preventing biotechnological risks. The aim of the protocol is to ensure an adequate level of protection for the transfer, handling & use of GMOs that may have adverse effects on the environment & human health, & specifically focusing on transboundary movements.	▪ Proposal for a Regulation of the European Parliament & the Council on transboundary movement of GMOs (1985)	▪ COM (2002) 85 Final	http://europa.eu.int/scadplus/leg/en/lvb/l28119.htm
	Interpretation Manual of European Union Habitats	Having in mind all the difficulties of classification, the Scientific Working Group, set up by the Habitats Committee (established by Directive 92/43/EEC), expressed in May 1992 the need to prepare a manual for the interpretation of Annex I. Following a call for proposals the Commission charged Professor Thanghe from the Université Libre de Bruxelles to prepare a draft manual.	▪ Report from the Directorate-General Environment on the Interpretation Manual of EU Habitats. (April 2003)	▪ DG Env Report – EUR 25	http://europa.eu.int/comm/environment/nature/nature_conservation/eu_enlargement/2004/pdf/habitats_im_en.pdf

Natura 2000 and Forests 'Challenges & Opportunities' – Interpretation Guide	This document was developed with extensive stakeholder consultation & is designed to give citizens a better understanding of Community legislation on conservation of our forests. It makes it clear that Natura 2000 is not opposed to economic activity in the forestry sector. It explains how sites are designated, & the practical consequences of designation. As another example of the EC openness & transparency policy, it recommends identifying the measures required to maintain biodiversity through discussion with stakeholders, & expressing the outcome of this process as formal management objectives. A separate section gives numerous examples of creative approaches to implementing Natura 2000 in forests & of the successful combination of forestry with nature conservation objectives.	Report from Nature & Biodiversity Unit of the EC Directorate-General Environment on Natura 2000 & Forests (2003)	DG Env Report	http://europa.eu.int/comm/environment/nature/nature_conservation/useful_info/documents_publications/pdf/n2kforest_en.pdf

Continued

Table 2.5.2 European Level Biodiversity Policy. *Continued*

Policy type	Programme Title	Policy Objective	Community Measure	Document Reference	Website
Miscellaneous	The Implementation of Directive 92/43/EEC on the Conservation of Natural Habitats & of Wild Fauna & Flora	This report provides a comprehensive overview of progress to date. While there have been serious delays in implementation, the recent progress provides ground for optimism with regard to the habitat protection requirements of the Directive. The situation concerning species protection measures is less positive and the report indicates clearly the need for higher priority in this area.	▪ Report from the Commission on the implementation of Directive 92/43/EEC on the conservation of natural habitats & of wild fauna & flora (05/01/04)	▪ COM (2003) 845	http://europa.eu.int/eur-lex/en/com/rpt/2003/com2003_0845en01.pdf
	LIFE-Nature: Communicating with Stakeholders & the General Public – *Best Practice Examples for Natura 2000*	The purpose of this report is threefold: ▪ to provide an overview of the key issues relating to communicating on Natura 2000. ▪ to provide a series of practical LIFE-Nature examples of successful communication techniques used by LIFE projects to communicate with different interest groups. ▪ to offer some basic guidelines and practical advice on how to communicate effectively on Natura 2000 with different audiences, based on the LIFE experiences.	▪ Report from the Directorate-General Environment on the Best Practice Examples of the Natura 2000 programme (2004)	▪ DG Env Report	http://europa.eu.int/comm/environment/nature/nature_conservation/natura_2000_network/communicating_natura_2000/pdf/comn2k_life.pdf

Table 2.5.3 UK Level Biodiversity Policy.

Policy Type	Document Title	Objective	Contents	Website
Strategies & Reports	Setting the Agenda for Biodiversity Research – Monitoring & Indicators (Oct 2000)	This was the 5th & final workshop in the series held as part of the Biodiversity Research Support Services contract managed by DETR on behalf of the UK Biodiversity Research Working Group (BRWG). The overall aims of the project were: ▪ To identify and prioritise cross-cutting research needed to support Government commitments under the UK Biodiversity Acton Plan (BAP). ▪ To facilitate the exchange of information on biodiversity research ▪ To establish links & collaboration with related activities ▪ To raise the profile of biodiversity issues among research sponsors & the scientific community.	▪ Section 1: Introduction; Section 2: Background; Section 3: Outcomes from themed sessions; Section 4: Workshop synthesis; Appendices.	http://www.ukbap.org.uk/Library/brwg05.pdf
	Our Countryside: The Future – A fair Deal for Rural England (Rural White Paper) (28/11/00)	The countryside is important to all of us. Town & country are interdependent & the needs of both have to be addressed together. But there are special problems in rural areas which require a direct response & that is the focus of this paper. We want to help build prosperous, sustainable and inclusive rural communities. We are addressing the needs of all who live & work in rural areas. We will maintain & improve basic local services whilst helping farmers & businesses adapt to change & support new enterprise. We will give local communities the chance to shape their own future.	Foreword; 10 ways we will make a difference; Intro; A new standard for rural services; Vital village services; Modern rural services; An affordable home; Flexible local transport; Market towns & a thriving local economy; A new future for farming; Conserving & enhancing our countryside; Restoring & maintaining wildlife diversity & the natural environment; Increasing enjoyment of the countryside; Local power for country towns & villages; Thinking rural; Reviewing progress; Conclusion; Annex: Helping rural business to succeed – sources of funding/advice.	http://www.defra.gov.uk/rural/ruralwp/whitepaper/default.htm

Continued

Table 2.5.3 UK Level Biodiversity Policy. *Continued*

Policy Type	Document Title	Objective	Contents	Website
Strategies & Reports	Countryside Survey 2000 – Accounting for Nature: Assessing Habitats in the UK Countryside (29/11/00)	Countryside Survey 2000 & the Northern Ireland Countryside Survey 2000 have been designed to provide detailed information about the habitats & landscape features that are important elements of our countryside. They can tell us about the 'stock' of these resources, that is how much of them we have & where they are to be found, & they can give us an insight into their condition based on the variety & abundance of the plant species associated with them. Using information from previous surveys, we can also gain an understanding of how the stock & condition of these habitats & landscape features are changing over time. We can build up a sort of balance sheet or an account of natural assets in the UK countryside. In this report we look in particular at the period between the last two surveys, 1990 & 1998.	Foreword; Executive summary; Chapter 1: Countryside survey; Chapter 2: The national picture; Chapter 3: Enclosed farmland – arable, horticultural, improved & neutral grassland; Chapter 4: Boundary & linear features; Chapter 5: Woodlands; Chapter 6: Mountain, moor, heath & down; Chapter 7: Rivers, streams & standing waters; Chapter 8: Developed land in rural areas; Chapter 9: Countryside change; Chapter 10: Countryside survey – where next?; Appendix A: Broad habitat definitions.	http://www.defra.gov.uk/wildlife-countryside/cs2000/index.htm
	What is Sustainability – Biodiversity, the Environment & SD (Sustainable Development): A Summary Guide for Companies (Dec 2000)	Common themes and organisational approaches are emerging with reference to the environmental aspect of sustainability, which are summarised in this paper.	What is sustainability? The UK Government context for sustainability; Sustainability in the context of the corporate sector; Sustainability, the environment & biodiversity; Eco-efficiency; Biodiversity; Corporate sustainability programmes; Performance indicators; 4 definitions of SD; Key publications & organisations.	http://www.earthwatch.org/europe/publications/what_is_sustain.pdf

Implementing EU Biodiversity Strategy: A UK Case Study (2000)	This paper offers a critical review of Europe's biodiversity policy, & examines the suitability of the EU's major policy response to the Convention on Biological Diversity, the Habitats Directive. Together with the Birds Directive, its aims are to create a network of designated areas (Natura 2000) to protect habitats & species of community-wide importance, on a bio-geographical basis. It is a biodiversity conservation strategy based on regulations & protected areas, which, although often criticised as being costly & insufficient, is one of the most long-standing & commonly used precautionary measures for nature protection.	Abstract; Introduction; Biodiversity conservation: the EU approach; Current EU biodiversity policies; The philosophy of EU biodiversity policy; Implementation of the Habitats Directive: UK case studies; Coastal management: the Cley case study; Wetland management: the Hickley case study; Towards a more flexible approach to biodiversity conservation; Deliberate & inclusionary processes (DIPs); Mitigation banking; Conclusion; References.	http://www.defra.gov.uk/wildlife-countryside/ewd/rrrpac/biodiv/bioresp.pdf
Sustaining the Variety of Life: 5 years of the UK Biodiversity Action Plan (BAP) (March 2001)	The report covered the first five years of the UK BAP. It contributed to the UK's reporting of its implementation of the CBD (Convention on Biodiversity) to the Conference of the Parties in 2002 (COP-6), which also covered implementation measures undertaken through mechanisms other than the UK BAP. For example the management of biodiversity in the overseas territories is the responsibility of the Overseas Territories' Governments and Administrations.	Ch 1: Introduction – where we are now; Ch 2: Milestones in the development of the UK Biodiversity Action Plan (BAP); Ch 3: The UK biodiversity policy framework; Ch 4: UK BAP – principles & objectives; Ch 5: The structure of delivery of the UK BAP; Ch 6: The country biodiversity groups; Ch 7: The habitat & species action plans; Ch 8: Progress with the implementation of the habitat & species action plans; Ch 9: Future direction for the habitat & species action plans; Ch 10: Local BAPs; Ch 11: The management of biodiversity; Ch 12: Biodiversity through sustainable development; Ch 13: Biodiversity & society; Ch 14: Large scale influences on biodiversity; Ch 15: Knowledge for biodiversity; App 1: Summary of conclusions & recommendations; App 2: The UK BAP; App 3: The 59 steps; App 4: The UKBG (UK Biodiversity Group) & country biodiversity groups; App 5: The habitat & species action plans; App 6: Lead partner reporting & the HAP/SAP database; App 7: Summary of research into the costs of the action plans; App 8: Glossary.	http://www.ukbap.org.uk/Library/library_5.htm

Continued

Table 2.5.3 UK Level Biodiversity Policy. *Continued*

Policy Type	Document Title	Objective	Contents	Website
Strategies & Reports	1st Report by the UK under Article 17 on the Implementation of Directive 92/43/EEC on the Conservation of Natural Habitats & of Wild Fauna & Flora (31/10/01)	This report summarises the UK's experience in implementing the European Community Directive 92/43/EEC on the Conservation of Natural Habitats & of Wild Fauna & Flora (the Habitats Directive) from 1994 to Dec 2000. The Directive has been implemented in Great Britain (England, Scotland & Wales) through the Conservation (Natural Habitat, etc.) Regulations 1994. The Conservation (Natural Habitat, etc.) Regulations (Northern Ireland) 1995 replicate the provisions in force in the GB Regulations by applying them to the separate legal system which exists in Northern Ireland.	Executive summary; Conservation of annex I habitats & the habitats of annex II species within Natura sites; Conservation of annex IV & annex V species in the UK; Other measures & additional provisions to implement the Directive; Final comments on implementation of the Directive; App 1: Organisations involved in compiling report; App 2: Organisations which contributed to, or were consulted on, proposals for site selection; App 3: No. of UK sites put forward as candidate SACs for habitats; App 4: No. of UK sites put forward as candidate SACs (Special Areas of Conservation) for species; App 5: Relationship between annex I habitats & UK biodiversity action plan priority habitats; App 6: Examples of community financing used to support UK implementation of Directive; App 7: Examples of research on species & habitats listed in annexes I, II, IV & V; App 8: Examples of candidate SAC conservation management & habitat restoration; App 9: Abbreviations; App 10: References; App 11: www addresses.	http://www.defra.gov.uk/wildlife-countryside/ewd/rrrpac/echabitats/habdir-full.pdf
	Literature Review of the Implications of Climate Change for Species, Habitats & the wider UK Countryside (13/02/01)	As a first step in identifying the implications of climate change for UK conservation policy, this review aims to collate and examine the range of literature available to understand the impacts of climate change in the UK for species and habitats. This review forms the basis of a report on the implications of climate change for nature conservation policy (Hossell *et al.*, 2000).	Introduction; Framework; UKCIP 98 scenarios; Climate change impact on species; Impacts of climate change in the wider countryside; Climate change impacts on habitats; Discussion; Conclusion; References; Appendix 1: Completed & ongoing UKCIP (UK Climate Impacts Programme) projects.	http://www.defra.gov.uk/wildlife-countryside/ewd/rrrpac/lreview/index.htm
	Science in Action for Biodiversity:	The Biodiversity Research Working Group (BRWG) which includes representatives from	Summary; Introduction; Identifying the research agenda; The key cross-cutting research	http://www.ukbap.org.uk/Library/BRWG06.pdf

A Report of the Biodiversity Research Working Group 1998–2001 (Nov 2001)	the Biodiversity Challenge Group, statutory conservation agencies, academia, the collections, Natural Environment Research Council & Government Departments was established in order to identify & prioritise the cross-cutting research needed to support Government commitments under the UK BAP. The group has progressed the definition of the cross-cutting research agenda through the organisation of five technical workshops, each addressing one cross-cutting theme. The workshops were attended by specialists from a range of government, research & voluntary organisations involved in biodiversity conservation & research. The group has analysed the outputs of the workshops to develop six cross-cutting research programmes. The BRWG review shows that there are a wide range of research needs, a large number of interested & responsible partners & limited resources.	programmes; Access to research information; Other related initiatives; Where next? Annexes.	
Rural White Paper Implementation Plan (11/10/02)	The Rural White Paper Implementation Plan was first published in March 2001, to set out the action on implementation and delivery of each of the measures announced in the White Paper, *Our Countryside: the future, a fair deal for rural England.*	Rural white paper overview of progress; A new standard for village services; Vital village services; Modern rural services; An affordable home; Flexible local transport; market towns & a thriving local economy; A new future for farming; Conserving & enhancing our countryside; Restoring & maintaining wildlife diversity & the natural environment; Increasing enjoyment of the countryside; Local power for country towns & villages; Thinking rural; Reviewing progress.	http://www.defra.gov.uk/rural/ruralwp/implementation/default.htm
UK Indicators for Sustainable Forestry (Oct 2002)	This document contains a set of UK Indicators of Sustainable Forestry. The indicators mostly provide information about the present state, & trends over time, of woodlands & their management, rather than measures of driving forces (pressures) or responses. They will enable the government, non-government organisations, the business community & society to track the contribution that our forests are making to our quality of life. They will inform of undesirable trends so that decisions can be made on how to act.	Introduction; Indicators; A. Woodland; B. Biodiversity; C. Condition of forest & environment; D. Timber & other forest products; E. People & forests; F. Economic aspects; Acronyms; Bibliography; Websites.	http://www.forestry.gov.uk/website/pdf.nsf/pdf/fullsfi.pdf/$FILE/fullsfi.pdf

Continued

Table 2.5.3 UK Level Biodiversity Policy. *Continued*

Policy Type	Document Title	Objective	Contents	Website
Strategies & Reports	Working with the Grain of Nature: A Biodiversity Strategy for England (24/10/02)	This biodiversity strategy for England builds upon the foundations already laid, but recognises that biodiversity must be 'mainstreamed' into all our activities – working the grain of nature through policies on agriculture, water, fisheries, woodland management & urban areas. The strategy also sets out a programme for the next 5 years for the main policy sectors, to make the changes necessary to conserve, enhance & work with the grain of nature & ecosystems rather than against them.	Foreword; Executive summary; Ch 1: Introduction; Ch 2: Biodiversity indicators measuring achievements; Ch 3: A holistic approach; Essay 1: How to manage non-native invasive species; Sustainable management by sector; Ch 4: Agriculture; Ch 5: Water & wetlands; Ch 6: Woodland & forestry; Ch 7: Towns, cities & development; Essay 2: The effects of climate change on biodiversity; Ch 8: The coasts & seas; The importance of people; Ch 9: Local & regional action; Essay 3: Biodiversity for recreation, health & well-being; Ch 10: The economics & funding of biodiversity; Essay 4: Biodiversity & sustainable tourism; Ch 11: The engagement of business; Ch 12: Promoting education & public understanding; Essay 5: Involving children & young people; Appendices.	http://www.defra.gov.uk/wildlife-countryside/ewd/biostrat/index.htm
	Directive on the Conservation of Wild Birds: 79/409/EEC – 7th Report by the UK under Article 12: Implementation of National Provisions 1999–2001 (07/11/02)	This is the 7th Triennial report for the UK on the implementation of Council Directive 79/409/EEC on the Conservation of Wild Birds. The statutes relevant to the Birds Directive are the Wildlife & Countryside Act 1981, the Wildlife (Northern Ireland) Order 1985, and the Nature Conservation & Amenity Lands (Northern Ireland) Order 1985 & the Countryside & Rights of Way Act 2000 (for England & Wales). The report also contains information on the UK's bird monitoring programmes & on scientific research undertaken during the reporting period which supports implementation of the Birds Directive.	Executive summary; Enforcement; SSSI (Site of Special Scientific Interest) legislation review; SPA (Special Protection Area) network: background, rationale & statistics; Recklessness in the countryside & rights of way; Capercaillie hunting ban; Lead poisoning of waterfowl/phasing out of lead shot; Derogation's report; Bird population monitoring & conservation; Rare breeding birds panel; App 1: Status of breeding population & assessment of change; App 2: List of sites classified as SPAs between 1999–01; App 3: List of sites on SPA network review 2001; App 4: Reports on the conservation of individual species; App 5: Agri-environmental schemes; App 6: Management agreements/schemes; App 7: Habitat restoration projects; App 8: Ornithological survey & monitoring of government involvement; App 9 Publications on the conservation of UK bird species.	http://www.defra.gov.uk/wildlife-countryside/ewd/rrrpac/wildbirds/report.pdf

UK Biodiversity Action Plan – tracking progress – results of 2002 reporting (June 2003)	The UK Biodiversity Action Plan established a planned approach with clear biological targets for 391 species and 45 habitats. Monitoring and reporting are crucial to assess progress, evaluate priorities and address emerging issues. At the end of 2002 the Lead Partners of the action plans and Local Biodiversity Action Plan officers reported on their progress. The 2002 reporting round was web-based and the first step towards fully web-based interactive reporting, supported by the Biodiversity Action Reporting System (BARS). This new approach proved a success. A high response rate was achieved with reports received for all 45 Habitat Action Plans, 364 (93%) Species Action Plans and 123 (76%) Local Biodiversity Action Plans.	Executive summary; Introduction & overview; Action plan progress; Species & habitats status; Species & habitats trends; Species & habitats targets; Links between lead partners & local BAPS; Action successes; Obstacles to progress; Sustainable development.	http://www.ukbap.org.uk/2002OnlineReport/2002Report.htm
Countryside Information for Policy – The Lessons from CS 2000 (Oct 2003)	This Report provides an assessment of CS 2000 from the perspective of users, & describes & discusses: ▪ the background to CS 2000; ▪ the characteristics & quality of data deriving from CS 2000, that might affect their use for policy & research applications; ▪ potential applications of CS 2000 in relation to policy & research; ▪ implications for future UK countryside survey.	Executive summary; Introduction; The CS 2000 process; Technical & scientific issues; CS 2000 for policy & research; Lessons for the future; Recommendations; References.	http://www.defra.gov.uk/wildlife-countryside/resprog/findings/cs2000review/cs2000report.pdf
Measuring the progress of the Biodiversity Strategy for England: Baseline Assessment (Dec 2003)	The purpose of these indicators is to help monitor the implementation of the England Biodiversity Strategy & to give some measure of progress towards the achievement of its vision. This first report on the indicators provides a baseline assessment and, where possible, the indicators also show trends over recent years. Some existing indicators are updated and published regularly and this will continue. However, the England Biodiversity Strategy set of indicators will be updated and published again in full alongside the full report of the strategy itself in 2006. By 2010 a time series for these indicators will be available to help assess England's contribution to the global and European targets for biodiversity.	Foreword; Ch 1: Introduction & summary; Ch 2: Headline indicators; Ch 3: Agriculture; Ch 4: Water & wetlands; Ch 5: Woodland & forestry; Ch 6: Towns, cities & development; Ch 7: The coasts & seas; Ch 8: Local & region action; Ch 9: The engagement of business; Ch 10: Promoting education & public understanding; Ch 11: The effects of climate change on biodiversity; Annexes.	http://www.defra.gov.uk/wildlife-countryside/ewd/biostrat/indicators031201.pdf

Continued

Table 2.5.3 UK Level Biodiversity Policy. *Continued*

Policy Type	Document Title	Objective	Contents	Website
Strategies & Reports	Review of Marine Nature Conservation – Working Group Report to Government (July 2004)	The Review of Marine Nature Conservation (RMNC) was established in 1999 to examine the effectiveness of the system for protecting nature conservation in the marine environment & develop practical & proportionate proposals for its improvement. This report to Government represents the Working Group's conclusions, based upon completing the recommendations from the Interim Report, a pilot study in the Irish Sea, & the reports of sub-groups created to consider issues of information provision & legislative mechanisms.	Executive summary; Preface; Background to the review; The wider policy context; Human activity & the marine environment; A marine nature conservation framework; Conservation objectives; Implementing the marine nature conservation framework through appropriate management of human activities; Assessing the impact of human activities; Monitoring; Effective communication & information provision; Enforcement; Governance; Prioritising action; Summary of recommendations; References; Appendices.	http://www.defra.gov.uk/wildlife-countryside/ewd/rmnc/pdf/rmnc-report-0704.pdf
	The Darwin Initiative – 7th Annual Report: Protecting Biological Diversity Around the World (Oct 2004)	The Darwin Initiative continues to support those countries which are rich in biodiversity but poor in financial resources to implement the Convention on Biological Diversity (CBD) through the funding of collaborative projects which draw on UK expertise.	Foreword; Chairman's report; Highlights of the year; Introduction; Objectives of Darwin's Initiative; Darwin Initiative projects; Monitoring & evaluation; Achievements & challenges; Abbreviations; Mid-term reviews; Recently completed projects; Current projects.	http://www.defra.gov.uk/environment/darwin/7threport/darwin7threport.pdf
	Rural Strategy 2004 (2004)	Working towards this aim requires both clarity of policy & effective delivery. At the start of this Parliament (Nov 2003), the Prime Minister set out his four principles of public service reform, which underpin the Government's approach to improving delivery. Applying these principles to the delivery of rural policy was a central & compelling message of Lord Haskins' review. This Strategy sets the framework for the Government's response.	Foreword; Overview; Economic & social regeneration; Social justice for all; Enhancing the value of the countryside; Summary of delivery reforms; A – New rural definition; B – Summary of evidence base; C – Government response to Lord Haskin's recommendations & implementation plan; D – Executive summary of funding streams report; E – Evaluation framework.	http://www.defra.gov.uk/rural/pdfs/strategy/rural_strategy_2004.pdf

Guidance Notes	Roots & Routes: Guidelines on Highways Works & Trees (14/07/99)	People are becoming increasingly aware of the quality of their surroundings. They want their landscapes softened by trees to create a pleasing & welcoming environment. These guidelines have been produced in response to that desire & they show how trees can be accommodated safely in our highways. That is: ▪ How trees may be incorporated in new projects, large or small; ▪ How existing trees should be protected during essential maintenance & repair works; ▪ How the problems associated with trees may be overcome & damage to the infrastructure minimised.	Consultation on draft guidance on highways works & trees; Highways works & trees; Contractors guide; Preventing damage to trees; Executive Summary; Chapter 1: Introduction; Chapter 2: Planning & design; Chapter 3: Highway construction & maintenance; Chapter 4: Strategy; Appendix A: Contract documentation & requirements; Appendix B: Highway works, trees & the law.	http://www.dft.gov.uk/stellent/groups/dft_roads/documents/pdf/dft_roads_pdf_505071.pdf
	Tree Preservation Orders: A Guide to the Law & Good Practice (17/04/00)	This guide sets out the government's policy advice on the tree preservation order system. It outlines the law as it currently stood in England, taking into account the Town & Country Planning (Trees) Regulations 1999. It also suggests ways in which local planning authorities can run the system in line with good administrative practice. Authorities are not required to follow the advice given.	Preface; Ch 1: Introduction to tree preservation orders (TPOs); Ch 2: Scope of TPOs; Ch 3: Making & confirming TPOs; Ch 4: Varying & revoking TPOs; Ch 5: Trees & development; Ch 6: Applications to carry out work under TPOs; Ch 7: Appeals against Local Plg Authority decisions; Ch 8: Modifying & revoking consents; Ch 9: Trees in conservation areas; Ch 10: Penalties; Ch 11: Tree replacement – enforcement; Ch 12: Appeals against tree replacement notices; Ch 13: Injunctions; Ch 14: Compensation; Annex 1: Contacts; Annex 2: Model regulation 3 notice; Annex 3: Model letter of confirmation; Annex 4: Code of practice for utility operators; Annex 5: Model tree work application form; Annex 6: Model acknowledgement of application; Annex 7: Model refusal notice; Annex 8: Model article 5 certificate; Annex 9: Model appeal statement; Annex 10: Model acknowledgement of section 211 notice; Annex 11: Model tree replacement notice; Index.	http://www.odpm.gov.uk/stellent/groups/odpm_urbanpolicy/documents/page/odpm_urbpol_607978.hcsp

Continued

Table 2.5.3 UK Level Biodiversity Policy. *Continued*

Policy Type	Document Title	Objective	Contents	Website
Guidance Notes	Managing Coastal Activities: A Guide for Local Authorities (2004)	This guide has been developed through consultation with local authorities, recreational representatives & users, harbour authorities, the Maritime & Coastguard Agency & others. Whilst focusing on local authorities in England, the guide should also be useful for harbour authorities & other bodies with coastal responsibilities. This guide will help them to: ▪ identify current issues about coastal activities; ▪ promote management solutions through voluntary action or self regulation; ▪ encourage partnership arrangements in managing recreational activities; ▪ encourage a strategic & coordinated approach to managing recreational activities; & ▪ promote & disseminate good practice in the planning & management of recreational activities at the coast.	Ministerial foreword; Introduction; Organisations with a role in the coast; Voluntary approaches; Byelaws; Designing a management scheme; Enforcement; Disseminating information; App 1 – Coastal recreational activities; App 2 – Government departments, agencies, Local Authorities & other organisations; App 3 – Information & guidance; App 4 – Other legislation.	http://www.defra.gov.uk/wildlife-countryside/issues/coastal/coastal-guidance.pdf

Table 2.5.4 Northern Ireland Level Biodiversity Policy.

Policy Type	Document Title	Objective	Contents	Website
Strategies & Reports	NI Biodiversity Strategy Proposals (June 1999)	These proposals for a NI biodiversity strategy have been drafted by a cross-sectoral group, the NI biodiversity group. They are intended to inform policy & promote action on biodiversity conservation in NI by: ▪ Identifying the main features of biodiversity in NI; ▪ Identifying the main issues affecting biodiversity, including institutional, political & legal issues; ▪ Proposing measures to support the conservation of biodiversity for the period 1999–2014.	Summary; People & biodiversity; The need for a NI biodiversity strategy; NI's biodiversity; Issues affecting biodiversity & draft recommendations; Priority habitats & species; Implementation & review; Bibliography; Appendix 1: Membership of the NI Biodiversity Group; Appendix 2: Acknowledgements; Appendix 3: Acronyms; Appendix 4: Legislation affecting biodiversity in NI; Appendix 5: Provisional list of species of conservation concern & priority species in NI; Appendix 6: List of species considered to be extinct in NI.	http://www.ehsni.gov.uk/pubs/publications/biodiversityproposals.pdf
	Biodiversity in NI – Recommendations to Govt for a Biodiversity Strategy (Oct 2000)	This Report to the Government sets out recommendations on how best to sustain biodiversity. It attempts to describe the importance of the diversity of habitats & species in both local & international terms. It looks at the everyday activities which affect biodiversity; & suggests the steps which should be taken to deal with those issues.	Foreword; Summary; Introduction; People & biodiversity; The need for a NI biodiversity strategy; NI biodiversity; Issues affecting biodiversity & recommendations; Priority habitats & species; Implementation & review; Bibliography; Appendices: NI biodiversity group (NIBG); Acknowledgements; Acronyms.	http://www.ehsni.gov.uk/pubs/publications/BiodiversityinNIsection1.pdf and so on with other sections.
	The NI Countryside Survey 2000 – Summary Report on Broad Habitats (Nov 2000)	Because of their species composition, the types & relative amounts of land cover such as woodland, bog, wetland, grassland & agricultural crops are key determinants of biodiversity at the landscape scale. Land uses such as farming, building, forestry & mineral extraction drive change & also influence biodiversity. For planning land use & ecosystem sustainability, it is important to know how much of each land cover type there currently is, what is the distribution of land cover in the landscape & what changes occur with time. The NI Countryside Survey 2000 was carried out to determine the current stock of land cover & field boundaries & to report on how they have changed over the past decade.	Summary; Introduction; NI countryside survey; Biodiversity reporting; Baseline survey; NI countryside survey 2000; Broad habitats; Habitat stock & change; Field boundaries; Agri-environmental measures; References.	http://www.ehsni.gov.uk/pubs/publications/nics2000report.pdf

Continued

Table 2.5.4 Northern Ireland Level Biodiversity Policy. *Continued*

Policy Type	Document Title	Objective	Contents	Website
Strategies & Reports	Partners in Protection – A Review of the Procedures for Notifying, Protection & Managing ASSIs in NI (March 2001)	The purpose of this consultation paper is to consider changes to legislation & procedures relating to the protection & management of ASSIs. The Department believes that better protection & management can be achieved through a combination of policy, administrative & financial changes, & by amending existing legislation. Various options for securing improvements are set out for discussion in this paper. In putting forward these options the Department is aware of possible human rights implications & these will be considered more fully in the framing of any legislative proposals & new procedures which might emerge as a result of this consultation.	Introduction & consultation; Site safeguard: the wider picture; Selection & notification; Protecting designated sites; Managing designated sites; Abbreviations.	http://www.ehsni.gov.uk/pubs/publications/partners_in_protection.pdf
	Habitat Change in the NI Countryside – Technical Report to the NI Countryside Survey 2000 (March 2002)	Quantitative information on the distribution, amount & type of wildlife habitats in the countryside & how they change with time, gives context to discussion on biodiversity & land use issues. It can also be used to indicate progress in meeting national & international obligations on wildlife conservation & sustainable development. Providing reliable information on habitats in different types of landscape is the purpose of the NI Countryside Survey 2000. This technical report describes the quantitative distribution of wildlife habitats recorded in 1998 & shows how they have changed since 1987.	Summary; NI countryside survey; Field & analytical methods; NI, lowlands & uplands change; Primary habitat change; Study area change; Field boundary change; UK broad habitats; References.	http://www.ehsni.gov.uk/pubs/publications/NICSTECH.PDF

The NI Biodiversity Strategy (Aug 2002)	Biodiversity – the variety of all living things – is under threat in NI as much as in any other part of the world. The Executive recognises the importance of taking action to protect & conserve our landscape, wildlife habitats & species & is fully committed to taking steps to protect & enhance biodiversity in NI. The first Programme for Government included a commitment to publish a Biodiversity Strategy. The purpose of this document is to meet that commitment & to set out how the Executive plans to protect & enhance biodiversity in NI over the period up to 2016.	Foreword; Introduction; Biodiversity in NI; The threats; Analysis of the issues; Our long-term goal; Mechanisms; All Ireland dimension; NI Departments working together; Additional resources; DOE responsibilities; DARD (Department of Agriculture and Rural Development) responsibilities; DCAL (Department of Culture Arts & Leisure) responsibilities; Knowledge, understanding & participation; Business & tourism; Policy background against which biodiversity framework will be actioned; Towards a sustainable future; Acronyms.	http://www.ehsni.gov.uk/pubs/publications/NIBS2002.pdf
The Habitats Regulations – A Guide for Competent Authorities (Sept 2002)	The Habitats Directive was transposed into Northern Ireland legislation by the Conservation (Natural Habitats, etc.) (Northern Ireland) Regulations 1995. These Regulations build on existing domestic law, chiefly the Nature Conservation & Amenity Lands (Northern Ireland) Order 1985, as amended in 1989, & the Wildlife (Northern Ireland) Order 1985. The Habitats Regulations place a statutory duty on all competent authorities to act in accordance with the Directive. Regulation 3(4) states that 'every Competent Authority in the exercise of any of its functions shall have regard to the requirements of the Habitats Directive so far as they may be affected by the exercise of those functions'.	Preface; Section 1: The Habitats Directive in NI; Introduction; The Habitats Directive; The Birds Directive; Natura 2000; European sites; Section 2: Competent authorities; Section 3: Assessment of implications for a European site; Section 4: Considerations of overriding public interest; Section 5: Compensatory measures; Section 6: Review of existing decisions & consents etc.; Section 7: Co-ordination where more than one competent authority is involved; Consideration of plans & projects affecting Natura 2000 sites; Appendix 1: Glossary.	http://www.ehsni.gov.uk/pubs/publications/Habitat_Regs_Guidance_Notes.pdf
Recent Changes in the Distribution of Red Squirrels in Northern Ireland (Mar 2003)	This report details the findings of a recent survey of squirrels within NI. The relative distribution of red & grey squirrels in the areas of interface in the distribution of the two species was examined. This report compares the current study with the results of O'Teangana (1999) who concluded that the ecological requirement of the red squirrel, i.e. the absence of the grey squirrel (Gurnell & Pepper, 1993), will not be met in Northern Ireland, as the latter disperses to occupy all suitable habitat throughout the region.	Executive summary; Introduction; Objectives of study; Methodology; Results (by County); General discussion; Forestry in NI; Future conservation & management of red squirrels in NI; Recommendations for future monitoring & research; Red squirrel reserve management; An examination of the coexistence of red & grey squirrels.	http://www.ehsni.gov.uk/pubs/publications/report.pdf

Continued

Table 2.5.4 Northern Ireland Level Biodiversity Policy. *Continued*

Policy Type	Document Title	Objective	Contents	Website
Strategies & Reports	NI Habitat Action Plan – Blanket Bog Final Draft (April 2003)	This plan encompasses all areas of blanket bog supporting semi-natural blanket bog vegetation, including intact surfaces, drained & cutover bog & whether or not it may be defined as 'active' (actively laying down peat). It excludes areas which no longer support such vegetation, except where the restoration of these areas is necessary for the protection and/or enhancement of adjacent bog.	Current status; Biological status; Links with other action plans; Current factors affecting the habitat; Current action; Legal status; Management, research & guidance; Action plan targets; Proposed action with lead agencies; Policy & legislation; Site safeguard & management; Advisory; International; Monitoring & research; Communications & publicity; Costing; References.	http://www.ehsni.gov.uk/pubs/publications/Blanket_Bog_Web_Version_April03.pdf
	Northern Ireland Habitat Action Plan – Lowland Heathland Final Draft (April 2003)	This Lowland Heathland Action Plan identifies specific targets and actions required to deliver Northern Ireland's contribution to the UK Lowland Heathland Action Plan, published in 1995 (UK Biodiversity Steering Group, 1995).	See NI Habitat Action Plan – Blanket Bog contents list above.	http://www.ehsni.gov.uk/pubs/publications/Lowland_Heath_Web_Version_April03.pdf
	Northern Ireland Habitat Action Plan – Lowland Raised Bog Final Draft (April 2003)	This Lowland Raised Bog Action Plan identifies specific targets and actions required to deliver Northern Ireland's contribution to the UK Lowland Raised Bog Action Plan, published in 1999 (UK Biodiversity Group, 1999).	See NI Habitat Action Plan – Blanket Bog contents list above.	http://www.ehsni.gov.uk/pubs/publications/Lowland_Raised_Bog_Web_Version_April03.pdf
	Northern Ireland Habitat Action Plan – Maerl Beds Final Draft (April 2003)	This maerl habitat action plan identifies specific targets and actions required to deliver Northern Ireland's contribution to the UK action plan. (UK Biodiversity Steering Group, 1999).	Current status; Physical & biological status; Links with other action plans; Current factors affecting the habitat; Current action; Legal status; Management, research & guidance; Action plan targets; Proposed action with lead agencies; Policy & legislation; Site safeguard & management; Advisory; International; Monitoring & research; Communications & publicity; Costing; References.	http://www.ehsni.gov.uk/pubs/publications/Maerl_Beds_Web_Version_April_03.pdf

Northern Ireland Habitat Action Plan – Montane Heath Final Draft (April 2003)	This plan recognises the importance of montane heath within NI where it is restricted in distribution. Montane heath is an important component of the mountain summits & ridges of NI, where below around 600m, it grades into upland heathland habitats. Upland heathland is subject to its own NI Habitat Action Plan & the requirements of this habitat should be taken into account during the implementation of this plan.	See NI Habitat Action Plan – Blanket Bog contents list above.	http://www.ehsni.gov.uk/pubs/publications/Montane_Heath_Web_Version_April03.pdf
Northern Ireland Habitat Action Plan – Mudflats Final Draft (April 2003)	This mudflats habitat action plan identifies specific targets and actions required to deliver Northern Ireland's contribution to the UK action plan. (UK Biodiversity Steering Group, 1999).	See NI Habitat Action Plan – Maerl Beds contents list above.	http://www.ehsni.gov.uk/pubs/publications/Mudflats_Web_Version_April_03.pdf
Northern Ireland Habitat Action Plan – Saline Lagoons Final Draft (April 2003)	This saline lagoons habitat action plan identifies specific targets and actions required to deliver Northern Ireland's contribution to the UK action plan, published in 1995. (UK Biodiversity Steering Group, 1995).	See NI Habitat Action Plan – Maerl Beds contents list above.	http://www.ehsni.gov.uk/pubs/publications/Saline_Lagoon_Web_Version_April_03.pdf
Northern Ireland Habitat Action Plan – Seagrass Beds Final Draft (April 2003)	This seagrass habitat action plan identifies specific targets and actions required to deliver Northern Ireland's contribution to the UK action plan, published in 1995. (UK Biodiversity Group, 1995).	See NI Habitat Action Plan – Maerl Beds contents list above.	http://www.ehsni.gov.uk/pubs/publications/Seagrass_Beds_Web_Version_April_03.pdf
Northern Ireland Habitat Action Plan – Sheltered Muddy Gravels Final Draft (April 2003)	This sheltered muddy gravels habitat action plan identifies specific targets and actions required to deliver Northern Ireland's contribution to the UK action plan. (UK Biodiversity Steering Group, 1999).	See NI Habitat Action Plan – Maerl Beds contents list above.	http://www.ehsni.gov.uk/pubs/publications/Sheltered_muddy_gravels_Web_Version_April_03.pdf
Northern Ireland Habitat Action Plan – Species-Rich Hedgerows Final Draft (April 2003)	This Species-rich Hedgerows Action Plan identifies specific targets and actions required to deliver Northern Ireland's contribution to the UK Ancient and/or Species-rich Hedgerows Action Plan published in 1995 (UK Biodiversity Steering Group, 1995).	See NI Habitat Action Plan – Blanket Bog contents list above.	http://www.ehsni.gov.uk/pubs/publications/Hedgerow_Web_Version_April_03.pdf

Continued

Table 2.5.4 Northern Ireland Level Biodiversity Policy. *Continued*

Policy Type	Document Title	Objective	Contents	Website
Strategies & Reports	Northern Ireland Habitat Action Plan – Tidal Rapids Final Draft (April 2003)	This tidal rapids habitat action plan identifies targets and actions required to deliver Northern Ireland's contribution to the UK action plan. (UK Biodiversity Steering Group, 1999).	See NI Habitat Action Plan – Maerl Beds contents list above.	http://www.ehsni.gov.uk/pubs/publications/Tidal_rapids_Web_Version_April_03.pdf
	Northern Ireland Habitat Action Plan – Upland Heath Final Draft (April 2003)	This Upland Heathland Action Plan identifies specific targets and actions required to deliver Northern Ireland's contribution to the UK Upland Heathland Action Plan, published in 1999 (UK Biodiversity Group, 1999). (See also Montane Heath Action Plan above.)	See NI Habitat Action Plan – Blanket Bog contents list above.	http://www.ehsni.gov.uk/pubs/publications/Upland_Heathland_Web_Version_April03.pdf
	Causeway Coast AONB (Area of Outstanding Natural Beauty) Summary Management Plan (April 2003)	The management plan provides the framework for the conservation & enjoyment of this wonderful coastline – a coastline that provides the setting & context to the Giant's Causeway, one of the most important scenic attractions on the island of Ireland. The AONB Management Plan is a plan prepared to guide the future protection, management & enjoyment of the special landscape of the Causeway Coast AONB. It highlights the unique qualities of the AONB, sets out a long-term holistic vision for the area & includes objectives & actions to help secure that vision. It also provides proposals on a management structure to ensure the implementation of the plan.	Introduction; Legislative background; Strategic Objectives; Implementing the plan.	http://www.ehsni.gov.uk/pubs/publications/EHS_Summary_Man_Plan_14-4-2003.pdf
	Natural Heritage Strategic Plan 2003 (Nov 2003)	This Strategic Plan sets out the direction which the Environment and Heritage Service plans to follow in its work on conserving the natural heritage of Northern Ireland. It has no fixed timetable; it seeks to define not so much the rate at which we will progress but the routes that we will take. For most purposes, however, it can be taken as looking 3 to 6 years ahead, or rather longer where we believe this is possible.	Introduction; Theme 1 – Conserving biodiversity; Theme 2 – Advice & influence; Theme 3 – Site safeguard; Theme 4 – Management of designated sites; Theme 5 – Sustainable use of the countryside; Theme 6 – Enjoyment of the countryside; Theme 7 – Working with others.	http://www.ehsni.gov.uk/pubs/publications/Nat_Her_Strat_PDF_72.pdf

	A Framework for Biodiversity Publicity & Promotion in Northern Ireland (March 2004)	This study is a thorough examination of the poor level of biodiversity awareness in Northern Ireland. The Strategy identifies a number of key messages that give greater resonance to the concept of biodiversity. Perhaps more importantly, it presents a robust vision for biodiversity publicity, developing a series of awareness targets that need to be met to make progress, whilst offering an action plan for the next five years to help deliver this.	Executive summary; Background & methodology; Biodiversity publicity & promotion in NI; Biodiversity publicity & promotion in England, Wales & Scotland; Baseline survey of awareness, knowledge & actions amongst the general public; Baseline survey of awareness, knowledge & actions in key sectors; Sectors of society – their potential influences on biodiversity & their attitudes; Biodiversity publicity & promotion – key issues; Towards a framework for biodiversity publicitiy & promotion; Proposals for a vision & key targets for biodiversity awareness in NI; Proposed action plan.	http://www.ehsni.gov.uk/pubs/publications/BiodiversityPublicityPromotionNI.pdf
	Northern Ireland Hare Survey 2004 (Sept 2004)	A survey of Irish hares in Northern Ireland was undertaken in spring 2004 & compared to a similar survey undertaken in spring 2002. The survey was based on observations of hares made at night using a spotlight while driving eight transects along roads. Hare density was estimated using Distance Sampling & populations were derived by multiplying density by land area. This survey provides evidence of a marked increase in hare numbers between 2002 & 2004, & indicates the potential of the species to increase rapidly from low densities. This is not, however, indicative of a general trend, since large annual fluctuations are characteristic of hare populations.	Executive summary; Introduction; Background to the survey; Methods; Results; Discussions; Acknowledgements; References.	http://www.ehsni.gov.uk/pubs/publications/QU04-01Irishharesurvey04.pdf
Consultation Papers	ASSIs in NI – Consultation on Proposals for their Future Management & Protection (Nov 2003)	The Department has published a consultation paper outlining its proposals for the future management and protection of Areas of Special Scientific Interest (ASSIs) in Northern Ireland. This consultation follows on from the earlier consultation 'Partners in Protection' which was issued in March 2001 & reviewed the key issues pertaining to the notification, protection & management of ASSIs.	Introduction; Context; The notification of ASSIs; Protecting from damage; Achieving better management; Public bodies; The next steps; Annexes.	http://www.ehsni.gov.uk/pubs/publications/ASSI_Proposals.pdf

Continued

Table 2.5.4 Northern Ireland Level Biodiversity Policy. *Continued*

Policy Type	Document Title	Objective	Contents	Website
Consultation Papers	Giant's Causeway & Causeway Coast World Heritage Site – Management Plan Public Consultation Draft (Sept 2004)	The plan provides a relatively detailed physical and historical description of the Site, its key features & its setting broadly based on the nomination document & other sources, as well as details on current ownership, management structures, designations & the planning context. The plan goes on to describe the Outstanding Universal Value of the site & other values at the Site & explores the issues facing the site & then identifies a series of management principles & objectives to guide the future strategic management of the site. It examines the approaches to monitoring change on & around the site & proposes a series of indicators that could be used to guide the long-term strategic management of the site & its setting & inform future reviews of the management plan.	Preface; Introduction; Site description; Ownership, management & planning; Justification for inscription & statement of values; Management issues, principles & objectives; Monitoring; Vision & action plan (2005–2011); Bibliography; Appendix 1 – reference points for the site.	http://www.ehsni.gov.uk/pubs/publications/WHS_condoc_web.pdf
	National Parks & other Protected Landscape Areas – A Discussion of Options for Establishing National Parks &	This discussion paper follows on from *Shared Horizons – Statement of Policy on Protected Landscapes in Northern Ireland*, which outlines the contexts & intentions for protected landscapes in more detail than this document. The primary purpose of this discussion paper is to stimulate debate about the need for new	Foreword; Introduction; Landscape protection in NI; National parks; Areas of outstanding natural beauty; Information, guidance & research in support of national parks & areas of outstanding natural beauty; Your views; Annex 1 – Protected landscapes background & examples from other countries; Annex 2 – Summary of management	http://www.ehsni.gov.uk/pubs/publications/National_Parks.pdf

	Managing Other Outstanding Landscapes in Northern Ireland (Oct 2004)	national park legislation in NI & to seek views which will assist the Department to prepare more detailed proposals to be the subject of public consultation in 2005. The paper also provides an opportunity to discuss the current arrangements for the management of AONBs (Areas of Outstanding Natural Beauty) & to determine the level of support for changes along similar lines to those introduced in England & Wales.	arrangements for AONB in NI; Annex 3 – Abbreviations; Annex 4 – List of questions.	
Guidance Notes	Guidelines for the Selection of Biological ASSIs in Northern Ireland (March 1999)	The purpose of the guidelines is: ▪ To provide a logical & consistent framework for the evaluation of sites throughout NI; ▪ To enable the scientific importance of each site to be seen in its regional, national or international context. These guidelines explain the rationale that underpins the selection of ASSIs & consider the application of this rationale by EHS using examples from a variety of habitats & species-groups.	Introduction; Chapter 1: Definitions & aims; Chapter 2: Selection criteria; Chapter 3: The process of site selection; Chapter 4: Boundary considerations; Chapter 5: Citation, notifying operations & conservation objectives; Chapter 6: The objectives & extent of the ASSI network; References	http://www.ehsni.gov.uk/pubs/publications/ASSIguidelines.pdf
	Nature in Your Neighbourhood: A Guide to Improving the Natural Environment & Biodiversity Where You Live (March 2004)	This guide sets out to encourage voluntary and community groups & individuals to participate in improving the quality of the natural environment.	Introduction; Foreword; What is biodiversity? Why include biodiversity when improving open space? How to run a community biodiversity project? Where are communities involved in biodiversity? Who can help?	http://www.ehsni.gov.uk/pubs/publications/NatureInYourNHood.pdf

3 Climate Change

3.1 Climate Change and the Built Environment

Climate Change is considered to be the most pervasively threatening and intractable of the global environmental problems and challenges to human well-being. It has a bearing on all people and their built environment to varying degrees and has attracted widespread attention from community levels to international political activity.

The concept of climate change is not easily understood by society, but its links with greenhouse gases, ozone layer depletion, global warming and seasonal variations are understood through observation. These issues have resonance with the use of renewable energy systems, business and financial decisions, geographical trends and complex scientific investigations. Indeed, there is now consensus that the greenhouse gas effect poses an unprecedented threat to the global climate. Whilst the consequences of such climate change may not be the same in all regions of the world, with uncertain and inexact impacts, it is clear that no country will be invulnerable. These impacts will undoubtedly stretch the resources of local and national governments, as well as the international agreements set up to address this complex matter.

Predictions of the effects of climate change across the built environment are dramatic with significant adverse consequences expected. Climate models have simulated several scenarios and these may be customised for a range of locations to fully enumerate the potential impacts. At a generic level it is clear that changes in climate will occur, and predictions for change in this century suggest headline elements such as:

- temperature increases (1.2–2.6°C);
- wetter climates (5–20%);
- rising sea-levels (0.3–1.0m);
- increased warm weather frequency; and
- changes in storm rainfall intensity and frequency.

This may result in flash floods and a need for greater flood protection and prevention, achieved by engineered construction projects and careful infrastructure planning. Climate change presents decision-makers with a set of formidable complications, including this inherent uncertainty, the potential for irreversible damage and costs, and need for long-term planning horizons to respond to the probable lengthy time lags between actions and their effects. On the short-term timescale, there is an urgent need for increased understanding of the physical impact of weather, well before it influences strategic decisions.

There are a number of international, national and local strategies which use tools such as legislation, taxes and incentives to address the reduction of the climate change effects in the sectors of greenhouse gas emission reduction, energy efficiency, pollution control, and environmental damage. The United Nations Framework Convention on Climate Change (1992) had the ultimate objective 'to achieve stabilisation of atmospheric concentrations of greenhouse gases at levels that would prevent dangerous anthropogenic (human-induced) interference with the climate system'. This keynote convention was supplemented by the detailed Kyoto Protocol, negotiated by more than 160 nations in December 1997, which aimed to reduce net emissions of six greenhouse gases, primarily carbon dioxide (CO_2), by 5.2% in the period 2008–2012. However, there remains a deficit in public awareness of the climate change effects and the development of the most appropriate strategies with associated political drivers. This is further complicated by the realisation that global warming may be caused by mechanisms other than the enhanced greenhouse gas effect, such as impact of solar wind on the atmosphere, cloud formation and water vapour movement, the nature of carbon cycle variation, and its link to periods of Earth warming and cooling over the last 400 000 years.

3.2 Key Climate Change Elements

The concept of climate change may be complex, but it can be demonstrated and evidenced by the visible presence, movement and combined effects of elements such as water, gases, droughts, flooding, pollution, diseases and desertification. Typically, global warming is caused by enhanced greenhouse gas emissions initiating a change in the Earth's temperature. This is done by unbalancing the energy entering the atmosphere from the Sun in the form of visible radiation or sunlight and the energy being constantly emitted from the Earth's surface to space. In addition, some of the infrared radiation emitted from the Earth is caught by the Earth's atmosphere and rebounds, resulting in further warming of the Earth's surface and its lower atmosphere. Together these are known as 'the greenhouse effect'.

Greenhouse gases such as water vapour, carbon dioxide, methane, ozone, nitrous oxide and chlorofluorocarbons occur naturally and are also observed on neighbouring planets. The enhanced greenhouse effect is caused by additional greenhouse gases being present in the atmosphere due to human activities such as the burning of fossil fuels and deforestation. Water vapour accounts for over 90% of the natural greenhouse effect, with the rest being due to carbon dioxide and methane, with the latter contributing to the warming of the atmosphere in an enhanced greenhouse mode. This water vapour is readily available in the form of surface water such as watercourses, lakes, seas and oceans, and in the form of ice, such as polar caps, glaciers, and snow-covered mountains. Hence, the Earth's water vapour could be increased through activities such as increased energy use on the Earth, external and internal influences on cloud formation and changes in the temperature of atmospheric gases. Pollution, industrial emissions and loss of 'carbon sinks', such as forests, also contribute to present and future global warming.

These varying climate change elements may result in flooding, storm surges, hurricanes and tsunamis, which in turn could have a dramatic and life-threatening effect on coastal zones, infrastructure, habitable land and local communities. Alteration to the Earth's temperature has several potential consequences on energy usage, human health, commercial and industrial decision-making and quality of life issues. In more detail, it has been observed that climate change will affect the physiology, survival and performance of many species of wildlife, due to temperature and humidity changes and the effects on competing species, predators, parasites and diseases. Similarly, there are political and human consequences from these complex and inter-dependent aspects of climate change. In this regard the international Kyoto Protocol (1997) aims to drive down the emission of carbon dioxide and other greenhouse gases whilst encouraging governments to co-operate with one another, improve energy efficiency, reform the energy and transportation sectors, promote renewable forms of energy, phase out inappropriate fiscal measures and market imperfections, limit methane emissions from waste management and energy systems, and protect forests and other 'carbon sinks'.

3.3 Sustainability Challenges for Climate Change

Developments in the latter part of the 20th century, both at International and European levels, have pushed issues related to climate change and sustainable development higher up national government agendas. While researchers have endeavoured to compile transparent data to support the alleged and perceived effects of climate change, industry has been drawn into the debate alongside politicians, scientists and engineers. Nations at the Earth Summit in Rio de Janeiro in 1992 agreed that reducing greenhouse emissions was vital, culminating in the UN Framework Convention on Climate Change (1992). Subsequently, the Kyoto Protocol set developed countries the target to cut emissions of a basket of six greenhouse gases by 5.2% below 1990 levels over the period 2008–2012. In recent years more specific measures have evolved to support this critical protocol, through the use of various mechanisms, policies, penalties and processes which help protect the environment, minimise health impacts and support a sound and viable economy.

There are clear recognisable links between carbon dioxide emissions and energy consumption in domestic, industrial, transport and public sectors. These sectors are indicative of and provide opportunities for sustainable development to be delivered, necessitating that the causes and effects of climate change are urgently addressed in tandem to reduce climate change impact on the Earth's sustainability. Climate change contributes to all four sustainable development objectives in the UK *Better Quality of Life* strategy. The following issues assist in contextualising climate change and indicate areas which need to be addressed to give a balanced and lasting output:

- *Effective protection of the environment* arises from reduction in greenhouse gas production in accordance with the Kyoto Protocol. Full consideration of biodiversity, air quality, wildlife habitat conservation and many related areas will arise from this gas-reduction campaign, while the reduction in ozone layer depletion can be achieved by control of CFC gas emissions. The secondary effects and benefits will be realised through improved energy efficiency, transport planning and filtration of air pollutants, while climate change research and modelling will assist in mitigating the effects of Global warming and the occurrence of flooding and reducing damage, injury or death from infrequent major incidents that arise from a combination of water, weather and wind movements.
- *Social progress which recognises the needs of everyone* arises from greater access to affordable energy and a reduction in the negative health impacts from fossil fuel emissions. The reduction in risk of vulnerability due to flooding adds to the social progress of communities and is aided by transparent and rigorous planning systems which do not support development on flood plains, and a drive towards increased public transport accessibility and frequency in order to encourage less private transportation use.
- *Prudent use of natural resources* can be achieved by a significant increase in renewable energy usage following on from a reduction in reliance on fossil fuel consumption. The development and construction of new energy schemes must contribute to waste minimisation, greater use of existing or recycled materials and a strong research-driven programme which provides a good understanding of a full range of renewable sources.
- *Maintenance of high and stable levels of economic growth and employment* arises from increased productivity at all stages of energy development and access to affordable energy production. This will require greater use of innovative processes, the application of green procurement, development of design build finance and operate (DBFO) projects through public private partnerships and full allocation of capital and recurrent expenditure. In this way, as a result of scenario building and modelling, using tools such as 'best practicable environmental option' and the Strategic Environmental Assessment, there can be an avoidance of fuel poverty levels.

The understanding, prediction, mitigation and trans-disciplinary aspects of climate change are central to the challenges in delivering sustainable development. The critical challenges which arise from climate change and contribute to this broader sustainability agenda include the following.

- *Energy consumption issues* which alter wealth development and impact on progress in the essential sectors of health, education and commerce through fuel poverty, community cohesion and inclusion, quality of life and their interactions.
- *Natural environment alteration*, producing a variety of forms of environmental damage and impacting on a range of species, habitats and ecosystems.
- *Community and government conflict* over energy systems and costs, damage to wildlife, threats or damage by flooding schemes, valuation of properties, and the undetermined effect of changing surface temperatures.

- *Resolution to the variable climate change impact* between developed and developing nations, resulting from inequity as developing nations face consequences from industrialised activity.
- *Stabilisation and reduction of CO_2 emissions* in industrialised countries, with supportive and complementary global monitoring and reporting mechanisms to track polluters. There is a central need for advocacy and political will to draw all nations into the Kyoto Protocol.
- *Melting polar ice caps and rising sea-levels* which require scientific understanding and engineered infrastructure responses to prevent greater flash storms and flooding to low lying areas.
- *A halt to environmental and economic deterioration* associated with human activity, caused by coal and oil power electricity, private transport overuse and deforestation.
- *Guarding against tropical country infrastructure and economic damage* due to drier seasons, drought and desertification and the spread of tropical diseases.
- *Emission trading* by enabling countries to burden share and sell excess capacity to countries producing more than their agreed limit; this also links to support for innovative technologies which help promote a low carbon economy.
- *Promotion of renewable energy sources*, including solar, wind, wave, hydropower and nuclear energy developments using research and development, demonstration projects and government financial incentives.

3.4 Climate Change Theme Index

Jurisdiction	Tier	Title	Date	Page No.
EUROPEAN	Council Directives	Establishing a Scheme for Greenhouse Gas Emission Allowance Trading *(Dir 2003/87/EC)*	2003	68
	Council Communication	The Energy Dimension of Climate Change *(COM (97) 196)*	1997	68
		Climate Change – Towards an EU post-Kyoto Strategy *(COM (98) 353)*	1998	68
		Greenhouse Gas Emissions Trading & Climatic Change Programme *(COM (2000) 87/88)*	2000	68
		Implementation of the European Climate Change Programme (ECCP) *(COM (2001) 580)*	2001	69
		The Implementation of the Criteria Listed in Annex III to Directive 2003/87/EC Establishing a Scheme for Greenhouse Gas Emission Allowance Trading within the Community *(COM (2003) 830)*	2003	69
		National Allocation Plans for the Allocation of Greenhouse Gas Emission Allowances *(COM (2004) 500)*	2004	69
	Council Decisions	Kyoto Protocol on Climate Change *(C Dec 2002/358/EC)*	2002	70
		Establishing Guidelines for the Monitoring & Reporting of Greenhouse Gas Emissions *(C (2004) 130)*	2004	70
		Mechanism for Monitoring Community Greenhouse Gas Emissions & for Implementing the Kyoto Protocol *(Dec No 280/2004)*	2004	70
	Miscellaneous	The Implementation of the Montreal Protocol on Substances that Deplete the Ozone Layer in the EU	2000	71
		Substances which Damage the Ozone Layer *(Reg No. 2037/2000 Amended by Reg No. 2038/2000; Reg No. 2039/2000)*	2000	71
		Report for A Monitoring Mechanism of Community Gas Emissions *(COM (2000) 749)*	2000	71
		EU Focus on Climate Change	2002	72
		Monitoring Mechanism of Community Greenhouse Gas Emissions *(COM (2002) 702)*	2002	72
		Proposal for a Decision for a Monitoring Mechanism of Community Greenhouse Gas Emissions & the Implementation of the Kyoto Protocol *(COM (2003) 51)*	2003	72
		2nd Report of the ECCP – Can we meet our Kyoto Targets	2003	72
		Proposal for a Directive Amending the Directive Establishing a Scheme for Greenhouse Gas Emission Allowance Trading in the Community, in Respect of the Kyoto Protocol Project Mechanisms *(COM (2003) 403)*	2003	73
		Proposal for Regulating Certain Fluorinated Gases *(COM (2003) 492)*	2003	73
		Control of Halon Exported for Critical Uses, the Export of Products & Equipment Containing Chloro-fluorocarbons & Controls on Bromochloro-methane *(Reg EC No. 1804/2003)*	2003	74

Jurisdiction	Tier	Title	Date	Page No.
EUROPEAN	Miscellaneous	A Monitoring Mechanism of Community Greenhouse Gas Emissions *(COM (2003) 735)*	2003	74
		Action on Climate Change Post 2012: A Stakeholder Consultation on the EU's Contribution to Shaping the Future Global Climate Change Regime	2004	74
UK	Strategies & Reports	Energy Efficiency in the Domestic Sector Technical Support Paper	1998	75
		Climate Change & UK Nature Conservation: A Review of the Impact of Climate Change on UK Species & Habitat Conservation Policy	2000	75
		Climate Change: The UK Programme	2001	76
		Draft Framework Document for the UK Emissions Trading Scheme	2001	76
		The Social Cost of Carbon Review – Background Paper	2003	77
		Climate Change & Local Communities – How Prepared are You?	2003	77
		Scenarios of Climate Change for Islands within the BIC Region – British-Irish Council	2003	77
		The Impacts of Climate Change: Implications for DEFRA	2003	78
		Energy White Paper – Our Energy Future: Creating a Low Carbon Economy	2003	78
		Options for a Low Carbon Future – Phase 2	2003	78
		Creating a Low Carbon Economy – 1st Annual report on the Implementation of the Energy White Paper	2004	79
		Scientific & Technical Aspects of Climate Change, Including Impacts & Adaptation & Associated Costs	2004	79
		UK International Priorities – the Energy Strategy	2004	79
NORTHERN IRELAND	Strategies & Reports	Environmental Monitoring – New Technologies for Monitoring: Status & Prospects	2001	80
		Climate Change & Nature Conservation in Britain & Ireland	2001	80
		EU Emissions Trading Scheme Guidance Note 1	2003	80
		Strangford Lough Ecological Change Investigation (SLECI) – Climate Change Desk Study	2004	81
		Climate Change Indicators for Northern Ireland	2004	81

3.5 Climate Change Policy Tables

Table 3.5.1 International Level Climate Change Policy.

Title	Venue	Purpose	Key Outcome(s)	Website
First Session of the Conference of the Parties – The Berlin Mandate (COP-1)	Berlin, Germany (1995)	At the very first session of the Conference of the Parties (COP-1 Berlin, 1995), the more than 120 countries that had already ratified the Convention found that the commitments referred to in Article 4.2(a) and (b) of the Climate Change Convention were inadequate to meet the overall goals of the Convention.	The Berlin Mandate which: ▪ Called for strengthening of the commitments made in 1992 & called for new proposals to be made in the form of a protocol at Kyoto ▪ Stated that quantified limits & reduction objectives & targets should be set for specific time frames (2005/2010/2015) ▪ Re-affirmed existing reporting obligations.	http://unfccc.int/resource/docs/cop1/07a01.pdf
Second Session of the Conference for the Parties – The Geneva Declaration (COP-2)	Geneva, Switzerland (1996)	The second session provided the Conference of the Parties with an opportunity to carry out an interim evaluation of the implementation of the United Nations Framework Convention on Climate Change and of the progress made in the negotiation process initiated at the first session with the adoption of the Berlin Mandate. Pursuant to that Mandate, the Parties were called upon, *inter alia*, to negotiate a protocol or another legal instrument in order to strengthen the commitments of Parties included in Annex I to the Convention (Annex I Parties) in Article 4.2(a) and (b) for adoption at the third session of the Conference of the Parties.	The Geneva Declaration which endorsed the IPCC (Inter-governmental Panel on Climate Change) Second Assessment Report as a basis for urgent action & instructed their representatives to accelerate negotiations on a legally binding protocol to be adopted by COP-3 scheduled for December 1997 in Kyoto, Japan.	http://www.ccsr.utokyo.ac.jp/unfccc1/pdfs/unfccc.int/resource/docs/cop2/15.pdf

Fourth Session of the Conference for the Parties – Buenos Aires Plan of Action (COP-4)	Buenos Aires, Argentina (1998)	A two-week meeting of 170 governments concluded by adopting a two year Plan of Action to reduce the risk of global climate change. The action plan was to accelerate work on the 1992 United Nations Framework Convention on Climate Change & prepare the way for industrialized countries to take future action under the 1997 Kyoto Protocol.	The Buenos Aires Plan of Action (BAPA) boosts work on transferring climate-friendly technologies to developing countries & addresses the special needs & concerns of countries affected by global warming & by the economic implications of response measures. It included deadlines on a number of important issues: ▪ Financial mechanisms – which will assist the developing world to respond to the challenges related to climate change. ▪ Further work on policies and measures – an issue introduced by the EU at a late stage in the Conference. ▪ Development and transfer of technologies. ▪ Rules governing the Kyoto Mechanisms with priority given to the Clean Development Mechanism. ▪ An undertaking to discuss supplementarity, ceilings, long term convergence and equity.	http://www.ccsr.utokyo.ac.jp/unfccc1/pdfs/unfccc.int/resource/docs/cop4/16a01.pdf
Fifth Session of the Conference for the Parties – Bonn Agreements (COP-5)	Bonn, Germany (1999)	During its last two days, COP-5 adopted 32 draft decisions and conclusions on, *inter alia*, the review of the implementation of commitments and other Framework Convention on Climate Change (FCCC) provisions and preparations for the first session of the COP serving as the Meeting of Parties to the Kyoto Protocol (COP/MOP-1). After a faltering COP in Buenos Aires, the process recovered vital momentum and began to gather determination and support for a self-imposed deadline for entry into force of the Protocol by 2002.	COP-5 emphasised the need for political will to fulfill the BAPA & allow for the Protocol to enter into force by Rio+10 in 2002. COP-5 adopted the draft decision on Part II of the guidelines (FCCC/CP/1999/L.3) & the addendum containing the guidelines (FCCC/CP/1999/L.3/Add.1). In the decision, the COP, *inter alia*: ▪ decides that Part II of the guidelines should be used for the preparation of third national communications; ▪ requests Annex I Parties to provide a detailed report on their activities in relation to systematic observation; and ▪ urges Annex II Parties to assist Parties with economies in transition (EITs) with technical aspects in preparing national communications.	http://www.iisd.ca/vol12/enb12123e.html

Continued

Table 3.5.1 International Level Climate Change Policy. *Continued*

Title	Venue	Purpose	Key Outcome(s)	Website
Sixth Session of the Conference for the Parties – The Hague (COP-6)	The Hague, Netherlands (2000)	The meeting aimed to set the operational details for commitments on reducing emissions of greenhouse gases under the 1997 Kyoto Protocol, and to reach agreement on actions to strengthen implementation of the UNFCCC itself. In attempting to achieve these goals, the meeting was intended to bring to a close more than two years of preparations and negotiations set out in the UNFCCC's 1998 Buenos Aires Plan of Action.	During the first week, delegates met in informal contact groups and other negotiating groups mandated by the UNFCCC subsidiary bodies. These meetings aimed to reduce differences on text for decisions on a range of issues related to the Protocol and the UNFCCC, including: ▪ the transfer of technology and capacity building to assist developing countries and countries with economies in transition; ▪ the adverse effects of climate change and the impact of implementation of response measures; ▪ best practice in domestic policies and measures to address greenhouse gas emissions ▪ the mechanisms outlined under the Protocol; ▪ a compliance system for the Protocol; and ▪ issues relating to the land use, land-use change and forestry (LULUCF) sector.	http://www.iisd.ca/vol12/enb12163e.html
Seventh Session of the Conference for the Parties – Marrakesh Accords (COP-7)	Marrakesh, Morocco (2001)	From 30 October to 6 November delegates met in negotiating groups, closed drafting groups and informal consultations in their attempt to resolve outstanding issues. These included: ▪ the mechanisms under the Protocol, ▪ a compliance system, accounting, reporting and review under Protocol Articles 5 (methodological issues), 7 (communication of information) and 8 (review of information), and	*The Marrakesh Ministerial Declaration* – reaffirmed that economic & social development, & poverty eradication were the first & overriding priorities of the developing countries. It also stressed the importance of capacity building, & developing innovative technologies through public & private sector involvement. *The Marrakesh Accords* – called upon Annex I Parties to provide funding to developing countries through these funds, increased GEF replenishment, & bilateral multilateral channels. However, these financial commitments were neither mandatory, nor were any	http://www.iisd.ca/vol12/enb12189e.html

		▪ land use, land-use change and forestry (LULUCF). Ongoing negotiations were also held on draft COP decisions relating to the least developed countries (LDCs), the Consultative Group of Experts (CGE), and input to the 2002 World Summit on Sustainable Development (WSSD).	funding levels specified according to a burden sharing formula. Some countries (Canada, EU, Iceland, New Zealand, Norway, & Switzerland) passed a political declaration, undertaking to provide an annual contribution of US$410 million by 2005 to developing countries.	
Eighth Session of the Conference for the Parties – Delhi Declaration on Climate Change & Sustainable Development (COP-8)	New Delhi, India (2002)	COP-8 Recognised that, as Africa is the region suffering the most from the combined impacts of climate change & poverty, development initiatives such as the New Partnership for Africa's Development (NEPAD) should be supported in the context of sustainable development. It also resolved that, in order to respond to the challenges faced now & in the future, climate change & its adverse effects should be addressed while meeting the requirements of sustainable development.	*Delhi Declaration on Climate Change & Sustainable Development* – The Declaration reiterated the importance of carrying out all existing international commitments under the United Nations Framework Convention on Climate Change. It also called for early ratification of the Kyoto Protocol. The Declaration further promoted less polluting energy and other innovative technologies. It urged governments to promote technological advances through research and development, to substantially increase renewable energy resources and to promote the transfer of technologies that can help reduce greenhouse gas emissions in major economic sectors including through public sector and market-oriented approaches.	http://unfccc.int/cop8/latest/1_cpl6rev1.pdf
Ninth Session of the Conference for the Parties – (COP-9)	Milan, Italy (2003)	The high-level political debate during the conference took place through three informal ministerial roundtables focusing on adaptation, mitigation, sustainable development, technology and assessment. Ministers agreed that climate change remains the most important global challenge to humanity and that its adverse effects are already a reality in all parts of the world. Many Parties, developed and developing, reported on activities to adapt to climate change and emphasised the moral duty to provide support to developing countries in these efforts.	Parties adopted numerous decisions & conclusions on various issues including: ▪ definitions & modalities for including afforestation & reforestation activites under the Clean Development Mechanism; ▪ good practice guidance on land use, land-use change & forestry (LULUCF); ▪ the Special Climate Change Fund (SCCF); and ▪ the Least Developed Countries (LDC) Fund.	http://www.iisd.ca/download/pdf/enb12231e.pdf

Continued

Table 3.5.1 International Level Climate Change Policy. *Continued*

Title	Venue	Purpose	Key Outcome(s)	Website
Inter-Governmental Panel on Climate Change (IPCC-4)	Colombo, Sri Lanka (2003)	The Expert Meeting held in Colombo on Climate Change and Sustainable Development was the first of two expert meetings approved by the IPCC Plenary in Paris, in February 2003. Sustainable development had been identified as a Cross Cutting Theme (CCT) for the Third Assessment Report (TAR), but was not developed fully therein.	Proceedings from the Expert Meeting of the Inter-governmental Panel on Climate Change. Fourth IPCC Assessment Report on Integrating Sustainable Development & Climate Change. (The other three Assessment Reports in 1990, 1995 and 2001 relate to wider issues of climate change.)	http://www.ipcc.ch/pub/colombo.pdf
Vienna Convention for the Protection of the Ozone Layer	Vienna, Austria (1985)	For the first time nations agreed in principle to tackle a global environmental problem before its effects were felt, or even scientifically proven.	Nations agreed to take 'appropriate measures . . . to protect human health and the environment against adverse effects resulting or likely to result from human activities which modify or are likely to modify the Ozone Layer' but the measures were unspecified. (See later Montreal Protocol, 1987.)	http://www.unep.org/ozone/vienna.shtml
UN Framework Climate Change Convention (UNFCCC)	Rio de Janeiro, Brazil (1992)	The ultimate objective of the Convention is 'to achieve stabilization of atmospheric concentrations of greenhouse gases at levels that would prevent dangerous anthropogenic (human-induced) interference with the climate system . . .'. (See later Kyoto Protocol, 1997.)	Parties to the UNFCCC have a number of commitments including: ▪ to submit a national inventory of emissions/removals of greenhouse gases (NGGI) ▪ to implement national programmes to mitigate climate change & adapt to its impacts ▪ to strengthen scientific & technical research & systematic observation related to the climate system, & promote the development & diffusion of relevant technologies ▪ to promote education programmes & public awareness about climate change & its likely effects ▪ to periodically submit comprehensive national reports on activities to implement commitments under the convention.	http://unfccc.int/resource/docs/convkp/conveng.pdf

Montreal Protocol on Substances that Deplete the Ozone Layer	Montreal, Canada (1987)	To further the issues discussed at the Vienna Convention & where possible set appropriate measures/targets.	The final agreement, which contains clauses to cover the special circumstances of several groups of countries, especially developing countries with low consumption rates who do not want the Protocol to hinder their development. The Protocol is constructively flexible: it can be tightened or 'adjusted', as the scientific evidence strengthened, without having to be completely renegotiated. Indeed, it sets the 'elimination' of ozone-depleting substances as its 'final objective.' (Subsequently amended at London 1990, Copenhagen 1992, Vienna 1995, Montreal 1997 & Beijing 1999.)	http://www.unep.org/ozone/pdf/Montreal-Protocol2000.pdf
Kyoto Protocol to the UN Framework Convention on Climate Change (COP-3)	Kyoto, Japan (1997)	The Kyoto Protocol, negotiated by more than 160 nations in December 1997, aims to reduce net emissions of certain greenhouse gases (primarily CO_2).	Protocol commits industrialised nations to legally binding reductions in emissions of six 'greenhouse gases'. The developed countries commit themselves to reducing their collective emissions of six key greenhouse gases by at least 5%. Each of the participating developed countries must decide how to meet its respective reduction goal during a five-year period (2008-2012); but specific ground rules remain to be worked out at future negotiating sessions. The Protocol encourages governments to co-operate with one another, improve energy efficiency, reform the energy and transportation sectors, promote renewable forms of energy, phase out inappropriate fiscal measures and market imperfections, limit methane emissions from waste management and energy systems, and protect forests and other carbon 'sinks'.	http://unfccc.int/resource/docs/convkp/kpeng.pdf

Table 3.5.2 European Level Climate Change Policy.

Policy Type	Programme Title	Policy Objective	Community Measure	Document Reference	Website
Council Directives	Establishing a Scheme for Greenhouse Gas Emission Allowance Trading	This Directive established a scheme for greenhouse gas emission allowance trading within the Community in order to promote reductions of greenhouse gas emissions in a cost-effective & economically efficient manner.	Dir 2003/87/EC of the European Parliament & of the Council establishing a scheme for greenhouse gas emission allowance trading within the Community, amending CD 96/61/EC (13/10/03).	▪ Off Jnl L275/32 25/10/03	http://europa.eu.int/eur-lex/pri/en/oj/dat/2003/l_275/l_27520031025en00320046.pdf
Council Communications	The Energy Dimension of Climate Change	To identify the policies to be implemented and the measures to be taken to meet the goals of reducing the emissions of greenhouse effect gases.	Commission Communication on the energy dimension of climate change (14/05/97).	▪ COM (1997) 196 Final	http://europa.eu.int/scadplus/leg/en/lvb/l27022.htm
	Climate Change: Community Post-Kyoto Strategy	To develop a Community strategy for achieving the targets laid down in the Kyoto Protocol & the Buenos Aires Action Plan.	Communication from the Commission to the Council & European Parliament on climate change – towards an EU post-Kyoto strategy (03/06/98).	▪ COM (1998) 353 Final	http://europa.eu.int/scadplus/leg/en/lvb/l28074.htm
			Communication from the commission on preparing for implementation of the Kyoto Protocol (19/05/99)	▪ COM (1999) 230 Final	
	Greenhouse Gas Emissions Trading & Climatic Change Programme	To launch a public debate with a view to implementing a possible Community policy on emissions trading before the Kyoto protocol is applied.	Green Paper on greenhouse gas emissions trading within the EU (2000).	▪ COM (2000) 87	http://europa.eu.int/scadplus/leg/en/lvb/l28109.htm
			Communication from the Commission to the Council & European Parliament on EU policies & measures to reduce greenhouse gas emissions: towards a European climate change programme (ECCP) (2000).	▪ COM (2000) 88	

Implementation of the European Climate Change Programme (ECCP)	To combat climate change by means of various cross-cutting measures & measures in the fields of energy, industry & transport in implementing the ECCP.	Communication from the Commission on the implementation of the 1st phase of the European Climate Change Programme (23/10/01).	■ COM (2001) 580	http://europa.eu.int/scadplus/leg/en/lvb/l28118.htm
The Implementation of the Criteria Listed in Annex III to Directive 2003/87/EC Establishing a Scheme for Greenhouse Gas Emission Allowance Trading within the Community.	The purpose of this guidance document is three-fold: ■ First, to assist Member States in drawing up their national allocation plans, by indicating the scope of interpretation of the Annex III criteria that the Commission deems acceptable; ■ Second, to support the Commission assessment of notified national allocation plans, pursuant to Article 9(3); ■ Third, to describe the circumstances under which *force majeure* is demonstrated.	Communication from the Commission on guidance to assist Member States in the implementation of the criteria listed in Annex III to Directive 2003/87/EC establishing a scheme for greenhouse gas emission allowance trading within the Community (07/01/04).	■ COM (2003) 830	http://europa.eu.int/eur-lex/en/com/cnc/2003/com2003_0830en01.pdf
National Allocation Plans for the Allocation of Greenhouse Gas Emission Allowances	This Communication sets out the Commission's assessment of 8 plans & is accompanied by decisions addressed to each Member State. One of the reasons why the Commission was given responsibility under the Directive to assess the national allocation plans is to ensure that the Directive's criteria have been correctly applied in the distribution of tradable allowances before trading starts. In a single EU internal market & a single EU emissions trading scheme it is important to guard against distortion of competition through an incorrect application of the Directive or Treaty provisions. This is the first time that the EU engages in EU wide emissions trading & the first trading period, from 2005–2007, has been designed as a 'learning phase'.	Communication from the Commission concerning national allocation plans for the allocation of greenhouse gas emission allowances of Austria, Denmark, Germany, Ireland, the Netherlands, Slovenia, Sweden & the UK in accordance with Dir 2003/87/EC (07/07/04).	■ COM (2004) 500	http://www.europa.eu.int/eur-lex/en/com/cnc/2004/com2004_0500en01.pdf

Continued

Table 3.5.2 European Level Climate Change Policy. *Continued*

Policy Type	Programme Title	Policy Objective	Community Measure	Document Reference	Website
Council Decisions	Kyoto Protocol on Climate Change	To tackle climate change by means of international action to reduce the emissions of certain greenhouse gases responsible for global warming.	Council Decision 2002/358/EC concerning the approval, on behalf of the European Community, of the Kyoto Protocol to the UN Framework Convention on Climate Change & the joint fulfilment of commitments thereunder (25/04/02).	▪ Off Jnl L130, 15/05/02	http://europa.eu.int/scadplus/leg/en/lvb/l28060.htm
	Establishing Guidelines for the Monitoring & Reporting of Greenhouse Gas Emissions	This Annex contains the general guidelines for the monitoring & reporting of emissions from the activities listed in Annex I to Directive 2003/87/EC of greenhouse gases specified in relation to those activities. Additional guidelines on activity specific emissions are set out in Annexes II–XI.	Council Decision C (2004) 130 establishing guidelines for the monitoring & reporting of greenhouse gas emissions pursuant to Dir 2003/87/EC (29/01/04)	▪ C (2004) 130 29/01/04	http://www.europa.eu.int/comm/environment/climat/pdf/c2004_130_en.pdf
	Mechanism for Monitoring Community Greenhouse Gas Emissions & for Implementing the Kyoto Protocol	This Decision establishes a mechanism for: ▪ monitoring all anthropogenic emissions by sources & removals by sinks of greenhouse gases not controlled by the Montreal Protocol on substances that deplete the ozone layer in the Member States; ▪ evaluating progress towards meeting commitments in respect of these emissions by sources & removals by sinks; ▪ implementing the UNFCCC (United Nations Framework Convention on Climate Change) & the Kyoto Protocol, as regards national programmes, greenhouse gas inventories, national systems and registries of the Community & its Member States, & the relevant procedures under the Kyoto Protocol; & ▪ ensuring the timeliness, completeness, accuracy, consistency, comparability & transparency of reporting by the Community & its Member States to the UNFCCC Secretariat.	Dec No 280/2004 concerning a mechanism for monitoring Community greenhouse gas emissions and for implementing the Kyoto Protocol (11/02/04)	▪ 19/02/04	http://europa.eu.int/eur-lex/pri/en/oj/dat/2004/l_049/l_04920040219en00010008.pdf

Miscellaneous	The implementation of the Montreal Protocol on Substances that Deplete the Ozone Layer in the EU	This study endeavours to provide systematic information about the implementation of the Montreal Protocol by the EU Member States. After providing information on the current international & European regulatory situation, case studies on the situation in each EU Member State are provided. These analyse the implementation of policies for the protection of the ozone layer in the respective country. Based on these case studies, a concluding assessment of the implementation of the Montreal Protocol in the EU Member States is provided. Differences in legislation & enforcement of regulations become visible as well as the achievements of Montreal Protocol implementation in the EU.	Final report to the European Commission, Environment DG & UK DETR (Feb 2000).	Final Report to EC & Env DG	http://www.europa.eu.int/comm/environment/ozone/implementation_report.pdf
	Substances which Damage the Ozone Layer	To limit & monitor the Community production, marketing & use of substances which deplete the ozone layer & exports of those substances to 3rd countries, in order to protect human health & the environment.	Regulation (EC) No. 2037/2000 of the European Parliament & the Council on substances that deplete the ozone layer (29/06/00). Amended by Reg (EC) No. 2038/2000 (28/09/00) and Reg (EC) No. 2039/2000 (28/09/00).	Off Jnl L244, 29/09/00	http://europa.eu.int/scadplus/leg/en/lvb/l28064.htm
	Report for a Monitoring Mechanism of Community Gas Emissions	This is the first progress report under Council Decision 99/296/EC amending Decision 93/389/EEC for a monitoring mechanism of Community CO_2 and other greenhouse gas emissions. The report presents the results of the evaluation process under the decision and assesses the actual and projected progress of Member States and the Community towards fulfilling their greenhouse gas (GHG) emission commitments under the United Nations Framework Convention on Climate Change (UNFCCC) and the Kyoto Protocol. The Convention was aimed at the stabilisation of greenhouse gases (at 1990 level) while the Kyoto Protocol set a target of –8% (from the 1990 level) for the EU.	Report under Council Decision 1999/296/EC for a monitoring mechanism of Community greenhouse gas emissions (22/11/00).	COM (2000) 749	http://europa.eu.int/eur-lex/en/com/rpt/2000/com2000_0749en01.pdf

Continued

Table 3.5.2 European Level Climate Change Policy. *Continued*

Policy Type	Programme Title	Policy Objective	Community Measure	Document Reference	Website
Miscellaneous	EU Focus on Climate Change	An overview of the EU position and action towards combating climate change. Includes reference to EU policies and discussion on the wider implications of climate change on sustainable development.	European Commission Directorate-General for the Environment Report (2002).	▪ EC DG Environment Report	http://www.europa.eu.int/comm/environment/climat/pdf/climate_focus_en.pdf
	Monitoring Mechanism of Community Greenhouse Gas Emissions	This is the third progress report under Council Decision 93/389/EEC as amended by Decision 99/296/EC for a monitoring mechanism of Community CO_2 & other greenhouse gas emissions. It assesses the actual & projected progress of Member States & the Community towards fulfilling their greenhouse gas emission commitments under the United Nations Framework Convention on Climate Change (UNFCCC) & the Kyoto Protocol. The Convention was aimed at returning greenhouse gas emissions to their 1990 level by 2000, while the Kyoto Protocol requires the EC to reduce greenhouse gas emission by 8% 2008–2012.	Report from the commission under Council Decision 93/389/EEC as amended by Decision 99/296/EC for a monitoring mechanism of Community greenhouse gas emissions (09/12/02).	▪ COM (2002) 702	http://europa.eu.int/eur-lex/en/com/rpt/2002/com2002_0702en01.pdf
	Proposal for a Decision for a Monitoring Mechanism of Community Greenhouse Gas Emissions & the Implementation of the Kyoto Protocol	This proposal for a Decision regarding a monitoring mechanism of Community greenhouse gas emissions & the implementation of the Kyoto Protocol replaces Council Decision 93/389/EEC for a monitoring mechanism of Community CO_2 and other greenhouse gas emissions, which established a mechanism for monitoring anthropogenic greenhouse gas emissions & evaluating progress towards meeting commitments in respect of these emissions.	Proposal for a Decision of the European Parliament & of the Council for a monitoring mechanism of Community greenhouse gas emissions & the implementation of the Kyoto Protocol (05/02/03).	▪ COM (2003) 51	http://europa.eu.int/eur-lex/en/com/pdf/2003/com2003_0051en01.pdf
	2nd Report of the ECCP – Can we meet	This report is a follow-up of the first ECCP report, published in June 2001. This report gives an overview of the work within the different working	European Climate Change Programme Report (April 2003).	▪ ECCP Report	http://europa.eu.int/comm/environment/climat/pdf/second_eccp_report.pdf

our Kyoto Targets?	groups of the European Climate Change Programme that were operational in 2002, in particular ▪ Joint Implementation (JI) and Clean Development Mechanism (CDM) & the link with emissions trading, ▪ Agriculture, ▪ Sinks in agricultural soils, ▪ Forest related sinks & ▪ Fluorinated gases. It also gives an overview of the follow-up work in terms of implementation of measures that were identified in the first phase of the ECCP.			
Proposal for a Directive Amending the Directive Establishing a Scheme for Greenhouse Gas Emission Allowance Trading in the Community, in Respect of the Kyoto Protocol Project Mechanisms	This proposal will boost JI & the CDM by providing additional incentives for business to engage in these mechanisms. It thereby promotes technology transfer to industrialised countries, for example Russia, & to developing countries while reducing the costs of meeting commitments under the EC emissions allowances trading scheme. The proposal sends a strong signal to other Parties to the Kyoto Protocol that the Community is fully committed to the Kyoto flexible mechanisms & the advantages that they entail for both the Community & other Parties. At the same time, industrialised countries have a responsibility to reduce their emissions of greenhouse gases through domestic measures, given their historic levels of emissions & current higher *per capita* emissions than developing countries.	Proposal for a Directive of the European Parliament & of the Council amending the Directive establishing a scheme for greenhouse gas emission allowance trading in the community, in respect of the Kyoto Protocol project mechanisms (23/07/03)	▪ COM (2003) 403	http://europa.eu.int/eur-lex/en/com/pdf/2003/com2003_0403en01.pdf
Proposal for Regulating Certain Fluorinated Gases	This proposal for a new EC Regulation on fluorinated gases is a key element of the first phase of the European Climate Change Programme. It will put in place a legislative framework to reduce emissions of hydrofluorocarbons (HFCs), perfluorocarbons (PFCs) and sulphur hexafluoride, which are powerful greenhouse gases covered by the Kyoto Protocol. The proposal includes provisions on the containment, reporting, marketing and use of fluorinated gases.	Proposal for a Regulation of the European Parliament & the Council on Certain Fluorinated Greenhouse Gases (11/08/03).	▪ COM (2003) 492	http://europa.eu.int/eur-lex/en/com/pdf/2003/com2003_0492en01.pdf

Continued

Table 3.5.2 European Level Climate Change Policy. *Continued*

Policy Type	Programme Title	Policy Objective	Community Measure	Document Reference	Website
Miscellaneous	Control of Halon Exported for Critical Uses, the Export of Products & Equipment Containing Chloro-fluorocarbons & Controls on Bromochloro-methane	In applying Regulation (EC) No 2037/2000 of the European Parliament & of the Council on substances that deplete the ozone layer, a number of issues have emerged that need to be addressed through amendments to that Regulation. These issues that relate to the effective & safe implementation of that Regulation have been discussed with the Member States in the Management Committee under that Regulation. This Regulation concerns four amendments to Regulation (EC) No 2037/2000.	Regulation (EC) No 1804/2003 of the European Parliament & of the Council amending Regulation (EC) No 2037/2000 as regards the control of halon exported for critical uses, the export of products and equipment containing chloro-fluorocarbons and controls on bromochloro-methane (22/09/03).	■ Off Jnl L265/1 16/10/03	http://europa.eu.int/eur-lex/pri/en/oj/dat/2003/l_265/l_26520031016en00010004.pdf
	A Monitoring Mechanism of Community Greenhouse Gas Emissions	This is the fourth progress report under Council Decision 93/389/EEC as amended by Decision 99/296/EC for a monitoring mechanism of Community carbon dioxide (CO_2) & other greenhouse gas emissions. It assesses the actual & projected progress of Member States & the Community towards fulfilling their greenhouse gas (GHG) emission commitments under the United Nations Framework Convention on Climate Change (UNFCCC) & the Kyoto Protocol.	Report from the Commission under Council Decision 93/389/EEC as amended by Dec 99/296/EC for a monitoring mechanism of Community Greenhouse Gas emissions (28/11/03).	■ COM (2003) 735	http://europa.eu.int/eur-lex/en/com/rpt/2003/com2003_0735en01.pdf
	Action on Climate Change Post 2012: A Stakeholder Consultation on the EU's Contribution to Shaping the Future Global Climate Change Regime	This paper provides a background to the stakeholder consultation on the EU's contribution to shaping the future global climate change regime. It outlines a range of elements & issues on which stakeholders may wish to comment, & lists a set of questions that stakeholders may wish to answer in their submissions. The elements & options set out in this paper aim to stimulate discussion & do not necessarily represent the Commission's position.	Background Commission Paper to the EU Stakeholder Conference (Sept 2004).	■ Commission Background Paper	http://www.europa.eu.int/comm/environment/climat/pdf/background_paper.pdf

Table 3.5.3 UK Level Climate Change Policy.

Policy Type	Document Title	Objective	Contents	Website
Strategies & Reports	Energy Efficiency in the Domestic Sector Technical Support Paper (09/11/98)	The purpose of this paper was to explain how the various energy efficiency figures in the Climate Change Consultation Paper have been estimated. The introductory section sought to explain the broad issues to the non-technical reader, while the remainder of the paper was aimed at technical experts for them to examine in detail the data sources and assumptions underlying the analysis. In this way, it was hoped to encourage an informed dialogue about future policies and programmes.	Summary; Section 1: Introduction; Section 2: Assumptions & data sources; Section 3: Detail of the measures considered; Section 4: Welfare costs & benefits calculation; Annex 1: Results by energy efficiency measure.	http://www.defra.gov.uk/environment/consult/climatechange/technical/index.htm
	Climate Change & UK Nature Conservation: A Review of the Impact of Climate Change on UK Species & Habitat Conservation Policy (20/12/00)	The aim of this study was: ▪ To review climate change impacts on species, habitats & networks, wider countryside, & current policy commitments. ▪ To consider & outline policy responses & a framework for implementing those policies; ▪ To recommend prioritised research & an early detection system.	Executive Summary; Chapter 1: Introduction; Chapter 2: Climate change; Chapter 3: Review of impacts on species & habitats; Chapter 4: Current conservation commitments; Chapter 5: Policy responses to climate change; Chapter 6: Detecting & monitoring climate change; Chapter 7: Research needs; Chapter 8: Overall conclusions; References; Appendix 1: Completed & ongoing UKCIP (UK Climate Impacts Programme) projects; Appendix 2: Species names; Appendix 3: Workshop reports; Appendix 4: Species & habitat matrices; Appendix 5: Adaptation matrices; Appendix 6: Evaluation matrices; Appendix 7: Derivation of vulnerability maps; Appendix 8: Monitoring networks & indicators.	http://www.defra.gov.uk/wildlifecountryside/climatechange/nature/pdf/climatechange.pdf

Continued

Table 3.5.3 UK Level Climate Change Policy. *Continued*

Policy Type	Document Title	Objective	Contents	Website
Strategies & Reports	Climate Change The UK Programme (02/02/01)	The UK's climate change programme sets out the Government's & the devolved administrations' approach to the challenge of climate change. It is designed to ensure that the UK's emissions remain on a downward path & to prepare the UK for making bigger cuts in the future. It aims to: ▪ Reflect the importance of tackling climate change; ▪ Take a balanced approach, with all sectors & all parts of the UK playing their part; ▪ Safeguard, & where possible enhance, the UK's competitiveness, tackle social exclusion & reduce harm to health; ▪ Focus on cost effective & flexible policy options which will work together to form an integrated package; & ▪ Take a long-term view, looking to targets beyond the Kyoto commitment period.	Foreword; Introduction; Section 1: The UK's strategy; Section 2: Delivering emission reductions; Section 3: Adapting to the impacts of climate change in the UK; Section 4: Action by the devolved administrations; Section 5: Annexes.	http://www.defra.gov.uk/environment/climatechange/cm4913/4913html/index.htm
	Draft Framework Document for the UK Emissions Trading Scheme (22/05/01)	The UK Government is introducing an emissions trading scheme to reduce greenhouse gas emissions as part of its Climate Change Programme. This will be one of the first national greenhouse gas trading schemes anywhere in the world. There is a broad consensus on the need for action to address climate change, which could have a serious impact on the world's economy, society and environment. Significant emissions reductions will be needed over the next century to avoid dangerous levels of climate change. Therefore, the Government wants to introduce policies that can reduce greenhouse gas emissions in the most efficient way possible, with emissions trading playing a central role.	Introduction; Part A: Framework for entry – Section 1: Preparing for entry into the scheme; Section 2: Entry into the scheme; Part B: Operating framework – Section 3: Targets & trading; Section 4: Reporting & compliance; Section 5: Project-based credits; Section 6: Links with International trading schemes & other trading-related policies; Section 7: Administration of the Scheme; Section 8: Future plans; Annex 1: Changes to the baseline list, baseline & targets.	http://www.defra.gov.uk/environment/climatechange/trading/uk/draft/

The Social Cost of Carbon Review – Background Paper (July 2003)	This paper provides a background to this review of the social cost of carbon. Section 2 provides a summary of the currently available estimates & explores the drivers in their variability. Section 3 looks at the main limitations of the current estimates, particularly the fact that they typically do not cover low-probability catastrophic impacts & 'socially contingent impacts' from climate change. Section 4 raises some important issues in terms of the use of social cost of carbon estimates in policy decision making, particularly the trade-offs between using broad ranges to reflect uncertainty & using more holistic decision-making frameworks to inform strategic climate change policy choices, & the benefits of having a consistent benchmark for the carbon externality. Section 5 attempts to draw some conclusions & to highlight key questions for the review.	Introduction; Available estimates & main drivers of variability; Limitations of the current estimates; Using SCC estimates in policy decision-making; Conclusions & key questions for a review; References.	http://www.defra.gov.uk/environment/climatechange/carbonseminar/background.pdf
Climate Change & Local Communities – How Prepared are You? (July 2003)	This report from the UK Climate Impacts Programme (UKCIP) is designed to provide practical help & support for local communities to ensure that their infrastructure & key services are resilient to the impacts of climate change.	Foreword; Why take account of climate change? Is climate changing? What is being done in the UK? How will climate change affect my council? Adapting to the effects of climate change; When do I need to take action? What are the impacts of climate change on the UK? Websites.	http://www.ukcip.org.uk/resources/publications/documents/Local_authority.pdf
Scenarios of Climate Change for Islands within the BIC Region – British–Irish Council (July 2003)	This report is a concrete example of the co-operation between the British–Irish Council on the issue of climate change. It provides new details on how the climate is likely to change across the BIC region. A key output of British–Irish Council activities are the BIC Climate Change Scenarios. These will help improve our understanding of the impacts of climate change in the different BIC administrations (particularly the islands), so that we can be in a better position to prepare to adapt to its consequences.	Foreword; Summary; Ch 1: Introduction to the climate change scenarios; Ch 2: Observations of trends & variability of climate; Ch 3: Future global climate change; Ch 4: Changes in seasonal climate; Ch 5: Changes in daily climate & extremes; Ch 6: Sea-level rise; Ch 7: Uncertainties in scenarios & some ways of managing them; Ch 8: Further work & development's in research; Annexes.	http://www3.britishirishcouncil.org/climatechange/bicpdf070703.pdf

Continued

Table 3.5.3 UK Level Climate Change Policy. *Continued*

Policy Type	Document Title	Objective	Contents	Website
Strategies & Reports	The Impacts of Climate Change: Implications for DEFRA (Jan 2003)	This report sets out the findings of a study to explore the potential implications of changing climate across the whole range of responsibilities of DEFRA.	Foreword; Summary; Introduction; Identifying priorities for DEFRA; Progress to date; Addressing the constraints on progress; Auditing progress; Annexes.	http://www.defra.gov.uk/environment/climatechange/impacts2/pdf/ccimpacts_defra.pdf
	Energy White Paper – Our Energy Future: Creating a Low Carbon Economy (Feb 2003)	This White Paper is a milestone in energy policy. It is based on the four pillars of the environment, energy reliability, affordable energy for the poorest, & competitive markets for our businesses, industries & households. The White Paper sets out a strategy for the long term, to give industry the confidence to invest to help deliver the goals to: ▪ Cut UK CO_2 emissions; ▪ Maintain the reliability of energy supplies; ▪ Promote competitive markets; and ▪ Ensure every home is adequately & affordably heated.	Foreword; Overview; Cleaner, smarter energy; The environment; Energy efficiency; Low carbon generation; Clean low carbon transport; Energy reliability; Productivity, competitiveness & innovation; Energy & the vulnerable; Delivery through partnership; Glossary; References.	http://www.dti.gov.uk/energy/whitepaper/ourenergyfuture.pdf
	Options for a Low Carbon Future – Phase 2 (Feb 2003)	This second phase of work on Long Term Low Carbon Options was commissioned by the DTI, with input from other government departments. It has made a more detailed analysis of the issues identified in the first study, & also examined issues relating to the phasing of emission reduction & the role of specific technologies. The results have helped inform the Energy White Paper.	Introduction; Approach – scenario assumptions; Economic implications of a low carbon future; Distribution of carbon abatement actions between sectors; Fuel costs, availability & infrastructure requirements; The importance of technology in supporting a low carbon future; Impact of different paths to a carbon future; References; Annexes.	http://www.dti.gov.uk/energy/whitepaper/phase2.pdf

Creating a Low Carbon Economy – 1st Annual report on the Implementation of the Energy White Paper (April 2004)	This report represents the first annual review of the energy white paper as obligated by the Sustainable Energy Act 2003. It provides an overview of the last 12 months and reinforces the commitments made in the Energy White Paper.	Foreword; The last 12 months & the way ahead: an overview; Reducing carbon emissions; Energy reliability; Competitive energy markets; Tackling fuel poverty; Annex: Other reports & plans published alongside this report.	http://www.dti.gov.uk/energy/sepn/annualreport/firstannualreportfull.pdf
Scientific & Technical Aspects of Climate Change, Including Impacts & Adaptation & Associated Costs (Sept 2004)	This paper reviews the evidence for climate change, its human causes, projections for future climate change, its likely impacts and related adaptation options, and the issues surrounding stabilisation of greenhouse gases in the atmosphere at an acceptable level.	Evidence for climate change & its effects; Evidence for human influence on climate change; Future climate change; Abrupt climate change & other high impact, low probability events; Potential impacts of climate change; Regional impacts of climate change; Adaptation to climate change in the UK; Developing countries; Limiting climate change & its effects; Confidence & uncertainty.	http://www.defra.gov.uk/environment/climatechange/pdf/cc-science-0904.pdf
UK International Priorities – the Energy Strategy (Oct 2004)	This document generated by the Foreign and Commonwealth Office (FCO) sets out how we propose to meet the international energy challenge over the next five to ten years. It builds on DTI's 2003 White Paper, *Our Energy Future creating a low carbon economy*, & on the FCO's White Paper, *UK International Priorities*, of Dec 2003. The Strategy will guide & inform the work of our officials on international energy issues at home & in our network of posts abroad, helping us to concentrate our efforts where they will be most effective.	Foreword; Introduction; Climate change; Energy security; Some major questions common to energy security & climate change; Energy for poverty reduction; Good governance in the energy sector; Promoting UK commercial interests; Linkages between various UK international energy objectives; Linkages between international energy objectives & other UK international priorities; Multi-lateral institutions & partners; UK working methods.	http://www.fco.gov.uk/Files/kfile/Energy_Report_281004.pdf

Table 3.5.4 Northern Ireland Level Climate Change Policy.

Policy Type	Document Title	Objective	Contents	Website
Strategies & Reports	Environmental Monitoring – New Technologies for Monitoring: Status & Prospects (Feb 2001)	The purpose of this paper is to: ▪ Report on current monitoring methods & technologies in use; ▪ Identify areas where new technologies are being researched or tested for future application; and ▪ Identify areas of common interest for North/South co-operation.	Part 1: Introduction; Part 2: Monitoring emissions to air & water; Part 3: Monitoring the aquatic environment; Part 4: Monitoring air quality & deposition; Part 5: Co-operation on technologies/methodologies for monitoring.	http://www.ehsni.gov.uk/pubs/publications/North_South_Min_Coun_Doc.pdf
	Climate Change & Nature Conservation in Britain & Ireland (Nov 2001)	The study Modelling Natural Resource Responses to Climate Change (MONARCH) evaluates the impacts of climate change on wildlife & geomorphological features in Britain & Ireland. It identifies the potential implications for species & habitats in four environments – terrestrial, freshwater, coastal & marine.	Introduction; A changing climate; What the study entailed; Bioclimatic classification; Impact models; Key findings; Terrestrial & freshwater habitats; Coastal & marine habitats; Implications of climate change for nature conservation policy & practice; At a glance – winners & losers of climatic space.	http://www.ehsni.gov.uk/pubs/publications/MONARCHSUMMABROCHURE.ART.PDF
	EU Emissions Trading Scheme (ETS) Guidance Note 1 (Dec 2003)	This guidance was drawn up on the basis of the draft EU ETS implementing the Greenhouse Gas Emissions Trading Scheme Regulations 2003 and are therefore subject to change pending further work on the regulations.	How do I know if I am included? Who should apply for a permit? How should I define the scope of the installation? When should capacities be aggregated to determine whether a scheduled activity is being carried out? How do I define a directly associated activity? Examples of the directly associated activity rule; If I have a directly associated activity on site do I need to account for the emissions from that	http://www.ehsni.gov.uk/pubs/publications/eu-ets-guidance01.pdf

			activity? I have stand-by generation or boiler capacity on site – should I include them in the aggregation? What is & isn't included in the definition of 'combustion installation'? What happens if there is a CHP (Combined Heat and Power) plant on site that supplies the process with electricity, but is operated by a different organisation? How do I find out what the rated thermal input of my combustion installation is? Further examples; Checklist; Annexes.	
	Strangford Lough Ecological Change Investigation (SLECI) – Climate Change Desk Study (Aug 2004)	This part of SLECI aimed to assess how climate change scenarios may have contributed to recent events in Strangford Lough & their potential future impacts on the Lough.	Introduction; Climate change in Northern Ireland; Predicted impacts of climate change in NI; Conclusions & recommendations for future work; References.	http://www.ehsni.gov.uk/pubs/publications/51ClimChge.pdf
	Climate Change Indicators for Northern Ireland (Sept 2004)	This document presents a series of indicators which have been carefully selected to be indicative of the changing climate in Northern Ireland. Information on each indicator is provided alongside data trends with an indication of how this can be linked back to climate change.	Introduction; Air temperature in NI; Number of hot & cold days per year; Seasonality of rainfall; Annual rainfall; Number of snow days; Position of the gulf stream; Length of growing season; Potato yields; Arrival date of the First swallow; First sighting of large white butterfly; Mean river flow in River Bush; Groundwater levels in Dunmurry & Killyglen; Sea level in NI; Appendices.	http://www.ehsni.gov.uk/pubs/publications/cci.pdf

4 Construction

4.1 Construction and the Built Environment

Construction is an essential element of the built environment, and can be described as the synergy between conceptual designs and development projects or how society imposes on the natural environment. Underlying the visual imposition of construction projects is the complex provision of infrastructural services which form the physical fabric of a nation. Highways, railways, water and sewerage services, energy, communications, ports and harbours add to the building stock and rely on efficient and effective provision of services as well as adequate funding to bring benefit to industrial and domestic societies.

The construction industry creates large revenue and provides direct employment to millions of people worldwide. In the UK between 10% and 15% of the nation's gross national product is typically generated by the construction industry, which has a profound impact on our daily lives. This industry represents a key indicator and driver of economic activity and wealth creation. Furthermore in order to remain competitive, the construction industry must continue to evolve and develop on business, technological and environmental fronts. The construction sector is multi-disciplinary, involving a range of professions and skills, and employing most of the built environment professionals in teams, as well as those from the financial and business sectors. While the industry is driven by key health and safety legislation, it is also seen as a measure of the nation's vitality and profitability, and addresses the needs of all communities, thereby becoming a vehicle for delivering social inclusion.

However, the construction industry has suffered from an image of inefficiency. This prompted an EC Communication in 1997 on *The Competitiveness of the Construction Industry* which identified four strategic objectives aimed at enhancing the sector through:

- development of a coherent quality policy;
- improvement of the regulatory environment;
- better provision of education and training; and
- reorientation and reinforcement of research and development.

In turn the UK Government delivered a *Rethinking Construction* report in 1998 which introduced the goals of quantifying the scope for:

- improving construction efficiency;
- deriving relevant quality/efficiency targets and performance measures which might be adopted by UK construction;
- examining the current practice and scope for improvement in innovation in products and processes;
- identifying specific actions and good practice which would help achieve more efficient construction in terms of quality, customer satisfaction, timeliness in delivery and value for money; and
- identifying projects to help demonstrate the improvements that can be achieved through the application of best practice.

These led to the *Movement for Innovation* demonstration projects show-cased through UK Rethinking Construction Centres, latterly re-named Centres of Construction Excellence.

4.2 Key Construction Elements

Construction is the tangible outcome from a range of processes. Clients perceive the need for a project, seeking advice from a range of disciplines such as planners, architects, civil and construction engineers to ensure that a preferred and optimal design is agreed within a cost budget. The selection and use of appropriate procurement processes will guide and inform the employment of individual companies or consortia to design and construct a project, with the added possibility of providing maintenance and operational functions.

The feasibility of each project is initially considered within the context of international and national strategies and conventions, and then becomes a matter for both technical and financial appraisals. This conceptual design approach of each individual project involves a breadth of considerations across several disciplines. Such an approach necessitates the study of a range of alternatives, the engagement of several professional experts, the application of legislation and the integration of all these issues into a weighted decision-making process to secure the optimum design. The final design is normally described in a set of working drawings, a schedule of works, a bill of quantities and the conditions of contract. The procurement method chosen will also have an influence on the selection of the design team and methodology deployed.

A wide range of procurement methods exist, from a simple 'no risk' approach of a lump sum tender by a contractor to complete a project within an agreed period, through to an 'admeasurement contract' in which a contractor is paid, in stage payments, for work completed against a set of working drawings and bill of quantities allowing for re-measurement of the finished project. Similarly 'design build finance and operate' projects require a multi-discipline consortium to take responsibility for all aspects of the design, construction and maintenance of the project over an extended period of up to 25 years. This latter approach has been driven by the *Rethinking Construction* initiative referred to earlier.

Construction methods are guided at a high level by health and safety legislation and should also be influenced by the wider sustainable development agenda. This has led to a greater need for and application of sustainable construction methods. In this respect the construction industry is challenged to mitigate or minimise environmental change created by development projects. This can be achieved by appropriate materials selection, recycling of materials, waste minimisation and adopting a whole life cycle approach to include maintenance, down-taking or demolition and site closure.

4.3 Sustainability Challenges for Construction

Construction challenges fall initially into both 'environmental protection' and 'economic growth' sectors as it is fully recognised that the industry has had significant impacts, both positive and negative, on the natural environment. In addition, sustainable construction requires detailed attention the 'prudent use of natural resources' and a full awareness of its impact on societies. Links between sustainable construction and the sustainable development objectives include:

- *Effective protection of the environment* is possible through controlling and minimising the impacts of construction acoustics, pollution and air quality, including potential damage to biodiversity. The re-use or decontamination of brown-field sites reverses environmental damage and enhances conservation. Construction site control and minimisation of energy usage, awareness of embodied energy, improved economy of materials transport and a reduction in water usage also improve environmental protection.

- *Social progress which recognises the needs of everyone* requires community consultation and feedback by the construction industry involving key stakeholder engagement, while adhering to health and safety legislation. Full attention to high quality design will ensure that many social issues are accounted for on the micro scale, issues such as accessibility for disabled and able-bodied persons, workers' training and development, affordability of shelter, quality of the internal environment, and building occupancy levels are important. Conflict can be minimised through close liaison between the clients, constructors and community groups and users.

- *Prudent use of natural resources* results in both greater industry efficiency and a restricted usage of natural materials. Practices such as materials recycling, waste minimisation, local product resourcing, land decontamination, and construction and demolition waste disposal make sound business sense and encourage good housekeeping. Application of the principles of 'lean construction' and life cycle analysis are important, while greater attention to renewable energy usage contributes to both financial and environmental efficiency.
- *Maintenance of high and stable levels of economic growth and employment* relates to increased levels of construction productivity which may involve innovative processes, the use of green procurement methods and the appropriate application of public private partnerships. The adoption of a whole life costing approach facilitates a fuller understanding of management and maintenance costs as well as capital and recurrent expenditure resulting in increased construction industry profitability.

Several ongoing issues and challenges exist for environmentalists, economists and strategists in developing the built environment through sustainable construction. This sector also maintains close relationships to the broader areas of climate change, urban development, waste management and social inclusion. Hence sustainable construction is seen as a key driver for the delivery of many of the principles of sustainable development. The critical challenges which arise for sustainable construction in contributing to this broader sustainability agenda include:

- *Creating a high quality construction product* which creates a lasting effect through ensuring a balance between 'value for money' and 'whole life costings'.
- *Improving construction procurement methods*, and facilitating a change in procurement structures, (especially the use of 'Design, Build, Finance and Operate' systems through Private Finance Initiatives or Performance Related Partnering) to increase construction quality and demonstrating an ability to learn from previous mistakes.
- *Using Life Cycle Analysis as a means of holistic management* to examine the whole life cycle of the building, including 'end of life' and maintenance period, while facilitating long-term decision-making with regard to materials and construction techniques in order to balance expensive techniques or materials against a building or project's life span and usage.
- *Recognising the impact of health and safety and social issues*, that the health and well-being of construction employees have a big impact on the effectiveness of the construction industry, and accounting for key health and safety risks such as exposure to construction vibrations, noise and asbestos.
- *Recognising the role of design in contributing to long-term project development*, that it is central to fully understanding the impact on the general public, respect for people, life-long learning and a full range of comfort factors. These could include the impact of the finished building product on potential occupiers, the quality of the internal environment through heating, lighting, ventilation and layout, and the quality of the external environment through landscape, streetscape, finishes, air quality and infrastructure provision.
- *Role of construction benchmarking* is important for monitoring and reporting as well as for disseminating best practice in the pursuit of construction efficiency. The use of Key Performance Indicators (KPIs) and Environmental Performance Indicators (EPIs) as performance measurements contributes to the 'Rethinking Construction' movement in terms of benchmarking and demonstration of sustainability projects and methods enabling performance comparisons. This provides the opportunity to showcase construction innovation in terms of construction companies, products and processes.
- *Using environmentally friendly construction materials* to assist in environmental protection and balancing the cost issues against long-term durability and maintenance-free arguments. The impact of the Construction Products Directive and the minimum standards of materials to be used in construction emphasises the use of local suppliers to support the local economy, avoiding costly embodied transportation energy and providing an Integrated Product Policy (IPP) approach to environmental management.
- *Attention to construction competitiveness*, which should help emphasise the importance of the education and training, and research and development aspects of the construction industry, and the operation of the industry within a regulatory environment.

- *The adoption of construction waste minimisation practices* requiring detailed evaluation of all construction and demolition waste activities against waste management statutory targets. The need for increased recycling and salvaging of waste products to reduce amounts sent to landfill sites is clear. Similarly the associated impact of aggregate and landfill taxes and related EC waste disposal directives, requires integrated site waste management plans and a greater understanding of the economic aspects of developing markets for construction waste material.
- *Application of 'lean construction' concepts* necessitating the management and improvement of construction processes and products to help minimise wastage, and the application of maintenance-free design and integration with the supply-chain management.
- *Integration of construction quality accreditation systems* to assist improved delivery, while providing a statement of capability for a company or supplier and acting as a marketing tool. This will require involvement with key construction organisations and programmes, such as the Confederation of House-builders, Constructing Excellence UK, Constructing Best Practice Programme, Considerate Constructors Scheme, Building Research Establishment (BRE) and the Civil Engineering Environmental Quality Assessment and Award Scheme (CEEQUAL).

4.4 Construction Theme Index

Jurisdiction	Tier	Title	Date	Page No.
EUROPEAN	Miscellaneous	Quality in the Construction Sector	1999	94
		Proposal for a Directive on the Energy Performance of Buildings *(COM (2001) 226)*	2001	94
		Sustainable Construction Final Report	2001	94
		Task Group 1 Final Report: Environmentally Friendly Construction Materials	2001	95
		Task Group 2 Final Report: Energy Efficiency in Buildings	2001	95
		Task Group 3: Construction & Demolition Waste	2001	96
		Benchmarking Pilot Study – Final Report	2001	96
		Education, Training & Image of the Construction Sector	2002	97
		Analysis of Life Cycle Analysis (LCA) Tools	2003	97
UK	Strategies & Reports	Rethinking Construction – The Report of the Construction Task Force on the Scope for Improving the Quality & Efficiency of UK Construction	1998	98
		Towards More Sustainable Construction: Green Guide for Managers on the Government Estate	1999	98
		Towards More Sustainable Construction: How Sustainable Construction Can Improve the Performance of Buildings & their Occupants	1999	99
		Building a Better Quality of Life: A Strategy for More Sustainable Construction	2000	99
		Achieving sustainability in Construction Procurement – Sustainability Action Plan	2000	99
		Better Public Buildings – A Proud Legacy for the Future	2000	99
		A Commitment to People 'Our Biggest Asset' – A Report from the Movement for Innovation's Working Group on Respect for People	2000	100
		The Movement for Innovation – Sustainability Working Group Report: Environmental Performance Indicators for Sustainable Construction	2000	100
		KPI Report for the Minister for Construction	2000	100
		Six Guiding Principles to Improve the Sustainability of the Housing Construction Industry	2001	101
		Building a Better Quality of Life: A Strategy for More Sustainable Construction – Report on Progress 2001	2001	101
		Rethinking the Construction Client – Guidelines for Construction Clients in the Public Sector	2001	101
		CE Marking Under the Construction Products Directive	2001	102
		Reputation, Risk & Reward: The Business Case for Sustainability in the UK Property Sector	2002	102
		Accelerating Change – A Report by the Strategic Forum for Construction	2002	102
		Rethinking Construction 2002 – Achievements, Next Steps, Getting involved	2002	103
		Building a Better East of England – A Review of the Construction Industry in the Area	2002	103
		Respect for People – Reaching the Standard	2002	104
		A Framework for Action – The Report of Rethinking Construction's Respect for People Working Group	2002	104

Jurisdiction	Tier	Title	Date	Page No.
UK	Strategies & Reports	Government Construction Procurement Guidance No. 7: Whole Life Costs	2002	105
		EU Legislation Study Report	2003	105
		Building on Success – Achieving Excellence in Construction	2003	105
		Demonstrations of Sustainability	2003	106
		Sustainable Construction Brief	2003	106
		Rethinking Construction Demonstrations Report: Demonstrating success through Rethinking Construction	2003	106
		Better Buildings Summit – Issues Paper	2003	107
		The UK Construction Industry: Progress Towards More Sustainable Construction 2000–2003 (The Sustainable Construction Task Group)	2003	107
		Timber Use for Construction in the UK	2004	107
		Health and Safety Best Practice Factsheet	2004	107
		Procurement Best Practice Factsheet	2004	108
		Sustainable Construction: An Introduction	2004	108
		Sustainable Construction: Constructors	2004	108
		Sustainable Construction: Suppliers	2004	109
		Sustainable Construction: Designers Best Practice Factsheet	2004	109
		Whole Life Costing Best Practice Factsheet	2004	109
		Sustainable Construction: Clients	2004	109
		Sustainability Accounting Best Practice Factsheet	2004	110
		Demonstrating Excellence – An Evolution of the Programme of Demonstrations	2004	110
		Lean Construction Best Practice Factsheet	2004	110
		Construction Products Industry Key Performance Indicators Handbook	2004	111
	Guidance Notes	A Guide to the Construction (Health, Safety & Welfare) Regulations 1996	1996	111
		Effective Teamwork: A Best Practice Guide for the Construction Industry	2004	111
NORTHERN IRELAND	Strategies & Reports	Building on Success – Proposals for Future Housing Policy	1996	112
		Measuring Construction Performance	2002	112
		Construction & Demolition Waste Survey 2001	2003	112
		Northern Ireland Construction Industry – Rethinking Construction Centre Northern Ireland	2004	113
		The Aggregates Levy (Northern Ireland Tax Credit) Regulations 2004	2004	113
		BPEO for the Management of Waste Asbestos	2004	113
	Guidance Notes	Central Procurement Directorate Procurement Guidance Note 03/04: Social Considerations in Public Procurement	2004	114
		Central Procurement Directorate: Procurement Guidance Note 04/04 – Integrating Environmental Considerations into Public Procurement	2004	114

4.5 Construction Policy Tables

Table 4.5.1 European Level Construction Policy.

Policy Type	Programme Title	Policy Objective	Community Measure	Document Reference	Website
Council Directives	The Performance of Heat Generators for Space Heating & the Production of Hot Water in New or Existing Non-Industrial Buildings	Member States shall take all necessary measures to ensure all new heat generators for space heating and/or the production of hot water in new/existing non-industrial buildings comply with minimum performance requirements. In the case of generators capable of using various forms of energy, the minimum performance requirements must relate to each form of energy used.	Council Directive 78/170/EEC on the performance of heat generators for space heating & the production of hot water in new or existing non-industrial buildings & on the insulation of heat & domestic hot water distribution in new non-industrial buildings (13/02/78). Amended by CD 82/885/EEC (10/12/82).	■ Off Jnl L52, 23/02/78 ■ Off Jnl L378, 31/12/82	http://europa.eu.int/smartapi/cgi/sga_doc?smartapi!celexapi!prod!CELEXnumdoc&lg=EN&numdoc=31978L0170&model=guichett
	Protection of Workers from the Risks Related to Exposure to Asbestos at Work	This Directive has as its aim the protection of workers against risks to their health, including the prevention of such risks, arising or likely to arise from exposure to asbestos at work. It lays down limit values & other specific requirements.	Council Directive 83/477/EEC on the protection of workers from the risks related to exposure to asbestos at work (second individual Directive within the meaning of Article 8 of Directive 80/1107/EEC) (19/09/83).	■ Off Jnl L263 24/09/83	http://www.europa.eu.int/smartapi/cgi/sga_doc?smartapi!celexplus!prod!CELEXnumdoc&lg=en&numdoc=31983L0477
	The Construction Products Directive	The CPD aims to break down technical barriers to trade in construction products between Member States in the European Economic Area (EEA). To achieve this the CPD provides for the following four main elements: ■ a system of harmonised technical specifications ■ an agreed system of attestation of conformity for each product family ■ a framework of notified bodies ■ the CE marking of products.	Council Directive 89/106/EEC on the approximation of laws, regulations and administrative provisions of the Member States relating to construction products (21/12/88). Amended by CD 93/68/EEC (22/07/93).	■ Off Jnl L40 ■ Off Jnl L220	http://europa.eu.int/comm/enterprise/construction/internal/cpd/cpd.htm

Energy Certification of Buildings	To limit carbon dioxide emissions and promote the rational use of energy in those economic sectors in the European Community which consume the most energy, namely the residential and tertiary sectors, in order to preserve the quality of the environment.	Council Directive 93/76/EEC to limit carbon dioxide emissions by improving energy efficiency (SAVE) (13/09/93).	Off Jnl L237, 22/09/93	http://europa.eu.int/scadplus/leg/en/lvb/l21252.htm
Minimum Health & Safety Requirements Regarding the Exposure of Workers to the Risks Arising from Physical Agents (Vibration)	This Directive lays down minimum requirements for the protection of workers from risks to their health and safety arising or likely to arise from exposure to mechanical vibration.	Directive 2002/44/EC of the European Parliament & of the Council on the minimum health & safety requirements regarding the exposure of workers to the risks arising from physical agents (vibration) (sixteenth individual Directive within the meaning of Article 16(1) of Directive 89/391/EEC) (25/06/02).	Off Jnl L177 06/07/02	http://europa.eu.int/eur-lex/pri/en/oj/dat/2002/l_177/l_17720020706en00130019.pdf
Minimum Health & Safety Requirements Regarding the Exposure of Workers to the Risks Arising from Physical Agents (Noise)	This Directive lays down minimum requirements for the protection of workers from risks to their health & safety arising or likely to arise from exposure to noise & in particular the risk to hearing.	Directive 2003/10/EC of the European Parliament & of the Council on the minimum health & safety requirements regarding the exposure of workers to the risks arising from physical agents (noise) (Seventeenth individual Directive within the meaning of Article 16(1) of Directive 89/391/EEC) (06/02/03).	Off Jnl L042 15/02/03	http://europa.eu.int/eur-lex/pri/en/oj/dat/2003/l_042/l_04220030215en00380044.pdf
Protection of Workers from the Risks Related to Exposure to Asbestos at Work (Amending CD 83/477/EEC)	This directive aims to protect workers against risks to their health, including the prevention of such risks, arising or likely to arise from exposure to asbestos at work. It lays down limit values & other specific requirements and amends CD 83/477/EEC.	Directive 2003/18/EC of the European Parliament & of the Council amending Council Directive 83/477/EEC on the protection of workers from the risks related to exposure to asbestos at work (27/03/03).	Off Jnl L097 15/04/03	http://europa.eu.int/eur-lex/pri/en/oj/dat/2003/l_097/l_09720030415en00480052.pdf

Continued

Table 4.5.1 European Level Construction Policy. *Continued*

Policy Type	Programme Title	Policy Objective	Community Measure	Document Reference	Website
Council Communication	The Competitiveness of the Construction Industry	The Communication identifies four strategic objectives aimed at enhancing the competitiveness of the sector: ▪ To develop a coherent quality policy for the sector ▪ To improve the regulatory environment ▪ To improve provision of education and training ▪ To reorient and reinforce research and development The Communication proposes more than 60 specific actions aimed at fulfilling these strategic objectives.	Communication from the Commission, DGIII, Brussels on the Competitiveness of the Construction Industry, COM (97) 539 final (04/11/97).	▪ COM (97) 539	http://europa.eu.int/comm/enterprise/construction/compcom/compcom.htm
Council Recommendation	The Rational Use of Energy by Promoting the Thermal Insulation of Buildings	Member States: ▪ Organise or sponsor specific information campaigns designed to stimulate public interest in improving inadequate or poor-quality thermal insulation systems in residential accommodation ▪ Assess the average extent of thermal insulation in existing residential accommodation ▪ Organise competitions or draw up development programmes to obtain the best original ideas for improving insulation in existing residential accommodation ▪ Adopt the harmonised Community reference standards for thermal insulation which are to be drawn up for use in the implementation of all laws, regulations and administrative measures	Council Recommendation 76/492/EEC on the rational use of energy by promoting the thermal insulation of buildings (04/05/76).	▪ Off Jnl L140, 28/05/76	http://europa.eu.int/smartapi/cgi/sga_doc?smartapi!celexapi!prod!CELEXnumdoc&lg=EN&numdoc=31976H0492&model=guichett

Rational Use of Energy in Industrial Undertakings	Member States to: ■ Take the necessary measures to mount campaigns to provide information & increase awareness, & to encourage, the achievement of energy savings; ■ Set up one or more public or approved bodies/departments with particular responsibility for these tasks; ■ Invite sectoral & general professional organisations as well as technical & scientific associations & consultancy services to arrange periodically for meetings & seminars; ■ Encourage industrial undertakings to: – Set up an energy department within their organisation, if they have not already done so, or call in an outside expert (see Annex) – Devote a chapter of an annual report to energy consumption, including any energy-saving measures taken or planned & the results obtained; ■ Collate all useful information concerning the results of the energy-saving measures taken by the industrial undertakings; ■ Inform the Commission regularly of the measures taken & of the results obtained or anticipated from the measures.	Council Recommendation 77/713/EEC on the rational use of energy in industrial undertakings (23/10/77).	■ Off Jnl L295, 18/11/77	http://europa.eu.int/smartapi/cgi/sga_doc?smartapi!celexapi!prod!CELEXnumdoc&lg=EN&numdoc=31977H0713&model=guichett

Continued

Table 4.5.1 European Level Construction Policy. *Continued*

Policy Type	Programme Title	Policy Objective	Community Measure	Document Reference	Website
Council Recommendation	The Reduction of Energy Requirements for Buildings in the Community	To pursue policies designed to save energy by reducing energy requirements through the improvement of the thermal efficiency of buildings, the Commission will organise appropriate consultations with the Member States within the context of the Energy Committee in order to: ■ prepare the reports ■ facilitate exchanges of information on matters relevant to the implementation of this recommendation which could prove helpful to Member States in the application of their programmes.	Council Recommendation 79/167/ECSC, EEC, Euratom on the reduction of energy requirements for buildings in the Community (05/02/79).	■ Off Jnl L37, 13/02/79	http://europa.eu.int/smartapi/cgi/sga_doc?smartapi!celexapi!prod!CELEXnumdoc&lg=EN&numdoc=31979H0167&model=guichett
Miscellaneous Reports	The Rational Use of Energy in the Building Sector	In the framework of this programme the Commission, in consultation with experts from the Member States, was to conduct a detailed assessment of the pilot projects relating to the upgrading of the energy efficiency of existing buildings already carried out by some Member States after thermal auditing in order to identify & compare the methods used. The Resolution emphasised the importance of the regulations in force in the Member States to improve the energy performance of new buildings and, if necessary, the importance of the introduction or reinforcement of such regulations.	Council Resolution on the rational use of energy in the building sector (15/03/85).	■ Off Jnl C78, 26/03/85	http://europa.eu.int/smartapi/cgi/sga_doc?smartapi!celexapi!prod!CELEXnumdoc&lg=EN&numdoc=31985Y0326(01)&model=guichett

Report to DG XI, European Commission: Construction & Demolition Waste Management Practices, & their Economic Impacts	Basic aim of the project was to investigate the quantities of construction and demolition waste (CDW) among the European Union as well as the measures which each Member State has taken to improve the re-use and the recycling of this waste stream. The report describes the best practices in this field as well as the economics of the CDW reuse and recycling.	The project was undertaken by Symonds Group (United Kingdom) in association with Argus (Germany), Cowi, Consulting Engineers and Planners (Denmark) and PRC Bouwcentrum (the Netherlands) (Feb 1999).	■ Report to DG XI, European Commission	http://europa.eu.int/comm/environment/waste/studies/cdw/c&dw_report.htm
Innovation in the Construction Sector	The aim of this study was to identify conditions that lead to successful innovation. The basis of the research was 17 case studies of organisations drawn from different parts of the construction sector. The organisations included product manufacturers, design firms, consultants, construction companies, specialists, research institutes, a university & a repeat order customer. The intention was not to regard their actions & achievements as representative of the sector as a whole – that would be impossible – but to use the findings to probe the reasons why construction has the reputation for a low propensity to innovate.	Final report of a study made by ECCREDI (European Council for Construction Research, Development and Innovation) (May 1999).	■ ECCREDI Report	http://europa.eu.int/comm/enterprise/construction/innov/innovhome.htm
Abnormally Low Tenders (ALTs) in the Construction Sector	The impact of abnormally low tenders is an inherent problem of the construction industry & this is investigated in the course of this report. Within the Report emphasis was placed on existing Directives & public procurement. The Working Group considered that this Report was only an interim statement of the views of the industry in relation to the prevention, detection and elimination of ALTs. Further work was necessary and the Recommendations contain the methodology by which more comprehensive and effective actions could be brought about.	Report of the ALT Working Group (May 1999).	■ Working Group Report	http://europa.eu.int/comm/enterprise/construction/alo/alohome.htm

Continued

Table 4.5.1 European Level Construction Policy. *Continued*

Policy Type	Programme Title	Policy Objective	Community Measure	Document Reference	Website
Miscellaneous Reports	Quality in the Construction Sector	The main objective of this study was to show what quality management meant for nine European companies in the construction sector & how they steered their organisation through the new challenges presented. Backgrounds to the companies involved in the study are presented in Appendix A.	The European Way to Excellence in Construction – Final report (Aug 1999).	▪ DG III Report	http://europa.eu.int/comm/enterprise/construction/quality/qualhome.htm
	Energy Performance in Buildings	To create a common framework to promote the improvement of the energy performance of buildings.	Proposal for a Directive of the European Parliament and of the Council on the energy performance of buildings (COM (2001) 226) (11/05/01).	▪ Off Jnl C213E, 31/07/01	http://europa.eu.int/scadplus/leg/en/lvb/l27042.htm
	Sustainable Construction Final Report	This report examines the implications of Article 6 of the Treaty of Amsterdam, in particular the requirements that 'environment protection requirements must be integrated into the definition and implementation of the Community policies and activities . . . , in particular with a view to promoting sustainable development'. This report will serve as a focal point for sustainable construction policies & strategies & will encourage most if not all Member States & Accession Countries, to develop their own national plans & strategies for a sustainable construction agenda.	Report from the EC working group for sustainable construction (20/05/01).	▪ Sustainable Construction Working Group Report	http://europa.eu.int/comm/enterprise/construction/suscon/finrepsus/susfin.htm

Task Group 1 Final Report: Environmentally Friendly Construction Materials (EFCM)	In June 1999 a Working Group on Sustainable Construction was set up with three Task Groups. The Task Group (TG) EFCM developed an approach which can be summarised as follows: ■ List the main environmental impacts of construction materials together with the main preventative and remedial measures already in place (see Annex 3); ■ Consider the use of a life cycle or IPP (Integrated Product Policy) approach to environmental management; ■ Using an IPP approach, analyse what the construction materials industry has achieved to date in improving its environmental performance; ■ Make recommendations on how industry can further improve its environmental performance by taking a holistic, life cycle/IPP approach to environmental management.	Final report of the working group on Environmentally Friendly Construction Materials (05/09/01).	■ TG EFCM Report	http://europa.eu.int/comm/enterprise/construction/suscon/tgs/tg1/efcmfin.htm
Task Group 2 Final Report: Energy Efficiency in Buildings	The objective of this Strategy Paper prepared by the second Task Group (TG EE) was to accelerate the installation of appropriate energy efficiency measures in all kinds of existing and new buildings, including specific actions for 'common and co-ordinated policies and measures' at community, Member State and industry level.	Final report of the Energy Efficiency working group (05/09/01).	■ TG EE Report	http://europa.eu.int/comm/enterprise/construction/suscon/tgs/tg2/eneffrep.htm

Continued

Table 4.5.1 European Level Construction Policy. *Continued*

Policy Type	Programme Title	Policy Objective	Community Measure	Document Reference	Website
Miscellaneous Reports	Task Group 3: Construction & Demolition Waste	The main function of the Task Group on Construction and Demolition Waste (C+DW) Management was to provide recommendations to the Working Group on Sustainable Construction on how to improve C+DW management. The total construction & demolition cycle can be divided into three steps, design & pre-construction, construction, & demolition. Each of these steps is discussed. In addition re-use, recycling, disposal and the issue of research and education are discussed in three further sections. Finally, the respective roles of industry, Member States and the European Commission are described and the Task Group's recommendations are assigned to each of these.	Final report of the Construction and Demolition Waste working group (05/09/01).	■ TG C and DW Report	http://europa.eu.int/comm/enterprise/construction/suscon/tgs/tg3/demfin.htm
	Benchmarking Pilot Study – Final Report	The overall aim of this exercise was to assess the practicality & feasibility of benchmarking on a pan-European scale – to be achieved by the fulfilment of six objectives: ■ Demonstrate the feasibility and benefits of benchmarking ■ Select and define Key Performance Indicators (KPIs) ■ Examine the feasibility of the collection and analysis of data at the individual enterprise level ■ Facilitate the comparison of performance ■ Determine best practice levels of performance ■ Provide the basis for a permanent system of benchmarking.	FIEC (European Construction Industry Federation) Pilot Study in co-operation with the Construction Benchmarking Centre (CBC) at the Building Research Establishment (BRE) in the United Kingdom (2001).	■ Pilot Study	http://europa.eu.int/comm/enterprise/construction/bench/benchpilot.htm

Education, Training & Image of the Construction Sector	The leading questions this report tried to answer in producing recommendations dealing with education, training & image are: ▪ how to attract the right people to ensure the quality of production in the construction sector; & ▪ how improvement of employment conditions & the image of the sector as well as improvement of education & training provisions at all levels of the sector can contribute to quality & thus to competitiveness.	European Commission's working group on Education, Training and Image of the Sector Report (July 2002).	▪ EC Working Group Report	http://europa.eu.int/comm/enterprise/construction/educim/eductra.htm
Analysis of Life Cycle Analysis (LCA) Tools	The overall objective of this study, was to provide the technical specification writers with specific information how & where to include environmental performance indicators & other key environmental aspects within the framework of European technical specifications for construction products. The recommendations of the study focused on construction products but did not explicitly exclude other products of the Construction Products Directive (CPD) & according to the Integrated Product Policy approach (IPP). (Green Paper on Integrated Product Policy (COM(2001)68 final)	European Commission (DG Enterprise, Construction Unit) Report (March 2003).	▪ DG III Report	http://europa.eu.int/comm/enterprise/construction/internal/essreq/environ/lcarep/lcafinrep.htm

Table 4.5.2 UK Level Construction Policy.

Policy Type	Document Title	Objective	Contents	Website
Strategies & Reports	Rethinking Construction – The Report of the Construction Task Force on the Scope for Improving the Quality & Efficiency of UK Construction (1998)	The Construction Task Force was set up by the Deputy Prime Minister against a background of deep concern in the industry and among its clients that the construction industry is under-achieving, both in terms of meeting its own needs and those of its clients. The Task Force's main goals were to: ▪ quantify the scope for improving construction efficiency and derive relevant quality and efficiency targets and performance measures which might be adopted by UK construction; ▪ examine current practice and the scope for improving it by innovation in products and processes; ▪ identify specific actions and good practice which would help achieve more efficient construction in terms of quality and customer satisfaction, timeliness in delivery and value for money; ▪ identify projects to help demonstrate the improvements that can be achieved through the application of best practice.	Foreword; Executive summary; Chapter 1: The need to improve; Chapter 2: Our ambition for UK construction; Chapter 3: Improving the project process; Chapter 4: Enabling improvement; Chapter 5: Improving house building; Chapter 6: The way forward.	http://www.constructingexcellence.org.uk/pdf/rethinking%20construction/rethinking_construction_report.pdf
	Towards More Sustainable Construction: Green Guide for Managers on the Government Estate (June 1999)	This guide was designed, as part of the move towards sustainable construction, to help managers on the government estate understand the basic principles of more environmentally conscious construction and refurbishment. It advises on good practice rather than legislative requirements.	Introduction; Green code for architecture; Procurement; Building location; Transport; Design & works contract; Demolition; Landscaping; Building materials (other than timber); Hazardous substances; Asbestos; Timber; Insulation (floors, walls, roofs & doors); Heating; Air conditioning, ventilation & refrigeration; Indoor air quality; Noise; Daylight; Windows; Lighting – internal; Lighting – external; Solar & other renewable energy sources; Electrical wiring; Electromagnetic fields; Water; Floor finishes & coverings; De-mountable office partitions; Office furniture; Catering & tea points; Fridges & freezers; Waste; Recycling facilities; Fire fighting; Security; Indoor plants; Cleaning; Environmental & energy audits; Legal compliance; Environmental management systems; Training ; Annexes.	http://www.sustainable-development.gov.uk/sdig/improving/partg/suscon/index.htm

Towards More Sustainable Construction: How Sustainable Construction Can Improve the Performance of Buildings & their Occupants (Dec 1999)	This guide is designed to help property managers understand the basic principles of green construction and refurbishment so that they can deal more effectively with designers, builders and other building professionals. It is not a checklist (every element of which can be incorporated into every project) and acknowledges that each project is different and will need its own solution.	Introduction; Radical ideas; Purpose of the guide; Model improvement framework; BREEAM (Building Research Establishment's Environmental Assessment Method); Eco-efficiency; Sustainable designs; Comfort; Construction industry; National pride; UK initiatives; Co-operation; Conclusion.	http://www.sustainable-development.gov.uk/sdig/improving/partg/suscon/notes/index.htm
Building a Better Quality of Life: A Strategy for More Sustainable Construction (April 2000)	This Strategy aims to provide a catalyst for change in construction across the United Kingdom. It identifies priority areas for action, and suggests indicators and targets to measure progress. It sets out action that the government has already taken and further initiatives that are planned, and highlights what others can do. The government will use the Strategy as a framework to guide its policies towards construction, and will encourage people involved in construction to do the same.	Foreword; Chapter 1: Introduction; Chapter 2: A sustainable economic base for construction; Chapter 3: Construction & its stakeholders; Chapter 4: Managing the environment & resources; Chapter 5: What the construction industry can do; Chapter 6: Action, measuring progress & reporting; Annex 1: Examples of current good practice; Annex 2: Examples of commitments to action; Annex 3: Sources of help & information.	http://www.dti.gov.uk/construction/sustain/bql/pdf/sus_cons.pdf
Achieving sustainability in Construction Procurement – Sustainability Action Plan (June 2000)	This strategy document sets out how government clients will take forward the sustainable development agenda through better procurement of new works, maintenance & refurbishment. This will deliver better value for money for occupiers, users & the public & will make clients and, in turn, suppliers fully aware of their responsibilities regarding sustainability.	Foreword; Executive summary; Background; Procurement on the basis of value for money; What is sustainable construction?; Themes for action; Implementation; Suggested further reading.	http://www.ogc.gov.uk/embedded_object.asp?docid=1039
Better Public Buildings – A Proud Legacy for the Future (Oct 2000)	This report shows how the government can work towards achieving a step change in the quality of building design in the public sector. Through the Achieving Excellence initiative, government organisations have started addressing their performance as clients by setting measurable targets and objectives with a strong focus on lifetime costs, quality and design, and, as this report shows, there are some outstanding examples for the rest of the public sector to learn from.	Foreword; Why good design matters; Why & how; Why our aspirations should be higher; Value for money; How our aspirations can be achieved; Continue; Stop; Start; Conclusion.	http://www.culture.gov.uk/PDF/architecture.pdf

Continued

Table 4.5.2 UK Level Construction Policy. *Continued*

Policy Type	Document Title	Objective	Contents	Website
Strategies & Reports	A Commitment to People 'Our Biggest Asset' – A Report from the Movement for Innovation's (M4I) Working Group on Respect for People (Nov 2000)	At the M4I's Conference in summer 1999 the industry was challenged to radically improve its performance on people issues. In response to this challenge the M4I set up a working group on respect for people. The working group was charged with identifying practical and effective ways in which the construction industry could improve its performance under this heading. Continuous improvement was not enough. Radical improvements needed to be identified which could be measured and tracked through a range of performance measures.	Summary; Introduction; The business case; Performance measures; Cross-cutting themes; Action themes – Diversity in the workplace; Site facilities & the working environment at site level; Health; Safety; Career development & lifelong learning; The working environment off-site; Next steps; Annex 1: Glossary; Annex 2: Membership of the working group & sub-groups; Annex 3: The investors in people standard.	http://www.constructingexcellence.org.uk//rc/publications/reports/rfp_report.pdf
	The Movement for Innovation – Sustainability Working Group Report: Environmental Performance Indicators for Sustainable Construction (2001)	In terms of sustainability, social and economic targets are reflected in several of the current Key Performance Indicators but, for some time, there has been a need for practical and quantifiable environmental targets that support the general desire to be 'green'. The six Environmental Performance Indicators (EPIs) produced by M4I's Sustainability Working Group are meant to be a start for the industry, so that it can get going on the process of measuring its environmental sustainability credentials project by project.	Introduction; The data – EPIs: Operational CO_2 emission (kg CO_2/m^2/yr); Embodied CO_2 (kg CO_2/m^2); Water (m^3/person/yr); Waste in the const. process (m^3/100m^2 floor area); Biodiversity; Transport; The benchmarks; What do we do now?; What can you do to improve?; Where to go for help?; Websites; Notes relating to the data; Acknowledgements; Appendix 1: Windrose diagrams; Tables of data; Appendix 2: Response form.	http://www.constructingexcellence.org.uk/pdf/document/m4i_epi_report2001.pdf
	KPI Report for the Minister for Construction (Jan 2000)	The purpose of the Key Performance Indicators (KPIs) is to enable measurement of project & organisational performance throughout the construction industry. This information can then be used for benchmarking purposes, & will be a key component of any organisation's move towards achieving best practice. While individual organisations have been measuring their performance for many years, there has been little consistency in the data, & the way it has been published. This report is another step in rectifying this deficiency, which builds on the foundation of the Construction Industry KPIs by detailing a comprehensive framework for measurement.	Foreword; Introduction; The purpose of KPIs; The KPI groups; The key project stages; Interpretation of the key project stages; The KPIs; The level of KPIs; The supply chain cube; The KPI definitions & guidance; Time; Cost; Quality; Client satisfaction; Change orders; Business performance; Health & safety; Appendix 1: Example of KPI calculations; Appendix 2: Quality register Appendix 3: Client satisfaction survey forms.	http://www.constructingexcellence.org.uk/pdf/document/kpireport2000.pdf

Six Guiding Principles to Improve the Sustainability of the Housing Construction Industry (July 2001)	There are Six Guiding Principles applying to improving the sustainability of the process and product of the housing construction industry. It is important to look at the issues in their global or national context in order to identify the roles & responsibilities that can have a positive impact on the sustainability of the housing construction industry.	Foreword; Executive summary; Introduction; Six guiding principles; Two Case studies; An indication to indicators; Signposts; Further Information.	http://www.constructingexcellence.org.uk/pdf/hforum/factor.pdf
Building a Better Quality of Life: A Strategy for More Sustainable Construction – Report on Progress 2001 (12/10/01)	This report reviews the work that has been done to take forward the sustainability agenda which was identified in Building a better quality of life. The report also notes other initiatives – from both government and within the industry – which have built on the sustainable principles set out in Building a Better Quality of Life.	Introduction: a year of progress; Industry response – Sustainable construction focus group; CIB sustainable construction task group; The movement for innovation; Confederation of construction clients; Sector sustainability strategies; Government response – Greening government; The government construction clients' panel; Regulation & fiscal charges; The construction best practice programmes; Local government task force; Housing forum; Research; Measuring progress & reporting – national indicators; Industry-level indicators; Other KPIs; Reporting; Forward look; Some useful links.	http://www.dti.gov.uk/construction/sustain/sustainableconstruction.htm
Rethinking the Construction Client – Guidelines for Construction Clients in the Public Sector (2001)	This report summarises the findings of the 12 regional rethinking the construction client events contributing to the national debate held in the latter part of 2001. Each event included workshop sessions where speakers were asked to discuss key topics with clients & their suppliers. This report is the result of analysing the workshop responses & provides advice & guidance to help organisations seeking to implement rethinking construction principles in their procurement.	What is rethinking construction?; Introduction – the guidelines; Background to report; Guideline 1: Traditional processes of selection should be changed as they don't lead to best value; Guideline 2: An integrated team should be formed before design starts & maintained throughout delivery; Guideline 3: Contracts should lead to mutual benefit for all parties & be based on a target & whole life cost approach; Guideline 4: Suppliers should be selected by best value & not by lowest price; Guideline 5: Performance measurement should be used to underpin continuous improvement within a collaborative working process; Guideline 6: Culture & processes should be changed so that collaborative rather than confrontational working is achieved; Where to go for help.	http://www.constructingexcellence.org.uk//rc/publications/reports/Rethinking_the_Construction_Client.pdf

Continued

Table 4.5.2 UK Level Construction Policy. *Continued*

Policy Type	Document Title	Objective	Contents	Website
Strategies & Reports	CE Marking Under the Construction Products Directive (2001)	This publication aims to explain the implications of CE (Conformité Européenne) marking under the Construction Products Directive (CPD) for manufacturers, specifiers, certification and test bodies and regulatory/enforcement authorities. The main features of the Directive are summarised as 'bullet points'. Which are then amplified in the following sections. The publication also deals with the transition phase from UK product specifications to the new European product specifications, focuses on the implications of CE marking for UK practitioners, and concludes with other sources of information.	Section 1: Introduction; Section 2: The framework of the CPD; Section 3: A system of harmonised technical specifications; Section 4: An agreed system of attestation of conformity for each product family; Section 5: A framework of notified bodies; Section 6: The CE marking of products; Section 7: Transitional arrangements & incorporation into national regulations; Section 8: Summary of implications for practitioners; Section 9: Further sources of information.	http://www.odpm.gov.uk/stellent/groups/odpm_buildreg/documents/pdf/odpm_breg_pdf_600433.pdf
	Reputation, Risk & Reward: The Business Case for Sustainability in the UK Property Sector (Jan 2002)	This paper is addressed primarily to chief executives, board members and senior managers in property development and property asset management; as their business partners and key suppliers, it should also be of interest to executives in the construction sector. The purpose of the paper is to demonstrate that sustainability is an issue that business leaders can no longer afford to ignore or treat as immaterial. This paper aims to demonstrate that sustainability issues are of critical and strategic importance to business. For companies in the property sector it will outline the business case for high-level strategies and policies designed to address the risks and opportunities posed by sustainability.	Summary; Introduction; The impacts of the property & construction sector; Changing business context; Sustainability linked drivers for business; Governmental & legislative pressure; Investor pressure & environmental, social & business probity; The increasing demands of clients; Benefits of good management of sustainability issues; Conclusion; References.	http://www.dti.gov.uk/construction/sustain/rrnr.pdf
	Accelerating Change – A Report by the Strategic	This publication considers the government's vision is for the UK construction industry to realise maximum value for all clients, end-users & stakeholders & exceed their expectations through the consistent	Foreword; Key measures to accelerate change; Vision; Chapter 1: Rethinking construction; Chapter 2; Progress since rethinking construction; Chapter 3: Strategic direction	http://www.constructingexcellence.org.uk//rc/publications/reports/AccChange.pdf

Forum for Construction (2002)	delivery of world class products & services. This vision needs to be supported by an education & training process that incorporates best practice & a systematic approach to continuing professional & personal development.	& targets; Chapter 4: Accelerating client leadership; Chapter 5: Accelerating supply side integration & integrated teams; Chapter 6: Accelerating culture change in 'people issues'; Chapter 7: Cross-cutting issues; Annex 1: Glossary & further information; Annex 2: Key steps a client must consider when faced with a business need; Annex 3: Publications & websites.	
Rethinking Construction 2002 – Achievements, Next Steps, Getting involved (June 2002)	Since the publication of the report, the rethinking construction agenda has been taken forward through a dynamic partnership between government, clients & industry. This has been given a further boost by the creation of the strategic forum for construction, that brings together all the key industry representatives in pursuit of improvement.	Introduction; 2002 a milestone; How can you get involved? Working together; Across the UK; Good for business; The 12 KPIs demonstrated; Who's who in rethinking construction; Making contact.	http://www.constructingexcellence.org.uk//rc/publications/reports/Rethinking_Construction_2002.pdf
Building a Better East of England – A Review of the Construction Industry in the Area (Sept 2002)	The overall aim of this report is to raise the profile, improve the image, & increase understanding of the construction industry amongst policy makers & the general public. The report identifies priorities for action in the region & provides a platform for future joint working with strategic partners, including the East of England Development Agency (EEDA). The objectives of this report are to: ▪ Demonstrate the importance of the construction industry to the regional economy & to the region as a whole ▪ Provide an audit of the current state of the construction industry in the region ▪ Identify restraints to the future success of the construction industry in the region ▪ Begin to identify the contribution the construction industry can make to help meet the objectives of the Regional Economic Strategy ▪ Identify future priorities for action by the Rethinking Construction Network for the East ▪ Assist in the development of partnerships with EEDA and the other organisations so as to address agreed priorities.	Foreword; Introduction; Executive summary; The region; The construction industry in the region; Statistical evidence; Construction's contribution to the regional economy; Commentary views from people within the industry; Learning & skills; Regional economy; Infrastructure; Communication; Sustainability; Conclusions & recommendations; Appendix 1: Questionnaire sent to construction organisations; Appendix 2: Questionnaire sent to construction businesses; Appendix 3: Rethinking construction demonstration projects in the region; Appendix 4: Stakeholder organisations in the East of England.	http://www.constructingexcellence.org.uk/pdf/document/building_a_better_east_of_england.pdf

Continued

Table 4.5.2 UK Level Construction Policy. *Continued*

Policy Type	Document Title	Objective	Contents	Website
Strategies & Reports	Respect for People – Reaching the Standard (Oct 2002)	During the development of the Respect for People toolkits, many managers recognised that the toolkits have a considerable value in addressing the people issues in quality standards. This report is a manager's tool that: ▪ Highlights key national standards affecting 'people' issues in construction and provides a brief overview of them ▪ Provides details on how the Respect for People Toolkits and Key Performance Indicators support achievement of these standards in sufficient detail to allow companies and projects to use them.	What is it for? How to use this tool? Benchmark index; Considerate constructors scheme; Construction clients charter; EFQM (European Foundation for Quality Management) (business) excellence model; Investors in people; ISO 9000–2000.	http://www.constructingexcellence.org.uk//rc/publications/reports/rfp_Reaching_the_Standard.pdf
	A Framework for Action – The Report of Rethinking Construction's Respect for People Working Group (2002)	Rethinking Construction identified 'a Commitment to People' as one of five drivers for change that must be addressed to make UK construction world-class. This report also identified seven targets for improvement – profitability, productivity, quality, safety, predictability, capital cost, & turnover & profits. The Task Group recognised that none of these could be achieved without attention to the people involved in construction activities. This report concludes the work of Rethinking Construction's Respect for People Working Group that started with the Ministerial challenge to the industry in 1998 & brings forward recommendations to support the industry in realising Respect for People.	Our recommendations; Foreword; Preface; Executive summary; Introduction; The framework; Action themes – Equality & diversity in the workplace; Working environment & conditions; Health; Safety; Career development & lifelong learning; Annex 1: Membership of the trial steering group & sub-groups; Annex 2: Trialling organisations & projects; Annex 3: Rethinking construction; Annex 4: Links to accelerating change; Annex 5: The respect for people trial; Glossary.	http://www.constructingexcellence.org.uk//rc/publications/reports/rfp_A_Framework_for_Action.pdf

Government Construction Procurement Guidance No.7: Whole Life Costs (2002)	Whole life costing requires the consideration of all costs of a facility over its life, so that investment at the beginning can be properly related to returns many years later. The capital cost will be a fraction of the lifetime cost and investment in the facility. This guidance sets out the main principles for preparing whole life cost models. This is the seventh in a series of nine guidance documents that translate the recommendations made in the Efficiency Unit Report *Construction Procurement by Government* into practical proposals for implementation. It addresses the recommendations made in the University of Bath School of Management Agile Construction Initiative report *Constructing the Government Client* and the recommendations made in the Construction Taskforce report *Rethinking Construction.*	Introduction; Scope; Business operations; Output functional specification; Elements to consider; Environment & sustainability; Risk analysis; Whole life cost model.	http://www.ogc.gov.uk/embedded_object.asp?docid=1576
EU Legislation Study Report (Jan 2003)	There is a wealth of EU legislative instruments, both in place and being considered, that are likely to have implications for sustainable construction in the UK. This report collates information on these legislative instruments and presents it in a format that can inform future policy decisions and be used by stakeholders in the construction industry.	Executive summary; Introduction; Findings; Access to environmental information; Building materials; Energy; Renewable energy sources; Energy efficiency in buildings; Energy efficient goods; Protection of the ozone layer; Planning & development; Urban development; Waste management; Packaging; Landfill; Waste electrical & electronic equipment; Composting; Pollution prevention; Hazardous waste; Water policy; Other matters; Dissemination & exploitation of results; Policy implications; Conclusions & recommendations; Appendices.	http://www.dti.gov.uk/construction/sustain/eu.pdf
Building on Success – Achieving Excellence in Construction (Feb 2003)	This document presents an overview of the achieving excellence in construction initiative launched in 1999, including the key findings of the previous 3 years and case study examples of excellence in practice.	Foreword; Introduction; Key findings from the last three years; Achievements; Case study: Highways Agency; Case study: Environment Agency; Case study: NHS Estates; Procurement strategies; Strategic targets; OGC support.	http://www.ogc.gov.uk/sdtoolkit/reference/achieving/AEbrochure.pdf

Continued

Table 4.5.2 UK Level Construction Policy. *Continued*

Policy Type	Document Title	Objective	Contents	Website
Strategies & Reports	Demonstrations of Sustainability (May 2003)	This report celebrates the success of the Rethinking Construction demonstrations in addressing sustainability. The demonstrations featured provide tangible evidence of the adoption of more sustainable practices by the construction industry. Innovations include improved community and user participation during planning and design, better treatment and training of construction staff and increased resource use efficiency in the construction process and finished product.	Executive summary; Foreword; Introduction; Economic sustainability; Social sustainability; Environmental sustainability; The future; Appendix 1 – Demonstrations we wish we had the space for; Appendix 2 – Planning & sustainability; Appendix 3 – Links.	http://www.constructingexcellence.org.uk/pdf/demonstrations_of_sustainability.pdf
	Sustainable Construction Brief (June 2003)	This brief provides a short outline on the key elements of sustainable construction including some documents, developments and good practice examples of sustainable construction in action.	What is sustainable construction? Some key documents; Sector strategies; The business case; Partners in innovation; Benchmarking performance; Government leading by example; Energy; Sustainable communities plan; The better buildings summit; Some links.	http://www.dti.gov.uk/construction/sustain/scb.pdf
	Rethinking Construction Demonstrations Report: Demonstrating success through Rethinking Construction (July 2003)	This report builds on the Rethinking Construction demonstrations programme with 134 active & 240 completed projects, including both large & small organisations within the project teams. There were some 101 brief case studies setting out the main lessons from completed projects available on the website, as well as 15 case studies on Respect for People. This report provides highlights of how these results have been achieved by giving specific examples from the demonstrations portfolio of 374 completed and active projects covering all sectors of construction, including refurbishment, repairs & maintenance & new build. The projects are located throughout the RDA regions, Wales, Northern Ireland & Scotland & represent a combined construction value of almost £7 billion.	Foreword; Executive summary; Introduction; Enhancing economic, social & environmental sustainability; Integrating the team to achieve a quality product; Promoting respect for people; The future of the demonstrations programme; Appendices.	http://www.constructingexcellence.org.uk/pdf/document/Demo_Report_2003.pdf

Better Buildings Summit – Issues Paper (Oct 2003)	This paper provided some context for current thinking on 'Better Buildings', including some best practice, looked at barriers to progress, and posed questions to stimulate deliberation and debate in the workshops and beyond. Although initiated by the Energy White Paper, the debate was not just about energy efficiency in buildings. A broader agenda of sustainability covering public and private development – homes, retail, offices, health, education and infrastructure – had to be addressed.	Introduction; Investment; Some best practice; Some barriers; Opportunities for change; Design; Some best practice; Some barriers; Opportunities for change; Building services & fabric; Some best practice; Some barriers; Opportunities for change; References.	http://www.dti.gov.uk/construction/sustain/betbuild.pdf
The UK Construction Industry: Progress Towards More Sustainable Construction 2000–2003 (The Sustainable Construction Task Group) (Oct 2003)	In 2000 'Towards Sustainability – A Strategy for the Construction Industry' identified a ladder, with each rung representing a key area in which progress had to be made to achieve sustainable construction. This report aims to: ▪ Account for progress up the ladder; ▪ Review the ladder; ▪ Highlight innovations that are ready for widespread uptake; ▪ Suggest priorities for next steps.	Introduction; Promoting awareness & educating people; Collecting information on sustainability initiatives & practical examples of sustainability in action; Monitoring & observing performance; Demonstrating a clear business case for more sustainable construction; Stakeholder dialogue; Spreading best practice; Setting & promoting targets; Establishing a voluntary code of reporting & reviewing performance; Learning from failures; At the top of the ladder? Conclusion; References.	http://www.dti.gov.uk/construction/sustain/sctg.pdf
Timber Use for Construction in the UK (Feb 2004)	The purpose of this report is to inform Government about, & to assist industry in the development of, timber products for the UK construction sector. Its primary focus is the economic activities between semi-finished timber products & end use, concentrating on the manufacture, distribution & use of timber for construction.	Executive summary; Introduction; The sector in context; The sawmilling industry; The sheets & boards industry; The wooden joinery industry; Markets & intermediaries; Conclusions & recommendations; Appendices.	http://www.dti.gov.uk/construction/help/timber_report_final_feb04.pdf
Health and Safety Best Practice Factsheet (15/03/04)	Health and safety is about identifying risks & eliminating or controlling them to prevent accidents & occupational ill-health. All construction work must comply with stringent UK health & safety legislation & this factsheet helps provide an overview of key health & safety information relevant to construction.	Who should read this factsheet? What is health and safety? Why are good health and safety processes important? Who is responsible for health and safety? Basic processes; What are the risks? Constructing a health and safety policy and programme; Further Assistance.	http://www.constructingexcellence.org.uk/pdf/fact_sheet/health_safety.pdf

Continued

Table 4.5.2 UK Level Construction Policy. *Continued*

Policy Type	Document Title	Objective	Contents	Website
Strategies & Reports	Procurement Best Practice Factsheet (15/03/04)	This factsheet is an introduction to construction procurement for occasional clients with a basic to intermediate understanding of the procurement process.	Who should read this factsheet? How to get started; Taking independent advice; Step 1 – Verifying need for the Construction Project; Step 2 Assessing the Options; Step 3 Developing Procurement Strategy; Traditional or design-bid build; Design–Build; Management Contracting; Step 4 Implementing Strategy; Step 5 Project Delivery; Step 6: Post Project Review.	http://www.constructingexcellence.org.uk/pdf/fact_sheet/procurement.pdf
	Sustainable Construction: An Introduction (15/03/04)	This factsheet provides an introduction to sustainable construction for companies across the industry, whether or not they have made progress on sustainability to date.	Who should read this factsheet? What is sustainable construction? The triple bottom line of sustainability; Why construct sustainably? Win business by responding to increasing client demand for sustainability; Manage your reputation; Attract & retain the best employees; Achieve cost savings through greater material efficiency; Reduce operational costs & resource consumption; Reduce transport costs; Case study examples; Sources of further information.	http://www.constructingexcellence.org.uk/pdf/fact_sheet/sus_construction.pdf
	Sustainable Construction: Constructors (15/03/04)	This factsheet complements the general sheet *Sustainable Construction: An Introduction*. It was written for organisations or individuals who work on construction sites. This includes builders, trades-people and landscapers. It outlines the aspects of sustainability to be considered, why they are important and what can be done to operate more sustainably.	Who should read this factsheet? What can sustainable construction mean for my business? What can I consider doing? Considerate Constructors Scheme Code of Practice headings and issues to address; Examples of constructors involved in more sustainable projects; Sources of further information.	http://www.constructingexcellence.org.uk/pdf/fact_sheet/sus_constructors.pdf

Sustainable Construction: Suppliers (15/03/04)	This factsheet complements the general sheet *Sustainable Construction: An Introduction*. Written for organisations or individuals who supply materials or systems for buildings and infrastructure, it outlines the aspects of sustainability that can be considered, why they are important and what can be done to operate in a more sustainable manner. The mini case studies towards the end of the document give a flavour of what some suppliers are offering.	Who should read this factsheet? What are the business benefits of sustainable construction? What should suppliers consider? In the quarry and factory; On the construction site; In finished buildings; End of life; Examples of suppliers involved in more sustainable projects; Sources of further information.	http://www.constructingexcellence.org.uk/pdf/fact_sheet/sus_suppliers.pdf
Sustainable Construction: Designers Best Practice Factsheet (18/03/04)	Written for designers this factsheet outlines aspects of sustainability to consider for more sustainable building design. The mini case studies towards the end of the document provide examples of sustainable designs.	What does sustainable construction mean for designers? What do I need to consider? Economic sustainability; Design Quality Indicator (DQI); Environmental sustainability; Social sustainability; Examples of more sustainable building designs; Sources of further information.	http://www.constructingexcellence.org.uk/pdf/fact_sheet/sus_designers.pdf
Whole Life Costing Best Practice Factsheet (01/04/04)	This factsheet introduces the principles of whole life costing for organisations that are new to the subject or in the early stages of working with whole life costing.	Who should read this factsheet? Who might be interested in Whole Life Costing? What is Whole Life Costing? Why does Whole Life Costing matter? What are the benefits of Whole Life Costing for companies? How do we get started? What are the basic steps in Whole Life Costing? What costs are taken into account? What other variables are important? Reporting formats & recording results; How important are maintenance & operational costs for different buildings? How Whole Life Costing analysis is used; References.	http://www.constructingexcellence.org.uk/pdf/fact_sheet/wholelife.pdf
Sustainable Construction: Clients (01/04/04)	Written for organisations or individuals who are about to procure a new property – a new building or a refurbishment – this fact sheet outlines the aspects of sustainability that should be considered. It applies to clients who procure new properties regularly and to more occasional clients. The mini case studies towards the end of this document provide examples of relevance to both of these groups.	What does sustainable construction mean for construction clients? What do I need to consider? Need & feasibility; Economic sustainability; Environmental sustainability; Social sustainability; Examples of clients who procured more sustainable buildings; Sources of further information.	http://www.constructingexcellence.org.uk/pdf/fact_sheet/sus_clients.pdf

Continued

Table 4.5.2 UK Level Construction Policy. *Continued*

Policy Type	Document Title	Objective	Contents	Website
Strategies & Reports	Sustainability Accounting Best Practice Factsheet (14/04/04)	Sustainability accounting provides a useful tool to identity, evaluate & manage social & environmental risks by identifying resource efficiency & cost savings & link improvements in social & environmental issues with financial opportunities. It also allows comparison & benchmarking of performance & identification of best practice.	What is Sustainability Accounting? Why is Sustainability Accounting Important? What are the benefits? Why sustainability accounting in the construction sector? How to do it? Preparation of the Direct Accounts; Preparation of the Indirect Accounts; Related resources; Case study at Great Western Hospital.	http://www.constructingexcellence.org.uk/pdf/fact_sheet/sus_accounting.pdf
	Demonstrating Excellence – An Evolution of the Programme of Demonstrations (June 2004)	This report provides a synopsis of the evolution of the demonstrations programme. It highlights the successes to date of demonstrating the future for the rest of the industry and, through the eyes of the industry leaders, it provides a glimpse of the future direction for this programme.	Background; Introduction; Traditional construction procurement; Early collaboration – design & build; Project partnering – a more integrated team; Strategic partnering alliances; Conclusions; Looking to the future; Demonstrating excellence the way forward; Appendices.	http://www.constructingexcellence.org.uk/pdf/bpknowledge/demonstrating_exc.pdf
	Lean Construction Best Practice Factsheet (28/10/04)	Factsheet detailing Lean construction as a philosophy based on the concepts of lean manufacturing. It is about managing & improving the construction process to deliver profitably what the customer needs.	What is lean construction? Lean manufacturing; The Lean Principles; Production & management principles; Design & product development; Critical success factors; Lean production; Supply chain management & supplier relationships; Applying lean thinking to construction; Design; Procurement; Production planning; Logistics; construction; Applying lean thinking in construction; Construction lean improvement programme; More detailed guidance.	http://www.constructingexcellence.org.uk/pdf/fact_sheet/lean.pdf

	Construction Products Industry Key Performance Indicators Handbook (2004)	The Construction Products Industry Key Performance Indicators (CPI KPIs) detailed in this handbook represent the third annual results for the construction products industry. Since their launch in 2002 when the first set of results were published the indicators underwent a review and in 2003 the industry's performance on a total of 15 indicators was published. These KPIs were developed following extensive dialogue with DTI, the industry's customers and manufacturers and suppliers.	Construction products association; Definitions; Purpose of CPI KPI; Measuring the company benchmark score; Dealing with 'flat' sections on graphs; Using the radar chart; Construction products industry key performance indicators, 2004 results; Secondary performance indicators; Sector specific – customer satisfaction; Appendices.	http://www.constprod.org.uk/download/network/CPI_KPI_2004_Handbook.pdf
Guidance Notes	A Guide to the Construction (Health, Safety & Welfare) Regulations 1996 (1996)	The Construction (Health, Safety & Welfare) Regulations 1996 were a set of modern & simplified construction regulations, which also included some important new provisions arising from the implementation of an EC Directive on construction (92/57/EEC). These Regulations completed the implementation of the Directive which was started by the introduction of the Construction (Design & Management) Regulations 1994.	Introduction; How the CHSW regulations affect you? Who has duties under the regulations? Safe places of work; Precautions against falls; Falling objects; Work on structures; Excavations, cofferdams & caissons; Prevention or avoidance of drowning; Traffic routes, vehicles, doors & gates; Prevention & control of emergencies; Welfare facilities; Training, inspection & reports; Related guidance & further information.	http://www.hse.gov.uk/pubns/indg220.pdf
	Effective Teamwork: A Best Practice Guide for the Construction Industry (Mar 2004)	This guide suggests ways in which construction teams can ensure they harness the collective energy of all their members. It has been written for those who lead, form or participate in construction teams & it applies to teams formed within organisations and those formed across two or more organisations. The guide discusses why teamwork is important & the key issues people face when working collaboratively. Overall it aims to help team members work better together & address many of the inter-personal issues such as trust, respect & leadership that influence how people collaborate successfully.	Introduction; What is teamwork; Selecting team members; Leadership; Team meetings; Chairing team meetings; The effective teamwork matrix; Team identity; Shared issues; Communication; Collaboration & participation; Issue negotiation & resolution; Reflection & self-assessment; Further reading.	http://www.constructingexcellence.org.uk/pdf/document/Teamwork_Guide.pdf

Table 4.5.3 Northern Ireland Level Construction Policy.

Policy Type	Document Title	Objective	Contents	Website
Strategies & Reports	Building on Success – Proposals for Future Housing Policy (Feb 1996)	This Review set out proposals for future government housing policies in NI. It summarised the initiatives taken in previous years & showed how these sought to meet the government's primary objective – to ensure that a decent home is within reach of every family. It outlined how the government proposed to respond to changing circumstances to ensure that this objective would continue to be met in the most effective, efficient & equitable way. The Review is set in the context of the White Paper on Housing Policies for England & Wales, published on 27 June 1995, & reflected the objectives & policies outlined in that Paper.	Introduction & summary; Background; Principles, objectives & success to date; Home ownership; The private rented sector; The social rented sector; Housing & regeneration; Delivery mechanisms; Housing management; Housing & the environment; Annex 1: Socio-economic analysis; Annex 2: Policy appraisal & fair treatment.	http://www.dsdni.gov.uk/publications/displayitemSections.asp?ID=239
	Measuring Construction Performance (June 2002)	The purpose of this paper is to identify the key factors involved in measuring construction performance. In line with the government's drive to modernise construction & gain value for money, the industry must recognise that it can only achieve these goals through the adoption of best practice processes, & by measuring performance against rigorous targets.	Introduction; Performance Measuring in Practice; Government's Initiative on Performance Measuring; The Theory & Practice of Performance Measuring; Conclusion; Annex I: Key Resources; Annex II: 5-4-7 Model; Annex III: Process Steps to Implement Performance Measurement.	http://www.engj.ulst.ac.uk/RCCNI/reports/kpi_report2.pdf
	Construction & Demolition Waste Survey 2001 (May 2003)	A survey of Construction & Demolition waste in Northern Ireland was commissioned by the Environment & Heritage Service (EHS) in March 2002. The aim of the project was to identify the quantity of construction & demolition (C & D) wastes arising in NI in 2001. The survey also sought to categorise the wastes by European Waste Classification (EWC) & to quantify the volumes reused and recycled.	Executive summary; Objective's & scope; Methodology; Survey results; Licensed landfills; Exempt landfills; C&D waste crushers/screeners; C&D companies; Waste breakdown by level 2 EXC category; Key findings.	http://www.ehsni.gov.uk/pubs/publications/ExecutiveSummary_constructionanddemolitionwastesurvey.pdf

Northern Ireland Construction Industry – Rethinking Construction Centre Northern Ireland (14/05/04)	In a period of rapid change & growth it became both confusing & cumbersome for industry stakeholders to keep abreast of the changing structures & standards. This report was to act as a key reference document on all new & essential aspects of the evolving modern construction industry.	Northern Ireland construction industry outlook; Government departments & agencies; Public procurement policy; Industry partnerships; Professional college; Contractors college; Specialist subcontractors college; Suppliers college; Construction industry training; Construction industry research; Annex I – key references.	http://www.engj.ulst.ac.uk/RCCNI/rccni_news/upload/nimap3.pdf
The Aggregates Levy (Northern Ireland Tax Credit) Regulations 2004 (2004)	Statutory Instrument 2004 No. 1959 states: this regulation applies to any eligible person within the meaning of paragraph (6) who on or after 1st April 2004 & before 1st April 2011 has commercially exploited within Northern Ireland taxable aggregate which has been extracted from a registered site in Northern Ireland & has accounted for the Aggregates Levy chargeable on that commercial exploitation.	Regulation text & Explanatory Note.	http://www.legislation.hmso.gov.uk/si/si2004/20041959.htm
BPEO (Best Practicable Environmental Option) for the Management of Waste Asbestos (Oct 2004)	This report describes various options for the management of asbestos waste in Northern Ireland based on a study carried out by Environmental Resources Management Limited (ERM) on behalf of the Environment and Heritage Service (EHS) in March 2004. The guidance is intended for use by EHS staff in assessing applications for treatment and disposal of asbestos waste and to inform operators on the approach to asbestos waste management in Northern Ireland.	Introduction; Asbestos; Legislative Requirements; The Northern Ireland Context; Treatment Options; Physical Treatment; Thermal Treatment Processes; Chemical Treatment Processes; Biological Processes; Summary of Treatment Technologies; BPEO; BPEO Decision Criteria; Sensitivity Analysis; Conclusions; Legislative Requirements.	http://www.ehsni.gov.uk/pubs/publications/BPEOAsbestos.pdf

Continued

Table 4.5.3 Northern Ireland Level Construction Policy. *Continued*

Policy Type	Document Title	Objective	Contents	Website
Guidance Notes	Central Procurement Directorate Procurement Guidance Note 03/04: Social Considerations in Public Procurement (2004)	The objective of this Guidance Note is to provide an outline of the range of possibilities for integrating social considerations into public procurement in Northern Ireland under the existing procurement rules.	Objective; Background; Selecting social policies for integration into public procurement; Subject Matter/Technical Specifications; Selection of Tenderers (Restricted Procedure); Tender Evaluation/Contract Award; Post Contract Award; Section 75 of the Northern Ireland Act 1998; Summary.	http://www.cpdni.gov.uk/social-integration-03.pdf
	Central Procurement Directorate: Procurement Guidance Note 04/04 – Integrating Environmental Considerations into Public Procurement (2004)	The purpose of this Guidance Note is to assist contracting authorities to identify the most effective ways of integrating environmental issues within the public procurement process while adhering to the UK Public Procurement Regulations & the EC Treaty. The guidance is based on the Office of Government Commerce (OGC) & the Department for Environment, Food & Rural Affairs (DEFRA) joint note on Environmental Issues in Purchasing (Information Note: 12/03) adapted to reflect Northern Ireland circumstances.	Introduction; Purpose; The Environmental Policy Context; Guidance; Scope; Best value for money policy; Legal compliance; Environmental procurement & sustainable development; Some key points for successful consideration of the environment in public procurement; Identifying a need & developing a business case; Specification; Selection stage; Tender evaluation; Contract management; Whole Life Costing; Glossary; Further information; Annex A – 'Quick Wins' for Greening Government Purchasing.	http://www.cpdni.gov.uk/pdf-environmental-notes.pdf

5 Energy

5.1 Energy and the Built Environment

Energy is considered to be essential to life in the developed world, but is accepted as a luxury in the developing world, due to inefficiency or non-existence of supply. Most of humanity's actions are based on energy supply, and the quest for reliable sources has exercised the minds and inventions of engineers and entrepreneurs down the years. The invention of Watt's steam engine in 1769 opened up the possibility for man to produce large amounts of energy on demand, laying the foundation for the Industrial Revolution and moving the energy source from human labour to fossil fuels.

The built environment has relied on a collection of fossil fuels throughout the nineteenth and twentieth centuries. During this period energy was derived initially from wood, and then more critically from coal, oil and natural gas (fossil fuels), being the breakdown products of plants millions of years old. The extraction and processing of these energy sources continues to demand the application of sophisticated technologies and increased levels of revenue due to the alleged shortage of fossil fuels. Predictions indicate that there is a diminishing supply of both oil and gas, with provisions due to run out as early as 2050 and probably by the end of this century. It should be noted that supply is strongly influenced by demand, with the world population on a steady growth curve. This alleged shortage is also influenced by the increased extraction costs of raw fuel, the need for more complex processing systems, and the development of alternative non-fossil or renewable fuel sources. Hence, the future of energy supply is not solely based on the availability of fossil fuels, but is part of a complex range of issues covering engineering, finance, change in energy consumption patterns and accurate investigations, with history showing that humanity can face up to the need to be creative and inventive. This lateral and essential approach addresses the principles of sustainable development and will require both global and local solutions.

While the oil and coal fields are being thoroughly examined, there is the additional threat of increased carbon dioxide emissions, leading to climate change and global warming. Gas remains a clean and cheap energy source with less environmental damage, but it requires large pipeline distribution systems. However, nuclear energy can provide a reliable alternative, through cleaving uranium molecules and reaping the heat energy by a process of energy fission. The security of nuclear energy has been undermined by well publicised accidents at Three Mile Island and Chernobyl. This has promoted energy supply into public debate with society displaying little confidence in nuclear energy generation. Therefore while the built environment can source energy at a reasonable cost society remains ambivalent to any energy crisis.

Renewable energy sources are perceived to provide viable alternatives to fossil fuels and nuclear power due to their infinite supply. Typical sources include the sun, wind, water and the Earth's internal heat, but renewable sources contribute less than 14% to global energy production. Hydro-electric power is currently the most productive of the renewable sources with almost 7%, while biomass, geothermal, wind and solar power constitute less than 15% of global energy supplies. Hence rapid growth in the development and use of renewable sources makes only a limited contribution to the global scene. This small share results from 'the renewables' not being fully competitive with fossil fuel sources.

Energy consumption and climate change are inextricably linked, as borne out by several meetings of the Conference of the Parties from 1995 to 2003, at which the United Framework on Climate Change and the 1997 Kyoto Protocol were addressed. Similarly, an International Conference for Renewable Energies 2004 addressed how the proportion of renewable energies used in industrialised and developing countries can be substantially increased, and how the markets for renewable energies can be better developed.

5.2 Key Energy Elements

Energy cannot be created or destroyed, for although its form may be changed its magnitude persists. This is governed by the 'First Law of Thermodynamics', better known as the 'Law of Conservation of Energy'. Energy is only valuable when transferred into a measurable such as temperature, heat, power or electricity. The efficiency of any item of equipment which requires energy will be judged on the amount of energy produced against the energy consumed, with efficiency ratings rarely above 90%. Energy can perform several types of work such as transporting goods and people, heating and cooling habitations, manufacturing and powering appliances and machines. The largest consumers in society are transportation, industry and buildings, the latter demanding up to 33% of global energy supplies through heating, refrigeration, ventilation, air conditioning and lighting.

Fossil fuels have been traditionally mined from the earth's subsurface or from deep sea locations, yielding the greenhouse gas, carbon dioxide, and other harmful emissions such as methane and sulphur dioxide. Power stations have relied on coal, oil and natural gas as the energy source to drive turbines and this is converted into electricity energy for distribution along high voltage cable grids and converted down to domestic low voltages. The development of renewable energy requires new technologies, providing much cleaner energy with minimal emissions. However a fluctuating supply cannot meet regular electricity demands, resulting in renewable energy sources having limited benefit when the supply barely meets the demand values. In contrast the more controversial nuclear energy supply is more reliable and cost effective.

The primary renewable energy source is the indirect solar energy conversion through wind energy and hydro-electricity, resulting from the differential heating of the earth's surface causing air to move about in the form of wind and precipitation to form as the air is lifted. The potential energy of water is created when it is evaporated by the sun's heat and released as rain, with this energy converted into kinetic or mechanical energy as it flows back to sea level, providing running water power (such as a river turbine) or hydropower (such as a storage reservoir). Similarly, wave energy can be derived from the movement of tides using fixed or floating wave power devices.

Photosynthesis produces energy from organic matter producing biomass when it is burned to generate steam for turbines. This process is being enhanced with the burning of fuels such as willow or agriculture sludge. Direct solar energy conversion can be achieved by thermal heating of buildings from sunlight and by photovoltaic cells which produce electricity when sunlight strikes a semiconductor material.

Efficient use of energy plays a key role in balancing the limited supply with increasing demands. Building energy efficiency can be improved through double glazing, window shading, loft and cavity wall insulation, the use of energy efficient appliances, thermostatic controls, and the use of natural heat and light, as well as regular servicing of equipment. In an industrial setting, there is the potential for co-generation of electricity by capturing the waste heat associated with a generation process and employing that heat to do other work, the most common type being Combined Heat and Power (CHP) units.

5.3 Sustainability Challenges for Energy

Energy supply sits alongside water supply in meeting one of society's basic needs by providing life through heat, warmth and energy. Energy also sits alongside climate change as a high level issue in addressing future trends and encouraging radical life-pattern changes to overcome both diminishing energy sources and the impact of pollution from energy production processes which cause global warming. The UK Government White Paper (2003) on creating a low carbon economy encapsulates these challenges by requiring energy producers, investors, business and consumers to have a clear, settled, long-term framework within which they can plan and make decisions with confidence, reflecting and reinforcing a wider commitment to sustainable development by challenging stakeholders to find ways to achieve simultaneously economic, social and environmental objectives. These high level challenges highlight the critical need to understand and build robust strategies and legislation to ensure the integrity of energy supplies and their efficient and effective use, and that these are synonymous with and contribute to all four objectives of sustainable development. The energy links with the sustainable development objectives are as follows:

- *Effective protection of the environment* arises from achieving a reduction in greenhouse gas production in accordance with the Kyoto Protocol, and an increased usage and reliance on renewable energy sources. Full consideration of biodiversity, air quality, wildlife habitat conservation and many related areas will arise from this emission-reduction campaign, while adoption of validated and secure nuclear energy systems can minimise both environmental risk and damage. Increased energy efficiency, transport planning and improved building management systems will all contribute to environmental protection.
- *Social progress which recognises the needs of everyone* implies greater access to affordable energy and a reduction in the negative health impacts from fossil fuel emissions. The reduction in and improved public perception of risk of vulnerability from nuclear energy processes will assist society in accepting alternatives to fossil fuel energy sources. Increased financial viability of renewable energy sources can help gain the confidence of communities, while greater awareness of energy efficiency devices can assist those in the fuel poverty bracket.
- *Prudent use of natural resources* can be achieved by a significant increase in renewable energy usage following on from a reduction in the reliance on fossil fuel consumption. The development and construction of new energy schemes must contribute to waste minimisation, greater use of existing or recycled materials and a strong research-driven programme which provides a full understanding of a range of renewable sources.
- *Maintenance of high and stable levels of economic growth and employment* comes from increased productivity at all stages of energy development and access to affordable energy production. This will also require greater use of innovative processes, the application of green procurement, the development of 'design, build finance and operate' projects through public-private partnerships and full allocation of capital and recurrent expenditure. At the same time it is necessary to incorporate whole life costing. This can assist low-income families in avoiding the fuel poverty trap.

The understanding, prediction, mitigation and multi-disciplinary aspects of managing energy supply and demand are central to the challenges in delivering sustainable development. The critical challenges which arise from energy management and contribute to this broader sustainability agenda include:

- *Increased energy consumption and its links to climate change,* which requires energy sources to move away from fossil fuels for electricity generation and the need to explore the role of nuclear energy, while sourcing cleaner fuels and reducing consumption patterns. Issues such as analysing embodied energy from buildings and the impact of transportation patterns within industrial processes must be examined alongside the recognition that increased fossil fuel consumption leads to increased carbon dioxide emissions and global warming.
- *Improved energy efficiency and conservation* is essential in light of a high proportion of energy consumption being wasted on space and water heating. This could be counterbalanced by energy consumption being dramatically reduced through the use of energy efficient devices (such as low energy light bulbs), electrical appliances with high energy ratings, double glazing and simple draught-proofing measures (reducing boiler temperatures), as well as the installation of loft and cavity wall insulation. There is potential to further influence energy efficiency through industrial regulatory measures such as emission monitoring and the application of fiscal measures to stimulate energy consumption changes.

- *Enhanced understanding of energy and its social implications*, through greater awareness of social responsibility and the consequences of poor energy efficiency to help stimulate reduced energy consumption. This must impact on fuel poverty where property occupiers are required to pay more than 10% of their household income to maintain adequate warmth, recognising the health-related implications. Fossil fuel price increases can exacerbate the fuel poverty trap and have global implications with the rising global population increasing energy demand.
- *Adoption of renewable energy sources and cleaner technologies* will enhance the future role of renewable energy, with fossil fuels having finite resources, and add to the importance of wind, solar, biomass, tidal, hydro-electric, wave and offshore resources as environment-friendly alternatives. There remains difficulty with the reliability of renewable sources in meeting peak demand, necessitating technological advancement in renewable technologies, the possible development of district heating and increased use of combined heat and power (CHP) cogeneration. This could lead to a carbon-free economy, utilising nuclear power or hydrogen as viable alternatives.
- *Development of integrated energy policies* will provide robust links to international agreements such as the Kyoto Protocol, the UN Climate Change Framework, the Montreal Protocol and the Vienna Convention, with further links to EU programmes such as Carnot, Altener, SAVE I & II and Synergy.
- *Securing energy markets and supply* recognises that an increasing energy demand cannot be met by a single source or supplier. This needs a combination of emission free energy sources such as nuclear power, hydropower, wind and solar energies. Other markets also need attention, including electricity generated from natural gas, oil and petroleum, confirming the importance of trans-European networks, market integration and an island-market approach.
- *Assessment and improvement of energy performance of buildings* can be assisted by the use of appropriate mechanisms such as energy management systems, ISO accreditation and Environmental Management Auditing Schemes, BREEAM (Building Research Establishment's Environmental Assessment Method) ratings, and building certification. In addition the use of energy saving devices can improve energy performance including sensor lighting, energy star office appliances, low energy light bulbs, insulation, double-glazing and appliances with energy labelling.

5.4 Energy Theme Index

Jurisdiction	Tier	Title	Date	Page No.
INTERNATIONAL	Conventions	Vienna Convention for the Protection of the Ozone Layer	1985	129
		UN Framework Climate Change Convention (UNFCC)	1992	129
	Protocols	Montreal Protocol on Substances that Deplete the Ozone Layer	1987	130
		Kyoto Protocol to the UN Framework Convention on Climate Change (COP-3)	1997	130
	Agreements	Energy Charter Treaty	1994	130
EUROPEAN	Council Directives	Minimum Stock Levels of Crude Oil &/or Petroleum Products *(CD 68/414/EEC Amended by CD 98/93/EC)*	1968 1998	131
		Measures to Mitigate the Effects of Difficulties in the Supply of Crude Oil & Petroleum Products *(CD 73/238/EEC)*	1973	131
		The Performance of Heat Generators for Space Heating & the Production of Hot Water in New or Existing Non-Industrial Buildings *(CD 78/170/EEC Amended by CD 82/885/EEC)*	1978 1982	131
		Price Transparency *(CD 90/377/EEC Amended by CD 93/87/EEC)*	1990 1993	132
		Energy Efficiency of Hot-Water Boilers *(CD 92/42/EEC Amended by CD 93/68/EEC)*	1992 1993	132
		Labelling Household Appliances *(CD 92/75/EEC)*	1992	132
		Energy Certification of Buildings *(CD 93/76/EEC)*	1993	132
		Energy Efficiency for Refrigerators *(Dir 96/57/EC)*	1996	133
		Dangers arising from Ionising Radiation *(CD 96/29/Euratom)*	1996	133
		Energy Efficiency of Ballasts for Fluorescent Lighting *(CD 2000/55/EC)*	2000	133
		Promotion of Electricity From Renewable Energy Sources *(Dir 2001/77/EC)*	2001	133
		The Energy Performance of Buildings *(Dir 2002/91/EC)*	2002	134
		Common Rules for the Internal Market in Electricity *(Dir 2003/54/EC Repealing Directive 96/92/EC)*	2003	134
		Common Rules for the Internal Market in Natural Gas *(Dir 2003/55/EC Repealing Directive 98/30/EC)*	2003	135
		Energy Labelling of Household Electric Refrigerators, Freezers & their Combinations *(Dir 2003/66/EC)*	2003	135
		Restructuring the Community Framework for the Taxation of Energy Products & Electricity *(Dir 2003/96/EC)*	2003	135
		The Promotion of Cogeneration Based on a Useful Heat Demand in the Internal Energy Market *(Dir 2004/8/EC amending Directive 92/42/EEC)*	2004	135
	Council Decisions	Community Targets for Reducing Consumption of Primary Sources of Energy *(C Dec 77/706/EEC)*	1977	136
		Set of Actions Relating to Trans-European Networks in the Energy Sector *(C Dec 96/391/EC)*	1996	136

Jurisdiction	Tier	Title	Date	Page No.
EUROPEAN	Council Decisions	The Energy Charter Protocol on Energy Efficiency & Related Environmental Aspects *(C Dec 98/181/EC, ECSC, Euratom)*	1998	136
		SYNERGY II Programme *(C Dec 1999/23/EC)*	1999	136
		Framework Programme for Actions in the Energy Sector *(C Dec 1999/21/EC)*	1999	137
		SURE Programme: Nuclear Safety *(C Dec 1999/25/Euratom)*	1999	137
		Thematic Programme: Energy, Environment & Sustainable Development *(C Dec 1999/170/EC)*	1999	137
		ETAP Programme: Multi-Annual Programme of Studies *(C Dec 1999/22/EC)*	1999	138
		Carnot Programme: Solid Fuels *(C Dec 1999/24/EC)*	1999	138
		Promotion of Renewable Energy – Altener Programme *(Dec No. 646/2000/EC)*	2000	138
		SAVE II Programme: Energy Efficiency *(C Dec 647/2000/EC)*	2000	138
		European Energy & Transport Forum *(C Dec 2001/546/EC)*	2001	139
		Office Appliances: Energy Star Programme *(C Dec 2001/469/EC)*	2001	139
		Co-Operation Between the US & Euratom in the Field of Fusion Energy	2001	139
		Laying down a Series of Guidelines for Trans-European Energy Networks *(Dec No. 1229/2003/EC repealing Decision No. 1254/96/EC)*	2003	140
		Adopting a Multi-annual Programme for Action in the Field of Energy: 'Intelligent Energy – Europe' (2003–2006) *(Dec No. 1230/2003/EC)*	2003	140
	Council Communications	The External Dimension of Trans-European Energy Networks *(COM (97) 125)*	1997	141
		Towards a Strategy for the Rational use of Energy *(COM (1998) 246)*	1998	141
		Integrating of Environmental Considerations in Community Energy Policy *(COM (1998) 571)*	1998	141
		The Northern Dimension of the European Energy Policy *(COM (99) 548)*	1999	141
		Energy Efficiency: Action Plan *(COM (2000) 247)*	2000	141
		Nuclear Safety in the Newly Independent States & Central/Eastern Europe *(COM (2000) 493)*	2000	141
		The Implementation of the Community Strategy & Action Plan on Renewable Energy Sources (1998–2000) *(COM (2001) 69)*	2001	142
		Euro-Mediterranean Co-operation on Energy & Transport *(COM (2001) 126)*	2001	142
		Global Partnership for Sustainable Development *(COM (2002) 82)*	2002	142
		Towards a European strategy for the security of energy supply *(COM (2002) 321)*	2002	143

Jurisdiction	Tier	Title	Date	Page No.
EUROPEAN	Council Communications	Energy Cooperation with the Developing Countries *(COM (2002) 408)*	2002	143
		The Promotion of Cogeneration Based on a Useful Heat Demand in the Internal Energy Market *(COM (2002) 415)*	2002	143
		The Development of Energy Policy for the Enlarged European Union, Its Neighbours and Partner Countries *(COM (2003) 262)*	2003	144
		The Share of Renewable Energy in the EU *(COM (2004) 366)*	2004	144
	Council Recommendation	The Rational Use of Energy by Promoting the Thermal Insulation of Buildings *(C Rec 76/492/EEC)*	1976	145
		The Rational Use of Energy in Urban Passenger Transport *(C Rec 76/495/EEC)*	1976	146
		Rational Use of Energy in Industrial Undertakings *(C Rec 77/713/EEC)*	1977	147
		The Reduction of Energy Requirements for Buildings in the Community *(C Rec 79/167/ECSC, EEC, Euratom)*	1979	148
		Investment in the Rational Use of Energy *(C Rec 82/604/EEC)*	1982	148
	Miscellaneous	A New Energy Policy Strategy for the Community	1974	148
		New Lines of Action by the Community in the Field of Energy Saving	1980	149
		The Improvement of Energy Saving Programmes in the Member States	1985	149
		The Rational Use of Energy in the Building Sector	1985	149
		Improving Energy Efficiency in Industrial Firms in the Member States	1986	150
		Shipments of Radioactive Substances Between Member States *(C Reg No. 1493/93)*	1993	150
		White Paper: An Energy Policy for the EU *(COM (95) 682)*	1995	150
		Renewable Energy: Green Paper *(COM (96) 576)*	1996	151
		Community Framework for the Taxation of Energy Products *(COM (97) 30)*	1997	151
		Community Strategy to Promote Combined Heat & Power	1997	151
		SYNERGY Programme: Co-operation with Third Countries *(C Reg No. 701/97 Extended by C Reg No. 2598/97)*	1997	151
		Renewable Energy Sources – White Paper Laying Down a Community Strategy & Action Plan *(COM (97) 599)*	1997	152
		Renewable Sources of Energy	1998	152
		European Energy Outlook to 2020 – Executive Summary	1999	152
		Green Paper towards a European strategy for the security of energy supply *(COM (2000) 769)*	2000	152

Jurisdiction	Tier	Title	Date	Page No.
EUROPEAN	Miscellaneous	Proposal for a Directive on the energy performance of buildings *(COM (2001) 226)*	2001	153
		Fostering The Use Of Clean Coal Technologies – The CARNOT Programme	2001	153
		A Community Energy Efficiency Labelling Programme for Office Equipment *(Reg (EC) No. 2422/2001)*	2001	153
		New Framework Programme Intelligent Energy for Europe *(COM (2002) 162)*	2002	154
		Energy – Let us Overcome our Dependence	2002	154
		European energy and transport – Trends to 2030	2003	154
		Proposal for a Directive on Establishing a Framework for the Setting of Eco-Design Requirements for Energy-Using Products *(COM (2003) 453)*	2003	155
		Market Development of Alternative Fuels – Report of the Alternative Fuels Contact Group	2003	155
		Study on Energy Supply Security & Geopolitics	2004	156
		Strategy Paper – Medium Term Vision for the Internal Electricity Market	2004	156
		The Share of Renewable Energy in the EU Country Profiles Overview of Renewable Energy Sources in the Enlarged European Union *(SEC (2004) 547)*	2004	157
		The Market for Solid Fuels in the Community in 2002 & 2003 *(SEC (2004) 692)*	2004	157
		Electricity From Renewable Energy Sources – Encouraging green electricity in Europe	2004	157
UK	Strategies & Reports	New & Renewable Energy: Prospects for the 21st Century: Conclusions in Response to the Public Consultation	1999	158
		UK Energy Sector Indicators	2000	158
		Energy Efficiency Commitment (2002–2005)	2001	158
		UK Fuel Poverty Strategy	2001	159
		The Royal Commission on Environmental Pollution 22nd Report: Energy – the Changing Climate	2001	159
		Social, Environmental & Security of Supply Policies in a Competitive Energy Market – A Review of Delivery Mechanisms in the UK	2001	160
		Domestic Energy Fact File: England, Scotland, Wales & Northern Ireland	2001	160
		The Government's Strategy for Combined Heat & Power to 2010 – Public Consultation Draft	2002	160
		Energy Consumption in the UK	2002	161
		The Energy Review: A Performance & Innovation Unit Report	2002	161
		Building Regulations & Historic Buildings – Balancing the Needs for Energy Conservation with those of Building Conservation: An Interim Guidance Note on the Application of Part L	2002	161
		Long-term Reduction in Greenhouse Gas Emissions in the UK	2002	162

Jurisdiction	Tier	Title	Date	Page no.
UK	Strategies & Reports	A Guide For UK Companies: Exporting Sustainable Energy Products & Services	2002	162
		Energy – Its Impact on the Environment & Society	2002	162
		Energy Paper 68: Energy projections for the UK	2002	163
		The World Offshore Renewable Energy Report 2002–2007	2002	163
		The UK Fuel Poverty Strategy – 1st Annual Progress Report	2003	164
		Our Energy Future: Creating a Low Carbon Economy – Energy White Paper	2003	164
		UK Energy Sector Indicators 2003	2003	164
		UK Energy in Brief	2003	164
		Digest of UK Energy Statistics	2003	165
		Energy – Its Impact on the Environment & Society: 2003 Update	2003	165
		Energy Market Competition in the EU & G7: The relative extent of energy market competition in the EU & G7	2003	165
		Review of Renewable Energy Development in Europe & the US	2003	166
		LNG Facilities & Inter-connectors: EU Legislation & Regulatory Regime – DTI/Ofgem Final Views	2003	166
		Offshore Wind – Economies of Scale, Engineering Resource & Load Factors	2003	166
		Renewable Energy Industry Gap Analysis: Summary Report	2004	167
		Energy Trends – Mar 2004	2004	167
		Creating a Low Carbon Economy – 1st Annual report on the Implementation of the Energy White Paper	2004	167
		The Government's Strategy for Combined Heat & Power to 2010	2004	168
		Energy Efficiency: The Government's Plan for Action	2004	168
		The UK Fuel Poverty Strategy 2nd Annual Progress Report: 2004	2004	168
		Creating a Low Carbon Economy – Progress on Regional Implementation of the Energy White Paper	2004	169
		Partial Regulatory Impact Assessment on a Proposal for a Directive on Energy End-Use Efficiency & Energy Services	2004	169
		UK Energy Sector Indicators 2004	2004	169
		Biomass as a Renewable Energy Source	2004	170
		Energy Trends – June 2004	2004	170
		Digest of United Kingdom Energy Statistics	2004	170
		UK Energy in Brief 2004	2004	170
		Energy – Its Impact on the Environment & Society – 2004 update	2004	171
		Energy Trends – Sept 2004	2004	171

Jurisdiction	Tier	Title	Date	Page No.
UK	Consultation Papers	Developing Energy Policy: Consultation for White Paper	2002	171
		Sustainable Energy – Response by the Sustainable Development Commission (SDC) to the Government's Energy Policy: Key Issues for Consultation	2002	172
		The Government's Draft Strategy for Combined Heat & Power to 2010 – Summary of Responses to the Public Consultation	2002	172
NORTHERN IRELAND	Strategies & Reports	Energy in the 90s & Beyond – An Energy Strategy for NI	1991	173
		North/South Energy Market Study: Final Report	2001	173
		Energy Bill	2002	173
		The Potential for the Use of Marine Current Energy in NI	2003	174
		A Study into the Economic Renewable Energy Resource in NI & the Ability of the Electricity Network to Accommodate Renewable Generation up to 2010	2003	174
		Action Renewables – A Study into the Renewable Energy Resource in the Six Counties of Northern Ireland: Generic Resource Opportunities	2004	174
		Energy – A strategic framework for Northern Ireland	2004	175
		All-Island Energy Market – A Development Framework	2004	175
	Consultation Papers	Domestic Energy Efficiency Scheme (DEES II) Proposals – A New Fuel Poverty Programme for warmer healthier homes in NI	2000	175
		Renewable Energy in NI – Realising the Potential	2001	176
		Towards a New Energy Strategy for NI – Consultation Document	2003	176

5.5 Energy Policy Tables

Table 5.5.1 International Level Energy Policy.

Title	Venue	Purpose	Key Outcome(s)	Website
First Session of the Conference of the Parties – The Berlin Mandate (COP-1)	Berlin, Germany (1995)	At the very first session of the Conference of the Parties (COP-1 – Berlin, 1995), the more than 120 countries that had already ratified the Convention found that the commitments referred to in Article 4.2(a) and (b) of the Climate Change Convention were inadequate to meet the overall goals of the Convention.	The Berlin Mandate which: ▪ Called for strengthening of the commitments made in 1992 & called for new proposals to be made in the form of a protocol at Kyoto ▪ Stated that quantified limits & reduction objectives & targets should be set for specific time frames (2005/2010/2015) ▪ Re-affirmed existing reporting obligations.	http://unfccc.int/resource/docs/cop1/07a01.pdf
Second Session of the Conference for the Parties – The Geneva Declaration (COP-2)	Geneva, Switzerland (1996)	The 2nd session provided the Conference of the Parties with an opportunity to carry out an interim evaluation of the implementation of the United Nations Framework Convention on Climate Change and of the progress made in the negotiation process initiated at the first session with the adoption of the Berlin Mandate. Pursuant to that Mandate, the Parties were called upon, *inter alia*, to negotiate a protocol or another legal instrument in order to strengthen the commitments of Parties included in Annex I to the Convention (Annex I Parties) in Article 4.2(a) and (b) for adoption at the third session of the Conference of the Parties.	The Geneva Declaration which endorsed the IPCC Second Assessment Report as a basis for urgent action & instructed their representatives to accelerate negotiations on a legally binding protocol to be adopted by COP-3 scheduled for December 1997 in Kyoto, Japan.	http://www.ccsr.u-tokyo.ac.jp/unfccc1/pdfs/unfccc.int/resource/docs/cop2/15.pdf

Continued

Table 5.5.1 International Level Energy Policy. *Continued*

Title	Venue	Purpose	Key Outcome(s)	Website
Fourth Session of the Conference for the Parties – Buenos Aires Plan of Action (COP-4)	Buenos Aires, Argentina (1998)	A two-week meeting of 170 governments concluded by adopting a 2 year Plan of Action to reduce the risk of global climate change. The action plan was to accelerate work on the 1992 United Nations Framework Convention on Climate Change (UNFCCC) & prepare the way for industrialised countries to take future action under the 1997 Kyoto Protocol.	The Buenos Aires Plan of Action (BAPA) boosts work on transferring climate-friendly technologies to developing countries & addresses the special needs & concerns of countries affected by global warming & by the economic implications of response measures. It includes deadlines on a number of important issues: ▪ Financial mechanisms (which will assist the developing world to respond to the challenges related to climate change). ▪ Further work on policies and measures (an issue introduced by the EU at a late stage in the Conference). ▪ Development and transfer of technologies. ▪ Rules governing the Kyoto Mechanisms with priority given to the Clean Development Mechanism. ▪ An undertaking to discuss supplementarity, ceilings, long term convergence and equity.	http://www.ccsr.u-tokyo.ac.jp/unfccc1/pdfs/unfccc.int/resource/docs/cop4/16a01.pdf
Fifth Session of the Conference for the Parties – Bonn Agreements (COP-5)	Bonn, Germany (1999)	During its last two days, COP-5 adopted 32 draft decisions and conclusions on, *inter alia*, the review of the implementation of commitments and other Framework Convention on Climate Change FCCC provisions and preparations for the first session of the COP serving as the Meeting of Parties to the Kyoto Protocol (COP/MOP-1). After a faltering COP in Buenos Aires, the process recovered vital momentum and began to gather determination and support for a self-imposed deadline for entry into force of the Protocol by 2002.	COP-5 emphasised the need for political will to fulfil the BAPA & allow for the Protocol to enter into force by Rio+10 in 2002. COP-5 adopted the draft decision on Part II of the guidelines (FCCC/CP/1999/L.3) & the addendum containing the guidelines (FCCC/CP/1999/L.3/Add.1). In the decision, the COP, *inter alia*: decides that Part II of the guidelines should be used for the preparation of third national communications; requests Annex I Parties to provide a detailed report on their activities in relation to systematic observation; and urges Annex I Parties to assist Parties with economies in transition (EITs) with technical aspects in preparing national communications.	http://www.iisd.ca/vol12/enb12123e.html

Sixth Session of the Conference for the Parties – The Hague (COP-6)	The Hague, Netherlands (2000)	The meeting aimed to set the operational details for commitments on reducing emissions of greenhouse gases under the 1997 Kyoto Protocol, and to reach agreement on actions to strengthen implementation of the UNFCCC itself. In attempting to achieve these goals, the meeting was intended to bring to a close more than two years of preparations and negotiations set out in the UNFCCC's 1998 Buenos Aires Plan of Action.	During the first week, delegates met in informal contact groups and other negotiating groups mandated by the UNFCCC subsidiary bodies. These meetings aimed to reduce differences on text for decisions on a range of issues related to the Protocol and the UNFCCC, including: the transfer of technology and capacity building to assist developing countries and countries with economies in transition; the adverse effects of climate change and the impact of implementation of response measures; best practice in domestic policies and measures to address greenhouse gas emissions; the mechanisms outlined under the Protocol; a compliance system for the Protocol; and issues relating to the land use, land-use change and forestry (LULUCF) sector.	http://www.iisd.ca/vol12/enb12163e.html
Seventh Session of the Conference for the Parties – Marrakesh Accords (COP-7)	Marrakesh, Morocco (2001)	From 30 October to 6 November delegates met in negotiating groups, closed drafting groups and informal consultations in their attempt to resolve outstanding issues. These included the mechanisms under the Protocol, a compliance system, accounting, reporting and review under Protocol Articles 5 (methodological issues), 7 (communication of information) and 8 (review of information), and land use, land-use change and forestry (LULUCF). Ongoing negotiations were also held on draft COP decisions relating to the least developed countries (LDCs), the Consultative Group of Experts (CGE), and input to the 2002 World Summit on Sustainable Development (WSSD).	*The Marrakesh Ministerial Declaration* – reaffirmed that economic & social development, & poverty eradication were the first & overriding priorities of the developing countries. It also stressed the importance of capacity building, & developing innovative technologies through public & private sector involvement. *The Marrakesh Accords* – called upon Annex I Parties to provide funding to developing countries through these funds, increased GEF replenishment, & bilateral multilateral channels. However, these financial commitments were neither mandatory, nor were any funding levels specified according to a burden sharing formula. Some countries (Canada, EU, Iceland, New Zealand, Norway, & Switzerland) have passed a political declaration, undertaking to provide an annual contribution of US$410 million by 2005 to developing countries.	http://www.iisd.ca/vol12/enb12189e.html

Continued

Table 5.5.1 International Level Energy Policy. *Continued*

Title	Venue	Purpose	Key Outcome(s)	Website
Eighth Session of the Conference for the Parties – Delhi Declaration on Climate Change & Sustainable Development (COP-8)	New Delhi, India (2002)	COP-8 Recognised that, as Africa is the region suffering the most from the combined impacts of climate change & poverty, development initiatives such as the New Partnership for Africa's Development (NEPAD) should be supported in the context of sustainable development. It also resolved that, in order to respond to the challenges faced now & in the future, climate change & its adverse effects should be addressed while meeting the requirements of sustainable development.	*Delhi Declaration on Climate Change & Sustainable Development* – The Declaration reiterates the importance of carrying out all existing international commitments under the United Nations Framework Convention on Climate Change. It also called for early ratification of the Kyoto Protocol. The Declaration further promoted less polluting energy and other innovative technologies. It urged governments to promote technological advances through research and development, to substantially increase renewable energy resources and to promote the transfer of technologies that can help reduce greenhouse gas emissions in major economic sectors including through public sector and market-oriented approaches.	http://unfccc.int/cop8/latest/1_cpl6rev1.pdf
Ninth Session of the Conference for the Parties – (COP-9)	Milan, Italy (2003)	The high-level political debate during the conference took place through three informal ministerial roundtables focusing on adaptation, mitigation, sustainable development, technology and assessment. Ministers agreed that climate change remains the most important global challenge to humanity and that its adverse effects are already a reality in all parts of the world. Many Parties, developed and developing, reported on activities to adapt to climate change and emphasised the moral duty to provide support to developing countries in these efforts.	Parties adopted numerous decisions & conclusions on various issues including: ▪ definitions & modalities for including afforestation & reforestation activites under the Clear Development Mechanism; ▪ good practice guidance on land use, land-use change & forestry (LULUCF); ▪ the Special Climate Change Fund (SCCF); and ▪ the Least Developed Countries (LDC) Fund.	http://www.iisd.ca/download/pdf/enb12231e.pdf

International Conference for Renewable Energies	Bonn, Germany (2004)	In June 2004, Bonn played host to guests from all over the world. 3,600 participants followed the German Government's invitation to attend the International Conference for Renewable Energies – *Renewables 2004* in order to address the two central issues: How can the proportion of renewable energies used in industrialised & developing countries be substantially increased? How can the markets for renewable energies be better developed?	The three outcomes of the Bonn conference: ▪ *The Political Declaration*, (containing shared political goals for an increased role of renewable energies & reflecting a joint vision of a sustainable energy future, which provides better and more equitable access to energy as well as increased energy efficiency). ▪ *A summary of the International Action Programme* (including actions & commitments by governments, international organisations & stakeholders). ▪ *The Policy Recommendations for Renewable Energies* (that can be of benefit to governments, international organisations & stakeholders as they develop new approaches & political strategies addressing the roles & responsibilities of key actors).	http://www.renewables2004.de/ http://www.renewables2004.de/pdf/conference_report.pdf
Vienna Convention for the Protection of the Ozone Layer	Vienna, Austria (1985)	For the first time nations agreed in principle to tackle a global environmental problem before its effects were felt, or even scientifically proven.	Nations agreed to take 'appropriate measures . . . to protect human health and the environment against adverse effects resulting or likely to result from human activities which modify or are likely to modify the Ozone Layer' but the measures were unspecified. (See later Montreal Protocol, 1987.)	http://www.unep.org/ozone/Treaties_and_Ratification/2A_vienna_convention.asp
UN Framework Climate Change Convention (UNFCCC)	Rio de Janeiro, Brazil (1992)	The ultimate objective of the Convention is 'to achieve stabilisation of atmospheric concentrations of greenhouse gases at levels that would prevent dangerous anthropogenic (human-induced) interference with the climate system . . .'. (See later Kyoto Protocol 1997.)	Parties to the UNFCCC have a no. of commitments including: ▪ to submit a national inventory of emissions/removals of greenhouse gases (NGGI) ▪ to implement national programs to mitigate climate change & adapt to its impacts ▪ to strengthen scientific & technical research & systematic observation related to the climate system, & promote the development & diffusion of relevant technologies ▪ to promote education programs & public awareness about climate change & its likely effects ▪ to periodically submit national reports on activities to implement commitments under the convention.	http://unfccc.int/resource/docs/convkp/conveng.pdf

Continued

Table 5.5.1 International Level Energy Policy. *Continued*

Title	Venue	Purpose	Key Outcome(s)	Website
Montreal Protocol on Substances that Deplete the Ozone Layer	Montreal, Canada (1987)	To further the issues discussed at the Vienna Convention & where possible sets appropriate measures/targets. (Subsequently amended at London 1990, Copenhagen 1992, Vienna 1995, Montreal 1997 & Beijing 1999.)	The final agreement, which contains clauses to cover the special circumstances of several groups of countries, especially developing countries with low consumption rates who do not want the Protocol to hinder their development. The Protocol is constructively flexible: it can be tightened or 'adjusted' as the scientific evidence strengthened, without having to be completely renegotiated. Indeed, it sets the 'elimination' of ozone-depleting substances as its 'final objective'.	http://www.unep.org/ozone/pdf/Montreal-Protocol2000.pdf
Kyoto Protocol to the UN Framework Convention on Climate Change (COP-3)	Kyoto, Japan (1997)	The Kyoto Protocol, negotiated by more than 160 nations in December 1997, aims to reduce net emissions of certain greenhouse gases (primarily CO_2).	Protocol commits industrialised nations to legally binding reductions in emissions of six 'greenhouse gases'. The developed countries commit themselves to reducing their collective emissions of six key greenhouse gases by at least 5%. Each of the participating developed countries must decide how to meet its respective reduction goal during a five-year period (2008–2012); but specific ground rules remain to be worked out at future negotiating sessions. The Protocol encourages governments to co-operate with one another, improve energy efficiency, reform the energy and transportation sectors, promote renewable forms of energy, phase out inappropriate fiscal measures and market imperfections, limit methane emissions from waste management and energy systems, and protect forests and other carbon 'sinks'.	http://unfccc.int/resource/docs/convkp/kpeng.pdf
Energy Charter Treaty	Lisbon, Portugal (1994)	The Energy Charter Treaty is the international community's most significant instrument for the promotion of co-operation in the energy sector. Following its entry into force on 16 April 1998, the Treaty, together with the related documents, provides an unprecedented legal basis for the creation of an open international energy market for the twenty-first Century.	The ECT is: ■ the first binding multilateral investment protection agreement; ■ the first multilateral agreement to cover both investment protection and trade; ■ the first application of transit rules to energy networks; ■ the first multilateral treaty to provide binding international dispute settlement as the general rule.	http://www.encharter.org/upload/1/TreatyBook-en.pdf

Table 5.5.2 European Level Energy Policy.

Policy Type	Programme Title	Policy Objective	Community Measure	Document Reference	Website
Council Directives	Minimum Stock Levels of Crude Oil &/or Petroleum Products	To ensure minimum stocks of crude oil and/or petroleum products in each Member State in order to improve the European Union's security of energy supplies.	Council Directive 68/414/EEC imposing an obligation on Member States of the EEC to maintain minimum stocks of crude oil and/or petroleum products (20/12/68). Amended by Council Directive 98/93/EC (14/12/98).	■ Off Jnl L308, 23/12/68 ■ Off Jnl L358, 31/12/98	http://europa.eu.int/scadplus/leg/en/lvb/l27045.htm
	Measures to Mitigate the Effects of Difficulties in the Supply of Crude Oil & Petroleum Products	Member States shall provide the competent authorities with the necessary powers in the event of difficulties arising in the supply of crude oil & petroleum products which might appreciably reduce the supply of these products & cause severe disruption. Those powers should enable the authorities to draw on emergency stocks, impose an obligation on MS to maintain minimum stocks of crude oil &/or petroleum products & distribute stocks to users.	Council Directive 73/238/EEC on measures to mitigate the effects of difficulties in the supply of crude oil & petroleum products (24/07/73).	■ Off Jnl L228, 16/08/73	http://europa.eu.int/smartapi/cgi/sga_doc?smartapi!celexapi!prod!CELEXnumdoc&lg=EN&numdoc=31973L0238&model=guichett
	The Performance of Heat Generators for Space Heating & the Production of Hot Water in New or Existing Non-Industrial Buildings	Member States shall take all necessary measures to ensure all new heat generators for space heating &/or the production of hot water in new/existing non-industrial buildings comply with minimum performance requirements. In the case of generators capable of using various forms of energy, the minimum performance requirements must relate to each form of energy used.	Council Directive 78/170/EEC on the performance of heat generators for space heating & the production of hot water in new or existing non-industrial buildings & on the insulation of heat & domestic hot water distribution in new non-industrial buildings (13/02/78). Amended by CD 82/885/EEC (10/12/82).	■ Off Jnl L52, 23/02/78 ■ Off Jnl L378, 31/12/82	http://europa.eu.int/smartapi/cgi/sga_doc?smartapi!celexapi!prod!CELEXnumdoc&lg=EN&numdoc=31978L0170&model=guichett

Continued

Table 5.5.2 European Level Energy Policy. *Continued*

Policy Type	Programme Title	Policy Objective	Community Measure	Document Reference	Website
Council Directives	Price Transparency	To improve the transparency of gas and electricity prices charged to industrial end-users in order to increase consumers' freedom of choice without creating obstacles to confidentiality.	Council Directive 90/377/EEC concerning a Community procedure to improve the transparency of gas and electricity prices charged to industrial end-users (29/06/90).	■ Off Jnl L185, 17/07/90	http://europa.eu.int/scadplus/leg/en/lvb/l27002.htm
			Amended by Commission Directive 93/87/EEC (22/10/93).	■ Off Jnl L277, 10/11/93	
	Energy Efficiency of Hot-Water Boilers	To ensure freedom of movement for new hot-water boilers in the internal market through the total harmonisation of the essential efficiency requirements that they must meet.	Council Directive 92/42/EEC on efficiency requirements for new hot-water boilers fired with liquid or gaseous fuels (21/05/92).	■ Off Jnl L167, 22/06/92	http://europa.eu.int/scadplus/leg/en/lvb/l21019.htm
			Amended by CD 93/68/EEC (22/07/93).	■ Off Jnl L268, 29/10/93	
	Labelling Household Appliances	To harmonise national measures relating to the publication of information on the consumption of energy and of other essential resources by household appliances, thereby allowing consumers to choose appliances on the basis of their energy efficiency.	Council Directive 92/75/EEC on the indication by labelling and standard product information of the consumption of energy and other resources by household appliances (22/09/92).	■ Off Jnl L297, 13/10/92	http://europa.eu.int/scadplus/leg/en/lvb/l32004.htm
	Energy Certification of Buildings	To limit carbon dioxide emissions and promote the rational use of energy in those economic sectors in the European Community which consume the most energy, namely the residential and tertiary sectors, in order to preserve the quality of the environment.	Council Directive 93/76/EEC to limit carbon dioxide emissions by improving energy efficiency (SAVE) (13/09/93).	■ Off Jnl L237, 22/09/93	http://europa.eu.int/scadplus/leg/en/lvb/l21252.htm

Energy Efficiency for Refrigerators	To lay down minimum energy efficiency standards for household refrigeration appliances sold in the European Union.	Directive 96/57/EC of the European Parliament and of the Council on energy efficiency requirements for household electric refrigerators, freezers and combinations thereof (03/09/96).	Off Jnl L236, 18/09/96	http://europa.eu.int/scadplus/leg/en/lvb/l21016.htm
Dangers arising from Ionising Radiation	To establish uniform basic safety standards to protect the health of workers and the general public against the dangers of ionising radiation.	Council Directive 96/29/Euratom laying down the basic safety standards for the protection of the health of workers and the general public against the dangers arising from ionising radiation (13/05/96).	Off Jnl L159, 29/06/96	http://europa.eu.int/scadplus/leg/en/cha/c11142.htm
Ballasts for Fluorescent Lighting	The purpose of the directive is to achieve cost-effective energy savings in fluorescent lighting, which would not otherwise be achieved with other measures. This directive covers only newly produced ballasts, which are responsible for high energy consumption and offer considerable potential for energy savings.	European Parliament and Council Directive 2000/55/EC, on energy efficiency requirements for ballasts for fluorescent lighting (18/09/00).	Off Jnl L279, 01/11/00	http://europa.eu.int/scadplus/leg/en/lvb/l27032.htm
Promotion of Electricity From Renewable Energy Sources	To create a Community framework which will facilitate a significant increase in the medium term in renewable generated electricity (RES-E).	Directive 2001/77/EC of the European Parliament and of the Council on the promotion of electricity from renewable energy sources in the internal electricity market (27/09/01).	Off Jnl L283, 27/10/01	http://europa.eu.int/scadplus/leg/en/lvb/l27035.htm

Continued

Table 5.5.2 European Level Energy Policy. *Continued*

Policy Type	Programme Title	Policy Objective	Community Measure	Document Reference	Website
Council Directives	The Energy Performance Of Buildings	The objective of this Directive is to promote the improvement of the energy performance of buildings within the Community, taking into account outdoor climatic & local conditions, as well as indoor climate requirements & cost-effectiveness. This Directive lays down requirements as regards: ▪ the general framework for a methodology of calculation of the integrated energy performance of buildings; ▪ the application of minimum requirements on the energy performance of new buildings; ▪ the application of minimum requirements on the energy performance of large existing buildings that are subject to major renovation; ▪ energy certification of buildings; & regular inspection of boilers & of air-conditioning systems in buildings & in addition an assessment of the heating installation in which the boilers are more than 15 years old.	Directive 2002/91/EC of the European Parliament and of the Council on the energy performance of buildings (16/12/02).	▪ Dir 2002/91/EC	http://europa.eu.int/eur-lex/pri/en/oj/dat/2003/l_001/l_00120030104en00650071.pdf
	Common Rules for the Internal Market in Electricity (Repealing Directive 96/92/EC)	This Directive establishes common rules for the generation, transmission, distribution & supply of electricity. It lays down the rules relating to the organisation & functioning of the electricity sector, access to the market, the criteria & procedures applicable to calls for tenders & the granting of authorisations & the operation of systems.	Directive 2003/54/EC of the European Parliament & of the Council concerning common rules for the internal market in electricity (repealing Directive 96/92/EC) (26/06/03).	▪ Off Jnl L176/37 15/07/03	http://europa.eu.int/eur-lex/pri/en/oj/dat/2003/l_176/l_17620030715en00370055.pdf

Common Rules for the Internal Market in Natural Gas (Repealing Directive 98/30/EC)	This Directive establishes common rules for the transmission, distribution, supply & storage of natural gas. It lays down the rules relating to the organisation & functioning of the natural gas sector, access to the market, the criteria & procedures applicable to the granting of authorisations for transmission, distribution, supply & storage of natural gas & the operation of systems.	Directive 2003/55/EC of the European Parliament & of the Council concerning common rules for the internal market in natural gas (repealing Directive 98/30/EC) (26/06/03).	Off Jnl L176/57 15/07/03	http://europa.eu.int/eur-lex/pri/en/oj/dat/2003/l_176/l_17620030715en00570078.pdf
Energy Labelling of Household Electric Refrigerators, Freezers & their Combinations	Electricity use by refrigerators, freezers and their combinations accounts for a significant part of total Community household energy demand. The further scope for a reduction of energy use by these appliances is substantial.	Commission Directive 2003/66/EC amending Directive 94/2/EC implementing Council Directive 92/75/EEC with regard to energy labelling of household electric refrigerators, freezers & their combinations (03/07/03).	Off Jnl L170/10 09/07/03	http://europa.eu.int/eur-lex/pri/en/oj/dat/2003/l_170/l_17020030709en00100014.pdf
Restructuring the Community Framework for the Taxation of Energy Products & Electricity	Member States shall impose taxation on energy products & electricity in accordance with this Directive.	Council Directive 2003/96 EC restructuring the Community framework for the taxation of energy products & electricity (27/10/03).	Off Jnl L283/51 31/10/03	http://europa.eu.int/eur-lex/pri/en/oj/dat/2003/l_283/l_28320031031en00510070.pdf
The Promotion of Cogeneration Based on a Useful Heat Demand in the Internal Energy Market (Amending Directive 92/42/EEC)	The purpose of this Directive is to increase energy efficiency & improve security of supply by creating a framework for promotion & development of high efficiency cogeneration of heat & power based on useful heat demand & primary energy savings in the internal energy market, taking into account the specific national circumstances especially concerning climatic & economic conditions.	Directive 2004/8/EC of the European Parliament & of the Council on the promotion of cogeneration based on a useful heat demand in the internal energy market (amending Directive 92/42/EEC) (11/02/04).	Off Jnl L52/50 21/02/04	http://europa.eu.int/eur-lex/pri/en/oj/dat/2004/l_052/l_05220040221en00500060.pdf

Continued

Table 5.5.2 European Level Energy Policy. *Continued*

Policy Type	Programme Title	Policy Objective	Community Measure	Document Reference	Website
Council Decisions	Community Targets for Reducing Consumption of Primary Sources of Energy	Where difficulties arise in the supply of crude oil or petroleum products in one or more Member States, the Commission may set a target for reducing consumption of petroleum products in the Community as a whole by up to 10% of normal consumption (applicable for a maximum of two months). MS shall without delay take all appropriate measures to reduce their consumption of petroleum products &/or energy consumption as a whole by at least their share of the reduction target.	Council Decision 77/706/EEC on the setting of a Community target for a reduction in the consumption of primary sources of energy in the event of difficulties in the supply of crude oil & petroleum products (07/11/77).	▪ Off Jnl L292, 16/11/77	http://europa.eu.int/smartapi/cgi/sga_doc?smartapi!celexapi!prod!CELEXnumdoc&lg=EN&numdoc=31977D0706&model=guichett
	Set of Actions Relating to Trans-European Networks in the Energy Sector	To create a more favourable context for the development of trans-European networks in the energy sector and speed up the realisation of projects of common interest already identified.	Council Decision 96/391/EC laying down a series of measures aimed at creating a more favourable context for the development of trans-European networks in the energy sector (28/03/96).	▪ Off Jnl L161, 29/06/96	http://europa.eu.int/scadplus/leg/en/lvb/l27012.htm
	The Energy Charter Protocol on Energy Efficiency & Related Environmental Aspects	The Energy Charter Treaty and the Energy Charter Protocol on energy efficiency & related environmental aspects are approved on behalf of the European Coal & Steel Community, the European Community & the European Atomic Energy Community.	Council & Commission Decision (98/181/EC, ECSC, Euratom) on the conclusion, by the European Communities, of the Energy Charter Treaty & the Energy Charter Protocol on energy efficiency & related environmental aspects (23/09/97).	▪ Off Jnl L69/1 09/03/98	http://europa.eu.int/eur-lex/pri/en/oj/dat/1998/l_069/l_06919980309en00010004.pdf
	SYNERGY II Programme	To improve energy efficiency, promote sustainable development, improve the	Council Decision 1999/23/EC adopting a multi-annual	▪ Off Jnl L7, 13/01/99	http://europa.eu.int/scadplus/leg/en/lvb/l27029.htm

	competitiveness of Community businesses and enhance security of supply by means of closer co-operation between the Community and third countries in the energy sector.	programme to promote international cooperation in the energy sector (1998–2002) – Synergy II programme (14/12/98).		
Framework Programme for Actions in the Energy Sector	To guarantee the coherence and efficiency of Community energy actions taken under the energy policy or under other Community policies.	Council Decision 1999/21/EC, adopting a Framework Programme for actions in the energy sector (1998–2002) and connected measures (14/12/98).	Off Jnl L7, 13/01/99	http://europa.eu.int/scadplus/leg/en/lvb/l27024.htm
SURE Programme: Nuclear Safety	To improve the safe transport of radioactive materials in the European Union and the safety of nuclear installations in countries participating in the TACIS (Technical Assistance to the Commonwealth of Independent States) programme by means of increased co-operation in the field of safeguards and industrial co-operation.	Council Decision 1999/25/Euratom adopting a multi-annual programme (1998 to 2002) of actions in the nuclear sector relating to the safe transport of radioactive materials and to safeguards and industrial co-operation to promote certain aspects of the safety of nuclear installations in the countries currently participating in the TACIS programme – SURE programme (14/12/98).	Off Jnl L7, 13/01/99	http://europa.eu.int/scadplus/leg/en/lvb/l27026.htm
Thematic Programme: Energy, Environment & Sustainable Development	Maintaining sustainable development to create new opportunities for citizens & companies, making the latter more efficient and less polluting. In other words, the objective is to reconcile economic growth & protection of the environment.	Council Decision 1999/170/EC adopting a specific programme for research, technological development and demonstration on energy, environment and sustainable development (1998 to 2002) (25/01/99). Council Decision 1999/175/Euratom adopting a research and training programme (Euratom) in the field of nuclear energy (1998 to 2002).	Off Jnl L64, 12/03/99	http://europa.eu.int/scadplus/leg/en/lvb/i23005.htm

Continued

Table 5.5.2 European Level Energy Policy. *Continued*

Policy Type	Programme Title	Policy Objective	Community Measure	Document Reference	Website
Council Decisions	ETAP Programme: Multi-Annual Programme of Studies (Environmental Technology Action Plan)	To promote a co-operative approach between the Community, the Member States, non-Community countries, international organisations & other interested parties to analysis of energy problems & trends at Community level.	Council Decision 1999/22/EC adopting a multi-annual programme of studies, analyses, forecasts and other related work in the energy sector (1998–2002) (ETAP programme) (14/12/98).	■ Off Jnl L7, 13/01/99	http://europa.eu.int/scadplus/leg/en/lvb/l27027.htm
	CARNOT Programme: Solid Fuels	To promote the use of clean and efficient solid fuel technologies.	Council Decision 1999/24/EC adopting a multi-annual programme of technological actions promoting the clean and efficient use of solid fuels (1998 to 2002) – CARNOT Programme (14/12/98).	■ Off Jnl L7, 13/01/99	http://europa.eu.int/scadplus/leg/en/lvb/l27030.htm
	Promotion of Renewable Energy – Altener Programme	The Altener programme follows on from the Altener I programme and the Altener II programme, which it replaces. The objective of Altener is to promote renewable energy sources in the European Union.	Decision No 646/2000/EC of the European Parliament and of the Council adopting a multi-annual programme for the promotion of renewable energy sources in the Community (Altener) (1998 to 2002) (28/02/00).	■ Off Jnl L79, 30/03/00	http://europa.eu.int/scadplus/leg/en/lvb/l27016b.htm
	SAVE II Programme: Energy Efficiency	On the basis of Article 174 of the Treaty, which stipulates that one objective of Community policy is to ensure prudent & rational utilisation of natural resources, the European Union is pursuing a programme to encourage rational & efficient use of energy resources. In the context of the Kyoto Protocol & the commitments to cut CO_2 emissions, energy efficiency is the cornerstone of Community action.	Council Decision No 647/2000/EC of the European Parliament and of the Council adopting a multi-annual programme for the promotion of energy efficiency (SAVE) (1998 to 2002) (28/02/00).	■ Off Jnl L79, 30/03/01	http://europa.eu.int/scadplus/leg/en/lvb/l27017.htm

European Energy & Transport Forum	To create a consultative committee consisting of qualified representatives of the energy & transport sectors whose role is to express an opinion on all Commission initiatives relating to energy & transport policy. The committee will also act as an observer commenting on competitiveness & changes in the structure of the sectors, taking into account environmental, social & safety considerations.	Commission Decision 2001/546/EC on setting up a consultative committee – European Energy & Transport Forum (11/07/01).	▪ COM (2001) 1843 ▪ Off Jnl L195, 19/07/01	http://europa.eu.int/scadplus/leg/en/lvb/l27044.htm
Office Appliances: Energy Star Programme	To co-ordinate the labelling of energy-efficient office equipment using the 'Energy Star' logo.	Council Decision 2001/469/EC concerning the conclusion on behalf of the European Community of the Agreement between the Government of the United States of America and the European Community on the co-ordination of energy-efficient labelling programs for office equipment (14/05/01).	▪ Off Jnl L172, 26/06/01	http://europa.eu.int/scadplus/leg/en/lvb/l27034.htm
Co-operation Between the US & Euratom in the Field of Fusion Energy	To continue and intensify cooperation between the European Atomic Energy Community (Euratom) and the United States in the field of fusion in order to develop the scientific understanding and technological capability underlying a fusion energy system.	Council Decision approving the conclusion, by the Commission, of the Agreement for co-operation between the European Atomic Energy Community represented by the Commission and the Department of Energy of the United States of America in the field of fusion energy research and development (08/03/01).	▪ Off Jnl L148, 01/06/01	http://europa.eu.int/scadplus/leg/en/lvb/l27043.htm

Continued

Table 5.5.2 European Level Energy Policy. *Continued*

Policy Type	Programme Title	Policy Objective	Community Measure	Document Reference	Website
Council Decisions	Laying down a Series of Guidelines for Trans-European Energy Networks (repealing Decision No 1254/96/EC)	This Decision defines the nature & scope of Community action to establish guidelines for trans-European energy networks. It establishes a series of guidelines covering the objectives, priorities and broad lines of action by the Community in respect of trans-European energy networks. These guidelines identify projects of common interest, including those which have priority, among trans-European electricity & natural gas networks.	Decision No 1229/2003/EC of the European Parliament & of the Council laying down a series of guidelines for trans-European energy networks (repealing Decision No 1254/96/EC) (26/06/03).	■ Off Jnl L176/11 15/07/03	http://europa.eu.int/eur-lex/pri/en/oj/dat/2003/l_176/l_17620030715en00110028.pdf
	Adopting a Multi-annual Programme for Action in the Field of Energy: *Intelligent Energy – Europe (2003–2006)*	The specific objectives of this programme are as follows: ■ to provide the elements needed for the promotion of energy efficiency, the increased use of renewable energy sources & energy diversification, such as through new developing & renewable energy sources, including in transport, the improvement of sustainability, the development of the potential of the regions & the preparation of the legislative measures needed to attain these strategic objectives; ■ to develop means & instruments, which may be used by the Commission & the Member States, to follow up, monitor & evaluate the impact of the measures adopted in the fields of energy efficiency & renewable energy sources, including the energy aspects of transport; ■ to promote efficient & intelligent patterns of energy production & consumption based on solid & sustainable foundations by raising awareness.	Dec No 1230/2003/EC of the European Parliament & of the Council adopting a multi-annual programme for action in the field of energy: *Intelligent Energy – Europe (2003–2006)* (26/06/03).	■ Off Jnl L176/29 15/07/03	http://europa.eu.int/eur-lex/pri/en/oj/dat/2003/l_176/l_17620030715en00290036.pdf

Council Communications	The External Dimension of Trans-European Energy Networks	To analyse natural gas and electricity interconnections with third countries and the development of these networks on the scale of the European continent, of the Mediterranean Basin and, where appropriate, Central Asia and the Middle East.	Communication from the Commission to the Council and the European Parliament: the external dimension of trans-European energy networks (1997).	COM (97) 125 Final	http://europa.eu.int/scadplus/leg/en/lvb/l27020.htm
	Towards a Strategy for the Rational use of Energy	To promote energy savings.	Communication from the Commission on Energy Efficiency in the European Community (29/04/98).	COM (1998) 246 Final	http://europa.eu.int/scadplus/leg/en/lvb/l27025.htm
	Integrating of Environmental Considerations in Community Energy Policy	To propose measures for further integrating of environmental considerations in Community energy policy and to review progress made so far.	Communication from the Commission: Strengthening environmental integration within Community energy policy (14/10/98).	COM (1998) 571 Final	http://europa.eu.int/scadplus/leg/en/lvb/l28071.htm
	The Northern Dimension of the European Energy Policy	To define and implement the concept of Northern Dimension in terms of European energy policy.	Commission Communication – Strengthening the Northern Dimension of European energy policy (08/11/99).	COM (99) 548	http://europa.eu.int/scadplus/leg/en/lvb/l27031.htm
	Energy Efficiency: Action Plan	To reduce energy consumption by improving energy efficiency in order to protect the environment, to increase security of supply and to establish a more sustainable energy policy.	Communication from the Commission to the Council, the European Parliament, the Economic and Social Committee and the Committee of the Regions. Action Plan to improve Energy Efficiency in the European Community (2000).	COM (2000) 247	http://europa.eu.int/scadplus/leg/en/lvb/l27033.htm
	Nuclear Safety in the Newly Independent States & Central/Eastern Europe	To evaluate the Commission's contribution towards the improvement of nuclear safety in the Newly Independent States (NIS) and in Central and Eastern Europe and to present proposals for future Community action.	Communication from the Commission to the Council and the European Parliament. Commission support to nuclear safety in the Newly Independent States and in Central and Eastern Europe (06/09/00).	COM (2000) 493	http://europa.eu.int/scadplus/leg/en/lvb/l27036.htm

Continued

Table 5.5.2 European Level Energy Policy. *Continued*

Policy Type	Programme Title	Policy Objective	Community Measure	Document Reference	Website
Council Communications	The Implementation of the Community Strategy & Action Plan on Renewable Energy Sources (1998–2000)	This Communication contains the first progress report on the implementation of the White Paper for a Community Strategy and Action Plan on Renewable Energy Sources. An updated version of the Action Plan is provided in Annex I.	Communication from the Commission to the Council, the European Parliament, the Economic & Social Committee & the Committee of the Regions on the implementation of the Community Strategy & Action Plan on Renewable Energy Sources (1998–2000) (2001).	▪ COM (2001) 69	http://www.europa.eu.int/smartapi/cgi/sga_doc?smartapi!celexplus!prod!DocNumber&lg=en&type_doc=COMfinal&an_doc=2001&nu_doc=69
	Euro-Mediterranean Co-Operation	To define the broad lines of Euro-Mediterranean cooperation in the field of transport and energy for the period 2000–2006 in order to inject new momentum into the transport sector and to boost the existing momentum in the energy sector.	Communication from the Commission to the Council and to the European Parliament, 'Enhancing Euro-Mediterranean cooperation on transport and energy' (07/03/01).	▪ COM (2001) 126	http://europa.eu.int/scadplus/leg/en/lvb/l27038.htm
	Global Partnership for Sustainable Development	To highlight how the European Union contributes to global sustainable development and the action aimed at establishing a global deal for sustainable development.	Communication from the Commission to the European Parliament, the Council, the Economic and Social Committee and the Committee of the Regions. Towards a global partnership for sustainable development (2002).	▪ COM (2002) 82	http://europa.eu.int/scadplus/leg/en/lvb/l28015.htm

Towards a European strategy for the security of energy supply	The Green Paper on the security of energy supply, adopted by the Commission more than a year before, opened up a debate on energy policy unprecedented in 30 years. In most of the Member States this debate revived discussion on national options in the energy field. It was used for reference in some third countries, like the United States (during the preparation of the Bush energy plan), Japan & Russia. Looking ahead to the next 20 to 30 years, the Green Paper drew attention to the structural weaknesses & geopolitical, social & environmental shortcomings of the EU's energy supply, notably as regards European commitments in the Kyoto Protocol.	Communication from the Commission to the Council & the European Parliament on the Final report on the Green Paper 'Towards a European strategy for the security of energy supply' (26/06/02).	COM(2002) 321	http://europa.eu.int/comm/energy_transport/livrevert/final/report_en.pdf
Energy Cooperation with the Developing Countries	This Communication first of all analyses the energy situation in the developing countries & then proposes a reference framework for energy cooperation with them. A series of operational recommendations to implement the cooperation flow from this analysis & the reference framework.	Communication from the Commission to the Council & the European Parliament – Energy cooperation with the developing countries (17/07/02).	COM (2002) 408	http://www.europa.eu.int/smartapi/cgi/sga_doc?smartapi!celexplus!prod!DocNumber&lg=en&type_doc=COMfinal&an_doc=2002&nu_doc=408 http://www.europa.eu.int/scadplus/leg/en/lvb/r12008.htm
The Promotion of Cogeneration Based on a Useful Heat Demand in the Internal Energy Market	The overriding objective of this proposal is to create a framework which can support & facilitate the installation & proper functioning of electrical cogeneration plants where a useful heat demand exists or is foreseen.	Proposal for a Directive of the European Parliament and of the Council on the promotion of cogeneration based on a useful heat demand in the internal energy market (22/07/02).	COM (2002) 415	http://europa.eu.int/eur-lex/en/com/pdf/2002/com2002_0415en01.pdf

Continued

Table 5.5.2 European Level Energy Policy. *Continued*

Policy Type	Programme Title	Policy Objective	Community Measure	Document Reference	Website
Council Communications	The Development of Energy Policy for the Enlarged European Union, its Neighbours and Partner Countries	This Communication is focused on energy relations of the enlarged EU with its neighbours & most important geographical partners in this sector. Although inspired by the recent Communication on a Wider Europe, its geographical scope is slightly different as it includes also South-East Europe – as an example of what might be envisaged on the regulation of the markets – & the Caspian Basin where appropriate. These two areas are of particular importance for the completeness of the Internal Energy Market & for the security of energy supplies of the EU.	Communication from the Commission to the Council and the European Parliament on the Development of Energy Policy for the Enlarged European Union, its Neighbours and Partner Countries (13/05/03).	▪ COM (2003) 262	http://europa.eu.int/comm/energy/ten-e/doc/com_2003_0262_en.pdf
	The Share of Renewable Energy in the EU	This Communication assesses the state of development of renewable energy in the EU. It serves three purposes: ▪ the formal report that the Commission is required to make under Article 3 of Directive 2001/77/EC, evaluating the progress made by the EU15 towards achieving national targets for 2010 for electricity from renewable energy sources;	Communication from the Commission to the Council & the European Parliament – The share of renewable energy in the EU (26/05/04).	▪ COM (2004) 366	http://www.europa.eu.int/comm/energy/res/legislation/country_profiles/com_2004_366_en.pdf

		■ assessment of the prospects for achieving the target of a 12% share of renewable energy in overall energy consumption in the EU15 in 2010 (including heating, electricity & transport), taking into account EU legislation since 2000 & other measures in renewable energy & energy efficiency; ■ proposals for concrete actions at national & Community level to ensure the achievement of EU renewable energy targets for 2010, in the context of the Bonn World Renewable Energy Conference (June 2004) &, building on this, the line to take on the scenario for 2020.			
Council Recommendations	The Rational Use of Energy by Promoting the Thermal Insulation of Buildings	That Member States: ■ Organise or sponsor specific information campaigns designed to stimulate public interest in improving inadequate or poor-quality thermal insulation systems in residential accommodation. ■ Assess the average extent of thermal insulation in existing residential accommodation ■ Organise competitions or draw up development programmes to obtain the best original ideas for improving insulation in existing residential accommodation ■ Adopt the harmonised Community reference standards for thermal insulation which are to be drawn up for use in the implementation of all laws, regulations & administrative measures.	Council Recommendation 76/492/EEC on the rational use of energy by promoting the thermal insulation of buildings (04/05/76).	■ Off Jnl L140, 28/05/76	http://europa.eu.int/smartapi/cgi/sga_doc?smartapi!celexapi!prod!CELEXnumdoc&lg=EN&numdoc=31976H0492&model=guichett

Continued

Table 5.5.2 European Level Energy Policy. *Continued*

Policy Type	Programme Title	Policy Objective	Community Measure	Document Reference	Website
Council Recommendations	The Rational Use of Energy in Urban Passenger Transport	That Member States: ■ Encourage the authorities responsible to promote frequent, convenient, regular, fast, reliable, comfortable urban public passenger transport services ■ Encourage research into the improvement of existing equipment & experimentation with public transport systems that will meet individual requirements more fully ■ Reduce commuting distances by encouraging construction of convenient parking areas near public transport termini & important stations & bus stops in urban & suburban areas ■ Examine the advisability of changing fiscal regulations which encourage commuting by private car and long-distance commuting by public transport ■ Encourage the highest load factor in private cars, by removing legislative & administrative barriers/regulations which prevent car-pooling on a cost-sharing basis & adopt the laws, regulations or administrative provisions to prevent insurance companies from introducing/maintaining contractual barriers to such a system ■ Encourage all measures making the flow of traffic smoother, especially the flow of public transport, e.g. by synchronisation/programming of traffic lights.	Council Recommendation 76/495/EEC on the rationa use of energy in urban passenger transport (04/05/76).	■ Off Jnl L140, 28/05/76	http://europa.eu.int/smartapi/cgi/sga_doc?smartapi!celexapi!prod!CELEXnumdoc&lg=EN&numdoc=31976H0495&model=guichett

Rational Use of Energy in Industrial Undertakings	That Member States: ▪ Take the necessary measures to mount campaigns to provide information & increase awareness, & to encourage the achievement of energy savings; and, to set up one or more public or approved bodies/departments with particular responsibility for these tasks ▪ Invite sectoral & general professional organisations as well as technical & scientific associations & consultancy services to arrange periodically for meetings & seminars ▪ Encourage industrial undertakings to set up an energy department within their organisation, if they have not already done so, or call in an outside expert the duties & tasks of such departments or persons could be similar to those proposed as a guide in the Annex to this recommendation and devote a chapter of an annual report to energy consumption, including any energy-saving measures taken or planned & the results obtained ▪ Collate all useful information concerning the results of the energy-saving measures taken by the industrial undertakings ▪ Inform the Commission regularly of the measures taken & of the results obtained or anticipated from the measures.	Council Recommendation 77/713/EEC on the rational use of energy in industrial undertakings (23/10/77).	▪ Off Jnl L295, 18/11/77	http://europa.eu.int/smartapi/cgi/sga_doc?smartapi!celexapi!prod!CELEXnumdoc&lg=EN&numdoc=31977H0713&model=guichett

Continued

Table 5.5.2 European Level Energy Policy. *Continued*

Policy Type	Programme Title	Policy Objective	Community Measure	Document Reference	Website
Council Recommendations	The Reduction of Energy Requirements for Buildings in the Community	To pursue, policies designed to save energy by reducing energy requirements through the improvement of the thermal efficiency of buildings the Commission will organise appropriate consultations with the Member States within the context of the Energy Committee in order to: ▪ Prepare the reports; ▪ Facilitate exchanges of information on matters relevant to the implementation of this recommendation which could prove helpful to Member States in the application of their programmes.	Council Recommendation 79/167/ECSC, EEC, Euratom on the reduction of energy requirements for buildings in the Community (05/02/79).	▪ Off Jnl L37, 13/02/79	http://europa.eu.int/smartapi/cgi/sga_doc?smartapi!celexapi!prod!CELEXnumdoc&lg=EN&numdoc=31979H0167&model=guichett
	Investment in the Rational Use of Energy	Member States should step-up their efforts to achieve a more rational use of energy by: ▪ improving the conditions necessary to enable the economic operators concerned to increase significantly their investment in this area; ▪ giving greater consideration over the whole range of investment to the requirement of rational use of energy.	Council Recommendation 82/604/EEC concerning the encouragement of investment in the rational use of energy (28/07/82).	▪ Off Jnl L247, 23/08/82	http://europa.eu.int/smartapi/cgi/sga_doc?smartapi!celexapi!prod!CELEXnumdoc&lg=EN&numdoc=31982H0604&model=guichett
Miscellaneous	A New Energy Policy Strategy for the Community	Affirms a political will to implement a Community energy policy – by preparing joint target figures, constituting guidelines for national policies & major indicators for energy producers & consumers. Confirms that the world-wide aspects of energy problems necessitate co-operation among the consumer countries & producer countries, in which the Community & the Member States intend to participate.	Council Resolution concerning a new energy policy strategy for the Community (17/09/74).	▪ Off Jnl C153, 09/07/75	http://europa.eu.int/smartapi/cgi/sga_doc?smartapi!celexapi!prod!CELEXnumdoc&lg=EN&numdoc=31975Y0709(01)&model=guichett

New Lines of Action by the Community in the Field of Energy Saving	Approves the objective of progressively reducing the average ratio for the whole of the Community between the rate of growth in gross primary energy consumption & the rate of growth in the gross domestic product to 0.77 or less by 1990. Agrees that Member States should, where this proves necessary, adapt their energy-saving programmes so that by the end of 1980 each Member State has an energy-saving programme covering all the main sectors of energy use & an appropriate energy-pricing policy. Agrees that work must be speeded up, particularly in international organisations, on the technical specifications of methods for measuring energy consumption & on performance standards, especially those relating to heat generators & domestic appliances.	Council Resolution concerning new lines of action by the Community in the field of energy saving (09/06/80).	Off Jnl C149, 18/06/80	http://europa.eu.int/smartapi/cgi/sga_doc?smartapi!celexapi!prod!CELEXnumdoc&lg=EN&numdoc=31980Y0618(02)&model=guichett
The Improvement of Energy Saving Programmes in the Member States	Invites Member States to pursue and where necessary increase their efforts to promote the more rational use of energy by the further development of integrated energy-saving policies.	Council Resolution on the improvement of energy saving programmes in the Member States (15/01/85).	Off Jnl C20, 22/01/85	http://europa.eu.int/smartapi/cgi/sga_doc?smartapi!celexapi!prod!CELEXnumdoc&lg=EN&numdoc=31985Y0122(01)&model=guichett
The Rational Use of Energy in the Building Sector	In the framework of this programme, the Commission in consultation with experts from the Member States will conduct a detailed assessment of the pilot projects relating to the upgrading of the energy efficiency of existing buildings already carried out by some Member States after thermal auditing in order to identify & compare methods used. Emphasises the importance of the regulations in force in the Member States to improve the energy performance of new buildings and, if necessary, of the introduction or reinforcement of such regulations.	Council Resolution on the rational use of energy in the building sector (15/03/85).	Off Jnl C78, 26/03/85	http://europa.eu.int/smartapi/cgi/sga_doc?smartapi!celexapi!prod!CELEXnumdoc&lg=EN&numdoc=31985Y0326(01)&model=guichett

Continued

Table 5.5.2 European Level Energy Policy. *Continued*

Policy Type	Programme Title	Policy Objective	Community Measure	Document Reference	Website
Miscellaneous	Improving Energy Efficiency in Industrial Firms in the Member States	There remains a large potential for energy saving in industry which could be achieved by introducing technologies whose sound economic profitability has been demonstrated. In the long run, the modernisation of productive plant will be the main source of energy savings, and this is likely to improve the competitiveness of European industry & promote job creation in undertakings.	Council Resolution on improving energy efficiency in industrial firms in the Member States (15/09/86).	■ Off Jnl C240, 24/09/86	http://europa.eu.int/smartapi/cgi/sga_doc?smartapi!celexapi!prod!CELEXnumdoc&lg=EN&numdoc=31986Y0924(01)&model=guichett
	Shipments of Radioactive Substances Between Member States	To set up a Community system for the declaration of shipments of radioactive substances between Member States, following the removal of frontier controls in the Community as from 1 January 1993, so that the competent authorities concerned receive the same level of information as before 1993 as regards radiation protection control.	Council Regulation (Euratom) No 1493/93 on shipments of radioactive substances between Member States (08/06/93).	■ Off Jnl L148, 19/06/93	http://europa.eu.int/scadplus/leg/en/lvb/l28099.htm
	An Energy Policy for the EU	The Commission feels that it is essential for the effectiveness of the policies conducted at national level, that the responsibilities deriving from energy policy should form part of the common aims that have been defined at Community level. The Commission is prepared to use all of the provisions of the Treaties in order to establish: ■ a framework for the discussion of energy policy that involves all of the public & private operators concerned;	White Paper: An Energy Policy for the EU (1995).	■ COM (95) 682	http://wire0.ises.org/wire/Publications/Whitepap.nsf/396e92819880db7dc125680f00443688/2EBE39B87F1DC00AC12567B100070C1A/$File/Whitepaper+RE.pdf

	■ a framework for consultation on energy policy guidelines & on activities in this area; &, finally; ■ a framework for co-operation with the Member States in order to achieve jointly-defined aims.			
Renewable Energy: Green Paper	Open up a debate on the most urgent and most important measures relating to renewable sources of energy, identifying the objectives, the obstacles and the means to be deployed.	Commission Green Paper on renewable sources of energy (20/11/96).	■ COM (96) 576	http://europa.eu.int/scadplus/leg/en/lvb/l27018.htm
Community Framework for the Taxation of Energy Products	To determine an overall tax system for the taxation of energy products, with a view to improving the functioning of the internal market, encouraging behaviour conducive to protection of the environment and promoting the greater use of the labour factor.	Proposal for a Council Directive restructuring the Community framework for the taxation of energy products (12/03/97).	■ COM (97) 30 ■ Off Jnl C139, 06/05/97	http://europa.eu.int/scadplus/leg/en/lvb/l27019.htm
Community Strategy to Promote Combined Heat & Power	The aim should be to establish a strategy that would lead to improved use of combined heat & power, taking into account the varying situations of the member states.	Council Resolution on a Community strategy to promote combined heat & power (18/12/97).	■ Off Jnl C4, 08/01/98	http://europa.eu.int/eur-lex/pri/en/oj/dat/1998/c_004/c_00419980108en00010002.pdf
SYNERGY Programme: Co-Operation with Third Countries	To develop a programme of assistance and co-operation with third countries for the formulation and implementation of energy policy.	Council Regulation (EC) No 701/97 amending a programme to promote international co-operation in the energy sector – Synergy programme (14/04/97) Extended by C Reg (EC) No. 2598/97 (18/12/97).	■ Off Jnl L104, 22/04/97 ■ Off Jnl L351, 23/12/97	http://europa.eu.int/scadplus/leg/en/lvb/l27013.htm

Continued

Table 5.5.2 European Level Energy Policy. *Continued*

Policy Type	Programme Title	Policy Objective	Community Measure	Document Reference	Website
Miscellaneous	White Paper Laying Down A Community Strategy & Action Plan	To attain, by 2010, a minimum penetration of 12% of renewable energy sources in the European Union.	Energy for the future: renewable energy sources – White Paper laying down a Community strategy and action plan (1997).	▪ COM (97) 599	http://europa.eu.int/scadplus/leg/en/lvb/l27023.htm
	Renewable Sources of Energy	Supports the need to promote a sustained & substantially increased use of renewable sources of energy throughout the Community in light of the valuable contribution renewables can make to environmental protection & the implementation of the commitments to the Kyoto Protocol.	Council Resolution on renewable sources of energy (08/06/98)	▪ Off Jnl C198, 24/06/98	http://europa.eu.int/eur-lex/pri/en/oj/dat/1998/c_198/c_19819980624en00010003.pdf
	European Energy Outlook to 2020 – Executive Summary	The purpose of this study is to present a consistent EU energy & energy related emissions outlook for the period to 2020. The derivation of the results & conclusions, which are presented here in the context of the 'Shared Analysis Project', has been based on quantitative analysis & on a process of communication with & feedback from a number of energy experts & organisations.	Shared Analysis Project (Nov 1999).	▪ Shared Analysis Project	http://europa.eu.int/comm/dgs/energy_transport/figures/energy_outlook_2020/execsum.pdf
	Security of Energy Supply – Green Paper	To launch a debate on the principal questions and risks linked to the future growth of European dependence on energy.	Green Paper towards a European strategy for the security of energy supply (29/11/00).	▪ COM (2000) 769	http://europa.eu.int/comm/energy_transport/doc-principal/pubfinal_en.pdf

Energy Performance in Buildings	To create a common framework to promote the improvement of the energy performance of buildings.	Proposal for a Directive of the European Parliament and of the Council on the energy performance of buildings (11/05/01).	■ COM (2001) 226 ■ Off Jnl C213E, 31/07/01	http://europa.eu.int/scadplus/leg/en/lvb/l27042.htm
Fostering the Use of Clean Coal Technologies – The CARNOT Programme	The promotion of existing & commercially available cleaner solid fuel technologies worldwide will not only enhance the sustainability of the European Union's industries based on energy but also ensure that third countries can be aware of those technologies which will be a benefit to & enable them to contribute significantly to the objectives of the Kyoto Protocol. Information on the cleaner use of solid fuels is especially important, since it is known that many countries will continue to use this cheap source of energy. This booklet demonstrates the way in which coal, through the whole chain from preparation to combustion, can help to contribute to sustainable development in many parts of the world.	Report to the Directorate-General for Energy & Transport (2001).	■ DG Energy & Transport Report	http://www.europa.eu.int/comm/energy/coal/fostering_en.pdf
A Community Energy Efficiency Labelling Programme for Office Equipment	This Regulation establishes the rules for the Community energy efficiency labelling programme for office equipment (the 'Energy Star programme') as defined in the Agreement between the Government of the United States of America & the European Community on the co-ordination of energy efficient labelling programmes for office equipment. Participation in the Energy Star programme is on a voluntary basis.	Regulation (EC) No 2422/2001 of the European Parliament & of the Council on a Community energy efficiency labelling programme for office equipment (06/11/01).	■ Off Jnl L332/1 15/12/01	http://europa.eu.int/eur-lex/pri/en/oj/dat/2001/l_332/l_33220011215en00010006.pdf

Continued

Table 5.5.2 European Level Energy Policy. *Continued*

Policy Type	Programme Title	Policy Objective	Community Measure	Document Reference	Website
Miscellaneous	New Framework Programme Intelligent Energy for Europe	To establish a new multi-annual programme for action in the field of energy reflecting the European Union's current objectives in this field, i.e. sustainable development & security of supply.	Proposal for a decision of the European Parliament and of the Council adopting a multi-annual programme for action in the field of energy: 'Intelligent Energy for Europe' Programme (2003–2006) (09/04/02).	▪ COM(2002) 162	http://europa.eu.int/scadplus/leg/en/lvb/l27046.htm
	Energy – Let us Overcome our Dependence	Important political decisions would have to be taken soon to ensure that European energy supply & consumption is more secure & more respectful of the environment. The purpose of this brochure was to provide readers with the knowledge they need to become aware of what is at stake regarding energy so that they may take part in the debate on energy launched by the Commission through a Green Paper on security of energy supply.	Report to the Directorate-General for Energy and Transport (2002).	▪ DG Energy & Transport Report	http://www.europa.eu.int/comm/energy_transport/livrevert/brochure/dep_en.pdf
	European energy and transport – Trends to 2030	The 'Trends to 2030' publication gives detailed projections on energy and transport for the current EU and the enlarged EU of 25. It also addresses wider European and world developments. These 'baseline' projections show how the future energy, transport and CO_2 emissions may unfold with a continuation of current trends and policies. They are built on a modelling approach encompassing both energy demand and supply.	Report from the Directorate-General for Energy and Transport (30/01/03).	▪ DG Energy & Transport Report	http://europa.eu.int/comm/dgs/energy_transport/figures/trends_2030/index_en.htm

Proposal for a Directive on Establishing a Framework for the Setting of Eco-Design Requirements for Energy-Using Products	This is a proposal for a framework Directive which will be followed by implementing measures establishing eco-design requirements adopted by the Commission assisted by a regulatory committee. Each implementing measure will be accompanied by an explanatory memorandum and an impact assessment & will be subject to a consultation procedure. The present text explains the underlying considerations & concepts of the framework Directive as well as the need for its establishment.	Proposal for a Directive of the European Parliament & of the Council on Establishing a Framework for the Setting of Eco-Design Requirements for Energy-Using Products – Amending Council Directive 92/42/EEC (01/08/03).	▪ COM (2003) 453 ▪ 2003/0172 (COD)	http://europa.eu.int/eur-lex/en/com/pdf/2003/com2003_0453en01.pdf
Market Development of Alternative Fuels – Report of the Alternative Fuels Contact Group	The EC, following the Communication on alternative fuels in 2001, set up a Stakeholder Contact Group in 2002 to advise on technical & economic status & developments of alternative fuels for road transport, with priority on natural gas & hydrogen, & on measures by which the Community could promote their use. Topic Groups on natural gas & hydrogen, respectively, & a Working Group on Biomass-to-Liquid fuels assessed these alternative fuels. The Contact Group published an Interim Report in Mar 2003, & now presents this report, offering the basis for a more detailed strategy on these fuels.	Stakeholder Contact Group Report (Dec 2003).	▪ Stakeholder Contact Group Report	http://europa.eu.int/comm/energy_transport/envir/2003_report_en.pdf

Continued

Table 5.5.2 European Level Energy Policy. *Continued*

Policy Type	Programme Title	Policy Objective	Community Measure	Document Reference	Website
Miscellaneous	Study on Energy Supply Security & Geopolitics	This study seeks to identify the main challenges with which the EU is confronted in search of strategies to maintain an adequate security of energy supply. These challenges emerge from a set of internal & external developments of geopolitical and economic origin. The internal context is the process of dual integration, involving the enlargement of the EU and a deepening of market integration. The external context involves a changing world, in which new relationships are emerging among the main powers, the US, Russia, the EU & possibly China & Japan.	Report prepared for the Directorate-General for Energy & Transport (Jan 2004).	▪ Directorate-General for Energy & Transport Report	http://www.europa.eu.int/comm/energy_transport/doc/2004_lv_ciep_report_en.pdf
	Strategy Paper – Medium Term Vision For The Internal Electricity Market	This document has been compiled in response to the request & with the co-operation of participants in the 'Florence' Forum on European Electricity Regulation. This subject was initially discussed at the ninth meeting of the Forum on 17–18 Oct 2002 at which an orientation document was produced. Several drafts of the full paper have since been prepared & participants in the Forum were consulted during Feb, July & Oct 2003 on its contents. Therefore although this paper is published as a document of the Commission services, there is a broad consensus within the industry regarding its content. Annexed to the document are a number of position papers that have been produced by various industry groups regarding the content of the paper.	Directorate-General for Energy & Transport Working Paper – Strategy Paper: Medium Term Vision For The Internal Electricity Market (01/03/04).	▪ Directorate-General for Energy & Transport Working Paper	http://www.europa.eu.int/comm/energy/electricity/florence/doc/florence_10/strategy_paper/strategy_paper_march_2004.pdf

The Share of Renewable Energy in the EU Country Profiles: Overview of Renewable Energy Sources in the Enlarged European Union	This Staff Working Document gives an overview of the different situations of renewable energy sources in the European Union. It includes part of the formal report that the Commission is required to make under Article 3 of Directive 2001/77/EC on electricity from renewable energy sources, and it completes the overall picture with information at a country level on the heat produced from renewable energies and biofuels in the transport sector. This Staff Working Document complements the Communication on 'The share of Renewable Energy sources in the EU'.	Commission Staff Working Document – The share of renewable energy in the EU Country Profiles: Overview of Renewable Energy Sources in the Enlarged European Union (26/05/04).	■ SEC (2004) 547	http://www.europa.eu.int/comm/energy/res/legislation/country_profiles/2004_0547_sec_country_profiles_en.pdf
The Market for Solid Fuels in the Community in 2002 & 2003	This working paper outlines the markets for solid fuels within the European Community for the period 2002–2003. Particular attention is focused on coal, coke, lignite & peat.	Commission Staff Working Paper – The market for solid fuels in the Community in 2002 & 2003 (01/06/04).	■ SEC (2004) 692	http://www.europa.eu.int/comm/energy/coal/pricing/sec_2004_0692_en.pdf
Electricity From Renewable Energy Sources – Encouraging green electricity in Europe	This publication explains the background & content of the EU's 2001 directive on the promotion of electricity produced from renewable energy sources in the internal electricity market, otherwise known as 'the RES-E directive'. The development of renewable energy sources is a central aim of European Union energy policy – reflecting the clear benefits that clean, sustainable & secure energy supplies will bring to current & future generations of Europeans, as well as other peoples worldwide.	Report for the Directorate-General for Energy and Transport (July 2004).	■ Directorate-General for Energy & Transport Report	http://www.europa.eu.int/comm/energy/res/publications/doc/2004_brochure_green_en.pdf

Table 5.5.3 UK Level Energy Policy.

Policy Type	Document Title	Objective	Contents	Website
Strategies & Reports	New & Renewable Energy: Prospects for the 21st Century: Conclusions in Response to the Public Consultation (March 1999)	The government has a manifesto commitment to 'a new & strong drive to develop renewable sources of energy'. It has undertaken a review of the status & prospects of renewables, including an examination of what would be necessary & practicable to achieve 10% of the UK electricity requirements from renewables by 2010 & what contribution renewables could make to reducing greenhouse gas emissions. This consultation paper reports the outcome of that review & possible ways forward in implementing the government's new drive for renewables.	Foreword; Summary; Section 1: Introduction; Section 2: Why support renewables; Section 3: Form of support; Section 4: Policy options for the future; Section 5: Responses to consultation; Annex 1: New & renewable energy policy review – terms of reference; Annex 2: Current status of renewables; Annex 3: potential future markets; Annex 4: Historic UK programmes; Annex 5: Introducing renewables – tackling the issues; Annex 6: Possible mechanisms to support market stimulation; Annex 7: Support programme to tackle other issues in the energy market.	http://www.dti.gov.uk/renew/condoc/energy.pdf
	UK Energy Sector Indicators (Dec 2000)	Readily available & financially accessible supplies of energy are fundamental to the functioning & growth of a modern industrial society. Energy policy usually starts from some broad principles aimed at ensuring that citizens & companies have access to the energy they need. In the UK the principles are brought together under the objective that ensures competitive energy markets while achieving safe, secure & sustainable energy supplies. The indicators in this publication are designed to throw light on the extent to which this objective is being achieved.	Introduction; Chapter 1: Energy in the Economy; Chapter 2: Oil, gas, coal & nuclear resources; Chapter 3: Primary energy supply; Chapter 4: Capacity utilisation; Chapter 5: Conversion efficiencies; Chapter 6: Energy use indicators; Chapter 7: Fuel prices; Chapter 8: Fuel poverty; Chapter 9: Competition in energy markets; Chapter 10: International comparisons of energy production & use; Chapter 11: Energy & the environment; Chapter 12: Standards of service.	http://www.dti.gov.uk/energy/inform/energy_indicators/2001/index.shtml
	Energy Efficiency Commitment (2002–2005) (August 2001)	This paper sets out & invites comments on the government's proposals for the Energy Efficiency Commitment (EEC) for 2002–05. Under the EEC commitment for 2002–05, electricity & gas suppliers will be required to achieve targets for the promotion of improvements in energy efficiency. There will be a focus on helping lower income consumers, including those in receipt of income & disability benefits.	Section 1: Introduction; Section 2: Summary of the Energy Efficiency Commitment 2002–05; Section 3: The Energy Efficiency Commitment 2002–05 description of the scheme; Section 4: Graphs – estimated costs & benefits; Annex 1: Explanatory notes on illustrative set of possible measures; Annex 2: Draft statutory order; Annex 3: Draft regulatory impact assessment.	http://www.defra.gov.uk/environment/consult/energy/eec0801/pdf/eec.pdf

UK Fuel Poverty Strategy (Nov 2001)	The Strategy sets out the approach of the Government & the devolved administrations to tackling fuel poverty. It focuses on practical action and on working together to get results. However, it should not be seen as the last word, but representing the start of the road to the end of fuel poverty in the UK. It focuses primarily on measures to improve energy efficiency & reduce the costs of fuel for fuel poor households, since the income measures which form part of a long-term solution are being addressed in wider poverty & social exclusion policies.	Foreword; Executive summary; Chapter 1: The causes & effects of fuel poverty; Chapter 2: The UK targets; Chapter 3: Common measures for tackling fuel poverty; Chapter 4: Tackling fuel poverty in England; Chapter 5: Tackling fuel poverty in Scotland; Chapter 6: Tackling fuel poverty in Wales; Chapter 7: Tackling fuel poverty in Northern Ireland; Chapter 8: Monitoring progress; Chapter 9: The way forward; Annex A: Energy efficiency measures; Annex B: Energy market measures; Annex C: Social inclusion measures; Annex D: English definition of a fuel poor household; Annex E: Fuel poverty monitoring; Annex F: Glossary of terms; Annex G: Bibliography.	http://www.dti.gov.uk/energy/consumers/fuel_poverty/strategy.shtml
The Royal Commission on Environmental Pollution 22nd Report: Energy – the Changing Climate (2001)	Part I of this report deals with the global context. The increased concentration of carbon dioxide is a global phenomenon which can be controlled only through concerted global action. The Report summarises current scientific knowledge about the natural processes involved, analyses the relationship between scenarios for future emissions & carbon dioxide concentrations, & considers the possible effects of climate change. It also discusses the actions the world community might take to limit the increase in the amount of carbon dioxide in the atmosphere. These include attempts to manage other aspects of the global carbon cycle, raising efficiency in order to limit the amounts of energy required, & large-scale replacement of fossil fuels by other energy sources.	Summary; Definitions & units; Chapter 1: The radical challenge; Chapter 2: Causes & effects of climate change; Chapter 3: Possible preventative measures; Chapter 4: Prospects for an effective global response; Chapter 5: The UK's present situation & policies; Chapter 6: Reducing energy use; Chapter 7: The alternatives to fossil fuels; Chapter 8: Patterns of energy supply & use; Chapter 9: Possible UK energy balances in 2050; Chapter 10: Adopting a long-term strategy; Recommendations; References; Appendixes.	http://www.rcep.org.uk/newenergy.htm

Continued

Table 5.5.3 UK Level Energy Policy. *Continued*

Policy Type	Document Title	Objective	Contents	Website
Strategies & Reports	Social, Environmental & Security of Supply Policies in a Competitive Energy Market – A Review of Delivery Mechanisms in the UK (May 2001)	This booklet outlines UK experience so far in using competitive energy markets to deliver social, environmental and security of supply policies. It highlights the benefits that have emerged from this approach and sets out the instruments the government has used to enhance policy delivery. It is intended to promote debate with other Member States about their own experiences and the mechanisms they have developed to deliver government policy in these areas.	Summary; Annex A: Social delivery; Annex B: Environmental delivery; Annex C: Security of supply delivery; Annex D Social, environmental & security of supply indicators; Bibliography.	http://www.dti.gov.uk/energy/publications/pdfs/socan.pdf
	Domestic Energy Fact File: England, Scotland, Wales & Northern Ireland (18/01/02)	This document shows comparative information for the main housing and energy efficiency criteria for each country of the UK. The information is useful for both fuel poverty and climate change purposes as it shows the indicative trends over the past decade. The information has been prepared by the Building Research Establishment using the BREHOMES model.	Summary; Introduction; Basic housing stock characteristics; Insulation measures; Central heating; Energy consumption; Fuel type; Expenditure; Conclusion; Further information; References.	http://www.defra.gov.uk/environment/energy/research/domestic/index.htm
	The Government's Strategy for Combined Heat & Power to 2010 – Public Consultation Draft (15/05/02)	This consultation document sets out & invites comments on the Government's strategy to meet the target for installed good quality CHP capacity of at least 10,000 MWe by 2010. This strategy describes the full range of measures, some recently announced, which are thought sufficient for the target to be met.	Foreword; Executive Summary; Section 1: Current situation: an overview of CHP in the UK; Section 2: The UK Strategy: support measures for CHP; Section 3: Monitoring progress towards the 2010 CHP target; Annex 1: Environmental impact of CHP; Annex 2: Glossary of terms; Annex 3: Sources of further information	http://www.defra.gov.uk/environment/consult/chpstrat/contents.htm

Energy Consumption in the UK (July 2002)	This publication looks at energy consumption & how it has changed over the last 30 years, with a focus on changes over the last ten years. It considers some of the drivers that have influenced energy consumption, updating some of the data that was published in Energy Paper 66, *Energy Consumption in the United Kingdom* in 1997. The main purpose of this publication is to provide data on energy consumption patterns. It contains data that is published in the annual Digest of UK Energy Statistics alongside information from other organisations that help us to understand some of the more detailed trends within the totals.	Introduction; Chapter 1: Overall energy consumption; Chapter 2: Transport energy consumption; Chapter 3: Domestic energy consumption; Chapter 4: Industrial energy consumption; Chapter 5: Service sector energy consumption; Annex: List of tables available on the internet.	http://www.dti.gov.uk/energy/inform/energy_consumption/ecuk.pdf
The Energy Review: A Performance & Innovation Unit Report (Feb 2002)	The review had three main objectives: ▪ to set out the objectives of energy policy, including the UK contribution to global policy initiatives, to 2050; ▪ to develop a framework for reconciling the trade-offs among the different objectives of energy policy; and ▪ to develop a vision and strategy for achieving these objectives and to identify the practical steps that need to be taken in the short and medium term, as well as the longer-term.	Executive summary; Ch 1 Introduction; Ch 2: The challenge ahead; Ch 3: Framework; Ch 4: Security in the energy system; Ch 5: Lessons from scenarios; Ch 6: Options for a low carbon economy; Ch 7: A programme for a low carbon future; Ch 8: Institutions; Ch 9: Concluding themes; Ch 10: Implementing the recommendations; Annexes.	http://www.number-10.gov.uk/su/energy/TheEnergyReview.PDF
Building Regulations & Historic Buildings – Balancing the Needs for Energy Conservation with those of Building Conservation: An Interim Guidance Note on the Application of Part L (Feb 2002)	This interim guidance note has been produced to help prevent conflicts between energy conservation policies enshrined in the revised building regulations & policies concerned with planning & the conservation of the historic environment. The right balance is needed between reducing energy use & greenhouse gas emissions, & conserving the national & local heritage. This is broadly in line with sound sustainability principles subscribed to nationally & developed locally through the agenda 21 programme.	Summary; Glossary; Preface; The context of the revisions to Part L; Historic building definitions & statutory protection; Principles of repair & alteration to historic buildings; Specific references to historic buildings in part L; Meeting the requirements of Part L; Historic buildings as environmental systems; Windows; Doors; Walls; Floors; Roofs; Building services; Forthcoming advice; References.	http://www.english-heritage.org.uk/Filestore/publications/pdf/free/interim_part_l_guidance.pdf

Continued

Table 5.5.3 UK Level Energy Policy. *Continued*

Policy Type	Document Title	Objective	Contents	Website
Strategies & Reports	Long-term Reduction in Greenhouse Gas Emissions in the UK (July 2002)	The report represents work by an Inter-departmental Analysts Group (IAG), comprising officials from DTI, DEFRA, DTLR, HM Treasury and the PIU. It considers the scale of the emission reduction required, taking into account potential future growth in energy demands and energy mix projected over the next fifty years. It illustrates the level of uncertainty in projecting so far into the future using a scenario approach. The report also considers options available to fill the estimated gap, their associated costs and the economic impact in relation to projected GDP growth to 2050. It does not make policy recommendations.	Executive summary; Introduction; Greenhouse gas emission projections to 2050; International context; Overview of cost issues; Options for reducing carbon emissions in the UK; Cost of reducing carbon emissions in the UK; General conclusions; Annexes.	http://www.dti.gov.uk/energy/greenhousegas/greenhouse.pdf
	A Guide For UK Companies: Exporting Sustainable Energy Products & Services (2002)	In a drive to grow the UK sustainable energy export market, the British Government actively promoted the export of British sustainable and renewable energy products and services to overseas markets. This guide was designed to signpost UK exporters to those organisations and government bodies which were available to help them export their products and services overseas.	Business link; Climate change projects office; DEFRA; DfID; DTI (sustainable energy programme); Export credits guarantee department; Foreign & Commonwealth office; Invest (NI); Joint environmental markets unit; Scottish development international; The carbon trust; Trade partners UK; Wales trade international; Useful contacts.	http://www.dti.gov.uk/energy/renewables/publications/pdfs/seecguide2002.pdf
	Energy – Its impact on the Environment & Society (July 2002)	This booklet looks at the social and environmental impact of the production & use of energy. Its aim was to increase awareness of the social & environmental aspects of energy use through the provision of a fuller range of analysis of the statistics that the DTI and others issue on a regular basis.	Introduction; Chapter 1: Emissions; Chapter 2: Energy production & supply; Chapter 3: Energy demand; Chapter 4: Competition & prices for domestic customers; Chapter 5: Service quality; Chapter 6: Fuel poverty.	http://www.dti.gov.uk/energy/environment/energy_impact/impact_booklet.pdf

	The booklet covers the evolution of the social and environmental impacts over previous years. It examines the key influences on the trends in the figures, including the interaction between economic, social & environmental trends. It looks at how government policies developed to limit the environmental damage that could be associated with the use of energy whilst, at the same time, ensuring that consumers could enjoy its benefits at an affordable cost.		
Energy Paper 68: Energy projections for the UK (July 2002)	This paper presents the results of an exercise to update the government's projections of future UK energy demand and energy-related emissions of carbon dioxide (CO_2) last published in Energy Paper 65 in 1995. It builds on work issued as a working paper in March 2000. The paper's projections provide a view of the possible future levels and composition of energy demand based on a set of different scenarios of growth in the economy and of world fossil fuel prices. Six core scenarios are considered. This approach provides a range of energy and emissions projections, comparable with Energy Paper 65. It also recognises that uncertainties are inherent in projecting the future path of energy demand.	Introduction; The Modelling Process; Energy Price, Industry Share and GDP Assumptions; Projections of Final Energy Demand; The Electricity Supply Industry; Primary Energy Demand; CO_2 and SO_2 Emission Projections; Sensitivity Analysis on CO_2 Emissions; Annexes.	http://www.dti.gov.uk/energy/inform/energy_projections/ep68_final.pdf
The World Offshore Renewable Energy Report 2002–2007 (Oct 2002)	This report summarises results of a study commissioned by Renewables UK from energy industry business analysts Douglas-Westwood Limited to meet a growing number of requests to quantify the prospects available to British industry in the offshore sector of the renewable energy industry.	Introduction; Summary & conclusions; Definitions; Offshore wind; Wave; Tidal/current stream; OTEC (Ocean Thermal Energy Conversion); Global market analysis & forecast; Appendices.	http://www.dti.gov.uk/energy/renewables/publications/pdfs/offshorereport.pdf

Continued

Table 5.5.3 UK Level Energy Policy. *Continued*

Policy Type	Document Title	Objective	Contents	Website
Strategies & Reports	The UK Fuel Poverty Strategy – 1st Annual Progress Report (Feb 2003)	This is the first progress report, providing information to better inform public concern about fuel poverty and the development of government policies and programmes aimed at tackling the problem. The English House Condition Survey, which provides the underpinning data on the majority of homes affected by fuel poverty, will now report annually, so regular and accurate measures of progress can be provided.	Foreword; Executive summary; Ch 1: Targets; Ch 2: Measures for tackling fuel poverty – actions over the past year; Ch 3: Progress on tackling fuel poverty in the UK; Ch 4: Progress towards meeting the target for England; Ch 5: Progress in the devolved administrations; Ch 6: The way forward.	http://www.defra.gov.uk/environment/energy/fuelpov/pdf/fuelpov_1stannrpt.pdf
	Our Energy Future: Creating a Low Carbon Economy – Energy White Paper (Feb 2003)	Energy producers, investors, business & consumers need a clear, settled, long-term framework within which they can plan & make decisions with confidence. The new energy policy set out in this white paper is designed to provide this. It reflects, & will reinforce, a wider commitment to sustainable development which challenges the DTI to find ways to achieve economic, social & environmental objectives at the same time.	Introduction; Section 1: Overview; Ch 1: Cleaner, smarter energy; Section 2: The low carbon economy; Ch 2: the environment; Ch 3: Energy efficiency; Ch 4: Low carbon generation; Ch 5: Clean low carbon transport; Section 3: Reliable, competitive & affordable supplies; Ch 6: Energy reliability; Ch 7: Productivity, competitiveness & innovation; Ch 8: Energy & the vulnerable; Section 4: Delivery through partnership; Ch 9: Annexes.	http://www.dti.gov.uk/energy/whitepaper/ourenergyfuture.pdf
	UK Energy Sector Indicators 2003 (March 2003)	This publication set out a range of indicators that could be used to monitor progress. These detailed indicators have been grouped into 12 sections covering different aspects of the energy sector.	Introduction: Energy in the economy, investment & productivity; Resources; Primary energy supply; Capacity utilisation; International comparisons of energy production & use; Fuel Prices; Competition in energy markets; Standards of service; Fuel poverty; Conversion efficiencies; Energy use indicators; Energy & the environment.	http://www.dti.gov.uk/energy/inform/energy_indicators/indicators.pdf
	UK Energy in Brief (July 2003)	This booklet summarised the latest statistics on energy production, consumption and prices in the United Kingdom. Figures were taken from the 2003 edition of the 'Digest of UK Energy Statistics', published on 31 July 2003.	Energy in the economy; Overall energy; Oil & gas production; Petroleum; Natural gas; Coal; Electricity; Combined heat & power; Nuclear power; Renewables; Prices; Expenditure; Fuel poverty; Environment; Energy efficiency; Contacts, conversion factors & definitions; References.	http://www.dti.gov.uk/energy/inform/energy_in_brief/energyinbrief2003.pdf

Digest of United Kingdom Energy Statistics (July 2003)	This issue of the Digest of United Kingdom Energy Statistics continues a series which commenced with the Ministry of Fuel & Power Statistical Digest for the years 1948 & 1949, published in 1950. This printed issue consists of seven chapters & four annexes. The first chapter deals with overall energy. The other chapters cover the specific fuels, combined heat & power & renewable sources of energy. The annexes cover major events in the energy industries, calorific values & conversion factors, a glossary of terms & further sources of information.	Introduction; Ch 1: Energy; Ch 2: Solid fuels & derived gases; Ch 3: Petroleum; Ch 4: Gas; Ch 5: Electricity; Ch 6: Combined heat & power; Ch 7: Renewable sources of energy; Annexes.	http://www.dti.gov.uk/energy/inform/dukes/dukes2003/dukes03.pdf
Energy – Its Impact on the Environment & Society: 2003 Update (July 2003)	This booklet gives an overview of the social and environmental impacts of the production and use of energy. It follows on from *Energy – Its Impact on the Environment and Society* issued in July 2002. Its aim is to increase awareness of the social and environmental aspects of energy use.	Introduction; Ch 1: Emissions; Ch 2: Energy production & supply; Ch 3: Energy demand; Ch 4: Competition & prices for domestic customers; Ch 5: Service quality; Ch 6: Fuel poverty.	http://www.dti.gov.uk/energy/environment/energy_impact/2003update.pdf
Energy Market Competition in the EU & G7: The relative extent of energy market competition in the EU & G7 (Sept 2003)	This report sets out a methodology for quantifying the competitiveness of energy markets in order to produce a ranking that will serve to provide this assessment. It adapts the approach taken by OXERA (Oxford Economic Research Associates Ltd.) in a previous study for the DTI & Dutch Ministry of Economic Affairs & is designed to be both analytically robust, drawing on well-identified indicators of market competitiveness from economic theory, & practical for the purpose of annual assessment. The report presents the first assessment of performance against the Public Service Agreement (PSA) target for 2002 to ensure the UK ranks in the top three most competitive energy markets in the EU and G7 in each year. This assessment indicates that the UK is likely to meet its target & has one of the most competitive energy markets in the EU & G7.	Executive summary; Introduction; Initial filter approach; Creating single comparators for multi-market countries; Listing of PSA countries after filtering; Detailed indicators – definition & scoring rules; Aggregation & weighting of indicators; Results; Sensitivity analysis; International price comparisons; Concluding remarks; Appendices.	http://www.dti.gov.uk/energy/gas_and_electricity/competitiveness_structure/oxera_report.pdf

Continued

Table 5.5.3 UK Level Energy Policy. *Continued*

Policy Type	Document Title	Objective	Contents	Website
Strategies & Reports	Review of Renewable Energy Development in Europe & the US (Oct 2003)	In this report a broad overview of progress with renewable technologies in Europe, the US & Japan is presented. From this review, a number of countries are highlighted for more in-depth analysis. The situation in the US is examined to contrast with the experience in Europe alongside six individual country case studies.	Introduction; Progress with renewable technologies in Europe, the US & Japan; Country study – Germany; Country study – Spain; Country study – Sweden; Country study – Austria; Country study – Netherlands; Country study – USA; Key findings from the case studies; References.	http://www.dti.gov.uk/energy/renewables/policy/iceptreview.pdf
	LNG (Liquitied Natural Gas) Facilities & Inter-connectors: EU Legislation & Regulatory Regime – DTI/Ofgem Final Views (Nov 2003)	This document gives the DTI & Ofgem's final views. These are based on the initial views paper, responses to that paper & further discussions with the European Commission, developers & other Member States regulators/governments currently or likely to be interconnected with GB. The document also sets out how Ofgem (Office of Gas and Electricity Markets) proposes to handle the process of providing early guidance (subject to certain legal caveats) to developers & handling exemption applications on the expectation that it will receive the relevant formal powers.	Introduction; Timetable; Background; Respondents views & the DTI & Ofgem's final views: proposed regulated & exempt regimes; Respondents views & the DTI & Ofgem's final views: exemptions criteria; Conclusions; Appendix 1: List of respondents to consultation.	http://www.dti.gov.uk/energy/publications/policy/lngfacilities.pdf
	Offshore Wind – Economies of Scale, Engineering Resource & Load Factors (Dec 2003)	The UK Dept of Trade and Industry and the Carbon Trust (DTI & CT) have contracted Garrad Hassan and Partners (GH) to carry out a review of specific subjects pertaining to the future development of offshore wind power in the UK. The scope of the work defined by DTI & CT requires appraisal of:	Introduction; Economics of offshore wind; Available engineering resource; Load factors; References; Appendix 1 capital cost budget breakdown.	http://www.dti.gov.uk/energy/renewables/policy/garradhassanoffshorewind.pdf

	■ economics of offshore wind power; ■ available engineering resource for the offshore wind industry; & ■ achievable load factors for UK offshore projects.		
Renewable Energy Industry Gap Analysis: Summary Report (Jan 2004)	Mott MacDonald, in association with the Bourton Group, was appointed by DTI's Renewables UK, Scottish Enterprise, Highlands & Islands Enterprise & the Scottish Executive to undertake a study of the current status of the UK renewable industry & to assess future potential. The aim of the study was to assist agencies in determining the most effective means of targeting their effort & resources. This document is a summary of a fuller report.	Overview; Main building blocks; Size of the renewable energy Industry at present; Future industry size; Gaps, Strengths & opportunities: Technologies; Gaps, Strengths & opportunities: R&D & Skills; Appendices.	http://www.dti.gov.uk/energy/renewables/publications/pdfs/renewgapreport.pdf
Energy Trends (March 2004)	Energy Trends and Quarterly Energy Prices are produced by the Department of Trade & Industry on a quarterly basis. Both periodicals are published concurrently in June, Sept, Dec & March. The March editions cover the fourth quarter of the previous year. Energy Trends includes information on energy as a whole & by individual fuels. The text & charts provide an analysis of the data in the tables. The tables are mainly in commodity balance format, as used in the DTI's annual Digest of UK Energy Statistics.	Introduction; Total energy; Solid fuels & derived gases; Oil & oil products; Gas; Electricity; Special features	http://www.dti.gov.uk/energy/inform/energy_trends/mar_04.pdf
Creating a Low Carbon Economy – 1st Annual report on the Implementation of the Energy White Paper (April 2004)	This report represents the first annual review of the energy white paper as obligated by the Sustainable Energy Act 2003. It provides an overview of the last 12 months and reinforces the commitments made in the energy white paper.	Foreword; the last 12 months & the way ahead: an overview; Reducing carbon emissions; Energy reliability; Competitive energy markets; Tackling fuel poverty; Annex: Other reports & plans published alongside this report.	http://www.dti.gov.uk/energy/sepn/annualreport/firstannualreportfull.pdf

Continued

Table 5.5.3 UK Level Energy Policy. *Continued*

Policy Type	Document Title	Objective	Contents	Website
Strategies & Reports	The Government's Strategy for Combined Heat & Power to 2010 (April 2004)	This Strategy document sets out a framework to support the growth of CHP capacity in the UK & to enable the CHP industry to meet the challenges ahead. The government announced, in 2000, a target of achieving at least 10,000 MWe of Good Quality CHP capacity by 2010 & the development of a Strategy to achieve it. A draft Strategy was published for consultation in May 2002 and in February 2003 the Energy White Paper, *Our energy future: creating a low carbon economy* reaffirmed the government's commitment to the target.	Foreword; Executive Summary; An Overview of CHP in the UK today; The UK Strategy: Support Measures for CHP; Fiscal Incentives; Grant Support; Regulatory Framework; Promotion of Innovation; Monitoring Progress Towards the 2010 CHP Target; Annexes.	http://www.defra.gov.uk/environment/energy/chp/pdf/chp-strategy.pdf
	Energy Efficiency: The Government's Plan for Action (April 2004)	This Plan sets out how commitments made in the Energy White Paper, particularly for period 2004–2010 will be delivered. It sets out, in Annex 1, progress against these commitments over the previous year, and it also identifies areas where further measures may be needed to reach energy policy goals. It sets the scene for the following decade, but more detailed work on the policies needed for 2010–2020 will begin in conjunction with the review of the Climate Change Programme later. The Plan does not cover transport measures, which will be covered by a forthcoming review of the 10-year transport plan.	Foreword; Executive Summary; Introduction; Implementing the Energy White Paper strategy; Households; Business; Public sector; Devolved Administrations; Cross-cutting issues; Delivering savings 2010–2020; Reporting and follow up; Annexes.	http://www.official-documents.co.uk/document/cm61/6168/6168.pdf
	The UK Fuel Poverty Strategy 2nd Annual Progress Report: 2004 (April 2004)	This was the government's second report on progress towards the targets set out in the UK Fuel Poverty Strategy. It provided an update on actions since the first progress report was published in 2003, and highlighted some of the key areas for action during the coming year.	Ministerial Foreword; Executive Summary; Targets; Measures for tackling fuel poverty; Progress on tackling Fuel Poverty in the UK; Progress towards meeting the target for England; Progress in the Devolved Administrations; The way forward.	http://www.defra.gov.uk/environment/energy/fuelpov/pdf/fuelpov_2nocover.pdf

Creating a Low Carbon Economy – Progress on Regional Implementation of the Energy White Paper (April 2004)	A number of regions had already made progress towards a regional approach to energy before the Energy White Paper was published. In all regions this is built on an existing, well-established base of sub-regional activity (not, for the most part, described in this report). This charts the progress that continued and accelerated since the publication of the White Paper.	Overview; The National Framework; The South East; East of England; London; Yorkshire and Humber; North East; West Midlands; East Midlands; North West; South West; Annex A: Regional Indicators; Annex B: Data Sources.	http://www.dti.gov.uk/energy/sepn/sepn_regional_annex.pdf
Partial Regulatory Impact Assessment on a Proposal for a Directive on Energy end-Use Efficiency & Energy Services (2004)	This document presents a partial regulatory impact assessment of a proposal for a Directive of the European Parliament & of the Council on Energy End-Use Efficiency & Energy Services. The Commission adopted a formal proposal 2003/0300 (COD), which was published on 10 December 2003 as a proposal to the Council & Parliament.	Introduction; Purpose & intended effect; Background; Risk assessment; Options; Benefits; Public sector target; Promotion of energy services; Qualification, Accreditation & certification of energy service providers; Availability of audits; Metering & billing; Business sectors affected; Issues of equity & fairness; Costs; Compliance costs for businesses, charities & voluntary organisations; General target; Promotion of energy services; Qualification, accreditation & certification of energy service providers; Metering & billing; Consultation with small business: Small firms impact test; Competition assessment; Enforcement & sanctions; Monitoring & review; Preliminary consultation; Summary & recommendations.	http://www.defra.gov.uk/corporate/consult/ec-energyservices/ria.pdf
UK Energy Sector Indicators 2004	In *UK Energy Sector Indicators 2003*, published at the same time as the Energy White Paper, views were sought on the most appropriate indicators to use to monitor progress annually, to help focus public attention on energy policy and give a broad overview of whether overall energy policy objectives were being delivered. As a result the structure of this booklet was amended this year and indicators are presented in three separate sections but set in a wider context. This remains an evolving set of indicators and work will continue to develop new indicators in a number of areas.	Introduction; Technical notes; Estimated data; Key indicators; Supporting indicators; Low carbon economy; Reliability; Competitiveness; Fuel poverty; Background indicators; Reliable supplies of energy; Industrial & business competitiveness; Social objectives; Environmental objectives.	http://www.dti.gov.uk/energy/inform/energy_indicators/indicators2004.pdf

Continued

Table 5.5.3 UK Level Energy Policy. *Continued*

Policy Type	Document Title	Objective	Contents	Website
Strategies & Reports	Biomass as a Renewable Energy Source (May 2004)	This study was carried out following the publication of the Energy White Paper, which accepted a number of the recommendations in the Twenty-second Report. This study now expands upon those recommendations and offers policy-based guidance on how to achieve them. A particular recommendation is that by 2050 up to 16 Gigawatts (about 12%) of the nation's energy should come from biomass. This would be a clear but not dominant role for biomass within a larger, diversified energy portfolio.	Introduction; Biomass fuels; Generation using biomass fuels; Meeting the target; Conclusions & recommendations; Appendices.	http://www.rcep.org.uk/biomass/Biomass%20Report.pdf
	Energy Trends (June 2004)	Energy Trends and Quarterly Energy Prices are produced by the Department of Trade & Industry on a quarterly basis. Both periodicals are published concurrently in June, Sept, Dec & March. The June editions cover the first quarter of the year. (See March 2004.)	Introduction; Total energy; Solid fuels & derived gases; Oil & oil products; Gas; Electricity; Special features.	http://www.dti.gov.uk/energy/inform/energy_trends/jun_04.pdf
	Digest of United Kingdom Energy Statistics (July 2004)	This issue of the Digest of United Kingdom Energy Statistics continues a series which commenced with the Ministry of Fuel & Power Statistical Digest for the years 1948 & 1949, published in 1950. The current issue updates the figures given in the Department of Trade and Industry's *Digest of United Kingdom Energy Statistics 2003*, published in July 2003. (See above)	Introduction; Energy; Solid fuels & derived gases; Petroleum; Gas; Electricity; Combined heat & power; Renewable sources of energy; Annexes.	http://www.dti.gov.uk/energy/inform/dukes/dukes2004/dukes04.pdf
	UK Energy in Brief 2004 (July 2004)	This booklet summarises the latest statistics on energy production, consumption and prices in the United Kingdom. Figures are taken from the 2004 edition of the *Digest of UK Energy Statistics*.	Key indicators for energy policy; Energy in the economy; Overall energy; Oil & gas production; Petroleum; Natural gas; Coal; Electricity; Combined heat & power; Nuclear power; Renewables; Prices; Expenditure; Energy efficiency; Contacts, Conversion factors & definitions; References.	http://www.dti.gov.uk/energy/inform/energy_in_brief/energyinbrief2004.pdf

	Energy – Its Impact on the Environment & Society – 2004 update (July 2004)	This booklet gives an overview of the social & environmental impacts of the production & use of energy. It follows on from the full version of *Energy – Its Impact on the Environment and Society* issued in July 2002 & the internet based 2003 annual update. Its aim is to increase awareness of the social & environmental aspects of energy use.	Introduction; Growth in greenhouse gas and other emissions; The environmental impact of the production of energy; Trends in energy demand and the drivers of its growth; Competition and prices in domestic energy markets; Quality of service that electricity and gas companies provide to domestic consumers and the factors that influence it; The extent of fuel poverty.	http://www.dti.gov.uk/energy/environment/energy_impact/2004update.pdf
	Energy Trends – Sept 2004 (Sept 2004)	Energy Trends and Quarterly Energy Prices are produced by the Department of Trade & Industry on a quarterly basis. Both periodicals are published concurrently in June, Sept, Dec & March. The September editions cover the second quarter of the year. (See March 2004.)	Introduction; Total energy; Solid fuels & derived gases; Oil & oil products; Gas; Electricity; Special features.	http://www.dti.gov.uk/energy/inform/energy_trends/sep_04.pdf
Consultation Papers	Developing Energy Policy: Consultation for White Paper (May 2002)	This consultation document was aimed primarily at energy stakeholders, i.e. consumer organisations, business, unions, environmental groups & others who had a particular expertise or interest in energy issues. The document invited comments on the main issues which government would need to consider in producing the 2003 white paper including its energy policy objectives, its underlying energy policy & the practical steps needed to implement these.	Section 1: Introduction; Section 2: Main themes for consultation; Section 3: How we will be conducting the review; Section 4: Deadlines & contracts; Annex A: Executive summary of the energy review Feb 2002. Annex B: Summary of recommendations from the energy review Feb 2002.	http://www.dti.gov.uk/energy/developep/energyconsreport.pdf

Continued

Table 5.5.3 UK Level Energy Policy. *Continued*

Policy Type	Document Title	Objective	Contents	Website
Strategies & Reports	Sustainable Energy – Response by the Sustainable Development Commission (SDC) to the Government's Energy Policy: Key Issues for Consultation (Sept 2002)	The response constituted the SDCs case for the government's forthcoming energy white paper to consist of proposals for a sustainable energy policy, itself geared to helping stimulate a low carbon economy.	Executive summary; Principal recommendations; Strategic; Operational; Main response; A low carbon economy; A sustainable framework methodology; Our vision for a low carbon economy; Sustainable energy strategy & targets; Key factors bearing on the case for a low carbon economy; Renewables; Nuclear; Coal, gas & oil; Innovation; Handling supply & demand side issues; Energy efficiency & CHP; Transport; Industry; Domestic; Government; Security; Climate change; International; Coherence; Economic dimension & role of financial/regulatory issues; Employment; Planning; Role of communications; conclusion; Annex 1: SDC's initial assessment of the use of economic instruments for achieving sustainability; Annex 2: SDC's six principles of sustainability.	http://www.sd-commission.org.uk/news/resource_download.php?attach_id=CCGG70X-CLUI9PA-E5H71YL-BSGU6GU
	The Government's Draft Strategy for Combined Heat & Power to 2010 – Summary of Responses to the Public Consultation (Nov 2002)	The draft Strategy was published for public consultation on 15 May 2002. It described a range of measures, some recently announced, considered by the government to be sufficient for the target of at least 10,000 MWe of CHP capacity in 2010 to be met. The Government received 80 responses to the draft Strategy consultation.	Executive Summary; The Consultation; Responses to issues covered in the consultation paper; Achieving the 2010 target for CHP; Concerns over the modelling work; UK energy market conditions, Frameworks & rules affecting CHP; Developing new technology; Other issues; CHP: The contribution to sustainable development; General comments & additional measures; Annex: List of respondents to the consultation.	http://www.defra.gov.uk/environment/consult/chpstrat/pdf/chpstrat_responses.pdf

Table 5.5.4 Northern Ireland Level Energy Policy.

Policy Type	Document Title	Objective	Contents	Website
Strategies & Reports	Energy in the 90s & Beyond – An Energy Strategy for NI (1991)	This energy strategy engaged collective effort & reflects new thinking. Efficiency of energy use was a prime consideration. It could reduce costs, help protect the environment & defer expenditure on expensive new power stations which consumers would have to pay for. The central premise was that energy use has consequences which could affect the environmental heritage & the means were available to produce more by using less. It was expected that those who supply & distribute energy sources, services & appliances should help. The document described the then Northern Ireland energy scene & how the strategic objectives were being & would be achieved taking into account economic growth & national & international influences.	Ministerial foreword; Introduction; Chapter 1: Present position & likely growth in energy usage; Chapter 2: Environmental & other influences; Chapter 3: The objectives & how they are being achieved; Chapter 4: Future Action; Chapter 5: Conclusion; Appendix: Energy efficiency in Practice.	http://www.detini.gov.uk/cgi-bin/downutildoc?id=49
	North/South Energy Market Study: Final Report (Aug 2001)	This final report is a summary of the key issues considered to be currently restricting trade between the north and south in each of the energy sectors, together with a number of options for addressing these issues with a view to the development of all-island markets. More detailed consideration of the issues and the options is contained within background reports submitted to the DPE and DETI.	Introduction & executive summary; Electricity; Renewables & CHP; Gas; Coal & oil; Indigenous fuels; Conclusions; Appendices.	http://www.detini.gov.uk/cgi-bin/downutildoc?id=36
	Energy Bill (18/06/02)	The purpose of this bill is to provide for the establishment & functions of the Gas & Electricity Markets Authority for Northern Ireland & an energy group of the General Consumer Council for Northern Ireland; to amend the legislation regulating the electricity & gas industries; & for connected purposes.	Part I: New regulatory arrangements; Part II: Objectives of regulation of electricity & gas; Part III: Functions of the general consumer council; Part IV: enforcement; Part V: Electricity from renewable sources; Part VI: Miscellaneous & supplementary; Schedules.	http://www.detini.gov.uk/cgi-bin/downutildoc?id=344

Continued

Table 5.5.4 Northern Ireland Level Energy Policy. *Continued*

Policy Type	Document Title	Objective	Contents	Website
Strategies & Reports	The Potential for the Use of Marine Current Energy in NI (June 2003)	This report was commissioned by the Department of Trade & Industry (DTI), the Department of Enterprise, Trade & Investment (DETI) Northern Ireland Electricity to assess the potential for generating power for NI from the kinetic energy of marine currents.	Executive Summary; Introduction; The state-of-the-art for tidal energy; Approach & method; Analysis; Proposal for future development; Conclusions & recommendations.	http://www.detini.gov.uk/cgi-bin/downutildoc?id=55
	A Study into the Economic Renewable Energy Resource in NI & the Ability of the Electricity Network to Accommodate Renewable Generation up to 2010 (June 2003)	The purpose of this study is to assess the realistic contribution that renewable energy sourced electricity, produced at an acceptable additional price to the end consumer, can make to the likely demand in Northern Ireland in 2010. The 'acceptable price' for the purpose of this study has been taken as 7p/kWh. This figure is calculated on the basis of a current average price of electricity in Northern Ireland of 4p/kWh plus a renewable 'premium' of 3p/kWh. This latter figure is based on the level of the 'buy-out' price set under the Renewables Obligation in Great Britain.	Executive Summary; Introduction; Resource availability & conversion technologies; Restrictions on recovery of resource; Summary of resource availability by 2010; Evaluation of resources & presentation of results; Conclusions.	http://www.detini.gov.uk/cgi-bin/downutildoc?id=73
	Action Renewables – A Study into the Renewable Energy Resource in the Six Counties of Northern Ireland: Generic Resource Opportunities (March 2004)	The purpose of the report is to enable potential developers, community groups, knowledgeable individuals and local authorities readily to identify the opportunity for sustainable energy developments in their region. The study extends the work carried out under the 2002 NIE/DETI renewables resource study which determined the extent and magnitude of renewable resources, the technical barriers to utilising those resources, the energy that could be made available from the resources by 2010 and the additional costs involved in supplying electricity to consumers from the resources.	Executive Summary; Introduction; Results of the County Studies – Offshore wind; Onshore wind; Biomass; Hydro-electricity; Tidal power; Municipal and agricultural wastes; Summary of resource; Antrim; Armagh; Down; Fermanagh; Londonderry; Tyrone; Appendices.	http://www.detini.gov.uk/cgi-bin/downutildoc?id=586

	Energy – A strategic framework for Northern Ireland (June 2004)	This Framework focuses on electricity, gas & renewable energy sources. The structure of the energy industry in Northern Ireland in relation to all these sources is very different now from that which existed 10 years ago. The primary objective is to achieve a competitive, sustainable, reliable energy market at the minimum cost necessary in an all-island, UK & European context.	Ministerial Foreword; Executive Summary; Overview; Policy goals; Reducing energy costs; Building competitive markets; Enhancing sustainability in energy; Reliable energy supplies; The future.	http://www.detini.gov.uk/cgi-bin/downutildoc?id=547
	All-Island Energy Market – A Development Framework (Nov 2004)	This paper sets out the policy context for action by the two governments, regulatory authorities & industry in creating an All-island Energy Market. It has been prepared by the Department of Communications, Marine & Natural Resources (DCMNR) & the Department of Enterprise, Trade & Investment (DETI), & the two regulatory authorities – Commission for Energy Regulation (CER) & the Northern Ireland Authority for Energy Regulation (NIAER). This work is set against the background of the European Union's single market for electricity & natural gas & the growing regionalisation of markets.	Ministerial Foreword; Introduction; Background; Policy Context; Definition; Benefits; Short to Medium Term Priorities; Implementation; Consultation; Contacts; Annex A Development Programme – All-island Energy Market.	http://www.detini.gov.uk/cgi-bin/downutildoc?id=812
Consultation Papers	Domestic Energy Efficiency Scheme (DEES II) Proposals – A New Fuel Poverty Programme for warmer healthier homes in NI (July 2000)	The proposals in this consultation paper set out a strategy for building on the achievements of DEES. The objective is to strengthen the scheme by broadening the kinds of assistance, which the scheme offers, and targeting those households most vulnerable to cold related ill health. In developing its proposals for the new NI scheme, the Department has taken account of fuel poverty schemes planned for England, Scotland & Wales. Following this consultation exercise more detailed work may be necessary to develop the NI scheme further.	Foreword; Executive summary; Effects of cold homes; Scale of fuel poverty in NI; Current assistance programmes; The current DEES; How the review was carried out; Proposed way ahead; The new programme; Operation of the new programme; What happens next; Annex A: Other action to help alleviate fuel poverty; Annex B: Measures to be offered under (DEES II) & (DEES II for over 60s).	http://www.dsdni.gov.uk/domestic_energy_efficiency_scheme_proposals.pdf

Continued

Table 5.5.4 Northern Ireland Level Energy Policy. *Continued*

Policy Type	Document Title	Objective	Contents	Website
Consultation Papers	Renewable Energy in NI – Realising the Potential (2001)	The aim of UK government policy for renewable energy is that it should make an increasing contribution to UK energy supplies in the years up to 2010 & beyond. To this end, targets have been set for 5% of licensed electricity supplies to be generated from renewable sources by 2005 & for 10% by 2010. Northern Ireland is required to contribute to the achievement of the UK targets. This Consultation Document sought to consider & invite views on the options for, & the implications of, increasing the amount of licensed electricity supplies to be generated from renewable sources in NI by 2010.	Executive summary; Introduction; Policy perspectives on increasing exploitation of renewable energy; Current uptake of renewable energy in NI – the potential for further exploitation & NI's contribution to the UK target; Potential benefits of a successful renewable energy industry in NI; Possible support mechanisms for renewable energy in NI; Weighing up the options – a possible way forward; Appendix 1: Summary of non-fossil fuel obligation orders in the UK; Appendix 2: details of NI non-fossil fuel obligation orders & non-grid connected renewable energy projects; Appendix 3: Comparison of NI's electricity process & rest of the UK & EU member states; Appendix 4: Second tier electricity supply licence holders in NI; Appendix 5: Glossary.	http://www.detini.gov.uk/cgi-bin/downutildoc?id=367
	Towards a New Energy Strategy for NI – Consultation Document (April 2003)	DETI published this consultation paper in order to seek final views on a new energy strategy for Northern Ireland for the following ten years. In order to provide a context for this strategy, the paper contained a short overview of the key energy sector developments that recently took place in Northern Ireland, its surrounding markets (Republic of Ireland and Great Britain) & at the European Union level. It also considered the challenges likely to be faced over the next ten years. Responses to this paper were to inform the preparation of a much shorter paper setting out the strategy for the next ten years. The final energy strategy was to be published later & reviewed at regular intervals. (See above: *Energy – A Strategic Framework for Northern Ireland* (June 2004).)	Foreword; Executive summary; Introduction; Outlook for the NI energy sector; Energy strategy objectives & challenges; Competition, cost reduction & energy process; All-island energy market & interconnection; Energy security & diversity; Environmental emission targets; The natural gas network; Renewable energy supply; Energy efficiency & fuel poverty; Further developments, recommendations & next steps; Appendix 1: Glossary; Appendix 2: Bibliography & websites.	http://www.detini.gov.uk/cgi-bin/downutildoc?id=281

6 Environment

6.1 Environment and the Built Environment

Environmental matters have exercised the minds of several generations to varying degrees since biblical times, and the diversity of approach has been evident in the move from a predominantly global agricultural society, in which environmental issues were extremely important, to the industrial society when environmental damage was often accepted in the pursuit of financial and community progress. In the 1980s, when there was an aspiration to move towards a sustainable society, a need evolved to balance simultaneous achievement of development and environmental goals. This was seen as the illuminating light of sustainable development and categorised the move away from singular environmental themes. The origins of this balanced approach can be traced back to the increased concern shown for the environment in the 1960s and 1970s, as doubts were raised about the potential of the world's resource base to sustain the established patterns of human activities. This was primarily attributed to the rapid growth of the built environment and its perceived negative impact on the natural environment.

The United Nations Conference on the Human Environment in Stockholm (1972) indicated that economic development, without proper regard to environmental constraints, was both wasteful and unsustainable. It also recognised that growing economies inevitably face some deterioration in environmental quality, necessitating a pro-active approach to the balance between environmental protection and economic growth, as well as careful attention to social issues. In this context, the environmental processes were seen as ecological and life support systems, genetic diversity and the sustainable use of species and ecosystems, while natural environmental products comprised the elements of soil, sea, fresh water and air providing harvests of energy, food and water.

Environmental concerns spread beyond the built environment, evidenced by the actions and lobbying of several environmentalists seeking conservation and preservation in areas such as bird life, natural habitat protection, air and soil quality issues, waters which support human and mammal life and protection of all types of species. While damage in these areas may arise from the impact of built developments, natural developments or disasters, there is need for protection through legislative and fiscal tools, alongside robust monitoring systems. This approach of restricting consumption of the natural environment is underpinned by education across the disciplines of science, history and geography, involving the professional response of scientists, engineers, financiers and planners. In addition, the use of appropriate monitoring and reporting to ensure environmental justice has been consolidated in the UN Economic Commission for Europe Aarhus Convention (1998) in which access to information, public participation in decision-making and access to justice in environmental matters are paramount. The UK government reinforced this by emphasising the importance of building trust within communities, increasing public authority accountability and devising better environmental policies.

6.2 Key Environment Elements

The environment is defined by the elements of land, air and water, and plants and animals which live on and feed off them in the natural environment, as well as the systems which support them. Concern for environmental issues arises when these elements are threatened or damaged, requiring man-made environmental systems to be developed to minimise this degradation. Hence, the elements of environmental protection are crucial as they address the monitoring of environmental support systems on a macro level, as well as both corporate and individual responsibilities at a micro level, in which there is a causal relationship with each environmental action. Table 1.1 in Chapter 1 highlights this range of environmental protection responsibilities including public health, the need for environmental stewardship for sustained protection, and the use of eco-labelling and purchasing by individuals. Corporate businesses have a different set of effects and actions in securing environmental protection, including corporate social responsibility, due attention to efficient resource management and the use of a whole life costing approach to financial decision-making.

These high level responsibilities require environmental reporting by corporations, stakeholders, government departments, communities and individuals. It is also important to ensure information is freely available for consultation and review, in the spirit of the Aarhus Convention, facilitating the opportunity for public participation in specific environmental matters. Environmental management and auditing systems play a key role in providing adequate reporting mechanisms, with ISO accreditation structures supplying the systems and performance benchmarks.

Environmental issues are driven through the planning process by requiring all major projects to have an environmental impact statement (CD 97/11/EC) carried out if they are considered to adversely impact on their surroundings. Furthermore site selection of the optimum project location and design can be informed by recently developed environmentally-driven processes such as the Best Practicable Environmental Option (BPEO) and Strategic Environmental Assessment (SEA). Additional local fiscal measures, such as the UK's 'aggregate tax' and the 'landfill tax' and the Republic of Ireland's 'plastic bag tax', can assist environmental protection, while higher level drivers such as the Kyoto Protocol and its climate change levy have environmental improvement as core values.

National and regional planning strategies and guidelines influence land use planning, including designation of specialist zones, such as Areas of Special Scientific Interest (ASSIs) and Areas of Outstanding Natural Beauty (AONBs), providing protection for heritage, archaeology and the wider environment. In turn, detailed environmental monitoring supports both planning guidelines and environmental reporting, and provides essential evidence of any environmental pollution incident in which legal charges may be applied in accordance with the 'polluter pays' principle.

Several professional groups interact with the natural environment and their environmental ethics should ensure that unjust environmental damage is minimised. For example planners govern the location and specification of a project, architects along with both civil and environmental engineers conceptualise and design, environmental scientists develop processes and identify endangered elements of the environment, while contractors build and deliver the finished product to be both assessed and used by public/private organisations or individuals. Hence, there is a need for both awareness and detailed education about many environmental elements across all groups. However, a sensitive and sustained support for the environment cannot be secured through education, good design and stringent fiscal measures alone. There needs to be a sense of natural and voluntary support, and this psychological aspect of human response also applies to other areas and activities which can minimise harm to the environment, such as water conservation, energy efficiency and waste management.

The need for global measurement and reporting on environmental performance was established by the *Agenda 21* document devised at the Earth Summit in Rio de Janeiro in 1992. This is further illustrated by the development of sustainable development indicators suites which have included key environmental elements, while the annual 'ecological footprint' of the World Wildlife Fund (WWF) gives a global assessment of damage to the natural environment.

6.3 Sustainability Challenges for the Environment

Environmental challenges fall within the 'protection of the environment' sector of sustainable development but also link to other sectors, as developments need to meet society's basic needs of food and shelter whilst resourcing supplies from the natural environment as well as financially accounting for its environmental impact. Although the natural environment is passive in its relationship with the sustainable development sectors, impressions on and the response from the environment contribute to all four objectives of sustainable development as follows:

- *Effective protection of the environment* is fundamental to the preservation of the Earth, and this can be temporarily or irrevocably damaged through irresponsible or unsustainable practices such as pollution to water sources, damage to the landscape, and extinction of vulnerable habitats and species. A host of human activities have the potential to damage the environment, with some agricultural processes and material extractive operations being potentially lethal.
- *Social progress which recognises the needs of everyone* relates directly to the attitude and behaviour of societies across several phases and actions, and most of these have the potential to damage the environment. Communities frequently have scant regard for the environment, so that social progress comes at both an environmental and financial cost. In contrast, responsible communities and businesses can make a significant voluntary response to environmental protection by good environmental stewardship, positively influencing peers, taking environmental ownership and respecting fiscal measures designed to minimise adverse environmental impacts.
- *Prudent use of natural resources* relates to the actions and reactions of members of society and business decision-makers. Procurement policies can request an emphasis on the use of recycled materials thereby reducing natural resource use, whilst large corporations and governments can adopt a general 'green approach'. Market developments can focus on a life cycle analysis approach to fully understand the materials flow and properly to allocate responsibility to any element throughout its lifespan. Reuse of materials may not always provide the best environmental solution, pointing to the need to apply appropriate decision-making tools. Pollution incidents requiring incident management to mitigate environmental damage, can arise from improper use of processes and can also occur due to unexpected events.
- *Maintenance of high and stable levels of economic growth and employment* may not necessarily lead to environmental damage, as the 'business case' for fully integrating environmental protection into project solutions needs to be developed for each individual situation. The use of Best Available Technology Not Entailing Excessive Cost (BATNEEC) can assist in a balanced approach, while the inclusion of environmental reporting in annual corporate reports demonstrates environmental awareness. Hence, informed business judgments can positively contribute to the environment, increase employment and prospects without depleting profit margins.

Several issues and challenges exist for environmentalists as well as economists and strategists in delivering sustainable development and enhancing environmental protection. This sector is integral with the broader sustainable development agenda and the critical challenges in securing and protecting the environment include:

- *Full use of Environmental Management Systems and Business Eco-Efficiency*, by appropriate application of EMAS (Eco-Management and Audit Scheme) or seeking accreditation through ISO9001, ISO14001 or ISO18001. These systems lend themselves to the use of environmental indicators, such as Environmental Performance Indicators, and the application of eco-purchasing or eco-labelling in many industry sectors.
- *Commitment to Environmental Reporting* is essential to achieve buy-in from businesses and the government to the environmental agenda. This involves company reporting on the environmental impacts of their operations, allowing freedom of access to environmental information, the monitoring of environmental impacts of EC Member States and facilitating public access to information and participation in environmental decision-making.
- *Application of Environmental Assessment methodologies* will ensure that the most appropriate solutions are achieved while accounting for a full range of potential environmental impacts, as required by European Directive 97/11/EC. Several assessment methods are available such as Environmental Impact Assessment (EIA), Best Practicable Environmental Option (BPEO), Regulatory Impact Assessment (RIA), Environmental Risk Assessment (ERA), and Strategic Environmental Assessment (SEA). Additional evaluation tools such as cost benefit and cost effectiveness analysis,

Life-Cycle Analysis (LCA) and Whole Life Costing (WLC) can supplement the assessments and complement environmental inspections in order to meet common compliance and implementation across Europe.

- *Adoption of Environmental Technologies* is necessary to enhance or protect the environment by using cleaner technology, reducing emissions, living within the environmental limits and contributing to the use of appropriate technology. This could be achieved by applying an Environmental Technologies Action Plan (ETAP), releasing untapped technological potential for improving the environment and contributing to competitiveness.
- *Full commitment to environmental liability responsibilities* by endorsing principles and practice such as 'polluter pays', the 'precautionary principle' and the 'proximity principle'. This can prevent serious environmental crime and combat environmental offences; however international consistency and co-operation is essential for effective delivery.
- *Consistency in local environmental performance, quality and management* is essential for delivering sustained environmental messages. Issues with potential local application include:
  - control of litter, refuse and fly-tipping;
  - impact of landfill usage and attitudes to disposal of harmful goods;
  - design and housekeeping of green spaces and streetscapes;
  - management of urban parks and city features;
  - development of cleaner, greener and safer play areas and multi-use zones;
  - protection of urban and rural landscapes, country parks, ASSIs and AONBs.
- *Encouragement of sustainable consumption and production* can be achieved through green purchasing and procurement policies, incentives to use recycled products and development of environmentally friendly business processes and products. The use of fiscal measures can support this approach, but it is as important to convince stakeholders of the broader benefits of environmental protection as it is to adopt an incremental approach to reducing the environmental impact of production and consumption of goods/services and the application of environmental product declaration schemes.

6.4 Environment Theme Index

Jurisdiction	Tier	Title	Date	Page No.
EUROPEAN	Council Decisions	The Recognition of the International Standard ISO 14001:1996 & the European Standard EN ISO 14001:1996, Establishing Specification for Environmental Management Systems *(C Dec 97/265/EC)*	1997	189
		The Review of the European Community Programme of Policy & Action in Relation to the Environment & Sustainable Development – 'Towards Sustainability' *(Dec 2179/98/EC)*	1998	189
		Allowing Voluntary Participation by Organisations in a Community Eco-Management & Audit Scheme (EMAS)	2001	190
		Laying Down the Sixth Community Environment Action Programme *(Dec No. 1600/2002/EC)*	2002	190
	Council Communications	Environmental Agreements *(COM (1996) 561)*	1996	191
		Environmental Taxes & Charges *(COM (97) 9)*	1997	191
		Europe-Asia Co-operation Strategy in the field of the Environment *(COM (97) 490)*	1997	191
		Partnership for Integration: A Strategy for Integrating Environment into EU Policies *(COM (1998) 333)*	1998	192
		Single Market & Environment *(COM (1999) 263)*	1999	192
		The Precautionary Principle *(COM (2000) 1)*	2000	192
		Bringing our Needs & Responsibilities Together: Integrating Environmental Issues with Economic Policy *(COM (2000) 576)*	2000	193
		The Sixth Environment Action Programme of the European Community 'Environment 2010: Our future, Our Choice' *(COM (2001) 31)*	2001	193
		Communication from the Commission on EU-Russia environmental co-operation *(COM (2001) 772)*	2001	193
		Pan-European Environmental Co-operation after the 2003 Kiev Conference *(COM (2003) 62)*	2003	193
		Developing an Action Plan for Environmental Technology *(COM (2003) 131)*	2003	194
		A European Environment & Health Strategy *(COM (2003) 338)*	2003	194
		Integrated Product Policy – Building on Environmental Life-Cycle Thinking *(COM (2003) 302)*	2003	195
		The 2003 Environment Policy Review: Consolidating the Environmental Pillar of Sustainable Development *(COM (2003) 745)*	2004	195
		Stimulating Technologies for Sustainable Development: An Environmental Technologies Action Plan for the European Union *(COM (2004) 38)*	2004	195
		The Integration of Environmental Aspects into European Standardisation *(COM (2004) 130)*	2004	196
		The European Environment & Health Action Plan 2004–2010 *(COM (2004) 416)*	2004	196

Jurisdiction	Tier	Title	Date	Page No.
EUROPEAN	Council Recommendations	Environmental Inspections: Minimum Criteria *(Rec 2001/331/EC)*	2001	196
		Allowing Voluntary Participation by Organisations in a Community Eco-Management & Audit Scheme (EMAS)	2001	197
		Allowing Voluntary Participation by Organisations in a Community Eco-Management & Audit Scheme (EMAS) Concerning the Selection & use of Environmental Performance Indicators	2003	197
	Miscellaneous	The Establishment of the EEA & the European Environment & Observation Network *(COM (95) 325)*	1995	198
		Proposal for a Council Directive on the assessment of the effects of certain plans and programmes on the environment *(COM (1996) 511)*	1996	198
		Economic Evaluation of Environmental Policies & Legislation	1998	198
		The Cologne Report on Environmental Integration – Mainstreaming of Environmental Policy *(SEC (99) 777)*	1999	199
		A Study on Environmental Reporting of Companies	1999	199
		Implementation & Enforcement of Community Environmental Law (1996–97) (1998–99) *(SEC (99) 592 & SEC (2000) 1219)*	1999 2000	199
		Environmental Liability – White Paper *(COM (2000) 66)*	2000	199
		Serious Environmental Crime *(CNS 2000/0801)*	2000	200
		Business, Eco-Efficiency & Sustainable Development – The role of Environmental Management Tools	2000	200
		Study on the relationship between environmental/energy taxation & employment creation	2000	200
		The Experience Gained in the Application of Council Directive 90/313/EEC on Freedom of Access to Information on the Environment *(COM (2000) 400)*	2000	201
		LIFE: A Financial Instrument for the Environment *(Reg No. 1655/2000)*	2000	201
		A Revised Community Eco-label Award Scheme *(Reg No. 1980/2000)*	2000	201
		Financial Assurance Issues of Environmental Liability	2000	201
		The Potential Economic Impact of Environmental Liability: The American & European Contexts	2000	202
		Indicators for Monitoring Integration of Environment & Sustainable Development in Enterprise Policy	2001	202
		LIFE – Environment in Action: 56 New Success Stories for Europe's Environment	2001	203
		Community Eco-Management & Audit Scheme (EMAS) *(Reg No. 761/2001)*	2001	203
		Protection of the Environment through Criminal Law *(COM (2001) 139)*	2001	203

Jurisdiction	Tier	Title	Date	Page No.
EUROPEAN	Miscellaneous	Study on the Economic & Environmental Implications of the use of Environmental Taxes & Charges in the EU & its Member States	2001	204
		SEA & Integration of the Environment into Strategic Decision-Making	2001	204
		The Eco-Management & Audit Scheme: A New Opportunity for Financial Institutions	2001	204
		An Analysis of the Preventive Effect of Environmental Liability: Environmental liability, location & emissions substitution – Evidence from the Toxic Release Inventory	2001	204
		Programme Promoting NGOs Active in the Field of Environmental Protection *(COM (2001) 337)*	2001	205
		Environmental Liability – Proposal for a Directive *(COM (2002) 17)*	2002	205
		Environmental Technology for Sustainable Development	2002	205
		EMAS 2000 – A Dynamic Instrument for Environmental Protection & Sustainable Development: Business Benefits & Eco-Management in the Mediterranean Region	2002	206
		The State of EMAS in the EU Eco-Management as a Tool for Sustainable Development	2002	206
		Evaluation of Environmental Product Declaration Schemes	2002	206
		Analysis of the 'Open List' of Environment-Related Headline Indicators *(COM (2002) 524)*	2002	207
		The Application and Effectiveness of the EIA Directive (*Directive 85/337/EEC as amended by Directive 97/11/EC*)	2002	207
		Europe's Environment: the Third Assessment	2003	208
		Review of the European Environment Agency (EEA) *(COM (2003) 800)*	2003	208
		A Sustainable Approach for the Environment – LIFE & the Community Eco-Management & Audit Scheme (EMAS)	2003	209
		Implementation of Directive 2001/42 on the Assessment of the Effects of Certain Plans & Programmes on the Environment	2003	209
		Promoting Environmental Technologies: Sectoral Analyses, Barriers & Measures	2004	209
UK	Strategies & Reports	Consumer Products & the Environment	1998	210
		Environmental Protection Act 1990: Code of Practice on Litter & Refuse	1999	210
		United Kingdom Response to European Commission White Paper on Environmental Liability	2000	211
		Green Spaces, Better Places	2001	211
		Extended Partial Regulatory Impact Assessment on Proposals for a Directive on Environmental Liability	2002	212
		Action Plan to Develop Organic Food & Farming in England	2002	212

Jurisdiction	Tier	Title	Date	Page No.
UK	Strategies & Reports	Living Places – Cleaner, Safer, Greener	2002	212
		Improving Urban Parks, Play Areas & Green Spaces	2002	212
		Paving the Way: How we Achieve Clean, Safe & Attractive Streets	2002	213
		The Strategy for Sustainable Farming & Food – Facing the Future	2002	213
		Farming and Food's Contribution to Sustainable Development – Economic & Statistical Analysis	2002	213
		Ecological Assessment of Wetlands, Scrub & Unimproved Grassland, Heath or Moorland for EIA Screening Decisions	2003	214
		The Annual Local Environmental Quality Survey of England 2002/2003	2003	214
		Implementation of the Recommendation (2001/331/EC) on the Minimum Criteria for Environmental Inspections in the UK: 2nd Report to the Commission	2003	214
		Policies for Sustainable Consumption	2003	215
		Changing Patterns UK Government Framework for Sustainable Consumption & Production	2003	215
		The Value of Public Space – How High Quality Parks & Public Spaces Create Economic, Social & Environmental Value	2004	215
		Sustainable Production & Consumption: Business Support Review Final Report	2004	216
		Action Plan to Develop Organic Food & Farming in England – Two Years On	2004	216
		Is the Grass Greener . . . ? Learning from International Innovations in Urban Green Space Management	2004	216
		The Environment in Your Pocket 2004 – Key Facts & Figures on the Environment of the United Kingdom	2004	217
		Reducing litter caused by 'food on the go' – A Voluntary Code of Practice for local partnerships	2004	217
	Consultation Papers	Living Places – Powers, Rights & Responsibilities	2002	217
		Living Places – Powers, Rights & Responsibilities Options for Reforming the Legislative Framework: Consultation Summary	2003	218
		Sustainable Consumption & Production Indicators – Joint DEFRA/DTI Consultation Paper on a Set of 'Decoupling' Indicators of Sustainable Development	2003	218
		Fly-Tipping Strategy – Consultation Document	2004	218
		Clean Neighbourhoods – Consultation Document	2004	219
	Guidance Notes	Guidelines for Environmental Risk Assessment & Management	2000	219
		Environmental Reporting – General Guidelines	2001	219
		EIA Guidelines – Environmental Impact Assessment for Use of Uncultivated Land or Semi-Natural Areas for Intensive Agricultural Purposes	2002	220

Jurisdiction	Tier	Title	Date	Page No.
UK	Guidance Notes	Green Claims – Practical Guidance: How To Make A Good Environmental Claim	2003	220
		Strategic National Guidance – The Decontamination of the Open Environment Exposed to Chemical, Biological, Radiological or Nuclear (CBRN) Substances or Material	2004	220
		Public Sector Food Procurement Initiative – Guidance for buyers & their Internal Customers: Advice for Public Sector Bodies on Integrating Sustainable Development into Food & Catering Services Contracts	2004	221
		A Guide to Producing Parks & Green Space Management Plans	2004	221
		Green Space Strategies: A Good Practice Guide	2004	221
NORTHERN IRELAND	Strategies & Reports	Framing the View – Window Frames for a Sustainable Future	2000	222
		Environmental Monitoring – New Technologies for Monitoring: Status & Prospects	2001	222
		Shared Horizons – Statement of Policy on Protected Landscapes in Northern Ireland	2003	222
		A Forward Programme for the Declaration of Areas of Special Scientific Interest in Northern Ireland	2003	223
		Areas of Special Scientific Interest in NI – Proposals for their Future Management & Protection	2003	223
		EHS Corporate & Business Plan 2003/2006	2003	223
		The Management of Country Parks – A Statement of Policy	2004	224
		EHS Corporate & Business Plan 2004–2006	2004	225
	Consultation Reports	Proposal for a Directive on Environmental Liability with regard to the Prevention & Remedying of Environmental Damage	2002	225
		The Draft Environmental Assessment of Plans & Programmes Regulations (Northern Ireland) 2004 – Consultation Paper	2004	225
	Guidance Notes	Tackling Fly Tipping – A Guidance Note & Response Protocol	1999	225
		Best Practicable Environmental Option (BPEO) – Decision-Makers Guide	2000	226
		Compliance Policy & General Guidance – ASSIs	2003	226

6.5 Environment Policy Tables

Table 6.5.1 International Level Environment Policy.

Title	Venue	Purpose	Key Outcome(s)	Website
First Euro-Environment Conference on Industry & Environmental Performance	Aalborg, Denmark (1998)	The conference aim was to establish itself as an event reporting on the status and situation of the international industrial environmental performance. The conference presentations and proceedings depicted a varied and highly active situation, leading to the general summarised conclusions. Detailed conclusions can be derived from the proceedings.	The key themes covered at this conference included: ▪ Environmental management; ▪ Environmentally sound product development; ▪ Self regulation versus public regulation; ▪ Cleaner technology & prevention; & ▪ Stakeholder management.	http://www.euro-environment.dk/page437.aspx http://www.euro-environment.dk/page438.aspx
Second Euro-Environment Conference on the International Political Environmental Agenda	Aalborg, Denmark (2000)	A key purpose of the Euro Environment 2000 conference was to create a forum where business, governments, academia & civil society could meet in an atmosphere of open and transparent dialogue. It emphasised the link between industrial policy setting & the environmental performance of individual companies, & therefore it presented visionary & strategic goals as well as practical results of the sustainability processes of individual businesses.	The key themes covered at this conference included: ▪ Business strategy & policy setting; ▪ Sustainability through the market; & ▪ Sustainability & financing.	http://www.euro-environment.dk/page442.aspx http://www.euro-environment.dk/page444.aspx
Third Euro-Environment Conference on Business, the Global Sustainability Agent	Aalborg, Denmark (2002)	The conference theme was 'Business the global Agent of Sustainability', indicating the active position business has to share with other stakeholders 'to make it happen'. This conference was one of the first opportunities for International business to meet its stakeholders after the Johannesburg Summit & the promises made there.	The key themes covered at this conference included: ▪ Corporate citizenship; ▪ Global governance; ▪ Global financing; ▪ Corporate globalisation; & ▪ Supply chain management.	http://www.euro-environment.dk/page448.aspx http://www.euro-environment.dk/page450.aspx

Seventh Biennial Conference and Trade Fair on Business and the Environment	Vancouver, Canada (2002)	Starting in 1998 'GLOBE' has grown to become a reputable event providing access to global environmental markets & senior-level corporate & government executives. In 2002 over 9,000 participants from 70 economies attended the event. The conference focused on how major developments in energy policy & sustainable development can be translated into technology solutions & business opportunities. These Conferences are biennial and further papers will regularly become available.	The conference had three main themes – corporate sustainability; environment & energy; & international business opportunities. The key objectives achieved are listed below: ▪ Accelerated Corporate Environmental Leadership; ▪ Facilitated Business Opportunities in the Environmental Industry; ▪ Enhanced SME Access to Global Environmental Markets; ▪ Defined Policy & Regulatory Trends; ▪ Provided Updates on Global Climate Change Negotiations; ▪ Attracted New Investment in Environmental Technology; ▪ United World Leaders to Discuss Global Environmental Issues.	http://www.globe2002.com/pdf/globe2002review.pdf http://www.globe2002.com/pdf/program.pdf
Convention on Access to Information, Public Participation in Decision-making & Access to Justice in Environmental Matters	Aarhus, Denmark (1998)	In order to contribute to the protection of the right of every person of present and future generations to live in an environment adequate to his or her health and well-being, each Party shall guarantee the rights of access to information, public participation in decision-making, and access to justice in environmental matters in accordance with the provisions of this Convention.	The Convention Treaty contains various articles & Annexes governing: ▪ The access to environmental information ▪ Collection & dissemination of environmental info ▪ Public participation in decisions on specific activities ▪ Public participation concerning plans, programmes & policies relating to the environment ▪ Public participation during the preparation of executive regulations &/or generally applicable legally binding normative instruments ▪ Access to justice ▪ Annex I – list of activities potentially harmful to the environment ▪ Annex II – arbitration in event of a dispute.	http://www.unece.org/env/pp/documents/cep43e.pdf

Table 6.5.2 European Level Environment Policy.

Policy Type	Programme Title	Policy Objective	Community Measure	Document Reference	Website
Council Directives	The Assessment of the Effects of Certain Public & Private Projects on the Environment	Member States shall adopt all measures necessary to ensure that, before consent is given, projects likely to have significant effects on the environment by virtue, *inter alia*, of their nature, size or location are made subject to an assessment with regard to their environmental effects.	Council Directive 85/337/EEC on the assessment of the effects of certain public and private projects on the environment (27/06/85).	▪ CD 85/337/EEC	http://europa.eu.int/comm/environment/eia/full-legal-text/85337.htm
	The Assessment of the Effects of Certain Public & Private Projects on the Environment (Amending CD 85/337/EEC)	This Directive amends CD 85/337/EEC and introduces provisions designed to clarify, supplement and improve the rules on the assessment procedure, in order to ensure that the Directive is applied in an increasingly harmonised and efficient manner.	Council Directive 97/11/EC amending Directive 85/337/EEC on the assessment of the effects of certain public and private projects on the environment (03/03/97).	▪ CD 97/11/EC	http://europa.eu.int/comm/environment/eia/full-legal-text/9711.htm
	Assessment of the Effects of Plans & Programmes on the Environment	To help integrate the environment into the preparation & adoption of plans & programmes liable to have significant effects on the environment, by subjecting them to an environmental assessment.	Directive 2001/42/EC of the European Parliament & of the Council on the assessment of the effects of certain plans & programmes on the environment (27/06/01).	▪ Off Jnl L197, 21/07/01	http://europa.eu.int/scadplus/leg/en/lvb/l28036.htm

Council Decisions	The Recognition of the International Standard ISO 14001:1996 & the European Standard EN ISO 14001:1996, Establishing Specification for Environmental Management Systems	For the purposes of Article 12 of Regulation (EEC) No 1836/93 the Commission hereby recognises that the international standard ISO 14001:1996 & the European standard EN ISO 14001:1996, establishing specification for environmental management systems, containing requirements corresponding to those of the above mentioned Regulation, as specified in the Annex to this Decision.	Commission Decision 97/265/EC on the recognition of the international standard ISO 14001:1996 & the European standard EN ISO 14001:1996, establishing specification for environmental management systems, in accordance with Article 12 of Council Regulation (EEC) No 1836/93, allowing voluntary participation by companies in the industrial sector in a Community Eco-Management & Audit Scheme (EMAS) (16/04/97).	Off Jnl L104 22/04/97	http://europa.eu.int/smartapi/cgi/sga_doc?smartapi!celexapi!prod!CELEXnumdoc&lg=EN&numdoc=31997D0265&model=guichett
	The Review of the European Community Programme of Policy & Action in Relation to the Environment & Sustainable Development – 'Towards Sustainability'	The objective of the fifth programme of action in relation to the environment was to transform patterns of growth in the Community in such a way as to promote sustainable development. The programme continued to tackle environmental problems (such as climate charge, aquatic pollution and waste management), but also aimed to establish new relations in the environmental sector.	Decision No 2179/98/EC of the European Parliament & of the Council on the review of the European Community programme of policy & action in relation to the environment & sustainable development 'Towards sustainability' (24/09/98).	Off Jnl L275/1 10/10/98	http://europa.eu.int/eur-lex/pri/en/oj/dat/1998/l_275/l_27519981010en00010013.pdf

Continued

Table 6.5.2 European Level Environment Policy. *Continued*

Policy Type	Programme Title	Policy Objective	Community Measure	Document Reference	Website
Council Decisions	Allowing Voluntary Participation by Organisations in a Community Eco-Management & Audit Scheme (EMAS)	This guidance has been prepared to assist organisations, environmental verifiers & competent bodies in deciding on the appropriate entity for registration as organisation to EMAS. The purpose of these requirements was to ensure that the organisation has management control & influence over its environmental aspects that have significant environmental impacts at all sites. Therefore participants were advised to have a clear & reasoned justification for selecting the sites or parts of sites of the organisation to be registered. By doing so they would be anticipating the requirements of the environmental statement & would be well placed to respond to possible queries, notably from verifiers & competent bodies but also from other interested parties.	Commission Decision on guidance for the implementation of Regulation (EC) No 761/2001 of the European Parliament & of the Council allowing voluntary participation by organisations in a Community Eco-Management & Audit Scheme (EMAS) (07/09/01).	▪ Off Jnl L247/24 17/09/01	http://europa.eu.int/eur-lex/pri/en/oj/dat/2001/l_247/l_24720010917en00240047.pdf
	Laying Down the Sixth Community Environment Action Programme	This Decision establishes a programme of Community action on the environment. It addresses the key environmental objectives & priorities based on an assessment of the state of the environment & of prevailing trends including emerging issues that require a lead from the Community. The Programme should promote the integration of environmental concerns in all Community policies & contribute to the achievement of sustainable development throughout the current & future enlarged Community. The Programme furthermore provides for continuous efforts to achieve environmental objectives & targets already established by the Community.	Decision No 1600/2002/EC of the European Parliament & of the Council laying down the Sixth Community Environment Action Programme (22/07/02).	▪ Off Jnl L242/1 10/09/02	http://europa.eu.int/eur-lex/pri/en/oj/dat/2002/l_242/l_24220020910en00010015.pdf

Council Communications	Environmental Agreements	Environmental Agreements can bring about effective measures in advance of legislation & thus reduce the volume of regulatory & administrative actions. This initiative was therefore related to the efforts of the Commission to simplify Community law & improve its quality. Agreements could be used as a supplement to legislation and might represent a particularly appropriate implementation tool. In a binding form, they are an effective & appropriate instrument for achieving general objectives set by Community Directives, in particular the implementation of reduction programmes. At Community level, such agreements enable the Commission to address environmental issues without resorting to the traditional regulatory instruments.	Communication from the Commission to the Council & the European Parliament on environmental agreements (27/11/96).	COM (1996) 561	http://europa.eu.int/comm/environment/docum/pdf/96561en.pdf
	Environmental Taxes & Charges	To promote the use of fiscal instruments by Member States to increase the efficiency of the environment policy & ensure that environmental taxes & charges are used in accordance with Community legislation.	Commission communication on environmental taxes & charges in the Single Market (26/03/97).	COM (97) 9	http://europa.eu.int/scadplus/leg/en/lvb/l28058.htm
	Europe–Asia Co-operation Strategy in the Field of the Environment	The goal is to enhance the efficiency of the co-operation between Europe & Asia in the field of the environment by promoting a proactive regional strategy.	Communication from the Commission to the Council, European Parliament & the Economic & Social Committee on a Europe–Asia Co-operation Strategy in the field of the environment (13/10/97).	COM (97) 490	http://europa.eu.int/scadplus/leg/en/lvb/l28067.htm

Continued

Table 6.5.2 European Level Environment Policy. *Continued*

Policy Type	Programme Title	Policy Objective	Community Measure	Document Reference	Website
Council Communications	Partnership for Integration: A Strategy for Integrating Environment into EU Policies	This Communication is the Commission's response to the request by the Luxembourg European Council for the submission of a strategy to implement the requirements of Article 6 of the consolidated EC-Treaty. This article requires that environmental protection is integrated into the definition & implementation of all Community policies & activities, in particular with a view to promoting sustainable development, established as a Community objective in Article 2 of the Treaty.	Commission communication to the European Parliament—Partnership for integration: A strategy for integrating environment into EU policies (1998).	▪ COM (1998) 333	http://europa.eu.int/comm/environment/docum/pdf/98333en.pdf
	Single Market & Environment	The Action Plan for the Single Market sets out the priority actions needed to improve its functioning; it takes account of the many dimensions of Single Market policy & includes among its four Strategic Targets that of 'Delivering a Single Market for the benefit of all citizens'. This emphasised that in delivering greater personal freedom & wider choice to consumers, the Single Market is also required to ensure a high level of protection of health, safety & the environment. The present Communication is a commitment to the Action Plan.	Communication from the Commission to the Council & the European Parliament on a single market & the environment (09/06/99).	▪ COM (1999) 263	http://europa.eu.int/comm/internal_market/en/update/general/263en.pdf
	The Precautionary Principle	To develop guidelines for the application of the precautionary principle so as to ensure a high level of protection of the environment & of human, animal & plant health whenever the available scientific data do not permit a complete evaluation of the risk. However also to prevent the precautionary principle from being used as a pretext for protectionist measures.	Communication from the Commission on the precautionary principle (02/02/00).	▪ COM (2000) 1	http://europa.eu.int/scadplus/leg/en/lvb/l32042.htm

Bringing our Needs & Responsibilities Together: Integrating Environmental Issues with Economic Policy	This Communication follows the presentation of strategies for integrating the environmental dimension into the Agriculture, Transport and Energy sectors. Building on the Commission's 1994 Communication on economic growth & the environment, it contributes towards the definition of a Community strategy to improve integration of the environmental & economic dimensions of sustainable development. In line with the conclusions of the Lisbon European Council, the strategy proposes no new process, but rather the use of existing instruments to achieve gradual but credible progress in integration.	Communication from the Commission to the Council & the European Parliament – Bringing our needs & responsibilities together: integrating environmental issues with economic policy. (20/09/00).	COM (2000) 576	http://europa.eu.int/eur-lex/en/com/cnc/2000/com2000_0576en01.pdf
The Sixth Environment Action Programme of the European Community 'Environment 2010: Our future, Our Choice'	This context has guided the strategic focus of the Sixth Environmental Action Programme, which effectively sets the environmental objectives & priorities that will be an integral part of the EC's strategy for sustainable development. The programme set out the major priorities & objectives for environment policy over the following five to ten years & details the measures to be taken.	Communication from the Commission to the Council, the European Parliament, the Economic & Social Committee & the Committee of the Regions on the sixth environment action programme of the European Community – 'Environment 2010: Our future, Our choice' (24/01/01).	COM (2001) 31	http://europa.eu.int/eur-lex/en/com/pdf/2001/en_501PC0031.pdf
Environmental Co-operation with Russia	To consolidate the environmental dialogue with Russia by devising a common strategic agenda & reinforced procedures.	Communication from the Commission on EU–Russia environmental co-operation (17/12/01).	COM (2001) 772	http://europa.eu.int/scadplus/leg/en/lvb/l28004.htm
Pan-European Environmental Co-operation after the 2003 Kiev Conference	This Communication sets out the views of the European Commission on how to pursue environmental co-operation between the enlarged EU and its neighbours in the coming years, as a contribution to the discussions at the Kiev Conference on the future development of such co-operation including the future of the 'Environment for Europe' process.	Communication from the Commission to the Council & the European Parliament on Pan-European environmental co-operation after the 2003 Kiev conference (06/02/03).	COM (2003) 62	http://europa.eu.int/eur-lex/en/com/cnc/2003/com2003_0062en01.pdf

Continued

Table 6.5.2 European Level Environment Policy. *Continued*

Policy Type	Programme Title	Policy Objective	Community Measure	Document Reference	Website
Council Communications	Developing an Action Plan for Environmental Technology	To deliver a strategic and concrete action plan to tackle obstacles to the development, take-up and use of environmental technologies requires the participation of all stakeholders throughout the whole policy chain. This Communication on the first findings facilitates this process as a starting point of the consultation with stakeholders. It was intended to put forward an action plan on environmental technology to be adopted by the Commission by the end of 2003.	Communication from the Commission – Developing an action plan for environmental technology (25/03/03). Resulting in *Stimulating Technologies for Sustainable Development* (see above – COM (2004) 38).	▪ COM (2003) 131	http://europa.eu.int/eur-lex/en/com/cnc/2003/com2003_0131en01.pdf
	A European Environment & Health Strategy	The added value of the proposed 'European Environment & Health Strategy' is the development of a Community System integrating information on the state of the environment, the ecosystem & human health. This renders the assessment of the overall environmental impact on human health more efficient by taking into account matters such as cocktail effects, combined exposure, cumulative effects, etc. The ultimate goal of the proposed strategy is to develop an environment & health 'cause–effect framework' that will provide the necessary information for the development of Community policy dealing with sources & the impact pathway of health stressors. To achieve this goal an integrated approach is needed within the framework of the Community's sustainable development policy.	Communication from the Commission to the Council, the European Parliament & the European Economic & Social Committee – A European Environment & Health Strategy (11/06/03).	▪ COM (2003) 338	http://www.europa.eu.int/smartapi/cgi/sga_doc?smartapi!celexplus!prod!DocNumber&lg=en&type_doc=COMfinal&an_doc=2003&nu_doc=338

Integrated Product Policy – Building on Environmental Life-Cycle Thinking	This strategy identified several actions in the environmental field, following the priority themes in the Sixth Environmental Action Programme. In the area of managing natural resources more responsibly the European Council agreed, 'that the EU Integrated Product Policy aimed at reducing resource use and the environmental impact of waste should be implemented in co-operation with business'.	Communication from the Commission to the Council & the European Parliament – Integrated product policy – building on environmental life-cycle thinking (18/06/03).	■ COM (2003) 302	http://europa.eu.int/eur-lex/en/com/cnc/2003/com2003_0302en01.pdf
The 2003 Environment Policy Review: Consolidating the Environmental Pillar of Sustainable Development	The main purpose of the Review was to report on developments since 2001 & highlight priority issues for the year to come in EU & national environmental policies. The Review was also aimed at monitoring the implementation of the EU's Sixth Environment Action Programme (6th EAP).	Communication from the Commission to the Council & the European Parliament – 2003 Environmental policy review: consolidating the environmental pillar of sustainable development (02/02/04).	■ COM (2003) 745	http://europa.eu.int/eur-lex/en/com/cnc/2003/com2003_0745en02.pdf
Stimulating Technologies for Sustainable Development: An Environmental Technologies Action Plan for the European Union	The Environmental Technologies Action Plan (ETAP) aimed to harness the full potential to reduce pressures on natural resources, improve the quality of life of European citizens & stimulate economic growth. It is an important means to implement the EU Sustainable Development Strategy & to pursue the Lisbon Strategy, while helping developing countries. It is based on the recognition that there is significant untapped technological potential for improving the environment while contributing to competitiveness & growth. The Action Plan sets out a series of measures to do this; measures that will require a concerted effort by the Commission, Member States & partners in the research community, industry & civil society.	Communication from the Commission to the Council & the European Parliament on Stimulating Technologies for Sustainable Development: An Environmental Technologies Action Plan for the European Union (28/01/04).	■ COM (2004) 38	http://www.europa.eu.int/eur-lex/en/com/cnc/2004/com2004_0038en01.pdf

Continued

Table 6.5.2 European Level Environment Policy. *Continued*

Policy Type	Programme Title	Policy Objective	Community Measure	Document Reference	Website
Council Communications	The Integration of Environmental Aspects into European Standardisation	The results showed that nearly all stakeholders felt that a Communication would be a useful contribution which would help them in their work. In particular, four key issues were identified, so these are specifically dealt with in this Communication: ▪ raising awareness & environmental thinking; ▪ setting priorities; ▪ enhancing stakeholder participation; ▪ using tools & offering incentives.	Communication from the Commission to the Council, the European Parliament & the European Economic & Social Committee (25/02/04).	▪ COM (2004) 130	http://europa.eu.int/eur-lex/en/com/cnc/2004/com2004_0130en01.pdf
	The European Environment & Health Action Plan 2004–2010	The action plan is based on three main themes: ▪ improving the information chain to understand the links between sources of pollution & health effects; ▪ filling the knowledge gap by strengthening research on environment & health & identifying emerging issues; ▪ reviewing policies & improving communication by developing the awareness-raising process, risk communication, training & education to give citizens the information they need to make better health choices, & to make sure that professionals in each field are alert to environment & health interactions.	Communication from the Commission to the Council, the European Parliament, the European Economic & Social Committee – 'The European Environment & Health Action Plan 2004–2010' [SEC(2004) 729] (09/06/04).	▪ COM (2004) 416	http://www.europa.eu.int/smartapi/cgi/sga_doc?smartapi!celexplus!prod!DocNumber&lg=en&type_doc=COMfinal&an_doc=2004&nu_doc=416
Council Recommendations	Environmental Inspections: Minimum Criteria	To ensure greater compliance, & a more uniform application & implementation, of Community legislation on the environment by providing for minimum criteria applicable to organising, carrying out, following up & publicising the results of environmental inspections in all Member States.	Recommendation 2001/331/EC of the European Parliament & of the Council providing for minimum criteria for environmental inspections in Member States (04/04/01).	▪ Off Jnl L118, 27/04/01	http://europa.eu.int/scadplus/leg/en/lvb/l28080.htm

Allowing Voluntary Participation by Organisations in a Community Eco-Management & Audit Scheme (EMAS)	EMAS is designed to help an organisation to manage & improve its environmental performance. This guidance was intended to assist the organisation in producing the environmental statement required by the scheme under Regulation (EC) No 761/2001 & identifies issues to be considered in preparing the statement in accordance with Annex III. In preparing this guidance consideration was given to the information needs of interested parties & how organisations can meet those needs. Openness, transparency & periodic provision of environmental information are key factors in differentiating EMAS from other schemes and are also important for the organisation in building confidence with interested parties.	Commission Recommendation on guidance for the implementation of Regulation (EC) No 761/2001 of the European Parliament & of the Council allowing voluntary participation by organisations in a Community eco-management & audit scheme (EMAS) (07/09/01).	Off Jnl L247/1 17/09/01	http://europa.eu.int/eur-lex/pri/en/oj/dat/2001/l_247/l_24720010917en00010023.pdf
Allowing Voluntary Participation by Organisations in a Community Eco-Management & Audit Scheme (EMAS) Concerning the Selection & use of Environmental Performance Indicators	Environmental performance indicators should be cost-effective & appropriate to the size & type of organisation, its needs & priorities. They should address primarily those environmental impacts that are most significant & which the company can influence by its operations, management, activities, products & services. They should also be sensitive enough to reflect significant changes in environmental impacts. In addition, organisations should make the optimum use of the environmental information they collect. To this end the indicators should fulfill the dual purpose of assisting the management of the organisation & providing information to stakeholders. Depending on an organisation's capabilities & resources, the use of environmental performance indicators may initially be limited to those aspects considered most relevant, with the initial scope being gradually widened over time. The indicators & measurement units provided in this guidance are given by way of example.	Commission Recommendation on guidance for the implementation of Regulation (EC) No 761/2001 of the European Parliament & of the Council allowing voluntary participation by organisations in a Community eco-management & audit scheme (EMAS) concerning the selection & use of environmental performance indicators (10/07/03).	Off Jnl L184/19 23/07/03	http://www.europa.eu.int/comm/environment/emas/pdf/guidance/guidance08_en.pdf

Continued

Table 6.5.2 European Level Environment Policy. *Continued*

Policy Type	Programme Title	Policy Objective	Community Measure	Document Reference	Website
Miscellaneous	European Environmental Agency	To provide the Community & Member States with reliable & comparable information on the state of the environment in co-operation with the European environment information & observation network.	Council Regulation (EEC) No. 1210/90 on the establishment of the EEA & the European environment & observation network (07/05/90).	▪ COM (95) 325 ▪ Off Jnl L120, 11/05/90	http://europa.eu.int/scadplus/leg/en/lvb/l28019.htm
			Amended by Council Regulation (EC) No. 933/1999	▪ Off Jnl L117, 05/05/99	
	Assessment of the Effects of Certain Plans & Programmes on the Environment	The objective of this Directive was to provide for a high level of protection of the environment by ensuring that an environmental assessment is carried out of certain plans and programmes and that the results of the assessment are taken into account during the preparation and adoption of such plans and programmes.	Proposal for a Council Directive on the assessment of the effects of certain plans and programmes on the environment (1996).	▪ COM (1996) 511	http://europa.eu.int/comm/environment/eia/full-legal-text/96pc511.htm
	Economic Evaluation of Environmental Policies & Legislation	The focus of this report was on the use of economic criteria for policy & regulatory analysis. The study was commissioned by DG III (Industry) of the European Commission to analyse the way in which cost-benefit analysis (CBA) & cost-effectiveness analysis (CEA) were currently used in the policy making & legislative processes by a range of mainly European countries.	Report for the Directorate General III (Industrial Affairs) on an economic evaluation of environmental policies & legislation (Sept 1998).	▪ DG III Report	http://europa.eu.int/comm/environment/enveco/others/environmental_policies_and_legislation.pdf

The Cologne Report on Environmental Integration – Mainstreaming of Environmental Policy	This report was part of the process launched at the Luxembourg Summit in December 1997. At the Cardiff Summit the European Council invited all relevant sectoral Councils to establish their own strategies for giving effect to environmental integration & sustainable development within their policy areas. The Transport, Energy & Agriculture Councils were invited to start this process. In Vienna, the Heads of State & Government invited the Council to develop further this work in other Community policies, particularly in the Development, Internal Market & Industry Councils. They also asked the Council to put emphasis on the environmental dimension of employment & enlargement.	Commission working paper addressed to the European Council on the Cologne report on environmental integration – Mainstreaming of environmental policy (1999).	▪ SEC (99) 777	http://europa.eu.int/comm/environment/docum/pdf/cologne.pdf
A Study on Environmental Reporting of Companies	The aim of the study was to identify disclosures that allow for standardisation & comparability of environmental information being presented in annual/financial reports. The report begins with a summary of the methodology. The analysis presents data relating to the current locations & types of environmental information found within the Annual Report & Financial Report.	Report by the Centre for Environmental Informatics, University of Sunderland to the DG Environment (1999).	▪ DG XI Report	http://europa.eu.int/comm/environment/enveco/industry_employment/envrep.pdf
Implementation & Enforcement of Community Environmental Law (1996–97) (1998–99)	To report on progress with implementing Community Environmental legislation. To improve awareness of Community Environmental legislation in the Member States & make its application more transparent.	Commission working document – second annual study on the implementation & monitoring of the application of Community Environmental Law (from Jan 1998–Dec 1999) (13/07/00)	▪ SEC (99) 592 ▪ SEC (2000) 1219	http://europa.eu.int/scadplus/leg/en/lvb/l28111.htm
Environmental Liability – White Paper	To set out the structure for a future EC environmental liability regime that aims at implementing the 'polluter pays' principle & to describe the key elements needed for making such a regime effective & practicable.	White paper on environmental liability (09/02/00).	▪ COM (2000) 66	http://europa.eu.int/scadplus/leg/en/lvb/l28107.htm

Continued

Table 6.5.2 European Level Environment Policy. *Continued*

Policy Type	Programme Title	Policy Objective	Community Measure	Document Reference	Website
Miscellaneous	Serious Environmental Crime	To protect the environment under criminal law & to establish police, criminal justice & administrative co-operation between Member States to combat serious environmental offences.	Initiative of the Kingdom of Denmark with a view to adopting a Council Framework Decision on combating serious environmental crime (11/02/00).	▪ CNS 2000/0801 ▪ Off Jnl C39, 11/02/00	http://europa.eu.int/scadplus/leg/en/lvb/l33120.htm
	Business, Eco-Efficiency & Sustainable Development – The role of Environmental Management Tools	The conference presentations & discussions covered all the three 'pillars' of sustainable development – economic development, environmental protection & social inclusion – with a focus on the role of environmental management tools in helping to achieve these objectives.	An International Workshop organised by INETI, Portuguese Directorate-General of Industry & the European Commission (Mar 2000).	▪ International Workshop	http://europa.eu.int/comm/enterprise/environment/reports_studies/reports/lisbon_3-00_final_report_ph.pdf
	Study on the relationship between environmental/energy taxation & employment creation	This report does not attempt to cover all the details of the double dividend literature. Instead, it presents & explains the main conclusions that may be drawn from this literature, paying particular attention to factors that could be relevant to the construction of empirical models. It first deals with some important general points that apply whether or not labour markets are in equilibrium. It then presents an analysis of the employment double dividend, dealing with the case of labour market disequilibrium, as this is generally assumed to be the situation in most of Europe. Finally it considers how the analysis has to be changed when the labour market is in equilibrium, & there is no involuntary unemployment.	A Report to the Directorate General for the Environment (April 2000).	▪ DG XI report	http://www.europa.eu.int/comm/environment/enveco/taxation/entaxemp.pdf

The Experience Gained in the Application of Council Directive 90/313/EEC on Freedom of Access to Information on the Environment	This Report is made by the Commission to the European Parliament & the Council in accordance with article 8 of Directive 90/313/EEC on freedom of access to information on the environment. It is made in the light of the experience gained by Member States in the operation of the Directive which has been the subject of reports made by them to the Commission in accordance with that article. It also takes into account reports made by non-governmental organisations (NGOs) active in the environmental sphere & developments in Community & international law.	Report from the Commission to the Council & European Parliament on the experience gained in the application of Council Directive 90/313/EEC on freedom of access to information on the environment (29/06/00).	■ COM (2000) 400	http://europa.eu.int/eur-lex/en/com/rpt/2000/com2000_0400en01.pdf
LIFE: A Financial Instrument for the Environment	To contribute to the development, implementation & updating of community environment policy & environmental legislation, especially as regards the integration of the environment into other policies, & to sustainable development in the Community.	Regulation (EEC) No 1655/2000 of the European Parliament & of the Council concerning the financial instrument for the environment (LIFE) (17/07/00).	■ Off Jnl L192, 28/7/00	http://europa.eu.int/scadplus/leg/en/lvb/l28021.htm
Eco-Label Award Scheme	To promote products with a reduced environmental impact compared with other products in the same product group.	Regulation (EC) No. 1980/2000 of the European Parliament & of the Council on a revised Community eco-label award scheme (17/07/00).	■ Off Jnl L237, 21/09/00	http://europa.eu.int/scadplus/leg/en/lvb/l28020.htm
Financial Assurance Issues of Environmental Liability	This study will try to achieve the two main goals, identified in the technical annex, being: ■ First, an analysis will be provided of how a liability & insurance regime (such as sketched out in the white paper on environmental liability) should be structured to provide an optimal assurance of environmental liability; ■ Second, a further analysis will be provided of other financial arrangements that could be advanced, other than traditional liability insurance, to cover damage caused to biodiversity & natural resources.	Report for the DG Environment on the financial assurance issues of environmental liability (01/12/00).	■ DG XI Report	http://europa.eu.int/comm/environment/liability/insurance_gen_finalrep.pdf

Continued

Table 6.5.2 European Level Environment Policy. *Continued*

Policy Type	Programme Title	Policy Objective	Community Measure	Document Reference	Website
Miscellaneous	The Potential Economic Impact of Environmental Liability: The American & European Contexts	This report will summarise the key findings from these earlier Superfund cost studies, & explain how these findings might apply in a European environmental liability regime. Because it is of particular interest, this report will discuss the distribution of Superfund costs per industry sector, comparing these costs with industry profit levels, or, in the absence of profitability data, value-added data. As the following discussion will indicate, in many ways it is very difficult to draw direct comparisons between the economic impacts of Superfund & the potential economic impacts of the proposed European environmental liability regime. But it is hoped that this report will at least create a framework to aid future debates.	Report for the Economic Analysis Unit of the DG Environment on the potential economic impact of environmental liability (Dec 2000).	■ DG XI Report	http://europa.eu.int/comm/environment/liability/competitiveness_finalrep.pdf
	Indicators for Monitoring Integration of Environment & Sustainable Development in Enterprise Policy	The aim of this study was to identify indicators that could support the formulation of an integration strategy & monitor the integration of environment & sustainable development into enterprise policy. The proposals & recommendations build on existing initiatives in the area of sustainable development indicators. They also take into consideration the approaches taken by integration strategies in other EU policy sectors (transport, agriculture etc.). The study also wanted like to contribute to the ongoing discussion about indicators for monitoring integration processes, which at that time lacked a harmonised methodology.	Report for the DG Enterprise on indicators for monitoring integration of environment & SD in enterprise policy (Feb 2001).	■ DG Enterprise Report	http://europa.eu.int/comm/enterprise/environment/reports_studies/studies/study99-502550_indicators-ph-finalreport010202.pdf

LIFE – Environment in Action: 56 New Success Stories for Europe's Environment	The key objective of LIFE-Environment is to implement Community environmental policies & legislation for the promotion of sustainable development. Sustainable development requires industry to find innovative solutions to reduce total life-cycle costs in terms of raw materials, energy consumption & environmental impact while reusing products at the end of their lives. LIFE-Environment provides support for both industry & local authorities. This book presents a number of successful projects financed by LIFE-Environment, projects which validate the idea that environment policy is about opportunities, solutions & success stories.	LIFE – Environment Report (Feb 2001).	LIFE Environment Report	http://europa.eu.int/comm/environment/life/infoproducts/successstories2001_en.pdf
Community Eco-Management & Audit Scheme (EMAS)	To promote continuous improvement of the environmental performance of European organisations, together with providing the public & interested parties with information.	Regulation (EEC) No. 761/2001 of the European Parliament & of the Council allowing voluntary participation by organisations in a Community eco-management & audit scheme (EMAS) (19/03/01).	Off Jnl L114, 24/04/01	http://europa.eu.int/scadplus/leg/en/lvb/l28022.htm
Protection of the Environment through Criminal Law	To ensure the effective application of Community law on protection of the environment; to establish a minimum standard for the constituent elements of criminal offences.	Proposal for a Directive of the European Parliament & of the Council on the protection of the environment through criminal law (2001).	COM (2001) 139	http://europa.eu.int/scadplus/leg/en/lvb/l33148.htm

Continued

Table 6.5.2 European Level Environment Policy. *Continued*

Policy Type	Programme Title	Policy Objective	Community Measure	Document Reference	Website
Miscellaneous	Study on the Economic & Environmental Implications of the use of Environmental Taxes & Charges in the EU & its Member States	This study explored whether taxes and charges have led to environmental benefits that the proponents of taxes would wish to see, & whether the concerns raised regarding competition are real. It therefore fulfils one of the Commission's intentions as noted in the 1997 Communication on Environmental Taxes & Charges.	Report for the DG Environment on the economic & environmental implications of the use of environmental taxes & charges in the EU & member states (April 2001).	▪ DG XI Report	http://europa.eu.int/comm/environment/enveco/taxation/ch1t4_overview.pdf
	SEA & Integration of the Environment into Strategic Decision-Making	The aim of the research was to evaluate the role of Strategic Environmental Assessment (SEA) in integrating the environment into strategic decision-making. The focus was on the way in which environmental considerations are included in policy, plan & programme decision-making in all sectors (including health, transport, education, defence etc.), rather than simply raising the profile of environmental policies within government & institutional agendas.	Directorate General Environment Report (May 2001).	▪ DG XI Report	http://europa.eu.int/comm/environment/eia/sea-studies-and-reports/sea_integration_main.pdf
	The Eco-Management & Audit Scheme: A New Opportunity for Financial Institutions	A brochure on the implications of EMAS in the financial sector this describes in detail the costs and benefits of the implementation of EMAS in the banking and insurance sector.	DG Environment Brochure (June 2001).	▪ DG XI Brochure	http://www.europa.eu.int/comm/environment/emas/pdf/finbroch/finbroch_en.pdf
	An Analysis of the Preventive Effect of	This paper examined the relationship between imposition of strict liability for environmental damages, and emissions of	Report for the DG Environment on an analysis of the preventive effect of	▪ DG XI Report	http://europa.eu.int/comm/environment/liability/preventive_final.pdf

Environmental Liability: Environmental liability, location & emissions substitution – Evidence from the Toxic Release Inventory	toxic chemicals into the environment. In particular, it explored whether the imposition of strict liability alters the relationship between emissions into the various media, due to possible differences in firms' exposure to damages from a given quantity of emissions into different media.	environmental liability (Oct 2001).		
Programme Promoting NGOs Active in the Field of Environmental Protection	To encourage the activities of European NGOs whose primary aim is to protect the environment & who contribute to developing & implementing European policy & legislation in the field of the environment.	Proposal presented by the Commission for a Decision of the European Parliament & of the Council laying down a community action programme promoting NGOs primarily active in the field of environmental protection (2001).	COM (2001) 337	http://europa.eu.int/scadplus/leg/en/lvb/l28041.htm
Environmental Liability – Proposal for a Directive	To adopt a community regime for the prevention & remedying of environmental damage.	Proposal for a Directive of the European Parliament & of the Council on environmental liability with regard to the prevention & remedying of environmental damage (23/01/02).	COM (2002) 17	http://europa.eu.int/scadplus/leg/en/lvb/l28120.htm
Environmental Technology for Sustainable Development	When meeting in Stockholm in March 2001, the European Council announced that it would review the contribution that the environment technology sector can make to promoting growth and employment, in the spring of 2002. The European Council in Gothenburg in June 2001 noted that the Commission would prepare a report on this issue. This Communication was a response to this commitment, setting the scene for the development of a future Action Plan.	Report From The Commission (13/03/02).	COM (2002) 122	http://europa.eu.int/eur-lex/en/com/rpt/2002/com2002_0122en01.pdf

Continued

Table 6.5.2 European Level Environment Policy. *Continued*

Policy Type	Programme Title	Policy Objective	Community Measure	Document Reference	Website
Miscellaneous	EMAS 2000 – A Dynamic Instrument for Environmental Protection & Sustainable Development: Business Benefits & Eco-management in the Mediterranean Region	A brochure specifically for the Mediterranean countries was developed, which explained the reasons to implement an environmental management system & described in detail the procedures to follow. This brochure also contained case studies & quotes from organisations that were EMAS registered.	DG Environment Brochure (2002).	▪ DG XI Brochure	http://www.europa.eu.int/comm/environment/emas/pdf/medbroch/emas_2000_en.pdf
	The State of EMAS in the EU Eco-Management as a Tool for Sustainable Development	This literature study aimed at giving an overview over the state of EMAS in the EU. It is based on studies from 11 Member States & other European countries as well as on publications of multiple international organisations.	DG Environment Report (June 2002).	▪ DG XI Report	http://europa.eu.int/comm/environment/emas/pdf/general/literature_study_020506.pdf
	Evaluation of Environmental Product Declaration Schemes	The aims of this project were to: ▪ Document and evaluate EPD systems (both national & industry) ▪ Compare them with each other & the current ISO TR 14025 ▪ Examine possible future directions for EPD systems in Europe ▪ Examine how best EPD systems should be integrated with other IPP tools.	EC DG Environment Final Report (Sept 2002).	▪ DG XI Report	http://europa.eu.int/comm/environment/ipp/pdf/epdstudy.pdf

Analysis of the 'Open List' of Environment-Related Headline Indicators	The analysis in this report has looked at existing known international data sources, at the data available from those sources, & at the definitions & methodologies applied, where these were accessible. Indicators have been considered feasible if they are based on sound science & up to date data is available from reliable sources, with a sufficient number of observations to identify trends. A further condition has been comparability across Member States &, as far as possible, other countries. Where the indicators or data are not particularly responsive, i.e. not able to change quickly in response to action & thus serve to monitor the effects of any policy measure or other change in circumstances, this has been pointed out. No explicit analysis of the policy relevance of the proposed indicators has been made.	Report from the Commission to the Council – Analysis of the 'Open List' of Environment-Related Headline Indicators (20/09/02).	COM (2002) 524	http://europa.eu.int/eur-lex/en/com/rpt/2002/com2002_0524en01.pdf
On the Application & Effectiveness of the EIA Directive (Directive 85/337/EEC as amended by Directive 97/11/EC) – How successful are the Member States in implementing the EIA Directive?	The European Commission, has prepared the following Five Year Report fulfilling the obligation found in Article 2 of Directive 97/11/EC & Article 11 paragraph 1 and 2 of Directive 85/337/EEC on the assessment of the effects of certain public & private projects on the environment (EIA Directive), which requires that 'five years after the notification of the Directive, the Commission shall send the European Parliament & the Council a report on its application & effectiveness. The report shall be based on exchange of information on experience gained in applying this Directive'.	Report from the Commission to the European Parliament & the Council On the Application & Effectiveness of the EIA Directive (Directive 85/337/EEC as amended by Directive 97/11/EC) – How successful are the Member States in implementing the EIA Directive? (2002).	DG XI Report	http://europa.eu.int/comm/environment/eia/report_en.pdf

Continued

Table 6.5.2 European Level Environment Policy. *Continued*

Policy Type	Programme Title	Policy Objective	Community Measure	Document Reference	Website
Miscellaneous	Europe's Environment: the Third Assessment	This report was prepared by the EEA for the Kiev Ministerial Conference in May 2003 as part of the 'Environment for Europe' process under the auspices of the UN Economic Commission for Europe (UNECE), with the support of the working group on environmental monitoring (WGEM). It was the third in a series, the first two published in 1995 & 1998 for the conferences in Sofia & Aarhus respectively. This third assessment differed in scope from previous reports by taking a more integrated approach both on environmental issues (e.g. combining inland & marine waters; assessing health & environment issues) & on the integration of environmental concerns into sectoral policies, reflecting policy developments in these areas.	European Environment Agency Report (June 2003).	▪ EEA Report	http://reports.eea.eu.int/environmental_assessment_report_2003_10/en/tab_content_RLR
	Review of the European Environment Agency (EEA)	The report reviewed the Agency's contribution to the Community's environment policy from 1994, when, its seat having been decided, it could start its operations, to 2003. It also formulates recommendations concerning EEA & its main stakeholders & partners with a view to improve the Agency's contribution to Community policy on the environment.	Report from the Commission to the Council on a review of the European Environment Agency (22/12/03).	▪ COM (2003) 800	http://europa.eu.int/eur-lex/en/com/rpt/2003/com2003_0800en01.pdf

A Sustainable Approach for the Environment – LIFE & the Community Eco-Management & Audit Scheme (EMAS)	This edition of *LIFE FOCUS* presented ten successful experiences selected from among the LIFE projects that have either prepared the way for EMAS or that have implemented it. There is no sign of this support abating, as 8 of the 109 LIFE-Environment projects adopted in 2002 are dedicated specifically to EMAS.	LIFE-Environment project report (2003).	▪ LIFE – Environment Report	http://europa.eu.int/comm/environment/life/infoproducts/emas_en.pdf
Implementation of Directive 2001/42 on the Assessment of the Effects of Certain Plans & Programmes on the Environment	The document is designed to help Member States, Acceding States & Candidate Countries understand fully the obligations contained in the Directive & assist them in transposing the Directive into their national law &, equally important, in creating or improving the procedures which will give effect to the legal obligations. It does not set out to explain how to carry out an environmental assessment although it does offer some practical advice on how certain requirements could be met.	Directorate General Environment Report (2003).	▪ DG XI Report	http://europa.eu.int/comm/environment/eia/030923_sea_guidance.pdf
Promoting Environmental Technologies: Sectoral Analyses, Barriers & Measures	The purpose of this report is to give an overview of the preparatory work carried out by the Sustainable Production & Consumption (SPC) Issue Group. At the same time, & in line with the essential character of the ETAP itself, it also aims at being a starting point for further initiatives by the private sector on the road to sustainable production patterns.	A Report from the Sustainable Production & Consumption Issue Group as a contribution to the Environmental Technologies Action Plan (2004).	▪ SPC Group Report	http://www.europa.eu.int/comm/environment/etap/pdfs/spc_finalreport_ipts.pdf

Table 6.5.3 UK Level Environment Policy.

Policy Type	Document Title	Objective	Contents	Website
Strategies & Reports	Consumer Products & the Environment (Oct 1998)	A key challenge in the drive towards development which is sustainable is how the environmental impacts associated with the production and consumption of the goods and services which meet human needs can be reduced. This paper tried to look in the round at approaches for doing that. The aim was to develop an integrated policy framework, helping the government to determine priorities and engage with partners in the action needed to deliver them.	Introduction; Section 1: Analysis; Section 2: Proposals on policy development & co-ordination; Section 3: Proposals & options on environmental information about products.	http://www.defra.gov.uk/environment/consult/consumerprod/index.htm
	Environmental Protection Act 1990: Code of Practice on Litter & Refuse (01/06/99)	The Environmental Protection Act 1990 imposes a duty on local authorities (and certain other landowners and occupiers) to keep specified land clear of litter and refuse so far as is practicable. This publication provides practical guidance on the discharge of these duties. The Code defines standards of cleanliness which are achievable in different types of location and under differing circumstances. It is concerned with how clean land is rather than how often it is swept. The Code does not therefore suggest cleaning frequencies. Rather it sets out how quickly differing types of land should be returned to a set cleanliness standard.	Introduction; Part 1: Code of practice on litter & refuse issued under section 89 of the EPA 1990; Part 2: Good practice advice; Chapter 1: Assessing the problem; Chapter 2: Taking action; Chapter 3: Involving the public; Chapter 4: Design against litter; Chapter 5: Making effective use of enforcement; Chapter 6: Beyond the litter duty; Annex A: Powers to tackle litter & refuse on private land.	http://www.defra.gov.uk/environment/localenv/litter/code/index.htm

United Kingdom Response to European Commission White Paper on Environmental Liability (Sept 2000)	Document stating that the UK fully supports the concept of environmental liability, & the need to ensure that those who cause damage to the environment should be made to pay for putting it right, affirming that the UK therefore welcomes the publication of the Commission's White Paper, & the consultation process which the Commission has started. The UK shall play an active & constructive part in helping to develop proposals, & judge the final package on its merits. While meeting the need to achieve equity, clarity and ease of effective implementation. It comments that important factors for safeguarding the environment will be clear responsibilities & workable definitions, and important factors for businesses will be the predictability of any regime, the extent to which it makes clear how they can manage their potential liability, & its compatibility with the national legal processes & traditions with which they are familiar.	Introduction; The case for EC action; Some general points; Comments on specific proposals; Definition of 'environmental damage' & related issues; Strict liability; Defences; Emissions explicitly allowed; Retroactivity; Dangerous activity; Valuation of biodiversity & scale of restoration to be required; Burden of proof; Role of public interest groups; Conclusion.	http://www.defra.gov.uk/environment/consult/liability/response/index.htm
Green Spaces, Better Places (15/11/01)	This final report of the Urban Green Spaces Taskforce had a simple message: now is the time for an urban renaissance with parks & diverse green spaces. The report set out a programme for national & local government to work in partnership with local communities, business, voluntary organisations & others to revitalise parks & green spaces. It summarised consideration of the issues, & set out conclusions & recommendations for reversing the well documented decline of urban parks & green spaces and to regenerate these priceless urban assets, & creating new & different ones which are attractive, accessible & sustainable.	Preface; Foreword; Executive summary; Part 1: Introduction; Section 1: Parks & green spaces & urban life; Part 2: Challenges & responses; Section 2: Issues & challenges; Section 3: Resources; Section 4: Building partnerships; Section 5: Supportive information & good practice; Part 3: Creating green space networks; Section 6: A strategic approach; Part 4: An urban renaissance with green spaces; Section 7: Making it happen; Section 8: Summary of recommendations; Annex 1: Acknowledgements; Annex 2: Responsibilities for urban green spaces at national level; Annex 3: Current green space initiatives & programmes; Annex 4: A national urban green spaces agency.	http://www.odpm.gov.uk/stellent/groups/odpm_urbanpolicy/documents/page/odpm_urbpol_607952.pdf

Continued

Table 6.5.3 UK Level Environment Policy. *Continued*

Policy Type	Document Title	Objective	Contents	Website
Strategies & Reports	Extended Partial Regulatory Impact Assessment on Proposals for a Directive on Environmental Liability (20/06/02)	This document constitutes an extended partial regulatory impact assessment of a proposal for a directive of the European Parliament & of the Council on environmental liability with regard to the prevention & remedying of environmental damage.	Executive summary; Proposed EU Directive on environmental liability; Purpose & intended effect of the measure; Options; The benefits of the options; Compliance costs for business, charities & voluntary organisations; Securing compliance; Competition assessment & impact on small businesses; Conclusions & recommendations.	http://www.defra.gov.uk/environment/consult/liability2/pdf/eu_env_liability_pria2.pdf
	Action Plan to Develop Organic Food & Farming in England (July 2002)	This Action Plan aims to identify what is required to ensure stable & strategic growth for the organic sector. It sets out a series of practical measures which the Government & the food & farming industry will take to encourage a sustainable organic farming & food sector in England.	Foreword; Introduction; Rationale; The Organic Action Plan Group; Priority action points; Areas for further work; Annex 1 Terms of reference and membership of Organic Farming and Food Action Plan team; Annex 2 Consolidated list of Action Plan Group's recommendations; Annex 3 Organic farming and the environment.	http://www.defra.gov.uk/farm/organic/actionplan/actionplan.pdf
	Living Places – Cleaner, Safer, Greener (31/10/02)	This document sets out the government's approach to making cleaner, safer, greener public spaces. It explains the need for new thinking to sort out the complex, overlapping responsibilities for public spaces, to remove the cumulative symptoms of degradation, to tackle the problems in deprived areas and to respond to changing times. It also identifies some key components that underpin successful schemes – committed leadership, strong partnerships, active community involvement, desire for quality and innovation and better communicating and sharing of ideas.	Ministerial foreword; Summary; Section 1: Setting the scene; Section 2: A constantly evolving landscape; Section 3: Urban parks & green spaces; Section 4: Continuous improvement; Annex 1: Funding & targets – delivering improvements; Annex 2: Government response to the recommendations of the Urban Green Spaces Taskforce; Annex 3: Rights & responsibilities – rationalising the legislative framework; Annex 4: Audit of main duties & powers.	http://www.odpm.gov.uk/stellent/groups/odpm_urbanpolicy/documents/downloadable/odpm_urbpol_607950.pdf
	Improving Urban Parks, Play Areas &	This research into urban green spaces was commissioned to inform the work of the Urban	Background; Key findings; Introduction; Definitions; Typology of urban green	http://www.odpm.gov.uk/stellent/groups/odpm_urbanpolicy/

Green Spaces (14/11/02)	Green Spaces Taskforce. The study looked at the uses of urban green spaces, barriers preventing use, the wider benefits to quality of life in urban areas & examples of innovative approaches to creation, management & maintenance.	spaces; Research methods; People & activities in urban green spaces; Non-users & barriers to use; The ideal urban green space; Adding up the benefits; Structures & mechanisms; Community involvement; Making things happen; A role in urban renewal; Overview & conclusions; Further information.	documents/page/odpm_urbpol_607951.hcsp
Paving the Way: How we Achieve Clean, Safe & Attractive Streets (2002)	The main aim of this report was to analyse why there was a failure to create & maintain quality streetscapes. It is not a design guide. Instead its main focus is the process of decision-making that has produced the kinds of streets we see in towns & cities. It endeavours to get behind the finished product to understand how & why it was created & how it is cared for – & by doing so, identify what the impediments are to the creation of good streetscapes.	Executive summary; Introduction; The quality of our streets; Case studies; The impediments to quality; Recommendations; Conclusions; Appendix 1: Case studies; Appendix 2: Management powers & duties for streets in England; Appendix 3: Lessons from abroad; Appendix 4: Legislation & guidance affecting the street; Appendix 5: Potential changes to guidance; Appendix 6: Consultees, project steering group & acknowledgements.	http://www.cabe.org.uk/pdf/Paving%20the%20way.pdf
The Strategy for Sustainable Farming & Food – Facing the Future (2002)	The challenge for the industry is how to create fresh prosperity, how to preserve and enhance the environment and contribute to stable rural communities, i.e. to rebuild a food and farming industry which is sustainable. The strategy for sustainable farming & food sets out how this agenda will be taken forward.	Prime Minister's foreword; Section 1 – The challenge of sustainability; Section 2 – A strategy for sustainability; Section 3 – Delivering change; Section 4 – Monitoring success; Section 5 – Related information; Annex A – Our approach to regulation.	http://www.defra.gov.uk/farm/sustain/newstrategy/strategy.pdf
Farming and Food's Contribution to Sustainable Development – Economic & Statistical Analysis (2002)	The structure of the paper is to look in turn at each of the three pillars of sustainable development and draw upon economic and statistical evidence to track the key changes in farming and food's contribution – including evidence upon a broad range of costs and benefits which result from current policies and practices. It acknowledges the significant interdependence between economic, environmental and social aspects and looks at competitiveness and the implications for farming's business prospects.	Section 1 – Farming & food's contribution to Sustainable Development: the current situation & future prospects; Section 2 – Using economic instruments to address the environmental impacts of agriculture; Section 3 – Sustainable farming & food strategy: a framework for evaluation & monitoring.	http://www.defra.gov.uk/farm/sustain/newstrategy/econ/contents.pdf http://www.defra.gov.uk/farm/sustain/newstrategy/econ/section1.pdf http://www.defra.gov.uk/farm/sustain/newstrategy/econ/section2.pdf http://www.defra.gov.uk/farm/sustain/newstrategy/econ/section3.pdf

Continued

Table 6.5.3 UK Level Environment Policy. *Continued*

Policy Type	Document Title	Objective	Contents	Website
Strategies & Reports	Ecological Assessment of Wetlands, Scrub & Unimproved Grassland, Heath or Moorland for EIA Screening Decisions (14/04/03)	The purpose of the Ecological Assessment, which will normally be undertaken by RDS (Rural Development Service) Ecologists, is two-fold: ▪ To determine whether any land included in the application falls within the scope of the regulations as uncultivated land or semi-natural habitat. This will be based on a field assessment; & ▪ To advise the RDS National Land Management Team whether any land falling within the scope of the regulations is of ecological significance. This will be based on a field assessment & on evidence submitted by statutory consultees.	Overview of ecological assessment procedure; Determination of unimproved grassland, heath & moorland; Determination of scrub; Determination of wetlands; Significance tests: all habitats; Special situations; References.	http://www.defra.gov.uk/environ/eia/field-guidance.pdf
	The Annual Local Environmental Quality Survey of England 2002/2003 (Sept 2003)	This is the second edition of the annual Local Environmental Quality Survey of England (LEQSE), which DEFRA has commissioned ENCAMS to develop & undertake on its behalf. ENCAMS is Environmental Campaigns Limited (the environmental charity formerly known as the Tidy Britain Group).	Foreword; Findings; Introduction; LEQ issues across landuse classes; Local environmental quality issues; Regional local environmental quality standards; Conclusions & priorities for action.	http://www.encams.org/information/publications/research/leqse2003.pdf
	Implementation of the Recommendation (2001/331/EC) on the Minimum Criteria for Environmental Inspections in the UK: Second Report to the Commission (Sept 2003)	The UK supports the principles & implementation of the Recommendation on minimum inspection criteria for environmental inspections, the purpose of which is to ensure that environmental inspection tasks should be carried out in Member States according to at least minimum criteria, thereby strengthening compliance with Community environmental law & contributing to a more consistent implementation & enforcement of that law.	Purpose of this report; Legislative scope of the recommendation; UK government & inspecting authorities; Co-operation & communication with other Member States; Data collection; Adoption of best practice as contained within recommendation for the implementation of future EU legislation; Annexes.	http://www.defra.gov.uk/environment/ea/mcei/second-report.pdf

Policies for Sustainable Consumption (Sept 2003)	The purpose of this document is to provide the Sustainable Development Commission with a more detailed guide to the complexity that lies behind this important policy debate. In particular, by highlighting some of the less obvious policy opportunities arising from a more detailed understanding of consumption, it aims to enhance the Commission's ability to exert a robust and creative influence on DEFRA's Strategy for Sustainable Consumption and Production.	Executive summary; Introduction; The sustainable consumption debate; Understanding consumption; Changing consumption patterns; Policy opportunities; References.	http://www.sd-commission.org.uk/news/resource_download_search.php?attach_id=OA492PF-KAOQ5IH-EAICDK6-H10C8GA
Changing Patterns UK Government Framework for Sustainable Consumption & Production (Sept 2003)	This strategy sets out a framework for how the government will take forward its commitment to sustainable consumption & production (SCP). The SCP framework set out in the current document develops this important element of sustainable development & will inform the forthcoming revision of the wider strategy. The social dimension of sustainable development, also relevant & important for SCP, is already addressed relatively fully in the Sustainable Development Strategy. The framework for SCP set out here is particularly concerned with the issues raised in the 'sustainable economy' chapter of the wider Sustainable Development Strategy. (see *Taking it On* consultation on p. 425)	Ministerial foreword; Executive summary; Sustainable consumption & production in context; Market framework & barriers; Policy & action under way; Next steps; Annexes.	http://www.defra.gov.uk/environment/business/scp/changing-patterns.pdf
The Value of Public Space – How High Quality Parks & Public Spaces Create Economic, Social & Environmental Value (Mar 2004)	Public space is all around us, a vital part of everyday urban life: an open-air living room, & outdoor leisure centre. This paper investigates how high quality parks & public space help create economic, social & environmental value.	Introduction; The Economic Value of Public Space; The Impact on Physical & Mental Health; The Benefits for Children & Young People; Reducing Crime & Fear of Crime; The Social Dimension of Public Space; Movement in & Between Spaces; Value from Biodiversity & Nature; References.	http://www.cabespace.org.uk/data/pdfs/TheValueofPublicSpace.pdf

Continued

Table 6.5.3 UK Level Environment Policy. *Continued*

Policy Type	Document Title	Objective	Contents	Website
Strategies & Reports	Sustainable Production & Consumption: Business Support Review Final Report (July 2004)	The government currently has a diverse range of programmes aimed at delivering resource efficiency in business (& the home). This complements other workstreams covering sustainable consumption, where the Sustainable Development Commission (SDC) & the Advisory Committee on Consumer Products & the Environment (ACCPE) have a particular interest. The following report provides an account of how this review of government programmes was conducted, & outlines the conclusions reached, including any lessons learned for shaping future support.	Foreword; Executive summary; Introduction; Conducting the review; Key findings; Conclusions & recommendations; Annexes.	http://www.defra.gov.uk/environment/business/scp/scp-bussupportrev.pdf
	Action Plan to Develop Organic Food & Farming in England – Two Years On (July 2004)	It is now two years since the launch, this progress report was produced with the Organic Action Plan Group, & it outlines progress & recommendations for further work, for all those involved in the production, preparation & sale of organic food.	Foreword; Part 1: Two Years On – The Action Plan to develop organic food and farming in England; Part 2: Action Plan progress report; Annex: Organic Food & Farming Action Plan Team; Terms of reference; Membership of the England Organic Food & Farming Action Plan Group.	http://www.defra.gov.uk/farm/organic/actionplan/actionplan2year.pdf
	Is the Grass Greener . . . ? Learning from International Innovations in Urban Green Space Management (July 2004)	This research from CABE Space brings together the experiences of 11 international cities, from Zurich to Melbourne in aspiring towards excellence in their green spaces & seeks lessons for practice in the UK. The report demonstrates what can be achieved & how the principles being pursued will lead to a transformation of towns & cities through the quality of their green spaces.	Introduction: the need for this report and who should read it; Why take an international view? How well is urban green space understood? What are the aspirations for urban green space? Who is responsible for urban green space management? How are management responsibilities co-ordinated and resourced? How is urban green space maintenance delivered? How are management practices applied to local contexts? What can we learn?	http://www.cabespace.org.uk/data/pdfs/grassgreenerreport.pdf

	The Environment in Your Pocket 2004 – Key Facts & Figures on the Environment of the United Kingdom (Oct 2004)	This is the eighth edition of our annual booklet of key environmental statistics, which includes eight of the 15 Government headline indicators of sustainable development. It is intended to be an easily accessible, handily sized, reference booklet, which will be particularly useful for ministers & officials across government, for students studying environmental topics & for members of the public with a general interest in environmental issues. It covers almost 60 key data series & focuses on providing trends over time, including performance against quantified targets & commitments set at the national & international level.	Introduction; Explanatory notes; Global atmosphere; Air quality; Inland water; Coastal & marine waters; Radioactivity; Noise complaints; Waste & recycling; Land; Wildlife; General.	http://www.defra.gov.uk/environment/statistics/eiyp/pdf/eiyp2004.pdf
	Reducing litter caused by 'food on the go' – A Voluntary Code of Practice for local partnerships (Nov 2004)	The aim of this Voluntary Code is not only to reduce levels of litter but also to promote the joint co-operation and community engagement that is key in driving up improvements in our local communities. It seeks to formulate local solutions to local problems noting that improvements will only come when issues affecting neighbourhoods are tackled in a mutually supportive way.	Ministerial Foreword; Introduction; Reducing Litter caused by 'Food on the Go' – A Voluntary Code of Practice for local partnerships; Framework; Step one – What particular issues are causing litter from products or around the business? Step two – The Solutions; Guidance on the Recommendations; Annexes.	http://www.defra.gov.uk/environment/localenv/litter/pdf/fastfoodcop.pdf
Consultation Papers	Living Places – Powers, Rights & Responsibilities (Oct 2002)	This consultation paper describes 27 options for reforming the legislative framework as part of the Government's strategy for clarifying & improving the statutory powers, rights and responsibilities associated with achieving cleaner & safer public spaces & local environments. The detail of the strategy & the context for the wider agenda are laid out in the accompanying publication *Living Places – Cleaner, Safer, Greener.*	Rights, powers, responsibilities – options for reforming the legislative framework; Main areas – views invited; Options summary; Realignment & extension of duties/powers; More robust/flexible powers; Reinforce LA strategic responsibility; Improved mechanisms for aggrieved persons to obtain redress; Improving communication: clearer guidance, working relationships & sharing best practice; Handling the consultation; Annex A: Audit of main duties/powers; Annex B: List of consultees.	http://www.defra.gov.uk/environment/consult/pubspace/pdf/living-places.pdf

Continued

Table 6.5.3 UK Level Environment Policy. *Continued*

Policy Type	Document Title	Objective	Contents	Website
Consultation Papers	Living Places – Powers, Rights & Responsibilities: Options for Reforming the Legislative Framework: Consultation Summary (July 2003)	The consultation produced options for delivering changes. Some of these options were contained in the consultation paper *Living Places – Powers, Rights, Responsibilities*. This paper now summarises the responses to that consultation exercise. It also details the comments received at regional workshops in the nine English regions conducted during the period March – May 2003.	Foreword; Background; General overview; Summary of options; Consultation responses; Regional workshops; Conclusions.	http://www.defra.gov.uk/environment/consult/pubspace/pdf/lp-responses.pdf
	Sustainable Consumption & Production Indicators – Joint DEFRA/DTI Consultation Paper on a Set of 'Decoupling' Indicators of Sustainable Development (Sept 2003)	This consultation document sets out the government's initial ideas for a 'basket' of 12 indicators by which the effectiveness of the planned actions for sustainable consumption & production can be measured & as a means for driving long-term improvements.	Introduction; Why have decoupling indicators? The concept of decoupling; The proposed basket of decoupling indicators & their relationship to other indicator sets; Updating; Targets & environmental limits; Presentation; Limitations; Aggregation; Gaps in data & selection criteria; Geographical coverage; Seeking views on proposals; Proposed basket of indicators; Consultation procedure; Annexes.	http://www.defra.gov.uk/corporate/consult/scp-indicators/consultdoc.pdf
	Fly-Tipping Strategy – Consultation Document (Feb 2004)	The government is developing this strategy to improve local environmental quality & deal with fly-tipping. This document also provides details of the other action being taken to help combat fly-tipping. It should also be noted that this strategy does not cover the issue of abandoned vehicles, since action is being taken separately to deal with these & they are covered by a different legislative framework.	Introduction; Tackling the illegal disposal of waste; The current situation; Potential future pressures on fly-tipping levels; Landfill directive; WEEE directive; Developing the fly-tipping strategy; Beginning to implement the strategy; More flexible & effective enforcement of the duty of care; Ensuring the main fly-tipping offence provides a greater deterrent; More robust powers for clearing waste from land; Consideration of the amount of construction, demolition & excavation waste being fly-tipped; Changes to secondary legislation;	http://www.defra.gov.uk/corporate/consult/flytipping-strategy/consultation.pdf

			Refinements to the way that stop, search & seizure powers work. Additional measures that do not need legislation; The consultation criteria.	
	Clean Neighbourhoods – Consultation Document (July 2004)	This document introduces a wide range of measures to bring about improvements to all aspects of our public space in a joined-up manner. It recognises the links that exist between local environmental quality & the wider quality of life agenda. It sets out powers that will greatly enhance the ability of those responsible for our public space to ensure it is clean, safe & green.	Foreword; Introduction; Communication strategy; Measures for consultation; Crime & disorder; Litter, graffiti & fly posting; Waste & contaminated land; Nuisance vehicles; Dogs; Fixed penalty notices; Nuisance; General public space; Appendices.	http://www.defra.gov.uk/corporate/consult/clean-neighbourhood/consultation.pdf
Guidance Notes	Guidelines for Environmental Risk Assessment & Management (Aug 2000)	The guidelines presented in this document provide a common framework for risk assessment as a key part of the process of appraisal for environmental decision-making. They build on the UK Department of the Environment's 1995 publication *A Guide to Risk Assessment and Risk Management for Environmental Protection*. The guidelines set out some basic principles which the Department of the Environment, Transport and the Regions (DETR) and the Environment Agency would normally intend to use in the assessment and management of environmental risks and which are recommended for all public-domain risk assessments. The guidelines provide decision-makers, practitioners and the public with a consistent language and approach for environmental risk assessment and management.	Foreword; Introduction to the guidelines; A framework for environmental risk assessment & management; The social aspects of risk; Problem formulation; Risk screening & prioritisation; Quantification & dealing with uncertainty; Evaluating the significance of a risk; Options appraisal & decision-making; Monitoring; Annexes.	http://www.defra.gov.uk/environment/risk/eramguide/index.htm
	Environmental Reporting – General Guidelines (Nov 2001)	These guidelines aim to help by cutting through some of the confusion & providing an introduction to environmental reporting. They explain how to produce an environmental report, outline its main contents & suggest key indicators to report against. They also include a possible incremental approach to help organisations plan staged improvements in their reporting. Although the guidelines are particularly aimed at organisations new to reporting, experienced reporters might also find them useful.	Introduction; The reporting process; The contents of an environmental report; Environmental performance indicators; Other issues.	http://www.defra.gov.uk/environment/envrp/general/pdf/envrptgen.pdf

Continued

Table 6.5.3 UK Level Environment Policy. *Continued*

Policy Type	Document Title	Objective	Contents	Website
Guidance Notes	EIA Guidelines – Environmental Impact Assessment for Use of Uncultivated Land or Semi-Natural Areas for Intensive Agricultural Purposes (01/02/02)	Environmental Impact Assessment (EIA) is a procedure for considering the potential environmental effects of land use change. The legislation has now been extended by bringing in EIA procedures for projects for the use of uncultivated land or semi-natural areas for intensive agricultural purposes. The regulations, while implementing specific European Community requirements, also contribute significantly to the wider aims of promoting sustainable agriculture; & a clearer understanding of good agricultural practice. The regulations also complement other environmental measures listed in Annex 1.	Introduction; How do these procedures link to other schemes & consent procedures? What types of land are covered? Definition of 'projects'; Procedures; Environmental statements; Appeals; Review; Annexes.	http://www.defra.gov.uk/environ/eia/guidelines.pdf
	Green Claims – Practical Guidance: How To Make A Good Environmental Claim (2003)	This guidance provides answers to some frequently asked questions & gives detailed advice & examples on how to make a good green claim; Giving clear, straightforward, environmental information, as outlined in this guidance, has benefits for both consumers & business alike. By providing information about the environmental effects & qualities of products, & how to use & dispose of them, environmental claims (sometimes called 'green' claims) help consumers to make informed buying choices. They also help to raise awareness of the issues, enhance consumer understanding & improve product standards overall. At the same time businesses can enhance their credentials & demonstrate that they are acting responsibly to consumers, other business partners & regulators.	Introduction; How to make a self-declared environmental claim; Existing environmental labelling & declaration schemes.	http://www.defra.gov.uk/environment/consumerprod/pdf/genericguide.pdf
	Strategic National Guidance – The Decontamination of the Open	The government has published this guidance to help local authorities, & others with responsibilities for protection of the public, to develop practical strategies for cleaning-up the open environment in	Introduction; Roles & responsibilities; Recovery principles & priorities; Factors influencing the recovery strategy; Priority actions for recovery; Management	http://www.defra.gov.uk/environment/risk/cbrn/cbrn-guidance.pdf

Environment Exposed to Chemical, Biological, Radiological or Nuclear (CBRN) Substances or Material (Mar 2004)	the event of releases of chemical, biological, radiological or nuclear (CBRN) substances, whatever their cause. It complements similar contingency planning guidance published by the Home Office. This guidance provides an agreed set of basic recovery principles & a shared understanding of the key issues that may need to be addressed.	approaches to environmental decontamination; Important local resources; Partner organisations; Annexes.	
Public Sector Food Procurement Initiative – Guidance for buyers & their Internal Customers: Advice for Public Sector Bodies on Integrating Sustainable Development into Food & Catering Services Contracts (May 2004)	The Public Sector Food Procurement Initiative (PSFPI) is designed to help the Government deliver its Sustainable Farming & Food Strategy (SFFS). This aims to achieve a world-class sustainable farming & food sector that contributes to a better environment & healthier & more prosperous communities. This guidance is to help buyers & their internal customers understand the issues & apply measures that are designed to facilitate implementation of this policy while adhering to good procurement practice & achieving value for money.	Part 1: Introduction; Part 2: Policy and Practice; Part 3: Action Sheet – Integrating sustainable development considerations; Part 4: Glossary.	http://www.defra.gov.uk/farm/sustain/procurement/pdf/foodprocure.pdf
A Guide to Producing Parks & Green Space Management Plans (May 2004)	This guide has been produced to enable anyone involved in the management of publicly accessible parks & green space to write management plans that help them to manage, maintain, develop & improve their green space in the most appropriate way.	Introduction to the guide; Who this guide is written for; How to use the guide; Part 1 – Planning the plan; Part 2 – Content & structure of the plan; Annex A – Heritage lottery fund management planning requirements.	http://www.cabespace.org.uk/data/pdfs/management_plan.pdf
Green Space Strategies: A Good Practice Guide (May 2004)	This guide is designed to steer authorities through the process of drawing up effective strategies based on clear assessments of stakeholders' needs & wishes. It will help provide a blueprint for working in partnership with other landowners & managers & with local communities to deliver excellent parks & green spaces now & in the future. It is aimed primarily at local government but its good practice advice will be useful to anyone with responsibility for the planning, design & maintenance of green spaces.	Foreword; Introduction; What is a green space strategy? Why prepare a green space strategy? What is involved in preparing a green space strategy? Checklist: What is included in a green space strategy? Annex 1 – Key references; Annex 2 – Green space audit methodology; Annex 3 – Community involvement; Annex 4 – Useful contacts.	http://www.cabespace.org.uk/data/pdfs/GSS_for_web.pdf

Table 6.5.4 Northern Ireland Level Environment Policy.

Policy Type	Document Title	Objective	Contents	Website
Strategies & Reports	Framing the View – Window Frames for a Sustainable Future (Jan 2000)	Window frames are an important environmental consideration, with the impact of modern building materials on the environment; and the waste & the toxins put into the atmosphere, rivers & the land with every piece of plastic developed. There is also the energy required to make a material with a limited life-span. This booklet examines some of the arguments but aims to inform rather than indoctrinate.	Introduction; Unplasticised Polyvinyl Chloride (PVC-u); The maintenance free myth; Price isn't everything; Wooden window frames: the long lasting old reliant; Wooden frames: part of Ireland's heritage; Extending the frame life: restoration; Life cycle considerations; PVC-u & the environment; Where does your expired PVC-u window frame go? Wood & the environment; Window frames & fire; Timber & PVC-u: the big picture; Conclusion.	http://www.ehsni.gov.uk/pubs/publications/Framing_the_View.pdf
	Environmental Monitoring – New Technologies for Monitoring: Status & Prospects (Feb 2001)	The purpose of this paper is to: ▪ Report on current monitoring methods & technologies in use; ▪ Identify areas where new technologies are being researched or tested for future application; and ▪ Identify areas of common interest for North/South co-operation.	Part 1: Introduction; Part 2: Monitoring emissions to air & water; Part 3: Monitoring the aquatic environment; Part 4: Monitoring air quality & deposition; Part 5: Co-operation on technologies/methodologies for monitoring.	http://www.ehsni.gov.uk/pubs/publications/North_South_Min_Coun_Doc.pdf
	Shared Horizons – Statement of Policy on Protected Landscapes in Northern Ireland (Feb 2003)	This statement sets out the issues associated with the protection & sustainable use of Northern Ireland's finest areas of rural landscape or countryside & indicates the way in which the Department of the Environment plans to address them. These areas fall within the internationally recognised description of 'Protected Landscapes'.	Introduction; Background to the designation & management of protected landscape areas; The designation of areas of outstanding natural beauty; Management of areas of outstanding natural beauty; National parks; Conclusion; Appendices.	http://www.ehsni.gov.uk/pubs/publications/SHARED_HORIZONS.pdf

A Forward Programme for the Declaration of Areas of Special Scientific Interest in Northern Ireland (Mar 2003)	What this paper aims to achieve is the establishment of a framework for taking the ASSI series towards substantive completion i.e. having declared more or less all of the area that can be reasonably predicted at this stage. It should therefore be noted that for some features further research will be required before a full complement of sites can be identified, & for others there may need to be a reassessment before the network is finalised.	Foreword; Executive Summary; Introduction; The Earth Science Programme; Biological Sites – Habitats; Biological Sites – Species; Discussion; References.	http://www.ehsni.gov.uk/pubs/publications/A_FORWARD_PROGRAMME_FOR_THE_DECLARATION_OF_ASSIs_IN_NI.pdf
Areas of Special Scientific Interest in NI – Proposals for their Future Management & Protection (Nov 2003)	Areas of Special Scientific Interest (ASSIs) are central to the delivery of nature conservation in Northern Ireland. Collectively they represent the best examples of natural habitats, sites which support rare or important species & earth science features present in Northern Ireland. Taking account of the many helpful comments & suggestions it received from previous consultations, the Department is now putting forward specific proposals for further consideration.	Introduction; Context; The notification of ASSIs; Protecting from damage; Achieving better management; Public bodies; The next steps; Annexes.	http://www.ehsni.gov.uk/pubs/publications/ASSI_Proposals.pdf
EHS Corporate & Business Plan 2003/2006 (Nov 2003)	The Corporate Plan covers the period up to 31 March 2006 & outlines the main themes of EHS (Environment and Heritage Service) work. These include protecting & improving the quality of air, water & land; promoting sustainable waste management; conserving biodiversity; protecting historic monuments & buildings of special architectural or historic interest; & raising awareness & appreciation of the environment & heritage.	Foreword; Introduction; Aims & objectives; key performance measures; The policy context; Corporate strategy; Business plan; Key performance targets & key management tasks; Resources; Key performance targets; Annexes.	http://www.ehsni.gov.uk/pubs/publications/CorporateBusinessPlan0306.pdf

Continued

Table 6.5.4 Northern Ireland Level Environment Policy. *Continued*

Policy Type	Document Title	Objective	Contents	Website
Strategies & Reports	The Management of Country Parks – A Statement of Policy (Mar 2004)	This document is not a statement of wider government policy on the provision of managed access to the countryside, but relates exclusively to the management by EHS of its own properties and the delivery of its countryside agenda to the visiting public. However others in the field of countryside recreation may find it includes some useful ideas which they may wish to incorporate into their own management regimes.	Executive Summary; Introduction; Historical Overview; A stimulating & safe environment in which to enjoy the countryside; Sound environmental principles with a presumption towards management to conserve the natural & historic heritage; Working in partnership with local communities & other stakeholders; A professional, friendly & efficient service consistent with the Customer Care Policy; A supportive working environment for staff to deliver the vision; Equality – a commitment to all; Best value for money; To provide a quality, safe environment for all to enjoy & appreciate countryside recreation; To provide appropriate countryside facilities; To provide exemplary habitat & heritage management to enhance biodiversity & to conserve the landscape & built environment; To provide enhanced appreciation of the countryside & opportunities to acquire the knowledge, values, attitudes & commitment needed to protect & improve the environment; To provide inspirational interpretation of the (natural & built) environment; To provide a positive 'interface' between EHS & the public; Appendices.	http://www.ehsni.gov.uk/pubs/publications/CPPolicyMarch04.pdf

	EHS Corporate & Business Plan 2004–2006 (July 2004)	The Business Plan for 2004/05 sets out the key targets, agreed with the Minister, & other targets which the EHS needs to achieve. Annual Reports will be published to present the agency's accounts & report on progress towards the achievement of its targets.	Introduction; Corporate plan 2004/06 aim & objectives; The policy context; Working with others; Corporate strategy; Business plan 2004/05; Targets 2004–05; Resources.	http://www.ehsni.gov.uk/pubs/publications/EHScorpbusplan0406.pdf
Consultation Papers	Proposal for a Directive on Environmental Liability with regard to the Prevention & Remedying of Environmental Damage (April 2002)	This consultation exercise invited comment on the European Commission's proposed Directive on Environmental Liability. It stated that the proposed Directive was a far-reaching initiative that would impact on a wide range of interests and, therefore, the Department needed to consult widely before forming a view on the Directive. The consultation response from Northern Ireland would help to inform the UK government's position ahead of negotiations on the final text of the Directive.	Consultation; Appendix A: Text of proposed directive; Appendix B: Summary & comparison with UK domestic liability arrangements.	http://www.ehsni.gov.uk/pubs/publications/Environmental_Liability_Consultation_Final.pdf
	The Draft Environmental Assessment of Plans & Programmes Regulations (Northern Ireland) 2004 – Consultation Paper (Mar 2004)	This consultation paper sets out the Department's intentions regarding the implementation of European Directive 2001/42 on the assessment of the effects of certain plans & programmes on the environment (known as the 'strategic environmental assessment' or SEA Directive).	Introduction; Background; Glossary; Scope of the directive; Exemptions; Who should carry out the SEA? The environmental report; Consultation; Decision-making; Monitoring; Guidance; Summary of the draft regulations; Following consultation; Annexes.	http://www.planningni.gov.uk/AreaPlans_Policy/SEA/SEA.pdf
Guidance Notes	Tackling Fly Tipping – A Guidance Note & Response Protocol (Mar 1999)	The purpose of this document is to provide operational guidelines, which may be referred to as a response protocol or best practice guidance, in respect of dealing with incidents of fly-tipping in Northern Ireland.	Introduction; Duties, powers & responsibilities in respect of fly-tipping; Duties/responsibilities & powers with regard to litter; A response strategy; An information strategy; Upon discovery fly-tipping; Appendices.	http://www.ehsni.gov.uk/pubs/publications/FlyTippingGuidance.pdf

Continued

Table 6.5.4 Northern Ireland Level Environment Policy. *Continued*

Policy Type	Document Title	Objective	Contents	Website
Guidance Notes	Best Practicable Environmental Option (BPEO) – Decision-Makers Guide (Sept 2000)	A guidance document on decision making for Best Practicable Environmental Option (BPEO) published by SNIFFER (Scottish Northern Ireland Forum for Environmental Research). This guidance document provides a stand-alone methodology to district councils in NI for determining BPEO, but may also be used as a complement to the SNIFFER guidance. Furthermore, there may be other approaches to identifying BPEO that the councils might choose to adopt.	Executive summary; Purpose; Introduction to BPEO; This guidance document; The BPEO process; Step 1: Define study objectives; Step 2: Identify decision criteria; Step 3: Develop options; Step 4: Apply constraints; Step 5: Evaluate options; Step 6: Weight decision criteria; Step 7: Generate final option scores; Step 8: Sensitivity analysis; Step 9: Create shortlist/iterate; Step 10: Identify the BPEO; Annex A: NI BPEO decision criteria; Annex B: Waste management methods; Annex C: BPEO practitioners' guide.	http://www.ehsni.gov.uk/pubs/publications/NI_BPEO_Guidance.pdf
	Compliance Policy & General Guidance – ASSIs (April 2003)	The paper outlines procedures which should be followed by officials involved in the investigation of possible offences & those required to consider evidence gathered through the investigative process and make recommendations on enforcement action to be taken. It embodies & takes cognisance of the general enforcement principles of the Enforcement Concordat that applies to Northern Ireland which requires enforcement activities to be carried out in a fair, equitable, practical & consistent manner.	Introduction; Compliance principles; Compliance action – general guidance; Prosecution action – enabling powers; Level of fines; Reasonable excuse within the legislation; Injunctions – prevention of damage; Investigation of possible offences; Preliminary investigation; Initial reports; Compliance action; Substantive investigation; Assessment by assistant director; Legal assessment by solicitors branch; Decision to prosecute; Key stages in the prosecution process – summary; Appointment of an investigation officer; Preparation for court: Annexes.	http://www.ehsni.gov.uk/pubs/publications/CompliancePolicy.pdf

7 Planning

7.1 Planning and the Built Environment

In terms of the built environment, the planning process plays a significant role in regulating the type, size, shape and spatial location of any development. It therefore becomes the strategic governing process by which any urban or rural setting spatially develops. The origins of the modern planning system lie in early public health and housing policies initiated to overcome some of the nineteenth century problems associated with the increase in population and rapid growth of towns and cities. Whilst the planning system has become more refined, to deal with the multitude of issues that affect modern society, it still retains that early focus of ensuring that development growth does not outstrip the ability of the infrastructure to keep pace.

Planning can be described as the process by which all new and existing spatial development is controlled, regulated and implemented in an urban or rural setting. Without a formal planning system any urban or rural location could become the victim of inappropriate, out of scale and highly intrusive development which is not only out of character with the area but also potentially damaging to its environment. It is then the task of the local planning authorities to ensure that the appropriate mechanisms are in place to guarantee that developments are in keeping with their immediate environment as well as their wider surrounding area.

The planning process can be instigated at many levels and stages depending on the anticipated impact of the development in question and the regulatory control necessary, including at project, city or urban, regional and national levels. Issues of regional or national importance are subject to a lengthy and onerous investigation usually involving either an Environmental Impact Assessment (EIA) or a Strategic Environmental Assessment (SEA) before approval, whilst smaller projects may only need to comply with local development plans and guidance before implementation. Adherence to pre-ordained plans and statutory guidance should ensure safe passage through the planning system, whilst failure to comply with the control mechanisms would result in enforcement proceedings being taken to secure compliance or the removal of the development.

The planning system puts in place the necessary statutory guidelines, which must be conformed to in meeting the basic need of 'adequate shelter for all'. In addition, modern planning systems have evolved to encompass issues as diverse as noise abatement, energy efficiency and the preservation of trees, offering the opportunity to impact on the character and design of the whole built environment. In this context the planning system is perfectly positioned to help instigate sustainable development through controlling developments that are granted planning permission and the imposition of sustainability compliant designs.

7.2 Key Planning Elements

The planning system is multi-faceted covering issues as varied as countryside protection to urban design. However whilst the issues or elements may be considered diverse, the common unifying premise is based on the need for development control. The following points provide a brief overview of the key elements, which form the basis of the modern planning system:

- *Countryside protection* – a key element of planning control involves ensuring that the countryside is not unduly damaged or compromised by new or existing developments. This includes introducing mechanisms such as Greenbelts, Areas of Outstanding Natural Beauty (AONB) and Areas of Special Scientific Interest (ASSI), which place severe restrictions on development occurring within these areas.
- *Housing* – the planning system sets the guidelines on where housing is located, the number of houses that can be constructed and the individual design details to be adhered to including building height, choice of materials, size and layout, and orientation. The planning system also played a key role in co-ordinating the removal of slum housing, a legacy of the industrial revolution when housing was constructed in an uncontrolled manner.
- *Conservation* – a key element of the planning system is the emphasis placed on conservation and the preservation of the existing townscape as a legacy of a bygone era, as well as maintaining rural character. This includes various mechanisms such as Listed Building Orders, Tree Preservation Orders and Conservation Areas, which place restrictions on the changes that can be made to buildings, trees and the townscape in certain areas.
- *Brown-field redevelopment and contaminated land* – where possible the planning system encourages the reuse of existing buildings or the redevelopment of brown-field land in preference to the green-field option. Similarly, the clean-up of contaminated land to bring such land back into viable use is an integral part of planning control and the drive towards sustainable development.
- *Urban design and urban form* – there is an increasing emphasis on ensuring that the location of buildings and their design contribute to the overall aesthetic quality of an area. This is promoted through the planning system in terms of building to appropriate densities, the compactness of urban form and the use of local materials which blend into their surroundings.
- *Retail and town centre development* – the planning system regulates out-of-town shopping complex applications to ensure that any such schemes take into account accessibility by public transport and the relative impact on town centre developments. The planning system also regulates the amount of car-parking space developed out-of-town and within town centres in a direct response to discouraging the use of private transport.
- *Regulatory documents* – various key documents form the basis of the regulatory control of the planning system. These include development plans, structure plans, local plans, guidance notes, planning policy statements, development control advice notes and planning applications.

Similarly, planning has an impact on other built environment areas, through the control of development, such as:

- *Energy* – increasingly developers are required to meet energy efficiency standards as laid down in the statutory guidance in terms of designing new build or redeveloped buildings.
- *Social Issues* – the planning system has a well-established method of ensuring that the general public get the opportunity to participate fully in offering opposition to undesirable development through planning appeals, third party appeals and public inquiries. This ensures that the public feel a sense of inclusion and can freely exchange their views on any planning application.
- *Transport* – the planning process has a direct impact at the strategic level on the placement of development in accordance with matters such as accessibility by public transport, car-parking provision, overall traffic management and infrastructure planning.
- *Urban Development* – the planning system has a direct impact on urban development through the restrictions it places on the market value of land. The planning controls which are placed on land can either adversely or positively impact on value by influencing the type, size and layout of development which can be erected on an individual building plot.

- *Waste* – the planning system will help implement the move away from landfill of waste towards sustainable waste management practices including the 'self-sufficiency' principle and the proximity principle of disposing of waste as close to the place of production as possible.

7.3 Sustainability Challenges for Planning

Planning is one of the key processes that can be used to help facilitate and implement sustainable development. If planning is to retain this key strategic role it will need to respond to the economic, environmental and social aspects of sustainability in equal measure. Whilst the planning system has obvious links to environmental and social issues it can also impact on the economic side of new or existing developments. In fact the planning process can exert influence on all four of the sustainable development objectives, as follows:

- *Effective protection of the environment* through the effective use and management of the natural and built heritage. Planning can help encourage environmental protection by preventing green-field development, ensuring mature trees are preserved, stipulating the need for environmental management systems in new builds to help reduce harmful emissions and by ensuring that spatial planning mitigates the need to travel by private transport.
- *Social progress which recognises the needs of everyone* is closely linked to planning through the need to secure a democratically accountable planning process, which encourages and facilitates general public participation. Mechanisms such as planning appeals and public inquiries help stimulate community involvement whilst instruments such as the EQIA (Equality Impact Assessment) help safeguard the views of various minority groups by promoting equality of opportunity for all.
- *Prudent use of natural resources* is a key part of the planning system as it governs the use of land, of which there is a finite amount. It is therefore important that the planning system controls the development of land and also helps increase productivity – achieved through increased densities, infill development or shared surface measures. Similarly, the planning system promotes the reuse of brown-field sites and buildings, thereby reducing the demand for green-field land and building materials such as virgin aggregates, as well as reducing the amount of demolition waste going to landfill.
- *Maintenance of high and stable levels of economic growth and employment* impacts on the planning system through the job creation associated with all disciplines within the construction industry and the ability of an area to attract inward investment. An area's strategic planning, its zoned development and the extent to which development has been effectively supplemented by the appropriate infrastructure planning all heavily influence economic growth. Planning also relates to economic growth through its impact on land values and the opportunity for the planning authorities to negotiate 'planning gain'.

The persuasive sustainability argument has placed new challenges on the planning system, which need to be responded to if society is to embrace the sustainable development aspiration. These challenges are varied and some require high level strategic action whilst others impact directly on the individuals when making decisions within the development process. The following list encompasses the key sustainability challenges that are likely to be faced, now and in the future, by the planning system:

- *Reconcile urban growth and limited land availability* – the increasing population trends are placing increasing pressure on green-field land. The planning system has a key role to play in reconciling urban growth and land availability through encouraging development on brown-field land and the reuse of existing buildings, thereby reducing pressure on greenbelt land.
- *Improve access to rural areas without compromising countryside protection* – there is a fine dividing line between securing environmental sustainability and rural protection and opening these areas up for more people to enjoy. The planning system has a responsibility to secure the protection of each rural area and its wildlife from any adverse human interference whilst ensuring that local communities can benefit from what the countryside has to offer.

- *Encourage sustainable settlement patterns* – one of the biggest threats to our countryside is the growth of urban sprawl and its associated environmental problems such as CO_2 emissions and pollution. The planning system will have to promote sustainable settlement patterns that encourage inner city living or home working to try and prevent suburbanisation and large-scale commuting to work.
- *Be mindful of flood threatened areas* – the planning system has a responsibility to minimise and safely control development in flood threatened areas, as the impacts of climate change are likely to increase the frequency of severe floods.
- *Secure continued community participation* – a key element of the modern planning system is its openness and ability to make the general public feel included in the process. It is very important that this trend is continued to help contribute to social sustainability.
- *Facilitate integrated infrastructure planning and spatial development* – the ability of the planning system to influence the location of future developments near main transport nodes or integrated with future infrastructure planning will make a significant contribution to achieving sustainable development.
- *Encourage high quality design* – high design standards in terms of the quality of materials selected and the ability of the designs to become adaptable for other uses and meet long term needs will become a key part of sustainable building. The planning system can help facilitate this through stipulating the design standards necessary for the achievement of planning permission.
- *Stipulate use of local materials and local suppliers (where possible)* – the economic sustainability of small to medium businesses requires support from local industry, to avoid or prevent the importing of goods and services from abroad, which not only takes wealth out of an area but also utilises high amounts of embodied energy and generates CO_2 emissions to transport them to the business location.

7.4 Planning Theme Index

Jurisdiction	Tier	Title	Date	Page No.
UK	Strategies & Reports	Planning for Communities of the Future	1998	238
		Planning for Sustainable Development – Summary	1998	238
		Training for Urban Design	2000	239
		Habitat UK National Report 2001	2001	239
		Planning Green Paper – Planning: Delivering a Fundamental Change	2001	239
		Delivering Affordable Housing Through Planning Policy	2002	240
		Sustainable Communities – Delivering through Planning	2002	240
		Making the System Work Better: Planning at Regional & Local Levels	2002	241
		Strategic Planning for Sustainable Waste Management	2002	241
		Contributing to Sustainable Communities – A New Approach to Planning Obligations	2003	241
		Community Involvement in Planning: The Government's Objectives	2004	242
	Consultation Papers	Planning for the Supply of Aggregates in England	2000	242
		Draft Planning Policy Statement 7: Sustainable Development in Rural Areas	2003	242
		Consultation on Draft Planning Policy Statement 22: Renewable Energy	2003	242
		Consultation on Draft Planning Policy Statement 6: Planning for Town Centres	2003	243
		Consultation Paper on Planning Policy Statement 1: Creating Sustainable Communities	2004	243
		A Draft Practical Guide to Strategic Environmental Assessment (SEA)	2004	243
		Consultation Paper on Planning Policy Statement 10: Planning for Sustainable Waste Management	2004	244
	Guidance Notes	Planning Policy Guidance Note 14: Development on Unstable Land	1990	244
		Planning Policy Guidance Note 18: Enforcing Planning Control	1991	244
		Planning Policy Guidance Note 19: Outdoor Advertisement Control	1992	245
		Planning Policy Guidance Note 20: Coastal Planning	1992	245
		Planning Policy Guidance Note 21: Tourism	1992	245
		Planning Policy Guidance Note 5: Simplified Planning Zones	1992	246
		Planning Policy Guidance Note 6: Town Centres & Retail Developments	1996	246
		Proposals for a Good Practice Guide on Sustainability Appraisal (Baker Associates Report)	1999	246
		Planning Policy Guidance Note 10: Planning & Waste Management	1999	247
		Planning Policy Guidance Note 12: Development Plans	2000	247
		Planning Policy Guidance Note 3: Housing	2000	247
		Controlling the Environmental Effects of Recycled & Secondary Aggregates Production: Good Practice Guide	2000	248
		By Design – Urban Design in the Planning System: Towards Better Practice	2000	248

Jurisdiction	Tier	Title	Date	Page No.
UK	Guidance Notes	Sustainability Appraisal of Regional Planning Guidance: Good Practice Guide	2000	248
		Tapping the Potential – Assessing Urban Housing Capacity: Towards Better Practice	2000	249
		Planning Policy Guidance Note 4: Industrial and Commercial Development and Small Firms	2001	249
		Planning Policy Guidance Note 2: Green Belts	2001	250
		Planning Policy Guidance Note 25: Development & Flood Risk	2001	250
		Planning Policy Guidance Note 1: General Policy and Principles	2001	251
		Planning Policy Guidance Note 8: Telecommunications	2001	251
		Planning Policy Guidance Note 16: Archaeology & Planning	2001	251
		Planning Policy Guidance Note 24: Planning & Noise	2001	252
		Better Places to Live by Design: A Companion Guide to PPG3	2001	252
		Planning Policy Guidance Note 9: Nature Conservation	2002	253
		Mineral Planning Guidance Note 1: General Considerations & the Development Plan System	2002	254
		Circular 06/1998 Planning and Affordable Housing	2002	254
		Planning Policy Guidance Note 17: Planning for Open Space, Sport & Recreation	2002	255
		Planning Policy Guidance Note 13: Transport	2002	255
		Planning Policy Guidance Note 15: Planning & the Historic Environment	2002	256
		Planning Circular 01/02: The Town & Country Planning (Residential Density) (London & South East England) Direction 2002	2002	256
		Monitoring Regional Planning Guidance	2002	256
		Planning & Access for Disabled People: A Good Practice Guide	2003	257
		The Strategic Environmental Assessment Directive: Guidance for Planning Authorities	2003	257
		Environmental Impact Assessment: Guide to Procedures	2004	258
		Note on EIA for Local Planning Authorities	2004	258
		Safer Places – The Planning System & Crime Prevention	2004	259
		Planning Policy Statement 7: Sustainable Development in Rural Areas	2004	259
		Planning Policy Statement 22: Renewable Energy	2004	260
		Planning Policy Statement 11: Regional Spatial Strategies	2004	260
		Planning Policy Statement 12: Local Development Frameworks	2004	261
		Planning Policy Statement 23: Planning & Pollution Control	2004	261
NORTHERN IRELAND	Strategies & Reports	The Planning Strategy for Rural Northern Ireland	1993	262
		Planning Policy Statement 3: Roads Consideration	1996	262
		Planning Policy Statement 5: Retailing & Town Centres	1996	263
		Planning Policy Statement 4: Industrial Development	1997	263

Jurisdiction	Tier	Title	Date	Page No.
NORTHERN IRELAND	Strategies & Reports	Planning Policy Statement 2: Planning & Nature Conservation	1997	264
		Your Permitted Development Rights & Environmental Assessment	1997	264
		Tree Preservation Orders	1997	264
		Environmental Assessment	1997	265
		Planning Policy Statement 1: General Principles	1998	265
		Planning Policy Statement 6: Planning, Archaeology & Built Heritage	1999	265
		Planning Policy Statement 8: Open Space, Sport & Recreation	1999	266
		Planning Policy Statement 9: The Enforcement of Planning Control	2000	266
		Planning Policy Statement 7: Quality Residential Developments	2001	267
		Belfast Metropolitan Area Plan 2015 (BMAP) Issues Paper	2001	267
		Planning Policy Statement 10: Telecommunications	2002	267
		Planning Policy Statement 11: Planning & Waste Management	2002	268
		Regulatory Impact Assessment of Third Party Appeals	2004	268
		Draft Belfast Metropolitan Area Plan 2015	2004	268
		Technical Supplements to the Draft Belfast Metropolitan Area Plan 2015	2004	268
	Consultation Papers	Draft Planning Policy Statement 12: Housing in Settlements	2002	269
		Public Consultation Draft – PPS3: Access, Movement & Parking	2002	269
		Draft Planning Policy Statement 13: Transportation & Land-use	2002	270
		Planning Policy Statement 4: Industry, Business & Distribution – Public Consultation Draft	2003	270
		Development Control Advice Note 11: Access for All, Designing for an Accessible Environment – Public Consultation Draft	2003	270
		Reforming Planning: Proposals to Amend Primary Planning Legislation in Northern Ireland – Public consultation paper	2004	271
		Public Consultation Draft – Planning Policy Statement 15 (PPS15) Planning & Flood Risk	2004	271
		Public Consultation Draft: Areas of Townscape Character – Addendum to PPS6 Planning, Archaeology & the Built Heritage	2004	271
	Guidance Notes	Development Control Advice Note 11: Access for People with Disabilities	1999	272
		Development Control Advice Note 10: Environmental Impact Assessment	1999	272
		Creating Places – Achieving Quality in Residential Developments	2000	273
		Improving the Quality of Housing Layouts in N.I.	2000	273
		Development Control Advice Note 8: Housing in Existing Urban Areas	2002	274
		Planning Policy Statement 8 – Open space, sport & outdoor recreation	2004	274

7.5 Planning Policy Tables

Table 7.5.1 Eurpopean Level Planning Policy.

Policy Type	Programme Title	Policy Objective	Community Measure	Document Reference	Website
Council Directive	The Assessment of the Effects of Certain Plans & Programmes on the Environment	This Directive provides for a high level of protection of the environment & contributes to the integration of environmental considerations into the preparation & adoption of plans & programmes with a view to promoting sustainable development by ensuring that an environmental assessment is carried out of plans & programmes likely to have significant effects on the environment.	Directive 2001/42/EC of the European Parliament & of the Council on the assessment of the effects of certain plans and programmes on the environment (27/06/01).	■ Off Jnl L197/30 21/07/01	http://europa.eu.int/eur-lex/pri/en/oj/dat/2001/l_197/l_19720010721en00300037.pdf
Council Communications	Towards an Urban Agenda in the EU	This Communication examined possibilities for improving urban development & for increasing the effectiveness of existing Community intervention in urban areas. The intention was not to develop European wide urban policies for matters best dealt with at a local or regional scale, but, since cities in the EU face a number of common problems, to look at opportunities at the European scale to share & facilitate potential solutions. This would not require additional powers at the European level, instead seeing what can be achieved through a more focused approach using existing instruments at national & Community level & enhanced co-operation & co-ordination at all levels.	Communication from the Commission on an urban agenda for the European Union (06/05/97).	■ COM (97) 197	http://europa.eu.int/comm/regional_policy/sources/docoffic/official/communic/pdf/urban/urban_197_en.pdf
	First Progress Report on Economic & Social Cohesion	This progress report on cohesion had two principal objectives: ■ to update the analysis of economic & social cohesion presented in the Second cohesion report published in January 2001, including an analysis of disparities in a Europe of 25 in the light of the potential enlargement with 10 new Member States;	Commission Communication – first progress report on economic & social cohesion (30/01/02).	■ COM (2002) 46	http://europa.eu.int/comm/regional_policy/sources/docoffic/official/reports/pdf/interim1/com_2002_046_en_acte.pdf

		■ to outline the state of the debate on future cohesion policy for the period after 2006, which began with the publication of the Second report & to prepare the next steps.			
Miscellaneous	First Official Draft of the European Spatial Development Perspective (ESDP)	The First official draft of the ESDP was intended as a common frame of reference to round off the period of reflection which the ministers of spatial planning had been undertaking since 1993. It addresses three main questions of common concern throughout the European continent: ■ Setting up a balanced & polycentric urban system; ■ Parity of access to infrastructure & knowledge & prudent management; and ■ Sustainable development of the natural & cultural heritage.	Summary report of ESDP (10/06/97).	■ Summary of Noordwijk document	http://www.kepemep-cree.org/PDF_Documents/European_Policies/Spatial_Planning/SDEC_up_to_1998/3-sdec-en.pdf
	Concept on the Establishment of the European Spatial Planning Observatory Network (ESPON)	A concept called *European Spatial Planning Observatory Network* (ESPON) was provided to Ministers in December 1997 to be ready for the implementation in the first half of 1998. It set out the tasks and areas of activity (Section II), the organisation (Section III), the financing (Section IV) and the launch of the ESPON (Section V).	Draft document for the meeting of the Committee for Spatial Development (25/11/97).	■ Concept Draft Report for CSD	http://europa.eu.int/comm/regional_policy/sources/docoffic/official/reports/pdf/orat_en.pdf
	The ESDP: A Strategy for Balanced & Sustainable Development of the EU – Synthesis report of the trans-national seminars	A series of trans-national seminars was organised by the European Commission during 1998 in different parts of the Union. This provided the opportunity for those directly involved in spatial planning at national & regional levels in Member States to give their views on the ESDP & how it might be refined & improved. This report synthesised the main ideas & suggestions from the trans-national seminars.	Report on the Trans-national Seminars on the ESDP (European Spatial Development Perspective) (1998).	■ EC Report	http://www.kepemep-cree.org/PDF_Documents/European_Policies/Spatial_Planning/SDEC_up_to_1998/2-session1-sdec_en.pdf

Continued

Table 7.5.1 Europpean Level Planning Policy. *Continued*

Policy Type	Programme Title	Policy Objective	Community Measure	Document Reference	Website
Miscellaneous	Report on Community Policies and Spatial Planning	This report was a first attempt by the services of the Commission to increase the awareness of the territorial dimension in the formulation of new policy guidelines as well as in the implementation of current Community policies. It was the result of eight months' joint work of representatives from 19 Directorates-General of the European Commission, EUROSTAT (the Statistical Office of the European Communities) & the EEA (European Environment Agency), and had three objectives: ▪ to provide a first answer to requests of the European Institutions & the Committee on Spatial Development to the Commission to establish a framework for co-ordinating Community policies; ▪ to contribute to the elaboration of the first final version of the ESDP which was envisaged in spring 1999; & ▪ to provide an input to the final ESDP Forum scheduled for February, 1999.	Working Document of the Commission Services (1998).	▪ Commission Report	http://europa.eu.int/comm/regional_policy/sources/docoffic/official/reports/pdf/coordfon/report_en.pdf
	Opinion of the Committee of the Regions on the European Spatial Development Perspective	The decision to frame the ESDP reflected the EU's determination to meet world economic challenges at a time when it was boosting its economic integration and recognised the growing role of local & regional authorities in spatial development. This determination went hand in hand with the three fundamental goals of economic & social cohesion, sustainable development, & competitive but mutually supportive regions.	Opinion of the Committee of the Regions on the European Spatial Development Perspective (1999/C93/07) (06/04/99).	▪ Off Jnl C93/37	http://www.kepemep-cree.org/PDF_documents/European_Policies/Spatial_Planning/SDEC_1999/AVIS-CR-EN.PDF

European Spatial Development Perspective: Towards Balanced & Sustainable Development of the Territory of the European Union	To define at Union level policy objectives & general principles of spatial development to ensure the sustainable balanced development of the European territory which respects its diversity.	Prepared by the Committee on Spatial Development – Potsdam Council (10/05/99).	ESDP report	http://europa.eu.int/scadplus/leg/en/lvb/g24401.htm http://europa.eu.int/comm/regional_policy/sources/docoffic/official/reports/pdf/sum_en.pdf
Multi-Level Governance: Linking & Networking the Various Regional & Local Levels	Sustainable & balanced development of the EU's territory was the subject area selected in order to examine new ways of linking various decision-making levels. The objective was to examine & propose co-operation mechanisms that would improve the ways in which the different levels of decision-making are interlinked. The underlying rationale is that powers are shared & not hierarchically organised. The challenge is to find ways of co-operating & organising new complementarities between the various levels, to provide them with better access to the decision-making process & also to help each level to strengthen the effectiveness, coherence & transparency of the whole system.	Working Group 4c Report (May 2001).	Working Group Report (Group 4c)	http://europa.eu.int/comm/governance/areas/group10/report_en.pdf
Implementation of Dir 2001/42 on the Assessment of the Effects of Certain Plans & Programmes on the Environment	The document is designed to help Member States, Acceding States & Candidate Countries understand fully the obligations contained in Dir 2001/42/EC & assist them in transposing the Directive into their national law &, equally important, in creating or improving the procedures which will give effect to the legal obligations. It does not set out to explain how to carry out an environmental assessment although it does offer some practical advice on how certain requirements could be met. In conjunction with national guidance prepared by Member States, it should also be of use to authorities which have to apply the Directive when preparing their plans & programmes.	Commission Guidance on implementing Dir 2001/42/EC (2001).	Commission Guidance	http://europa.eu.int/comm/environment/eia/030923_sea_guidance.pdf

Table 7.5.2 UK Level Planning Policy.

Policy Type	Document Title	Objective	Contents	Website
Strategies & Reports	Planning for Communities of the Future (23/02/98)	This document was the first major step in the drive to make towns & cities more attractive & sustainable, & so reduce the need for development to spread into the countryside. It was to ensure that the best use is made of the many thousands of previously developed (or 'brown-field') sites around the country with local authorities, English Partnerships & others taking a lead role in that work. Where some development is still needed outside urban areas there should be steps to ensure that it is carried out in the most sustainable way to ensure the continuing vitality of rural life, combined with an active approach to the protection of the countryside.	Foreword; Introduction; The context; Household growth; Housing provision & patterns of development; The green paper consultation; The economics of household growth & housing supply; The way forward; A new way to assess & cascade the housing figures; Reusing previously developed land; The options – the government's preferred approach; Other options – housing policy; Improving the planning tools; References.	http://www.odpm.gov.uk/stellent/groups/odpm_planning/documents/pdf/odpm_plan_pdf_606125.pdf
	Planning for Sustainable Development – Summary (Oct 1998)	The Guide concentrated on those planning policy areas where sustainable development raised new issues for planners, or required a new or revised approach. It also suggested a systematic method to assist local authorities in integrating sustainable development into their development plans. As part of this, the longer term and secondary effects of development needed to be recognised and that good planning also depends on effective participation by the local community. The Guide offered suggestions on how individuals and communities could be helped to play their part in the process.	Ch 1: Purpose & principles of the guide; Ch 2: Realising the potential of existing urban areas; Ch 3: Growing new urban areas; Ch 4: Revitalising & enhancing the countryside; Ch 5: Incorporating other sustainability issues; Ch 6: Methodology for integrating sustainable development into plans.	http://www.odpm.gov.uk/stellent/groups/odpm_planning/documents/pdf/odpm_plan_pdf_606393.pdf

Training for Urban Design (Oct 2000)	The terms of reference for this project were 'to clarify the "baseline" on urban design skills in terms of the training received by those who influence urban design'. The report was prepared in a very short space of time relying on published sources including the professional guidelines for accredited courses by the UDAL professions, giving information about the content & structure of accredited courses, the specialised urban design courses, & provision for continuing professional development taken from the professional institutions, university handbooks, prospectuses & the web pages.	Executive summary; Introduction; The nature of urban design; Urban design & initial professional education; Post initial professional education & urban design; Findings & conclusions; References.	http://www.odpm.gov.uk/stellent/groups/odpm_planning/documents/pdf/odpm_plan_pdf_606435.pdf
Habitat UK National Report 2001 (15/10/01)	This report reviewed UK policies and actions at home and overseas, demonstrating progress contributing to the achievement of shelter for all and the development of sustainable cities. It showed how common problems result from similar causes, even if solutions may vary.	Foreword; Introduction – UK action in a global context; International urban development; Adequate shelter for all; Sustainable cities & settlements; Economic growth & poverty elimination; Equality & social cohesion; Urban governance; UK implementation; Key references; Glossary & list of acronyms.	http://www.odpm.gov.uk/stellent/groups/odpm_planning/documents/pdf/odpm_plan_pdf_606539.pdf
Planning Green Paper – Planning: Delivering a Fundamental Change (12/12/01)	A Green Paper with the objectives of: ■ good planning to deliver sustainable development, to harness growth to build a better future, and ■ a better, simpler, faster, more accessible system that serves both business and the community.	Foreword; Chapter 1: The planning system we need: our objectives for the planning system; Chapter 2: The need for change; Chapter 3: How the planning system works; Chapter 4: A fundamental change for plans; Chapter 5: A fundamental change in development control; Chapter 6: Fundamental change at national, regional & local level; Sending in your views; Appendix.	http://www.odpm.gov.uk/stellent/groups/odpm_planning/documents/pdf/odpm_plan_pdf_605832.pdf

Continued

Table 7.5.2 UK Level Planning Policy. *Continued*

Policy Type	Document Title	Objective	Contents	Website
Strategies & Reports	Delivering Affordable Housing Through Planning Policy (Mar 2002)	This research report was an important contribution to the Government's on-going work to address the need for more affordable housing. The research was carried out during 2000 & 2001 & examined the effectiveness of current planning policy guidance on affordable housing (set out in Planning Policy Guidance note 3: Housing & Circular 6/98, *Planning & Affordable Housing* (see below)). The report addressed a range of issues in respect of the use & implementation of the policy across five English regions. Based on the findings, it included examples of, & recommendations for, better practice in using existing policy guidance to best effect to deliver affordable housing.	Preface; Executive Summary; Chapter 1 – Introduction; Chapter 2 – Definition of Affordable Housing; Chapter 3 – Housing Needs Assessment; Chapter 4 – Significance of Affordable Housing Provision; Chapter 5 – Thresholds & Targets; Chapter 6 – Supplementary Planning Guidance; Chapter 7 – Section 106 Agreements; Chapter 8 – Economics of Development; Chapter 9 – Use of Commuted Sums; Chapter 10 – Working with Registered Social Landlords; Chapter 11 – Monitoring of Delivery; Chapter 12 – Corporate & Collaborative Working; Chapter 13 – Rural Issues; Chapter 14 – Towards Better Practice; Appendices.	http://www.odpm.gov.uk/stellent/groups/odpm_planning/documents/pdf/odpm_plan_pdf_606099.pdf
	Sustainable Communities – Delivering through Planning (18/07/02)	This paper set out government plans for transforming the planning system to be faster, fairer and more predictable and to contribute to the delivery of government objectives.	Our aims; Where are we now; What happens next; Planning for a purpose; Making the system work better at the national level; Improving national planning policy; Improving government performance; Major infrastructure projects; Making the system work better – planning at regional & local level; Regional spatial strategies; Local development frameworks; Making the system work better at local level – improved implementation; Overcoming the obstacles to land assembly; Encourage high quality development in the right places to meet local needs; Give greater certainty over what development may be permitted; Planning obligations; More efficient processing of planning applications; Appeal process improvements; The role of statutory consultees; Helping communities; Enforcement; The need for culture change; Best value; Tools for the job; Skills; Resources; Taking things forward; Conclusions.	http://www.odpm.gov.uk/stellent/groups/odpm_planning/documents/pdf/odpm_plan_pdf_605857.pdf

Making the System Work Better: Planning at Regional & Local Levels (July 2002)	The above Planning Policy Statement, *Sustainable Communities – Delivering through Planning*, set out the Government's plans for reform of the planning system. This was in the light of responses received to the Green Paper *Planning: delivering a fundamental change* (see above), including the recommendations of the Transport, Local Government & Regions Select Committee. That statement included a summary of the key changes to the proposals for reform of the plan making system at regional & local levels. This paper provides some additional detail on those proposals.	Introduction; Regional Spatial Strategies; Preparing the Regional Spatial Strategy; Local Development Frameworks; Nature of the Local Development Frameworks; The Local Development Framework Scheme; The Statement of Community Involvement; Preparing Local Development Frameworks; Testing Local Development Frameworks; Adopting Local Development Frameworks; Monitoring and review of Local Development Frameworks; Minerals and Waste Development Frameworks; Transitional arrangements.	http://www.odpm.gov.uk/stellent/groups/odpm_planning/documents/pdf/odpm_plan_pdf_605840.pdf
Strategic Planning for Sustainable Waste Management (Oct 2002)	This Guidance seeks to address the limitations of BPEO (Best Practicable Environmental Option) assessment. It sets out a methodology for appraising strategic waste planning options that takes account of environmental, socio-economic & implementation issues – as well as those normally addressed by BPEO assessments. The step by step methodology is intended to provide greater transparency in decision-making, as well as a robust & comprehensive approach. The Guidance is aimed primarily at RTABs (Regional Technical Advisory Bodies), to assist their work at the regional level, but may also assist strategic planning authorities engaged in planning for waste management.	Preface; Executive summary; Introduction; Identifying & agreeing appraisal criteria; Developing strategic waste planning options; Data collection; Appraising strategic waste planning options; Ranking & valuing performance; Weighting indicators; Sensitivity testing & option refinement; Appendices.	http://www.odpm.gov.uk/stellent/groups/odpm_planning/documents/pdf/odpm_plan_pdf_606386.pdf
Contributing to Sustainable Communities – A New Approach to Planning Obligations (Nov 2003)	This consultation presents proposals for reforming and improving planning obligations (also known as Section 106 agreements). The objective is to provide greater transparency and certainty for all stakeholders and help enable developers and local planning authorities to avoid the delays that sometimes occur under the current system. But the proposals also recognise that the ability to use negotiation to tailor contributions to the circumstances of an individual site is important and flexibility is often essential to securing agreement on planning obligations.	Introduction; What are planning obligations; Objectives for a new approach to planning obligations; Our proposals for improving planning obligations; Proposal to establish a new optional planning charge; Annexes.	http://www.odpm.gov.uk/stellent/groups/odpm_planning/documents/pdf/odpm_plan_pdf_025543.pdf

Continued

Table 7.5.2 UK Level Planning Policy. *Continued*

Policy Type	Document Title	Objective	Contents	Website
Strategies & Reports	Community Involvement in Planning: The Government's Objectives (Feb 2004)	The paper sets out: ▪ The importance of greater community involvement in planning and how this fits with the Government's broader agenda. ▪ The principles underpinning this approach. ▪ How the reform programme will strengthen community involvement.	Introduction; Community Involvement in Planning – Delivering Sustainable Communities; Community Involvement in Planning: The Principles; Community Involvement in Planning: How it Happens; Conclusion; Annex A – Planning Processes: Opportunities for Community Involvement; Annex B – Related Legislation and Policy Initiatives; Bibliography.	http://www.odpm.gov.uk/stellent/groups/odpm_planning/documents/page/odpm_plan_027497.pdf
Consultation Papers	Planning for the Supply of Aggregates in England (23/10/00)	This consultation paper identifies key issues for, and principles of, planning for the supply of aggregates in England, to stimulate discussion on how they should be best addressed. The paper is not exhaustive so comment can be given on other relevant matters. Following the analysis of the responses, a draft revision of MPG6 will be prepared for general consultation.	Summary; Introduction; Events since 1994; Key policy issues; Operational framework; Regulatory impact assessment; Conclusion; References; Annex A: Consolidated list of consultation questions; Annex B: Background information on aggregates & their supply; Annex C: Structure of the 1994 issue of MPG6 (Mineral Planning Guidance Note 6); Annex D: Options for estimation of demand for aggregate minerals.	http://www.odpm.gov.uk/stellent/groups/odpm_planning/documents/pdf/odpm_plan_pdf_605804.pdf
	Draft Planning Policy Statement 7: Sustainable Development in Rural Areas (PPS7) (Nov 2003)	The draft sets out broad policy objectives relevant to rural areas in England, & proposed planning policies that will help deliver these objectives. These policies are firmly based on the principles of sustainable development & the need to protect the wider, largely undeveloped countryside for the benefit of all.	Part 1: Introduction; Part 2: Draft PPS7 – the government's proposals; Part 3: Initial/partial regulatory impact assessment; Part 4: Statement of consultation.	http://www.odpm.gov.uk/stellent/groups/odpm_planning/documents/pdf/odpm_plan_pdf_024014.pdf
	Consultation on Draft New Planning Policy Statement 22: Renewable Energy (PPS22) (Nov 2003)	The draft sets out the Government's broad policy objectives relevant to renewable energy in England, and its proposed planning policies that will help deliver these objectives. These policies are firmly based on the principles set out in the Government's recent Energy White Paper.	Part 1: Introduction; Review of PPG (Planning Policy Guidance); Draft PPS22 – the government's proposals; Relationship with other policies & policy reviews; Part 2: Draft PPS22 renewable energy; The government's objectives; National planning policies; Regional targets; Policies in regional planning guidance & development plans; Locational	http://www.odpm.gov.uk/stellent/groups/odpm_planning/documents/pdf/odpm_plan_pdf_025517.pdf

		considerations; Other considerations; Part 3: Initial/partial regulatory impact assessment on new draft PPS22 renewable energy; Purpose & intended effect; Background & options; Part 4: Statement of consultation.	
Consultation on Draft Planning Policy Statement 6: Planning for Town Centres (Dec 2003)	This consultation paper seeks comments on a draft of a new Planning Policy Statement (PPS) 6: Planning for Town Centres. It is the Government's intention that this PPS (and accompanying guidance to be drafted) should, in due course, replace *Planning Policy Guidance Note 6 (PPG6): Town Centres and Retail Developments* (June 1996).	Ch 1: Introduction; Ch 2: PPS6: Planning for town centres; The Government aims; Positive planning for centres: a plan-led approach; Development control; Monitoring & review; Annex A: Glossary; Annex B: Implementation & community involvement; Ch 3: Partial regulatory impact assessment; Ch 4: Consultation criteria.	http://www.odpm.gov.uk/stellent/groups/odpm_planning/documents/pdf/odpm_plan_pdf_026232.pdf
Consultation Paper on Planning Policy Statement 1: Creating Sustainable Communities (Feb 2004)	This consultation paper seeks comments on a draft of a new Planning Policy Statement (PPS) 1 – *Creating Sustainable Communities*. This PPS will replace Planning Policy Guidance Note 1: *Policy and Principles* (PPG1). The Government is committed to a planning system which creates sustainable communities & delivers sustainable development. Planning has a critical role in supporting the Government's wider economic, social & environmental objectives, for example, the Government's objectives for balancing housing markets and for achieving sustainable improvements in the economic performance of all English regions.	Creating Sustainable Communities; Sustainable Development; Planning for Sustainable Development; Delivering Sustainable Development; Integrating the Four Aims of Sustainable Development; Sustainable Development and Design; Spatial Planning; Community Involvement in Planning; Principles of Community Involvement in Planning; Annex A – Sustainable Communities; Annex B – The Plan Led System; Annex C – Design; Bibliography; Initial/Partial Regulatory Impact Assessment; Consultation criteria.	http://www.odpm.gov.uk/stellent/groups/odpm_planning/documents/page/odpm_plan_027494.pdf
A Draft Practical Guide to Strategic Environmental Assessment (SEA) (July 2004)	This consultation paper presents a draft Practical Guide to European Directive 2001/42/EC 'on the assessment of the effects of certain plans & programmes on the environment', known as the Strategic Environmental Assessment or SEA Directive. The purpose of the Practical Guide is to provide key facts together with guidance on how to comply with the Directive in an environmental assessment of a plan or programme. The Guide is intended to apply to all plans & programmes in the UK which fall within the scope of the Directive.	Introduction; Background & context; SEA & consultation; SEA & sustainable development; Stages of SEA; Indicative list of plans & programmes subject to the SEA Directive; FAQs on SEA Directive; EC Directives & other International instruments related to the SEA Directive; Glossary; Appendices.	http://www.odpm.gov.uk/stellent/groups/odpm_planning/documents/page/odpm_plan_029817.pdf

Continued

Table 7.5.2 UK Level Planning Policy. *Continued*

Policy Type	Document Title	Objective	Contents	Website
Consultation Papers	Consultation Paper on Planning Policy Statement 10: Planning for Sustainable Waste Management (Dec 2004)	This consultation seeks views & comments on the draft of the new Planning Policy Statement (PPS) 10 *Planning for Sustainable Waste Management*. The intention is that PPS10, together with an accompanying practice guide should, in due course, replace Planning Policy Guidance Note 10 (PPG10) *Planning & Waste Management*, published in 1999.	Introduction; Proposed Planning Policy Statement; Planning for Sustainable Waste Management; Draft framework of practice guide; Partial Regulatory Impact Assessment; Non-Technical Summary of the Environment Report; List of consultation documents & availability; Consultation questions; Statement of consultation.	http://www.odpm.gov.uk/stellent/groups/odpm_planning/documents/pdf/odpm_plan_pdf_033314.pdf
Guidance Notes	Planning Policy Guidance Note 14: Development on Unstable Land (April 1990)	These guidelines explain briefly the effects of instability on development and land use. Consideration is given to the responsibilities of the various parties to development and the need for instability to be taken into account in the planning process is emphasised. Ways in which instability might be treated in development plans and in considering applications for planning permission are outlined. The role of expert advice is emphasised and Appendices examine and explain the different causes of instability and indicate some possible sources of information.	Introduction; Purpose of the guidelines; Effects of instability & the need for guidance; Causes of instability; Responsibilities of the different parties to development; Planning control; The need for expert advice; Conclusion; Appendix A: Causes for instability; Appendix B: Sources of information.	http://www.odpm.gov.uk/stellent/groups/odpm_control/documents/contentservertemplate/odpm_index.hcst?n=3416&l=3
	Planning Policy Guidance Note 18: Enforcing Planning Control (Dec 1991)	New and substantially improved powers to enforce planning control were given to local planning authorities (LPAs) by the Planning & Compensation Act 1991. The enforcement provisions of the Act were based on the main recommendations of the report, entitled *Enforcing Planning Control* (HMSO, February 1989). The report also recommended that then current Ministerial policy guidance about enforcement should be revised, taking account of the concern expressed about certain aspects of the current guidance. This Note gave revised guidance.	The new enforcement regime; The general approach to enforcement; Where development is carried out without permission; Where unauthorised development can be made acceptable by the imposition of conditions; Where the unauthorised development is unacceptable on the site but relocation is feasible; Where the unauthorised development is unacceptable & relocation is not feasible; Where the unauthorised development is unacceptable & immediate remedial action is required; Unauthorised development by small businesses or self-employed people; Authorised development by private householders; Enforcement of planning control over mineral working; The organisation of their enforcement functions by LPAs; Cancellation of advice.	http://www.odpm.gov.uk/stellent/groups/odpm_planning/documents/page/odpm_plan_606903.pdf

Planning Policy Guidance Note 19: Outdoor Advertisement Control (March 1992)	All advertisements affect the appearance of the building, structure or place where they are displayed. The main purpose of the advertisement control system is to help everyone involved in the display of outdoor advertising to contribute positively to the appearance of an attractive and cared-for environment in cities, towns and the countryside.	Introduction; The role of guidance on design; Advertisement control by local planning authorities; Criteria for dealing with advertisement applications; Considerations of 'amenity'; Considerations of 'public safety'; Advertisement control policies; Reasons for advertisement control decisions; Advertisement control conditions; Advertisements in conservation areas; Listed buildings & ancient monuments; Areas of special control of advertisements & other special areas; The Secretary of State's direction-making power; The power to discontinue 'deemed consent' advertisements; The operation of the advertisement control system; Advertisement applications & appeals to the Secretary of State; Annex 1: Criteria for deciding applications & appeals involving poster sites.	http://www.odpm.gov.uk/stellent/groups/odpm_planning/documents/page/odpm_plan_606904.pdf
Planning Policy Guidance Note 20: Coastal Planning (Sept 1992)	This note covers planning policy for the coastal areas of England & Wales. It sets the general context for policy (Chapter 1) & identifies planning policies for the coast (Chapter 2). Policies for development that require a coastal location are presented in Chapter 3. Guidance is then given on how these policies should be reflected in development plans (Chapter 4).	Introduction; Character of the coastline; The coastal zone; Designated areas; International dimension; Existing planning policy guidance; Heritage coasts; Planning policies for the coast; Types of coast; Conservation policies; Policies for development; Policies for risk; Environmental assessment; Policies for improving the environment; Policies for development that require a coastal location; The coastal zone & development plans; Annex 1: Note on definitions; Annex 2: Central government departments, agencies & statutory bodies with responsibilities & interests in the coastal zone; Annex 3: Coastal planning contract points.	http://www.odpm.gov.uk/stellent/groups/odpm_planning/documents/page/odpm_plan_606907.pdf
Planning Policy Guidance Note 21: Tourism (Nov 1992)	This PPG outlines the economic significance of tourism & its environmental impact, & therefore its importance in land-use planning. It explains how the needs of tourism should be dealt with in development plans & in development control.	Introduction; Tourism & planning; Trends in tourism; Development plans; Tourism development; Consultation Cancellations; Annex A: Hotels; Annex B: Holiday & touring caravans; Annex C: Seasonal & holiday occupancy conditions; Annex D: Useful addresses.	http://www.odpm.gov.uk/stellent/groups/odpm_planning/documents/page/odpm_plan_606908.pdf

Continued

Table 7.5.2 UK Level Planning Policy. *Continued*

Policy Type	Document Title	Objective	Contents	Website
Guidance Notes	Planning Policy Guidance Note 5: Simplified Planning Zones (Nov 1992)	This one covers simplified planning zones (SPZs), the procedures for preparation of which have been streamlined by the Planning and Compensation Act 1991 and new regulations. It builds on previous advice updated to reflect changes in the procedures. The guidance note outlines the general nature and role of SPZs. Detailed guidance is given in Annexes A and B.	Nature of SPZs; General benefits of SPZs; Promotion of particular areas; Initiative in making SPZ schemes; Relationship to development plans; More detailed guidance; Cancellation of previous advice; Annex A: Use, content & effect of SPZs; Annex B: Guidance on SPZ procedures.	http://www.odpm.gov.uk/stellent/groups/odpm_planning/documents/page/odpm_plan_606914.pdf
	Planning Policy Guidance Note 6: Town Centres & Retail Developments (June 1996)	The Government's objectives are: ▪ To sustain/enhance the vitality & viability of town centres; ▪ To focus development, especially retail development, in locations where the proximity of businesses facilitates competition from which all consumers are able to benefit; ▪ To maximise the opportunity to use means of transport other than the car; ▪ To maintain an efficient, competitive & innovative retail sector; & ▪ To ensure the availability of a wide range of shops, employment, services & facilities to which people have easy access by a choice of means of transport.	Objectives & approach; Town centres; Planning for retail developments; Assessing new retail developments; Annex A: Glossary; Annex B: Development plans; Annex C: Town centre management; Annex D: Amusement centres; Annex E: Traffic management for town centres.	http://www.odpm.gov.uk/stellent/groups/odpm_planning/documents/page/odpm_plan_606915.pdf
	Proposals for a Good Practice Guide on Sustainability Appraisal (Baker Associates Report) (Aug 1999)	This document set out guidelines for RPBs (Regional Planning Bodies) to use as they developed their sustainability appraisals. In so doing, it drew on the practical experience of a wide range of practitioners working to make sense of a difficult, yet potentially rewarding, task. It did not claim to be definitive, but	Preface; Ch 1: Introduction; Ch 2: Approach to sustainability appraisal; Ch 3: Setting appraisal criteria; Ch 4: Sustainability scoping of RPG (Regional Growth Plan); Ch 5: Appraisal of spatial strategy; Ch 6: Appraisal of policies; Ch 7: Recording & reporting findings; Ch 8: Monitoring & evaluation; Ch 9 Implementing	http://www.odpm.gov.uk/stellent/groups/odpm_planning/documents/pdf/odpm_plan_pdf_606943.pdf

	aimed to develop the broad scope of a logical & consistent methodology, acknowledging that it would expand & improve as relevant practice developed over time.	the approach; Annex 1: Sustainable development & baseline studies; Bibliography.	
Planning Policy Guidance Note 10: Planning & Waste Management (12/10/99)	This guidance note provides advice about how the land-use planning system should contribute to sustainable waste management through the provision of the required waste management facilities in England and explains how this provision is regulated under the statutory planning and waste management systems.	Preface; Introduction; The waste planning framework – European legislation & regional planning guidance; The main organisations & their responsibilities; The relationship between the planning system & waste management licensing; Development plans; Development control; Planning control; Planning considerations & conditions; Environmental impact assessment & site investigation; Monitoring & enforcement; Annex A: Planning considerations & planning conditions; Annex B: Developing regional strategies for waste management & the role of regional technical advisory bodies; Annex C: County matters; Annex D: Information on waste.	http://www.odpm.gov.uk/stellent/groups/odpm_planning/documents/pdf/odpm_plan_pdf_606925.pdf
Planning Policy Guidance Note 12: Development Plans (18/01/00)	This Planning Policy Guidance note was a revision of PPG12 published in February 1992. It was refocused to provide a more strategic overview of the role & importance of development plans within the planning system. It outlined key issues on plan content & procedures. It did not attempt to summarise planning policies for particular issues or subjects, set out elsewhere in Regional Planning Guidance, other PPGs, mineral planning guidance & circulars.	Foreword; Chapter 1: Development plans in the planning system; Chapter 2: Improving the process of plan preparation; Chapter 3: Plan content & level of detail; Chapter 4: Sustainable development Chapter 5: Integration of transport & land use policies; Chapter 6: Other content & procedural issues; Annex A: Types & forms of plans; Annex B: New procedures for local plans & UDPs; Annex C: Consultees for development plans.	http://www.odpm.gov.uk/stellent/groups/odpm_planning/documents/pdf/odpm_plan_pdf_606929.pdf
Planning Policy Guidance Note 3: Housing (07/03/00)	This PPG provides guidance on a range of issues relating to the provision of housing. It replaces the 1992 version of PPG3. Circular 6/98 *Planning and Affordable Housing* continues to apply, within the framework of policy set out in this guidance.	Guidance note; The government's objectives; Widening housing opportunity & choice; Maintaining a supply of housing; Creating sustainable residential environments; Monitoring & review; Annex A: London; Annex B: Exception housing in rural areas; Annex C: Definitions; Annex D: Bibliography.	http://www.odpm.gov.uk/stellent/groups/odpm_planning/documents/pdf/odpm_plan_pdf_606933.pdf

Continued

Table 7.5.2 UK Level Planning Policy. *Continued*

Policy Type	Document Title	Objective	Contents	Website
Guidance Notes	Controlling the Environmental Effects of Recycled & Secondary Aggregates Production Good Practice Guide (22/03/00)	Government policy aims to minimise the loss of land & the environmental impacts associated with primary extraction of minerals & waste disposal by increasing the amount of recycled & secondary materials used as aggregate in construction. The policy is set out in Minerals Planning Guidance note 6. The policy is that aggregates, & products made from aggregates, should be recycled wherever possible &, where technically, economically & environmentally acceptable, mineral & construction wastes should be used instead of primary materials. This document addresses this problem by providing good practice guidance on siting considerations, management controls & physical measures to avoid or reduce the local environmental impacts of producing recycled & secondary aggregates.	Preface; Executive Summary; Chapter 1: Introduction; Chapter 2: Overview of nature & scale of recycled secondary aggregates production; Chapter 3: Location of recycled & secondary aggregates production; Chapter 4: Land take, development & operations involved; Chapter 5: Planning & regulatory framework; Chapter 6: Overview of environmental impacts; Chapter 7: Impacts of land take & ancillary development; Chapter 8: Dust; Chapter 9: Noise; Chapter 10: Additional transport impacts; Chapter 11: Water pollution & land contamination; Chapter 12: Vibration; Chapter 13: Gaseous emissions & odour; Chapter 14: Planning for recycled & secondary aggregates production; Chapter 15: Controlling environmental effects.	http://www.odpm.gov.uk/stellent/groups/odpm_planning/documents/pdf/odpm_plan_pdf_606242.pdf
	By Design – Urban Design in the Planning System: Towards Better Practice (27/06/00)	This guide will help to encourage better design and is intended as a companion to the PPGs. It has been written to stimulate thinking about urban design, not to tell the reader how to design. The central message is that careful assessments of places, well-drafted policies, well-designed proposals, robust decision-making and a collaborative approach are needed throughout the country if better urban space is to be created.	Foreword; The need for better urban design; Thinking about urban design; Urban design & the planning toolkit; Raising standards in urban design; Appendices – Is design working for you?; Glossary; Further reading; Useful contacts.	http://www.odpm.gov.uk/stellent/groups/odpm_planning/documents/pdf/odpm_plan_pdf_605981.pdf
	Sustainability Appraisal of Regional Planning Guidance: Good Practice Guide (Oct 2000)	This document sets out guidelines which RPBs (Regional Planning Bodies) can use when preparing their sustainability appraisals. It is based on Baker Associates' report. It refines that guidance, taking account of feedback from a series of regional seminars, and written	Preface; Ch 1: Introduction; Ch 2: Approach to sustainability appraisal; Ch 3: Setting appraisal criteria; Ch 4: Sustainability scoping of RPG; Ch 5: Appraisal of spatial strategy; Ch 6: Appraisal of policies; Ch 7: Recording & reporting findings; Ch 8: Monitoring &	http://www.odpm.gov.uk/stellent/groups/odpm_planning/documents/pdf/odpm_plan_pdf_606126.pdf

	responses on the feasibility of applying this guide in each region.	evaluation; Ch 9: Implementing the approach; Annex 1: Sustainable development & baseline studies; Bibliography.	
Tapping the Potential – Assessing Urban Housing Capacity: Towards Better Practice (14/12/00)	In order to make the best use of the potential to recycle land & buildings in any one location, PPG3 has for the first time set out the expectation that all local planning authorities should undertake urban housing capacity studies. These studies are now at the heart of the planning for housing process, & form the basis for both the sequential approach and the managed release of sites. This short guide to better practice is based on a programme of research with local planning authorities. It draws on examples of good practice from the level of the region through to the district. As appropriate, reference is made to the studies which informed the guide so to as allow readers to refer to them for further information. It also cross refers to the National Land Use Database which local authorities are updating.	Section 1: Why assess urban housing capacity?; Section 2: Identifying capacity resources; Section 3: Surveying the capacity; Section 4: Assessing the yield; Section 5: Discounting potential; Annex A: Useful data sources; Annex B: Summaries of main case studies examined.	http://www.odpm.gov.uk/stellent/groups/odpm_control/documents/contentservertemplate/odpm_index.hcst?n=2321&l=2
Planning Policy Guidance Note 4: Industrial and Commercial Development and Small Firms (05/03/01)	The planning system plays an important role integrating environmental and economic objectives. Development plans provide the policy framework, weighing the importance of industrial and commercial development with that of maintaining and improving environmental quality. The principles of sustainable development require the responsible use of man-made and natural resources by all concerned in a way that ensures that future generations are not worse off. Careful attention to environmental issues makes good economic sense for business and industry.	Introduction; Development plans; Locational factors; Development control – a positive approach; Helping small firms through the planning system; Mixed uses; Conservation & heritage; Reuse of urban land Speculative development; Environmental assessment; The imposition of conditions; Deciding whether permission is needed; Enforcement action; Cancellation of advice.	http://www.odpm.gov.uk/stellent/groups/odpm_planning/documents/pdf/odpm_plan_pdf_606913.pdf

Continued

Table 7.5.2 UK Level Planning Policy. *Continued*

Policy Type	Document Title	Objective	Contents	Website
Guidance Notes	Planning Policy Guidance Note 2: Green Belts (02/07/01)	This PPG replaces the 1988 version of PPG2, and advice in Circulars. It: ▪ States the general intentions of Green Belt policy, including its contribution to sustainable development objectives; ▪ Reaffirms the specific purposes of including land in Green Belts, with slight modifications; ▪ Gives policy a more positive thrust by specifying for the first time objectives for the use of land in Green Belts; ▪ Confirms that Green Belts must be protected as far as can be seen ahead, advises on defining boundaries & on safeguarding land for longer-term development needs; ▪ Maintains the presumption against inappropriate development within Green Belts; & ▪ Refines the categories of appropriate development, including making provision for the future of major existing developed sites & revising policy on the re-use of buildings.	Foreword; Introduction; Designation of green belts; Control over development; Cancellation of advice; Map – approved green belt based on structure plans & local plans (September 1993); Annex A: Land management; Annex B: Safeguarded land; Annex C: Future of major developed sites; Annex D: Re-use of buildings – additional advice; Annex 5: Further guidance from other PPGs & circulars.	http://www.odpm.gov.uk/stellent/groups/odpm_planning/documents/pdf/odpm_plan_pdf_606905.pdf
	Planning Policy Guidance Note 25: Development & Flood Risk (17/07/01)	This guidance sets out the Government's policy on the positive role of land-use planning in achieving these aims and the lead responsibilities of local planning authorities and other bodies. It provides guidance to planning authorities, developers, the public and the Environment Agency on a wide range of planning and other measures so that decisions on plans and applications in areas of both existing and future development take full account of flood risk, whether inland or on the coast.	Foreword; Introduction; Purpose & scope of guidance; Sustainable development & the precautionary principle; Responsibilities of the parties in the development process; Flooding & land-use planning – general considerations; Regional planning & development plans; Development control; Further information & advice; Conclusion; Annex A: Causes of flooding & the impact of climate change; Annex B: Publications relevant to flooding; Annex C: Some relevant statutory & non-statutory pans; Annex D: Role of the	http://www.odpm.gov.uk/stellent/groups/odpm_planning/documents/pdf/odpm_plan_pdf_606931.pdf

		Environment Agency & other operating authorities; Annex E: Sustainable drainage systems; Annex F: Guidance on requirements for undertaking a flood risk assessment; Annex G: Warning notices/signs at caravan/camping sites & other developments at risk of flooding.	
Planning Policy Guidance Note 1: General Policy and Principles (22/08/01)	This Note begins by discussing sustainable development, mixed use and design, three themes which underpin the government's approach to the planning system. It then provides a number of policy messages in relation to various particular land uses. More detailed guidance is given in the other Government planning policy notes and good practice guides listed in Annex B. Finally, it sets out the operational principles to be observed in the planning system.	Introduction; The government's approach & planning – sustainable development, mixed use & design; Key policy objectives; Planning for industry & commerce; Landuse & transport; Planning for housing; The importance of town centres; Rural areas; Conserving the historic environment; Access; Operational principles; Cancellation of advice; Annex A: Handling design issues; Annex B: Government statements of planning policy; Annex C: Legislation; Annex D: Other procedural issues.	http://www.odpm.gov.uk/stellent/groups/odpm_planning/documents/pdf/odpm_plan_pdf_606895.pdf
Planning Policy Guidance Note 8: Telecommunications (22/08/01)	This PPG replaces PPG8 of December 1992 *Telecommunications* and Circular 4/99 *Planning for Telecommunications*. It gives guidance on planning for telecommunications development – including radio masts and towers, antennas of all kinds, radio equipment housing, public call boxes, cabinets, poles and overhead wires.	Preface; Planning policy; General policy; Technical limitations; Discussion & consultation; Pre-rollout discussions; Pre-application discussions; Application consultation & publicity; Environmental considerations; Mast & site sharing; Design; Health considerations; Other development; Appendix: Supporting guidance; Annex 1: Guidance on prior approval procedures; Note A: Extract from schedule 2 to the Telecommunications Act 1984; Annex 2: The control of radio interference.	http://www.odpm.gov.uk/stellent/groups/odpm_planning/documents/pdf/odpm_plan_pdf_606918.pdf
Planning Policy Guidance Note 16: Archaeology & Planning (22/08/01)	This sets out the Secretary of State's policy on archaeological remains on land, & how they should be preserved or recorded both in an urban setting & in the countryside. It gives advice on the handling of archaeological remains & discoveries under the development plan & control systems, including the weight to be given to them in planning decisions & the use of planning conditions.	Introduction; The importance of archaeology; Advice on the handling of archaeological matters in the planning process; Annex 1: Key bodies & organisations; Annex 2: Contact addresses for County archaeological officers & SMRs (Sites and Monuments Records); Annex 3: Legislative arrangements; Annex 4: Secretary of State's criteria for scheduling; Annex 5: Ancient Monuments (class consent) Order 1981.	http://www.odpm.gov.uk/stellent/groups/odpm_planning/documents/pdf/odpm_plan_pdf_606901.pdf

Continued

Table 7.5.2 UK Level Planning Policy. *Continued*

Policy Type	Document Title	Objective	Contents	Website
Guidance Notes	Planning Policy Guidance Note 24: Planning & Noise (22/08/01)	The aim of this guidance is to provide advice on how the planning system can be used to minimise the adverse impact of noise without placing unreasonable restrictions on development or adding unduly to the costs & administrative burdens of business. It builds upon the principles established in Circular 10/73 *Planning & Noise*, & takes account of the recommendations of the Noise Review Working Party 1990. It outlines some of the main considerations which local planning authorities should take into account in drawing up development plan policies & when determining planning applications for development which will either generate noise or be exposed to existing noise sources.	Foreword; Introduction; General principles; Noise policies in development plans; Noise exposure categories for residential development; Development control; Environmental assessment; Other statutory controls; Cancellation of advice; Glossary; Annex 1: Noise exposure categories for dwellings; Annex 2: Noise exposure categories – explanation of noise levels; Annex 3: Detailed guidance on the assessment of noise from different sources; Annex 4: Examples of planning; Annex 5: Specifying noise limits; Annex 6: Insulation of buildings against external noise; Annex 7: More information on other noise control regimes; Annex 8: Statutory instruments.	http://www.odpm.gov.uk/stellent/groups/odpm_planning/documents/pdf/odpm_plan_pdf_606912.pdf
	Better Places to Live by Design: A Companion Guide to PPG3 (11/09/01)	The purpose of this guide is to help deliver the step-change in quality required to break the mould of mediocrity that has characterised so much new housing development. It does not set out new policy, rather it is a companion to PPG3 & should be read alongside it. It is a guide to better practice & aims to prompt greater attention to the principles of good design.	Foreword; Chapter 1 – The need for better housing design; Chapter 2: Understanding the context; Chapter 3: Creating a movement framework – Part 1: Creating connections, routes & functions, walking & cycling & public transport; Part 2: Managing the traffic, servicing the home; Chapter 4: Housing mix & neighbourhood; Chapter 5: Housing layout & urban form – Part 1: Importance of structure & perimeter blocks; Part 2: Other block structures, street widths & enclosure; Part 3:	http://www.odpm.gov.uk/stellent/groups/odpm_control/documents/contentservertemplate/odpm_index.hcst?n=2325&l=2

		Solar orientation, public, private & communal space, designing for privacy, creating a feeling of safety; Part 4: Accommodating car-parking; Chapter 6: Space in & around the home; Chapter 7: Thoroughness in design; Appendices – Case studies; Further reading.	
Planning Policy Guidance Note 9: Nature Conservation (18/02/02)	The Government's objectives for nature conservation are to ensure that its policies contribute to the conservation of the abundance and diversity of British wildlife and its habitats, or minimise the adverse effects on wildlife where conflict of interest is unavoidable, and to meet its international responsibilities and obligations for nature conservation. This Planning Policy Guidance note (PPG) sets out the principles and policies that apply to the integration of nature conservation priorities and land-use planning.	Summary; Introduction; Statutory framework; International obligations & interests; The role of English Nature; Designated sites; Nature conservation outside designated sites; Nature conservation & land-use planning; Regional & strategic planning guidance; Structure plans; Local plans; Nature conservation & development control; SSSIs (Sites of Special Scientific Interest); Special procedures to protect sites of particular importance; Environmental assessment; Minerals matters including peat extraction; Protection of species; Lawful development certificates; Cancellation of advice; Annex A: Designated sites; Annex B: Classified & potential special protection areas & listed Ramsar sites Sept 1994; Annex C: Development control for special protection areas & special areas of conservation; Annex D: Local authority action under the Countryside Act 1968; Annex E: The Ramsar convention on wetlands of international importance; Annex F: The birds directive on the conservation of wild birds (79/409/EEC); Annex G: The habitats directive on the conservation of natural habitats & of wild fauna & flora (92/43/EEC); Annex H: Addresses.	http://www.odpm.gov.uk/stellent/groups/odpm_planning/documents/pdf/odpm_plan_pdf_606919.pdf

Continued

Table 7.5.2 UK Level Planning Policy. *Continued*

Policy Type	Document Title	Objective	Contents	Website
Guidance Notes	Mineral Planning Guidance Note 1: General Considerations & the Development Plan System (20/02/02)	This Guidance Note sets out the Government's policy on minerals and planning issues and provides advice to mineral planning authorities and the minerals industry on policies and the operation of the planning system with regard to minerals. Mineral planning authorities must take its contents into account in preparing their development plans. The guidance may also be material to decisions on individual planning applications and appeals.	Introduction; The development plan & minerals; Policy considerations for minerals planning; Sustainable Development; Safeguarding; Ensuring supply; Landbanks & continuity of production; Areas for future working; National parks & AONBs (Areas of Outstanding Natural Beauty); SSSIs & NNRs (National Nature Reserves); Other environmentally important areas; Agricultural land; Forestry; Water environment; Archaeology, listed buildings & the historic environment; Green belt; Coastal planning; Land instability; Development control policies; Environmental impacts; Transport; Recycling plants; Restoration, aftercare & after-use; Mineral waste & forms of waste disposal; Environmental management; Annex A: General information on minerals planning; Annex B Advice in individual minerals; Annex C: Figures 1/tables 1&2; Annex D: Legislation relevant to minerals planning; Annex E: Current PPGs; Annex F: References.	http://www.odpm.gov.uk/stellent/groups/odpm_planning/documents/pdf/odpm_plan_pdf_606843.pdf
	Circular 06/1998 Planning and Affordable Housing (16/05/02)	This Circular supplements PPG3, by amplifying the Government's preferred approach to planning and affordable housing. It aims to provide a clearer framework for preparing plan policies, and practical advice to local planning authorities on how they should encourage the supply of affordable housing in appropriate circumstances through negotiation with developers and others.	Introduction; Assessment of need for affordable housing; Affordable housing policies in local plans; Securing affordable housing & controlling occupancy; Rural sites; Involvement of a registered social landlord; Content & discharge of conditions & obligations; Possible approaches to the use of conditions & planning obligations in the range of general situations; Situations where a registered social landlord is involved; Situations where no social landlord is involved; Endnotes.	http://www.odpm.gov.uk/stellent/groups/odpm_planning/documents/pdf/odpm_plan_pdf_606806.pdf

Planning Policy Guidance Note 17: Planning for Open Space, Sport & Recreation (24/07/02)	Open spaces, sport & recreation all underpin people's quality of life. Well-designed & implemented planning policies for open space, sport & recreation are therefore fundamental to delivering broader Government objectives. These include: ▪ Supporting an urban renaissance – local networks of high quality & well managed & maintained open spaces, sports & recreational facilities help create urban environments that are attractive, clean & safe. ▪ Supporting a rural renewal – the countryside can provide opportunities for recreation & visitors can play an important role in the regeneration of the economies of rural areas. ▪ Promotion of social inclusion & community cohesion – well planned & maintained open spaces & good quality sports & recreational facilities can play a major part in improving people's sense of well being in the place they live. ▪ Health & well-being – open spaces, sports & recreational facilities have a vital role to play in promoting healthy living & preventing illness ▪ Promoting more sustainable development – by ensuring that open space, sports & recreational facilities (particularly in urban areas) are easily accessible by walking & cycling & that more heavily used or intensive sports & recreational facilities are planned for locations well served by public transport.	Planning objectives; National planning policies; Assessment of needs & opportunities; Setting local standards; Maintaining an adequate supply of open space & sports & recreational facilities; Planning for new open space & sports & recreational facilities; Planning obligations; Annex: definitions	http://www.odpm.gov.uk/stellent/groups/odpm_planning/documents/pdf/odpm_plan_pdf_606902.pdf
Planning Policy Guidance Note 13: Transport (11/10/02)	This guidance sets out the circumstances where it is appropriate to change the emphasis and priorities in provision between different transport modes in pursuit of wider Government objectives. The car will continue to have an important part to play and for some journeys, particularly in rural areas, it will remain the only real option for travel.	Introduction & summary; Planning policies; Managing travel demand; Implementation; Annex A: London; Annex B: Planning for transport; Annex C: Transport infrastructure; Annex D: Maximum parking standards; Annex E: Park & ride in the green belt.	http://www.odpm.gov.uk/stellent/groups/odpm_planning/documents/pdf/odpm_plan_pdf_606896.pdf

Continued

Table 7.5.2 UK Level Planning Policy. *Continued*

Policy Type	Document Title	Objective	Contents	Website
Guidance Notes	Planning Policy Guidance Note 15: Planning & the Historic Environment (11/10/02)	This PPG provides a full statement of Government policies for the identification and protection of historic buildings, conservation areas, and other elements of the historic environment. It explains the role played by the planning system in their protection. It complements the guidance on archaeology and planning given in PPG 16.	Introduction; Part 1: Planning & conservation; Development plans & development control; Listed building control; Conservation areas; Transport & traffic management; Part 2: Identifying & recording the historic environment; The upkeep & repair of historic buildings; Churches & the ecclesiastical exemption; Annex A: The legislation & main heritage bodies; Annex B: Listed building control procedures; Annex C: Guidance on alterations to listed buildings; Annex D: Bibliography.	http://www.odpm.gov.uk/stellent/groups/odpm_planning/documents/pdf/odpm_plan_pdf_606900.pdf
	Planning Circular 01/02: The Town & Country Planning (Residential Density) (London & South East England) Direction 2002 (31/10/02)	In the Statement, the Deputy Prime Minister announced he would intervene in planning applications for housing that involve a density of less than 30 dwellings per hectare net. The Town and Country Planning (Residential Density) (London and South East England) Direction 2002 gives effect to this announcement. A copy of the Direction, which comes into force on 2 December 2002, forms the Annex to this Circular.	The Town and Country Planning (Residential Density) (London and South East England) Direction 2002.	http://www.odpm.gov.uk/stellent/groups/odpm_planning/documents/pdf/odpm_plan_pdf_606794.pdf
	Monitoring Regional Planning Guidance (Nov 2002)	This guide has been produced in order to assist Regional Planning Bodies (RPBs) & others involved in the Regional Planning Guidance (RPG) process to develop a robust framework for performance review & monitoring of RPG. It takes forward an earlier draft guide prepared for the then DETR. It sets out a step by step approach to developing appropriate monitoring frameworks & provides advice on formulating targets & indicators. A series of Topic Guides at Annex. A provide more detail on setting targets & indicators to deliver policy objectives, with examples.	Ch 1: How to use this guide; Ch 2: Context & basic principles of regional monitoring; Ch 3: Key stages in the monitoring & review process; Ch 4: The objectives led approach to monitoring & review; Ch 5: Developing targets for RPG; Ch 6: Developing indicators for RPG; Glossary; Annex A – topic guides.	http://www.odpm.gov.uk/stellent/groups/odpm_planning/documents/pdf/odpm_plan_pdf_606395.pdf

Planning & Access for Disabled People: A Good Practice Guide (March 2003)	This good practice guide, stemming from a recommendation of the Disability Rights Task Force, describes how all those involved in the development process can play their part in delivering physical environments which can be used by everyone. It encourages local planning authorities & developers to consider access for disabled people, & stresses the importance of early consultation when formulating development plans & preparing planning applications. The guide explains the relevant legislation & policy frameworks, shows how local planning authorities can put in place appropriate planning policies & development control processes, & suggests ways in which these can be implemented & enforced effectively. It pinpoints the role of developers & occupiers & underlines the benefits to them in providing environments which are accessible & inclusive.	Foreword; Part 1: A summary of good practice points; Part 2: Understanding the issues; Part 3: Local planning authorities – appropriate planning policy; Part 4: Local planning authorities – effective development control; Part 5: Combining the interests of all statutory authorities; Part 6: Good practice by developers, occupiers & owners; Appendices; Glossary; Useful contacts; References.	http://www.odpm.gov.uk/stellent/groups/odpm_planning/documents/pdf/odpm_plan_pdf_609460.pdf
The Strategic Environmental Assessment Directive: Guidance for Planning Authorities (Oct 2003)	This publication presents guidance on how to carry out environmental assessment of English land-use and spatial plans in accordance with European Directive 2001/42/EC on the assessment of the effects of certain plans and programmes on the environment, known as the 'Strategic Environmental Assessment' or SEA Directive.	Ch 1: Introduction; Ch 2: Background & context; Ch 3: Stages of SEA. App 1: Application of the SEA Directive to plans & programmes; App 2: Other relevant plans, programmes & objectives; App 3: Devising SEA & sustainability appraisal objectives & indicators; App 4: Sources of baseline data; App 5: Presenting baseline information; App 6: Assessing secondary, cumulative & synergistic effects; App 7: Identifying & comparing alternatives; App 8: Prediction & evaluation; Quality assurance checklist; References; The SEA Directive.	http://www.odpm.gov.uk/stellent/groups/odpm_planning/documents/page/odpm_plan_026670.pdf

Continued

Table 7.5.2 UK Level Planning Policy. *Continued*

Policy Type	Document Title	Objective	Contents	Website
Guidance Notes	Environmental Impact Assessment: Guide to Procedures (Jan 2004)	Environmental impact assessment (EIA) is an important procedure for ensuring that the likely effects of new development on the environment are fully understood & taken into account before the development is allowed to go ahead. This booklet, which is intended primarily for developers & their advisers, explains how European Community (EC) requirements for the environmental impact assessment of major projects have been incorporated into consent procedures in the UK. It revises the booklet *Environmental Assessment: A Guide to the Procedures*, first published in 1989, to take account of the requirements of Directive 97/11/EC, which was adopted on 3 March 1997 & came into effect on 14 March 1999.	Introduction; Environmental impact assessment & projects which require planning permission; Preparing an environmental statement: The planning procedures; Arrangements for other projects; Appendices.	http://www.odpm.gov.uk/stellent/groups/odpm_planning/documents/pdf/odpm_plan_pdf_026667.pdf
	Note on EIA for Local Planning Authorities (Jan 2004)	This note, in the form of answers to frequently asked questions, offers a brief & basic introduction to the Directive, the Regulations and planning authority responsibilities. It does not offer definitive guidance & is not a substitute for the Regulations or for guidance published in the official Departmental Circular (DETR Circular 02/99, *Environmental Impact Assessment*).	Background; What do the Regulations require? Do the Regulations apply to all applications for planning permission? What are the lessons of these cases? Where EIA is required, what information has to be included in the Environmental Statement? Is there a standard format for an ES? What action does the planning authority have to take? Who has to carry out the screening opinion? What factors are taken into consideration when making a screening opinion? Does the screening opinion have to give reasons for the decision? Can the authority change its screening opinion? Scoping; Once a scoping opinion is issued can I request further information? Lessons from Gillespie; Evaluating the Environmental Statement; Does this also apply to applications	http://www.odpm.gov.uk/stellent/groups/odpm_planning/documents/pdf/odpm_plan_pdf_026668.pdf

		for outline planning permission where some matters may be reserved for later determination? EIA and Unauthorised Development; Domestic challenges; Complaints to the European Commission.	
Safer Places – The Planning System & Crime Prevention (April 2004)	This guide is not solely about crime prevention. It is about making places that are safer, but also better in a number of other ways. This means it is concerned with the promotion of safe, sustainable & attractive environments that meet the full set of planning objectives. In short, it is about good planning in general, & its particular role in tackling crime & the fear of crime. The guide's aim is to show how good planning can contribute to crime prevention & the creation of safer places & hence to well-designed, sustainable communities.	Foreword; Safe & sustainable communities; The attributes of safe, sustainable places; The planning & crime prevention toolkit; Annex 1 – case studies; Annex 2 – Reconciling evidence of what works, knowledge of crime reduction & community safety principles & values; Annex 3 – Evidence base & further reading; Annex 4 – summary of checklists; Annex 5 – Glossary.	http://www.odpm.gov.uk/stellent/groups/odpm_planning/documents/page/odpm_plan_028449.pdf
Planning Policy Statement 7: Sustainable Development in Rural Areas (Aug 2004)	Planning Policy Statements (PPS) set out the Government's national policies on different aspects of land use planning in England. The policies in this statement apply to the rural areas, including country towns & villages & the wider, largely undeveloped countryside up to the fringes of larger urban areas. These policies complement, but do not replace or overrule, other national planning policies & should be read in conjunction with other relevant statements of national planning policy. This PPS replaces Planning Policy Guidance (PPG) Note 7, *The Countryside – Environmental Quality & Economic & Social Development*, published in February 1997.	The Government's Objectives; National Planning Policies; Key Principles; Sustainable Rural Communities, Economic Development; The Countryside; Agriculture, Farm Diversification, Equine-Related Activities & Forestry; Tourism & Leisure; Annex A – Agricultural & other Occupational Dwellings.	http://www.odpm.gov.uk/stellent/groups/odpm_planning/documents/downloadable/odpm_plan_030185.pdf

Continued

Table 7.5.2 UK Level Planning Policy. *Continued*

Policy Type	Document Title	Objective	Contents	Website
Guidance Notes	Planning Policy Statement 22: Renewable Energy (Aug 2004)	Planning Policy Statements (PPS) set out the Government's national policies for different aspects of land use planning in England. This PPS replaces Planning Policy Guidance Note 22 (PPG22) issued in 1993, the annexes issued in 1994 and the photo-voltaics annex issued in 2002. The policies set out in this statement will need to be taken into account by regional planning bodies & the Mayor of London in the preparation of regional spatial strategies (or the Spatial Development Strategy in London), & by local planning authorities in the preparation of local development documents. They may also be material to decisions on individual planning applications. National policies set out in other planning policy statements or PPGs may also be relevant to consideration of planning for renewable energy.	Introduction; The Government's Objectives; National Planning Policies.	http://www.odpm.gov.uk/stellent/groups/odpm_planning/documents/pdf/odpm_plan_pdf_030334.pdf
	Planning Policy Statement 11: Regional Spatial Strategies (Sept 2004)	Planning Policy Statements (PPSs) set out the Government's national policies on different aspects of planning. The policies in this statement apply throughout England apart from London where the Mayor is responsible for preparing a spatial development strategy. These policies complement, but do not replace or overrule, other national planning policies & should be read in conjunction with other relevant statements of national planning policy.	Chapter 1: What is a Regional Spatial Strategy? Chapter 2: Requirements for preparing a Regional Spatial Strategy Revision; Chapter 3: Implementation, Monitoring and Review; Annex A: Policy and Guidance on topics to be covered in a RSS; Annex B: Regional Transport Strategies; Annex C: Examinations-in-Public into Revised RSSs; Appendix (i) Indicative Timetable; Appendix (ii) Panel Secretary; Appendix (iii) Panel Assistant Annex D: Partnership Working & Community Involvement in the RSS Process; Glossary; List of Abbreviations/Acronyms.	http://www.odpm.gov.uk/stellent/groups/odpm_planning/documents/page/odpm_plan_030921.pdf

Planning Policy Statement 12: Local Development Frameworks (Sept 2004)	Planning Policy Statements (PPS) set out the Government's national policies on different aspects of planning. The policies in this statement apply throughout England & focus on procedural policy & the process of preparing local development documents. These policies complement, but will not overrule, other national planning policies & should be read in conjunction with other relevant statements of national planning policy.	Chapter 1 Development plans in the planning system; Chapter 2 Plan content and level of detail; Chapter 3 Involvement appraisal and management of the process; Chapter 4 Preparation of a local development document; Chapter 5 Transferring to the new system; Annex A: Illustrative material; Annex B: Policy Advice and integration with other strategies; Annex C: Making representations on submitted development plan documents; Annex D: Independent Examination; Annex E: Consultees; Annex F: Complaints procedure.	http://www.odpm.gov.uk/stellent/groups/odpm_planning/documents/page/odpm_plan_030922.pdf
Planning Policy Statement 23: Planning & Pollution Control (Nov 2004)	Planning Policy Statements (PPSs) set out the Government's core policies and principles on the most important aspects of land use planning. The policies in this statement & the advice in the accompanying Annexes (Annex 1: *Pollution Control, Air and Water Quality*; & Annex 2: *Development on Land Affected by Contamination*) should be taken into account by Regional Planning Bodies (RPBs) & Local Planning Authorities (LPAs) in preparing Regional Spatial Strategies (RSSs) & Local Development Documents (LDDs) – referred to in this Statement as 'development plans'.	Introduction; Sustainable Development; The Precautionary Principle; National & International Obligations; Government policies on planning & pollution control; Development Plans; Pre-application Discussions; Development Control; Government policies on land affected by contamination; Development Plans; Pre-application Discussions; Development Control; Conclusion; Appendix A. Matters for consideration in preparing; Local development documents & taking decisions on individual planning applications.	http://www.odpm.gov.uk/stellent/groups/odpm_planning/documents/downloadable/odpm_plan_032633.pdf

Table 7.5.3 Northern Ireland Level Planning Policy.

Policy Type	Document Title	Objective	Contents	Website
Strategies & Reports	The Planning Strategy for Rural Northern Ireland (1993)	The Planning Strategy for Rural Northern Ireland covers all of the towns, villages and countryside of Northern Ireland outside Belfast (& adjoining built up areas) and Londonderry. It provides a comprehensive framework for the preparation of development plans for 19 individual District Council areas across the region. The Strategy establishes the objectives & the policies for land use & development appropriate to the particular circumstances of Northern Ireland & which need to be considered on a scale wider than the individual District Council Area. It provides a basis for co-ordinating decisions in both the public & the private sectors & sets out regional policies for the control of development, to ensure a consistent approach to rural planning matters.	Introduction; The Context – Background; Consultation responses; Issues; The strategy – Planning strategy; Strategic policies; Regional planning policies; Housing; Agriculture; Industry & commerce; Minerals; Tourism, sport & recreation; Conservation; Public services & utilities; Design principles; Green belts & countryside policy areas; The coast; Implementation; Appendices.	http://www.planningni.gov.uk/AreaPlans_Policy/Strategies/PSRNI/psrni_1.htm
	Planning Policy Statement 3: Roads Consideration (May 1996)	This PPS sets out the roads matters which will be taken into account in determining planning applications for such development. In order to properly evaluate the traffic impact of development proposals & to determine what infrastructure improvements may be necessary, a Traffic Impact Assessment (TIA) may be required in support of the planning application. A TIA for a large-scale development may establish the acceptability of the development proposal from a traffic/roads view & in some circumstances may establish a requirement for the developer to undertake works to the road system.	Introduction; Accesses; Parking; Layout of residential roads; Main traffic routes; The protection of road lines; Off-site road improvements; Developer contributions; Annex A: Planning conditions requiring works to the roads system; Annex B: Planning agreements.	http://www.doeni.gov.uk/planning/Planning_Policy_Statements/Planning_Policy_Statement_3/PPS3.pdf

Planning Policy Statement 5: Retailing & Town Centres (June 1996)	This Planning Policy Statement sets out the Department's policy for town centres and retail developments for all of Northern Ireland.	Introduction; Objectives & approach; Planning strategy for NI; Development plans; Town centres; Vitality & viability of town centres; Diversity of use in town centres; Physical environment in town centres; Access to town centres; Planning for retail development; Planning for local shopping; Assessment of retail development; Assessment of major retail proposals; Environmental assessment; PPSs; Development control advice notes; Glossary.	http://www.doeni.gov.uk/planning/Planning_Policy_Statements/Planning_Policy_Statement_5/PPS%205.pdf
Planning Policy Statement 4: Industrial Development (March 1997)	This statement is relevant to all uses of land involving the carrying out of any industrial process or research and development into industrial products or processes. It does not, however, deal with other business activities such as retailing, offices, mineral extraction, waste disposal or agriculture.	Context; Scope; Importance of industrial development; Sustainable development; Transport considerations; Targeting social need; Development plans; Development control; Home-working & other ancillary ind uses; General criteria for assessing ind dev proposals; Development incompatible with existing industry; Industries of an offensive or hazardous nature; Environmental assessment; Control of hazardous substances; Retention of industrial land & buildings; Non-industrial Uses on zoned industrial land; Enterprises on farm holdings; Small off-farm rural enterprises; Rural business use class; Major Industrial Projects in the countryside; Expansion of rural enterprises; Unauthorised development.	http://www.doeni.gov.uk/planning/Planning_Policy_Statements/Planning_Policy_Statement_4/PPS%204.pdf

Continued

Table 7.5.3 Northern Ireland Level Planning Policy. *Continued*

Policy Type	Document Title	Objective	Contents	Website
Strategies & Reports	Planning Policy Statement 2: Planning & Nature Conservation (June 1997)	Planning & Nature Conservation (PPS2) states the Department's planning policy for nature conservation for the whole of NI. Such policy is strongly guided by Government policy on nature conservation & by the statutory framework. This PPS describes the statutory framework that helps to protect designated areas, including those of international importance, & advises on the treatment of nature conservation issues in development plans. In addition, it outlines the criteria which the Planning Service will employ when processing planning applications which might affect nature conservation interests & to which developers should have regard when preparing proposals.	Introduction; Aims, approach & objectives; Statutory framework; Legislation; International obligations; Policy on designated sites; Development plans; Countryside assessments; Environmental appraisal; Designated areas; Outside designated areas; Nature conservation & development control; Development affecting sites of international nature conservation importance; Development affecting sites of national nature conservation importance; Development affecting recognised sites of local nature conservation importance; Development affecting other sites of local conservation importance; Trees & woodland; Protection of species; Peatlands; Environmental assessment; Annex 1: International treaties, conventions & directives; Annex 2: Nature conservation & planning legislation in NI; Annex 3: Proposed & designated sites of international importance in NI; Annex 4: Consideration of development proposals affecting international sites; Annex 5: Permitted development rights in European sites; Annex 6: Bibliography; Annex 7: Useful addresses.	http://www.doeni.gov.uk/planning/Planning_Policy_Statements/Planning_Policy_Statement_2/PPS2.pdf
	Your Permitted Development Rights & Environmental Assessment (June 1997)	This Leaflet explains how permitted development rights in Northern Ireland may be affected by Environmental Assessment.	What is environmental assessment? What type of development is subject to EA? Who decides whether EA is needed? Which permitted development proposals are likely to require EA? How can you judge whether a project is likely to have significant environmental effects? What if EA is required? What are the benefits of EA?	http://www.doeni.gov.uk/planning/Information_Leaflets/Permitted_Development_and_EA/pd&ea.htm
	Tree Preservation Orders (June 1997)	This leaflet explains the purpose of Tree Preservation Orders. It also discusses the instances when a TPO is needed & the procedures which must be followed to meet their requirements.	Introduction; Why protect trees? What do tree preservation orders do? How do we decide when a tree preservation order is needed? What are the procedures? What if I object? Can protected trees ever be cut down? What happens if I cut down a protected tree without permission? Where can I get information?	http://www.doeni.gov.uk/planning/Information_Leaflets/TPO_Leaflet/treepo.htm

Environmental Assessment (June 1997)	This leaflet answers some basic questions about the requirements for environmental assessment in respect of development which requires planning permission under the planning procedures in Northern Ireland. In general similar provisions apply in Scotland, England & Wales.	Introduction; What is environmental assessment (EA)? What types of development are subject to EA as part of the planning procedures? In the case of schedule 2 projects, how will 'significance' be assessed? Who decides whether EA is necessary? What are the advantages of EA? How will EA help the public? How can I find out whether EA is needed for particular development? How do I find out? What is in the ES? What about development that is not subject to EA? Will EA affect my permitted development rights?	http://www.doeni.gov.uk/planning/Information_Leaflets/EA_Leaflet/Environmental%20Assessment.htm
Planning Policy Statement 1: General Principles (March 1998)	This Statement sets out the general principles that the Department observes in formulating planning policies, making development plans & exercising control of development. The Statement also sets out the key themes that underlie the Department's overall approach to planning across the whole range of land-use topics.	Introduction; The Department's approach to planning; Sustainable development; Quality development; Design considerations; Mixed uses; Policy publications; Regional strategic framework; A planning strategy for rural NI; PPSs; Development plans; Supplementary planning guidance; Development control; Annex 1: PPSs; Annex 2: Development control advice notes; Annex 3: Design; Annex 4: Statutory framework.	http://www.doeni.gov.uk/planning/Planning_Policy_Statements/Planning_Policy_Statement_1/pps1.pdf
Planning Policy Statement 6: Planning, Archaeology & Built Heritage (March 1999)	This PPS sets out the Department's planning policies for the protection & conservation of archaeological remains & features of the built heritage. It embodies the Government's commitment to sustainable development & environmental stewardship.	Introduction; Preamble; Planning & conservation; Identifying the archaeological & built heritage & the role of development plans; Archaeological sites & monuments; World heritage sites; Historic parks, gardens & demesnes; Listed buildings; Conservation areas; Industrial heritage; Non-listed vernacular buildings; Other aspects of planning control; Annex A: Transport & traffic management; Annex B: Legislative arrangements for archaeological sites; Annex C: The selection of listed buildings; Annex D: The upkeep & repair of listed buildings; Annex E: Guidelines for the alteration, extension & maintenance of listed buildings; Annex F: Useful addresses.	http://www.doeni.gov.uk/planning/Planning_Policy_Statements/Planning_Policy_Statement_6/PPS%206.htm

Continued

Table 7.5.3 Northern Ireland Level Planning Policy. *Continued*

Policy Type	Document Title	Objective	Contents	Website
Strategies & Reports	Planning Policy Statement 8: Open Space, Sport & Recreation (March 1999)	This PPS sets out the Department's planning policies for the protection of open space, the provision of new areas of open space in association with residential development & the use of land for sport & physical recreation. It embodies the Government's commitment to sustainable development & to the promotion of a more active & healthy lifestyle.	Preamble; Introduction; Policy context; Health & welfare; Sustainable development; Draft regional strategic framework; Policy objectives; Background information; The role of development plans; Planning policies – Policy OS1: Protection of open space; Policy OS2: Open space in new residential developments; Policy OS3: Outdoor recreation in the countryside; Policy OS4: Intensive sports facilities; Policy OS5: Noise generating sports; Policy OS6: Development assoc. with water sports; Policy OS7: The floodlighting of recreational facilities; Annex A: Key bodies which make a valuable contribution to development of sport, recreation & open space facilities; Annex B: Minimum standards of outdoor playing space; Annex C: Divisional planning offices.	http://www.doeni.gov.uk/planning/Planning_Policy_Statements/Planning_Policy_Statement_8/Draft%20PPS%208%20Open%20Space,%20Sport%20&%20Recreation.pdf
	Planning Policy Statement 9: The Enforcement of Planning Control (March 2000)	This Planning Policy Statement sets out the general policy approach that the Department will follow in taking enforcement action against unauthorised development in Northern Ireland. It also provides advice on the Department's approach to the enforcement of additional aspects of planning control over listed buildings, conservation areas, hazardous substances, trees & advertisements. Nothing in the Statement should be taken as condoning any breach of planning law.	Introduction; A breach of planning control; The general approach of the department to enforcement; Where acceptable but unauthorised development has been carried out; Where unauthorised development can be made acceptable through the imposition of conditions; Where unauthorised development is unacceptable; Where unauthorised development is unacceptable & urgent remedial action is required; Unauthorised development by private householders; Enforcement of planning control over waste disposal; Listed buildings & conservation areas; Hazardous substances; Trees; Advertisements; Appeals; Annex 1: Statutory time limits on enforcement action for a breach of planning control; Annex 2: Planning service offices.	http://www.doeni.gov.uk/planning/Planning_Policy_Statements/Planning_Policy_Statement_9/PPS%209%20Enforcement.pdf

Planning Policy Statement 7: Quality Residential Developments (June 2001)	This PPS sets out the Department's planning policies for achieving quality in new residential development & advises on the treatment of this issue in development plans. It embodies the Government's commitment to sustainable development & the Quality Initiative.	Preamble; Policy context; Introduction; The quality initiative; Regional development strategy; Planning policy statements; Moving forward; The NI transport policy statement; Supplementary planning guidance for residential development; Policy objectives; Development plans; Infrastructure requirements; Local design requirements; Concept master plans as supplementary planning guidance; Requirements for local neighbourhood facilities; Policies for establishing residential areas; Planning policies – Policy QD1: Quality in new residential development; Policy QD2: Design concept statements, concept master plans & comprehensive planning; Annex A: Local community development guidelines for urban & rural areas; Annex B: PPSs; Annex C: Planning service offices.	http://www.doeni.gov.uk/planning/Planning_Policy_Statements/Planning_Policy_Statement_7/PPS_7_Final.pdf
Belfast Metropolitan Area Plan 2015 (BMAP) Issues Paper (07/12/01)	A metropolitan plan which encompasses the entire built-up area & its immediate hinterland permits a strategic examination of the many inter-related issues – social, economic & environmental & transport which needs to be addressed on a metropolitan wide basis. It provides a major opportunity to influence the quality of the environment, both urban & rural, now and in the future.	Part 1: Introduction; Part 2: Recent trends; Part 3: Guiding principles; Part 4: Plan-wide issues; Part 5: Local issues; Part 6: Villages & small settlements; Part 7: The countryside; Part 8: Next steps; Appendix 1: Settlements in BMAP rural hinterland; Appendix 2: Existing conservation areas & areas of townscape character.	http://www.doeni.gov.uk/planning/Area_Plans/Belfast_Metropolitan_Area_Plan/BMAP2015.pdf
Planning Policy Statement 10: Telecommunications (April 2002)	This PPS sets out the Department's planning policies for telecommunications development. It embodies the Government's commitment to facilitate the growth of new & existing telecommunications systems whilst keeping the environmental impact to a minimum. The PPS also addresses health issues associated with telecommunications development.	Preamble; Introduction; Policy & legislative context; background information; Policy objectives; Development plans; Planning policies – Policy TEL1: Control of telecommunications development; Policy TEL2: Development & interference with television broadcasting services; Annex A: Accommodating telecommunications infrastructure in new development; Annex B: The radio communications agency & the control of radio interference; Annex C: Telecommunications systems Glossary.	http://www.doeni.gov.uk/planning/Planning_Policy_Statements/Planning_Policy_Statement_10/PPS_10.pdf

Continued

Table 7.5.3 Northern Ireland Level Planning Policy. *Continued*

Policy Type	Document Title	Objective	Contents	Website
Strategies & Reports	Planning Policy Statement 11: Planning & Waste Management (Dec 2002)	This PPS sets out the Department's planning policies for the development of waste management facilities. It seeks to promote the highest environmental standards in development proposals for waste management facilities & includes guidance on the issues likely to be considered in the determination of planning applications. In addition, it explains the relationship between the planning system & authorities responsible for the regulation & management of waste.	Preamble; Introduction; The relationship between the planning & pollution control regimes; Policy objectives; Development plans; Planning policies; Environmental impact; Waste collection & treatment facilities; Waste disposal; Land improvement; Development in the vicinity of waste management facilities; Annex A: Planning conditions; Annex B: Waste minimisation in new development; Annex D: European legislation; Annex C: Agricultural permitted development; Glossary.	http://www.doeni.gov.uk/planning/Planning_Policy_Statements/Planning_Policy_Statement_11/PPS11_DECEMBER_2002.pdf
	Regulatory Impact Assessment of Third Party Appeals (June 2004)	The objective of this RIA is to explore the risks, costs & benefits that are likely to accrue from the possible introduction of a third party appeals process. In line with RIA guidelines, costs and benefits are quantified where possible & where this is not possible detailed qualitative descriptions are be given. As a preliminary objective & in order to facilitate the above, the RIA attempts to identify what features a third party appeal system in Northern Ireland would or should have.	Objective, background & risk assessment; Consideration of options; Benefits & costs; Summary & interpretation of cost–benefit analysis; Third party appeals consultations completed; Appendix 1 – Staff & accommodation cost calculations.	http://www.planningni.gov.uk/Corporate_Services/Publications/TPA_RIA/TPA_RIA_Report.pdf
	Draft Belfast Metropolitan Area Plan 2015 (Nov 2004)	The purpose of the Plan is to inform the general public, statutory authorities, developers, & other interested bodies of the policy framework & land-use proposals that will be used to guide development decisions within the BMA over the Plan period. The Plan will help to give effect to the Regional Development Strategy (RDS).	Part 1 Introduction; Part 2 The Plan Strategy; Part 3 Strategic Plan Framework; Settlements; Housing; Employment; Transportation; Retailing; Offices; The urban environment; The natural environment; Countryside & coast; Open space, sport & outdoor recreation; Tourism; Public services & utilities; Education, health, community & cultural facilities; Appendices.	http://www.planningni.gov.uk/AreaPlans_Policy/Plans/BMA/draft_plan/00contents/00Contents.htm
	Technical Supplements to the Draft Belfast Metropolitan Area	These technical supplements underpin the draft metropolitan area plan and provide detailed information on a number of key issues such as – population growth & housing needs;	Population & housing; Employment; Transportation; Retailing; Offices; Urban environment; Open space, sport & outdoor recreation; Tourism; Public services & utilities;	http://www.planningni.gov.uk/AreaPlans_Policy/Plans/BMA/draft_plan/00contents/00ContentsTechSupps.htm

	Plan 2015 (Nov 2004)	employment trends; links to Belfast metropolitan transport plan; shopper, retailer & pedestrian surveys; office market appraisals; urban design & urban form; open space & accessibility; sustainable tourism; water, sewerage, drainage, waste disposal, electricity, natural gas, fire & telecommunications; education, health & community/cultural facilities; natural–built heritage & landscape assessment; collection of baseline data and assessment of impacts.	Education, health, community & cultural facilities; Countryside assessment; Strategic environmental assessment; Equality impact assessment; Report on public consultation.	
Consultation Papers	Draft Planning Policy Statement 12: Housing in Settlements (PPS12) (Nov 2002)	This Planning Policy Statement (PPS) has been prepared in accordance with the DRD's (Department for Regional Development) statutory function of providing policy guidance & advice under the Strategic Planning (Northern Ireland) Order 1999. Its purpose is to provide guidance in the form of strategic planning policy to assist the implementation of the RDS. This PPS flows directly from the vision, spatial strategy & strategic planning guidelines contained in the RDS. It is one of the key mechanisms for the implementation of the RDS (Regional Development Strategy).	Preamble; Policy context; Policy objectives; Development Plans; Process for allocating housing land; Measures to be contained in development plans; Implementing, monitoring & reviewing the development plan; Regional housing policies; Monitoring & review; Glossary; Appendices.	http://www.drdni.gov.uk/shapingourfuture/pdf/reports/PPS%2012.pdf
	Public Consultation Draft – PPS3: Access, Movement & Parking (AMP) (Dec 2002)	The draft PPS sets out the Department's planning policies for vehicular and pedestrian access, the protection of transport routes, transport assessment, and parking. It forms an important element in the integration of transport and land-use planning and embodies the Programme for Government's commitments to the provision of a modern, safe, sustainable transport system, the improvement of mobility for those who are socially excluded or whose mobility is impaired and the promotion of healthier living and improved road safety.	Preamble; Introduction; Policy context; Policy objectives; Development plans; Planning policies; Policy AMP1: Creating an accessible environment; Policy AMP2: Access to public roads; Policy AMP3: Access to protected routes; Policy AMP4: Route protection; Policy AMP5: Disused transport routes; Policy AMP6: Transport assessment; Policy AMP7: Parking & servicing in new development; Policy AMP8: Cycle provision in new development; Policy AMP9: Design of car-parking; Policy AMP10: Development of public/private carparks; Policy AMP11: Temporary car-parking; Annex A: Parking standards; Annex B: Equality impact assessment screening.	http://www.doeni.gov.uk/planning/Planning_Policy_Statements/Planning_Policy_Statement_3/PPS3%20Public%20Consultation%20Draft%20Nov%2002.pdf

Continued

Table 7.5.3 Northern Ireland Level Planning Policy. *Continued*

Policy Type	Document Title	Objective	Contents	Website
Consultation Papers	Draft Planning Policy Statement 13: Transportation & Land-use (PPS13) (Dec 2002)	This Planning Policy Statement (PPS) has been prepared in accordance with the DRD's (Department for Regional Development) statutory function of providing policy guidance and advice under the Strategic Planning (Northern Ireland) Order 1999. Its purpose is to provide guidance in the form of strategic planning policy to assist the implementation of the RDS. This PPS flows directly from the vision, spatial strategy and strategic planning guidelines contained in the RDS. It is one of the key mechanisms for the implementation of the RDS. This statement also recognises the significance of the Regional Transportation Strategy for Northern Ireland 2002–2012 (RTS) in delivering the transportation vision of the RDS (Regional Development Strategy).	Preamble; Policy context; Policy objectives; The process of integrating transportation & land-use planning; Land-use allocation; Locational criteria; The role of development plans; Monitoring & review; Glossary; Appendices.	http://www.drdni.gov.uk/shapingourfuture/pdf/PPS%2013.pdf
	Planning Policy Statement 4: Industry, Business & Distribution – Public Consultation Draft (Jan 2003)	This Planning Policy Statement, PPS4 sets out the Department's planning policies for industrial development & other defined business uses & indicates how economic & employment growth associated with such uses can be accommodated & promoted in development plans. It seeks to facilitate & accommodate economic growth in ways compatible with social & environmental objectives & sustainable development.	Preamble; Policy context; Policy objectives; The role of development plans; Planning policies; Annex 1 – Home-working; Annex 2: Screening for equality impact assessment.	http://www.planningni.gov.uk/AreaPlans_Policy/PPS/pps4/PPS4_Revised.pdf
	Development Control Advice Note 11: Access for All, Designing for an Accessible Environment –	This Advice Note has been drawn up following discussions with Disability Action, the Equality Commission & the Chief Building Control Officers Group. It seeks to increase awareness among building owners & managers, architects, surveyors & developers about their	Introduction; The relationship between planning control & building control; Legislation & codes of practice; Access audits & action plans; Planning policy; Improving access to historic buildings & sites; Design guidance; Annexes.	http://www.planningni.gov.uk/AreaPlans_Policy/Supplementary/DCAN/dcan11/dcan11.pdf

Public Consultation Draft (July 2003)	statutory & legal obligations regarding the needs of people with disabilities. Information about undertaking an access audit is included together with advice regarding improved access to historic buildings & sites.		
Reforming Planning: Proposals to Amend Primary Planning Legislation in Northern Ireland – Public consultation paper (Aug 2004)	This Consultation Paper seeks views on a series of legislative proposals affecting the development control process, enforcement & the preparation of development plans. The paper also outlines the current position on the issues of introducing a third party right of appeal in the planning system & making it a criminal offence to proceed with development without planning permission.	Foreword; Executive summary; Introduction; Development control; Enforcement; Preparing development plans; Following this consultation; Responding to this consultation paper; Annexes.	http://www.planningni.gov.uk/Corporate_Services/Publications/Reforming_Planning/E_Consultation_Paper.pdf
Public Consultation Draft – Planning Policy Statement 15 (PPS15) Planning & Flood Risk (Dec 2004)	This PPS sets out the Department's planning policies to reduce flood risk to people, property & the environment. It embodies the Government's commitment to sustainable development & the conservation of biodiversity. It adopts a precautionary approach to decision-making taking account of climate change, & promotes an approach to development supportive to the well-being & safety of people.	Introduction; Policy context; Flood risk – raising awareness; The precautionary principle; Policy objectives; The role of development plans; Development control considerations; Planning policies; Annexes.	http://www.planningni.gov.uk/AreaPlans_Policy/PPS/pps15/PPS15_draft.pdf
Public consultation Draft: Areas of Townscape Character – Addendum to PPS6 Planning, Archaeology & the Built Heritage (Dec 2004)	This document is an addendum to Planning Policy Statement 6 *Planning, Archaeology & the Built Heritage* & must be read in conjunction with the policies contained within that PPS. The addendum provides additional planning policies relating specifically to Areas of Townscape Character (ATC) for new development, the control of advertisements & demolition of buildings. It embodies the Government's commitment to sustainable development & environmental stewardship.	Preamble; Introduction; Planning policy; Policy ATC 1 – New development; Policy ATC 2 – The control of advertisements; Policy ATC 3 – Demolition; Annexes.	http://www.planningni.gov.uk/AreaPlans_Policy/PPS/pps6/PPS6_Addendum.pdf

Continued

Table 7.5.3 Northern Ireland Level Planning Policy. *Continued*

Policy Type	Document Title	Objective	Contents	Website
Guidance Notes	Development Control Advice Note 11: Access for People with Disabilities (March 1999)	The purpose of this Advice Note is to give general guidance to intending developers, their professional advisors & agents. This is designed to provide advice on the planning criteria to be applied in regard to the provision of facilities for disabled people in the case of developments to which the public are to have access. It should be stressed that the note is not a specific statement of Departmental policy but rather one of advice & guidance.	Introduction; Legislative background; Planning guidelines; Car-parking provision; Changes in level; Street furniture; Further information.	http://www.doeni.gov.uk/planning/Development_Control-Advice_Notes/Development_Control_Advice_Note_11/DCAN11.html
	Development Control Advice Note 10: Environmental Impact Assessment (Aug 1999)	The purpose of this Advice Note is to give general guidance to intending developers, their professional advisors & agents, on the operation of the Planning (Environmental Impact Assessment) Regulations (Northern Ireland) 1999. Any legal views expressed in this note have no statutory force & should not be relied upon as an authoritative interpretation of the law. The Regulations implement EC Directive 85/337/EEC as amended by Directive 97/11/EC on the assessment of the impact of certain public & private projects on the environment insofar as it applies to development under the Planning (Northern Ireland) Order 1991. They incorporate the requirements of Directive 97/11/EC into the planning system in Northern Ireland & consolidate the Planning (Assessment of Environmental Effects) Regulations (Northern Ireland) 1989 & subsequent amending regulations.	Introduction; Aims; Background; The regulations; Schedule 1 developments; Schedule 2 developments; The screening process; Establishing whether a development requires EIA; Indicative criteria & thresholds; Outline planning applications; Procedures for establishing whether or not EIA is required; Pre-planning application procedures; Planning applications not accompanied by an Environmental statement (ES); Planning applications submitted with an ES; Permitted development; Crown development; Time periods for decisions; The role of consultees; ES: their adequacy & publicity; Determining the planning application; Securing mitigation measures; Publicising decisions on EIA applications; Unauthorised development; Transboundary cases; Financial implications; An illustrative guide to the thresholds system; Annex A: Indicative thresholds & criteria for identification of Schedule 2 development requiring EIA; Annex B: Consultees; Annex C: Selection criteria for screening Schedule 2 developments; Annex D: Information to be included in an ES.	http://www.doeni.gov.uk/planning/Development_Control-Advice_Notes/Development_Control_Advice_Note_10/Dcan%2010%20final.pdf

Creating Places – Achieving Quality in Residential Developments (May 2000)	This guide describes the contributions to quality & sustainability that developers in Northern Ireland will be expected to make through the design of new residential developments. It seeks to ensure that what is designed & built today will be cherished by both present & future generations. The guide is for use by all those involved in the design of new residential developments & the rejuvenation of existing housing areas – primarily house-builders, architects, landscape architects, urban designers, planners & road engineers.	Foreword; Preamble; Introduction; Quality & sustainability through design; Part 1: Context for overall design character; Respecting the local setting; Responding to the site; Creating attractive surroundings & places to live; Part 2: Main elements of design; Landscape design; Open space; Local neighbourhood facilities; Dwellings; Pedestrian & cycle routes; Bus routes; Routes for cars & service vehicles; Parking provision; Provision for services; Part 3: Detailed design requirements; Planting; Footways, verges, footpaths & cycle tracks; Special facilities for buses; Carriageways; Visibility; Speed restraints; Shared surfaces; Parking spaces: numbers & dimensions; Utilities' services; Appendix A: Definitions; Appendix B: Information required; Appendix C: Adoption & maintenance; Appendix D: Planning & Roads service offices; Appendix E: Public consultation; Appendix F: Regional planning policy & publications; Further reading; Acknowledgements.	http://www.doeni.gov.uk/planning/Guidance/Creating_Places/Creating_Places.pdf
Improving the Quality of Housing Layouts in N.I. (2000)	Since the introduction of the Quality Initiative to Northern Ireland, developers, urban designers, architects, landscape architects, planners and road engineers have collectively made considerable progress towards improving the quality of new residential environments. There is now common understanding of the need to adopt a more holistic design-led approach. The production of this document confirms the will to work together and strive for the highest possible standards in external residential environments.	Foreword; Preface; Introduction; Landscaping; House types; Roads standards; Planning standards; Land values; The design process; Site appraisal; Concept statement; Sketch plan; Notes.	http://www.doeni.gov.uk/planning/Guidance/Improving_the_Quality_of_Housing_Layouts.pdf

Continued

Table 7.5.3 Northern Ireland Level Planning Policy. *Continued*

Policy Type	Document Title	Objective	Contents	Website
Guidance Notes	Development Control Advice Note 8: Housing in Existing Urban Areas (June 2002)	The purpose of this advice note is to provide advice which will help to ensure that urban & environmental quality is maintained, amenity preserved, & privacy respected when proposals are being considered for new housing development within existing urban areas.	Executive summary; Introduction; Policy context; Regional development strategy; PPS1 – General principles; PPS7 – Quality residential environment; PPS12 – Housing in settlements; Supplementary planning guidance; A design-led approach to housing in existing urban areas; New housing in established residential areas; Maintaining environment quality; Appreciating the context; Building on local character; Protecting the characteristics which contribute to local character; Types of proposals for new residential development in existing urban areas; Demolition & redevelopment; Backland development; Conversion/Extension of houses; Living over the shop (LOTS); Appendix 1: Need for equality impact assessment; Appendix 2: New TSN implications; Appendix 3: List of development control advice notes.	http://www.doeni.gov.uk/planning/Development_Control-Advice_Notes/Development_Control_Advice_Note_8/DCAN_8.pdf
	Planning Policy Statement 8 – Open space, sport & outdoor recreation (Feb 2004)	This PPS sets out the Department's planning policies for the protection of open space, the provision of new areas of open space in association with residential development & the use of land for sport & outdoor recreation, & advises on the treatment of these issues in development plans. It embodies the Government's commitment to sustainable development, to the promotion of a more active and healthy lifestyle & to the conservation of biodiversity.	Introduction; Policy context; Policy objectives; The role of development plans; Planning policies; Policy OS1 – Protection of open space; Policy OS2 – Public open space in new residential development; Policy OS3 – Outdoor recreation in the countryside; Policy OS4 – Intensive sports facilities; Policy OS5 – Noise generating sports & outdoor recreational activities; Policy OS6 – Development of facilities ancillary to water sports; Policy OS7 – The floodlighting of sports & outside recreational facilities; Annexes.	http://www.planningni.gov.uk/AreaPlans_Policy/PPS/pps8/PPS8.pdf

8 Pollution

8.1 Pollution and the Built Environment

The causes and effects of pollution in the built environment are inextricably linked with environmental matters and securing the appropriate levels of protection. Hence pollution is a direct result of many other activities, but its sources, control, effects and implications have been extensively highlighted in both strategies and in wide-ranging monitoring programmes. The International Pollution Prevention Summit in 2000 at Montreal, Canada, provided an opportunity for national and regional roundtables to explore how pollution prevention could be better understood and implemented as a result of partnerships with other organisations and agencies. This resulted in action plans aimed at fostering change, expanding education and awareness, financing appropriate work and research, and national-level policy development. This event also saw the launch of the Global Information Network, a cross-regional programme linking innovation, ideas, practitioners and decision-makers.

Pollution is essentially an anthropogenic process which can be characterised as a negative impact on life's essentials that are provided by the natural environment. It is typified by water pollution, air pollution and land contamination, primarily through waste disposal. Typical examples of such pollution are sewage, smog, hazardous waste, fertilisers and domestic waste, with this pollution only becoming a problem when the natural assimilative ability of the ecosystem in question is reached. Pollution tends to be more prevalent in large population centres such as mega-cities, often referred to as mega-slums, where widespread technology and industrial development, supplemented by other incidents such as major oil spills or nuclear accidents, can prevent the local ecosystems from sustaining a reasonable quality of life. Pollution prevention is recognised as the key response, alongside better production methods, an improved health and safety commitment, improved monitoring and environmental justice similar to the 'polluter pays' principle.

The impact of pollution falls unequally on humanity across the Earth with clear disparities between developing and developed nations, with animals, birds, crops and all living organisms affected. Therefore the actions of communities and industry in use of or interaction with the essentials of water, air and land are influenced by their conduct in respect of key human responsibilities such as energy consumption, urban and rural planning, transport planning and travel patterns, water conservation and usage rates and waste management practices, with environmental damage limitation or protection as the overarching consideration.

The European Directive CD 96/61/EC on *Integrated Pollution Prevention and Control* (IPPC) established an important structure to help prevent or minimise air, water and soil pollution caused by emissions from industrial installations in the European Community, with a view to achieving a high level of environmental protection. Several other European Directives and Communications dealt with pollution streams in specific media or locations such as water and air. In turn the Environment Agencies in the UK developed a series of Pollution Prevention Guidance notes, known as PPGs, providing practical advice to help avoid the causes of pollution, minimise waste outputs and comply with legal requirements. The PPGs recognise that the necessary preventive measures cost less if identified at an early design stage with the potential to save money in the long-term. In contrast, the fines for failing to comply with the relevant regulations or the costs of cleaning up pollution, normally recovered from the polluter, can be very high.

Several relevant environmental and pollution-related principles have been established as guidelines including the following:

- *Polluter Pays Principle* – that the polluter should bear full costs of all measures, encourages a rational approach to costing and avoids distortion of internal trade and investment.
- *Proximity Principle* – that disposal of waste be close to the place of generation, minimising environmental damage by waste movement controls, acts as a planning tool and raises awareness in communities that waste is their problem.
- *Self Sufficiency Principle* – that nations be self sufficient, driven by CD 259/93/EC *Waste Shipments Regulations* and the Basel Convention (1989) for control of hazardous wastes across boundaries, and that prior consent is needed for waste movement encourages responsible planning and development.

8.2 Key Pollution Elements

The pollution history of the Earth can be described through a sequence of changes in the nature and causes of pollution:

- Pollution caused by *bacteria* living and developing in decaying and putrefying materials and stagnant water associated with settlements of people which are growing in size;
- Pollution associated with *artisans* involving small-scale craft industries and activities such as tanneries, potteries, soap manufacture, bone burning and glue making;
- *Industrial* pollution from large scale and pervasive polluting organisations in major centres;
- Pollution from *fundamental* and in-built activities across whole regions such as the desiccation and subsequent salination of sea catchments;
- *Proximity* pollution in which large quantities of chemicals are applied to land as fertilisers and biocides, with an environmental cost, and possibly physical tax, to be paid later;
- *Accidental* pollution resulting from major accidents which are neither foreseen nor can be calculated.

These pollution processes or causes, resulting primarily from human activity, can be measured in a broad set of 'component indicators of the biosphere'. These cover measures of:

- deforested area;
- terrestrial vertebrate diversity;
- carbon dioxide releases;
- population size;
- water abstractions or withdrawals; and
- nitrogen releases.

It is pertinent to note that pollution, as expressed in these component indicators, has changed more in a single population generation since 1950 than in the whole of human history.

Air pollution exists in gaseous and particulate forms. Gaseous pollutants occur in primary forms such as sulphur dioxide, oxides of nitrogen and carbon monoxide, probably emitted from industrial sources. Secondary gaseous air pollutants occur in the form of ozones and products of photochemical reactions as a result of subsequent chemical processes in the atmosphere, involving primary pollutants and sunlight. Particulate pollutants consist of very small solid or liquid-suspended droplets, such as dust, smoke and aerosolic salt, and contain a wide range of soluble and insoluble components. These air pollutants are toxic and can affect humans, animals and vegetation to varying degrees.

Water pollution has been understood for much longer than air pollution, as it has been recognised as a cause of human, animal and fish infection, disease and death. The main groups of water pollutants and their causes are listed below:

- Sewage and other oxygen-demanding wastes, via human and animal excreta;
- Infectious and disease-transmitting agents, via water-borne activities;
- Organic chemicals, via industrial and agricultural effluents;
- Chemical, mineral and heavy metal substances, via materials extractive processes which intercept with the water table;

- Sediments and water turbidity, via materials movement in watercourses;
- Radioactive substances, via industrial sectoral deposits;
- Acid rain affecting rain forests, via air pollutants and subsequently via surface water;
- Thermal pollution, via heat emissions into bodies of water from turbines and power generating units.

Water pollutants can be derived from single point or diffuse sources. The former will be readily identified by tracing back to the source, while the latter may be caused by a combination of actions and will require chemical and biological forensic analysis.

Land pollution can be categorised into contamination and land-blight zones, with possibly both types occurring at a single location. Contamination can arise from a range of industrial and commercial activities in which undesirable deposits or spoil heaps have been abandoned in pursuit of progress and profit. The subsequent need for de-contamination therefore arises from potential or actual polluting incidents or as a result of the enforcement of environmental legislation. Pollution of the natural environment can occur when poor design and planning processes are in operation, resulting in domestic or industrial buildings and an infrastructure provision which deploys an inappropriate use of land and the absence of good design principles. This latter aspect of pollution is seen as a soft issue but has strong resonance with many other themes which need to be in tune and support of each other in order to deliver sustainable development.

Several elements of legislation are in place to encourage or drive pollution reduction and these need to have appropriate penalties attached in order to satisfy the 'polluter pays' principle. However, the use of guidelines, engagement with professional groups and their advice, and a psychological desire to eliminate pollution are all assisting progress towards environmental protection. Reference to Chapter 6 clarifies the process tools which are used to assist with holistic decision-making and the identification of links between different activities and pollution.

8.3 Sustainability Challenges for Pollution

Pollution challenges fall mainly within the 'protection of the environment' sector of sustainable development, with discrete links to the other objectives, as housing and industrial developments are necessary to meet society's basic food and shelter needs, and these can produce some pollutants. Likewise this pollution comes at an environmental and a financial cost. Pollution can arise as a side effect of human activities rather than there being a primary cause, but has strong implications for sustainable development. The interactions between pollution and the four sustainable development objectives are as follows:

- *Effective protection of the environment* is a definitive message for the reduction or eradication of all sources of pollution. To remove pollutants from the air, water and on land requires a determined international effort, with a high level of funding for new processes, a cohort of highly experienced staff and a strong legislative structure. However this lofty goal is probably not possible as it would imbalance the other sustainable development objectives. Hence, progress can only be made incrementally by tackling the task of reducing high-risk pollutants and building up a stronger database of the causes and effects of each potentially polluting activity.

- *Social progress which recognises the needs of everyone* relates directly to the attitude and behaviour of society across several phases and actions, and has a direct relationship to pollution incidents as many of these arise from human-based activity. In turn, a society's quality of life, health and well-being relate to the presence of pollutants in the air, in surface and drinking water and the land on which communities depend for shelter, food and recreation. As pollution is a secondary outcome, society will only enjoy a pollution-free environment when the primary sources are eliminated. However, communities frequently have scant regard for the environment, so that social progress comes at both an environmental and financial cost, requiring society to both understand and make voluntary responses towards environmental protection.

- *Prudent use of natural resources* encapsulates the contribution made by members of society and business decision-makers as they are encouraged to procure and use goods which have a minimal polluting effect. Emphasis on recycled materials will reduce natural resource use, while a 'green approach' to all domestic and industrial practices will reduce pollution. Pollution incidents can arise from improper use of processes but can also occur due to unexpected events, requiring responsive incident management to mitigate accidental pollution and environmental damage.

- *Maintenance of high and stable levels of economic growth and employment* may not necessarily cause pollution and environmental damage, for pollution may still arise in areas of low socio-economic status. The use of Best Practicable Environmental Option (BPEO) and Best Available Technology Not Entailing Excessive Cost (BATNEEC) assist in a balanced approach, while the inclusion of environmental reporting in annual corporate reports demonstrates environmental awareness, and specific reporting on pollution incidents. Increased employment and profitability can result from more efficient and cleaner businesses.

Several issues and challenges exist for a range of built environment professionals in reducing pollution and enhancing environmental protection to deliver sustainable development. This pollution theme is integral with the broader sustainable development agenda and the critical challenges in pollution control and prevention include:

- *Full use of Environmental Management Systems and Business Eco-Efficiency*, through the appropriate application of EMS, Environmental Management Auditing Systems or accreditation to ISO9001, ISO14001 or ISO18001 standards. Specific application of the Integrated Pollution Prevention and Control legislation should identify optimum solutions and facilitate greenhouse gas emissions allowance trading schemes, permit emissions ceilings and limit values to be set, demand air quality monitoring, and give support to the Clean Air for Europe (CAFE) Programme.
- *Commitment to Environmental Reporting* is essential to achieve buy-in from businesses and the government to underpin pollution reduction and environmental protection. This involves companies reporting on the environmental impacts of their operations and allowing freedom of access to environmental information. There is also the monitoring of environmental impacts of EC Member States and the facilitation of public access to information and participation in environmental decision-making.
- *Application of Environmental Assessment Methodologies* will ensure that the most appropriate solutions are achieved while accounting for a full range of potential polluting and environmental impacts, as required by European Directive 97/11/EC. Several assessment methods are available such as Environmental Impact Assessment (EIA), Best Practicable Environmental Option (BPEO), Regulatory Impact Assessment (RIA), Environmental Risk Assessment (ERA), and Strategic Environmental Assessment (SEA).
- *Adoption of Environmental Technologies* which enhance or protect the environment by using cleaner technology, reducing emissions, living within environmental limits and contributing to the use of appropriate technology. A specific challenge arises in the handling of pollution from hazardous chemicals, such as pesticides and surface run-off causing eutrophication, as well as the presence of persistent organic pollutants and volatile organic compounds.
- *Investigation into the links between air quality and other key activities,* such as greenhouse gas emissions and global warming relationships, by studying links between human health and air pollution, links with transport and exhaust emissions, links with waste incineration, links with CFC gases, ozone layer depletion and skin cancer, and wider links to acid rain and possible building deterioration and crop destruction.
- *Investigation into the links between noise pollution and other key activities,* by studying issues of permissible sound levels for vehicles, airports and household appliances, noise complaints issues and social impact on local communities.
- *Investigation into the links between land contamination and other key activities,* such as soil protection from erosion and contamination, by studying links to crop production, links between land contamination and human health, relationships between PVC in landfill pollution and performance, issues around land remediation solutions, due costs of pollutant removal and reuse of land after remediation.
- *Full commitment to environmental liability responsibilities* by endorsing principles and practice such as 'polluter pays', the 'precautionary principle', 'self sufficiency' and the 'proximity principle'. These can prevent serious environmental crime, assist in identifying polluters and combat environmental offences, but international consistency and co-operation is essential for effective delivery.

- *Consistency in local environmental performance, quality and management* is essential for delivering sustained environmental messages. Issues with potential local application include control of litter, refuse and fly-tipping, impact of landfill usage and attitudes to disposal of harmful goods, design and housekeeping of green spaces and streetscapes, management of urban parks and city features, development of cleaner, greener and safer play areas and multi-use zones, protection of urban and rural landscapes, country parks, ASSIs and AONBs.

8.4 Pollution Theme Index

Jurisdiction	Tier	Title	Date	Page No.
INTERNATIONAL	Conference	International Pollution Prevention Summit	2000	286
	Conventions	Convention on the Prevention of Marine Pollution by Dumping Waste & Other Matter (London Dumping Convention)	1972	286
		International Convention for the Prevention of Pollution from Ships (MARPOL 73/78)	1973	286
		Convention on Long-Range Transboundary Air Pollution (UNECE)	1979	287
		Vienna Convention for the Protection of the Ozone Layer	1985	287
		Convention on Oil Pollution Preparedness, Response & Co-operation (OPRC)	1990	287
		Convention on Environmental Impact Assessment in a Transboundary Context	1991	288
		UN Framework Climate Change Convention (UNFCCC)	1992	288
		Convention on the Prior Informed Consents (PIC) Procedure for certain Hazardous Chemicals & Pesticides in International Trade	1998	289
		Stockholm Convention on Persistent Organic Pollutants (POPs)	2001	289
	Protocols	Montreal Protocol on Substances that Deplete the Ozone Layer	1987	289
		Sofia Protocol Concerning the Control of Emissions of Nitrogen Oxides	1988	290
		Geneva Protocol concerning the Control of Emissions of Volatile Organic Compounds	1991	290
		Oslo Protocol on Further Reduction of Sulphur Emissions	1994	291
		Kyoto Protocol to the UN Framework Convention on Climate Change (COP-3)	1997	291
EUROPEAN	Council Directives	Motor Vehicles with Trailers – Permissible Sound Level *(CD 70/157/EEC Amended by Com Dir 73/350/EEC; CD 77/212/EEC; Com Dir 81/334/EEC; CD 84/372/EEC; CD 84/424/EEC; CD 87/354/EEC; Com Dir 89/491/EEC; CD 92/97/EEC; Com Dir 96/20/EC; Com Dir 99/101/EC)*	1970 1973 1977 1981 1984 1987 1989 1992 1996 1999	292

Jurisdiction	Tier	Title	Date	Page No.
EUROPEAN	Council Directives	Motor Vehicles with Trailers: Petrol & Diesel Engines *(CD 70/220/EEC Amended by CD 74/290/EEC; Com Dir 77/102/EEC; Com Dir 78/665/EEC; CD 83/351/EEC; CD 88/76/EEC; Com Dir 89/491/EEC; CD 89/491/EEC; CD 91/441/EEC; CD 93/59/EEC; CD 94/12/EC; Com Dir 96/44/EC; CD 96/69/EC; CD 98/69/EC; Com Dir 99/102/EC; CD 2001/1/EC; CD 2001/100/EC)*	1970 1974 1977 1978 1983 1988 1989 1991 1993 1994 1996 1998 1999 2001	293
		Two or Three Wheeled Motor Vehicles – Permissible Sound Level of Motorcycles *(CD 78/1015/EEC Amended by CD 87/56/EEC; CD 89/235/EEC)*	1978 1987 1989	293
		Limit Value for Lead in the Air *(CD 82/884/EEC Amended by CD 90/656/EEC; CD 91/692/EEC)*	1982 1990 1991	294
		Quality Standards for Nitrogen Dioxide *(CD 85/203/EEC)*	1985	294
		Airborne Noise Emitted by Household Appliances *(CD 86/594/EEC)*	1986	294
		Tower Cranes – Sound Levels *(CD 87/405/EEC)*	1987	294
		Motor Vehicles with Trailers: Emission of Gaseous Pollutants from Diesel Engines *(CD 88/77/EEC Amended by CD 91/542/EEC; CD 96/1/EC; CD 1999/96/EC)*	1988 1991 1996 1999	294
		Pollution from Large Combustion Plants *(CD 88/609/EEC Amended by CD 94/66/EC; Dir 2001/80/EC)*	1988 1994 2001	295
		Air Pollution by Ozone *(CD 92/72/EEC)*	1992	295
		Sulphur Content of Certain Liquid Fuels *(CD 93/12/EEC Amended by Dir 98/70/EC; CD 99/32/EC)*	1993 1998 1999	295
		Volatile Organic Compounds (VOCs) Resulting from the Storage of Petrol *(CD 94/63/EC)*	1994	295
		Integrated Pollution Prevention & Control: Directive IPPC *(CD 96/61/EC)*	1996	296
		Management & Quality of Ambient Air *(CD 96/62/EC)* (Air Quality Framework Directive)	1996	296
		Non-Road Mobile Machinery: Gaseous Pollutants *(CD 97/68/EC Amended by CD 2001/63/EC; CD 2002/88/EC; CD 2004/26/EC)*	1997 2001 2002 2004	296
		Quality of Petrol & Diesel Fuels *(Dir 98/70/EC Amended by Dir 2003/17/EC)*	1998 2003	296
		Volatile Organic Compounds (VOCs) Resulting from Certain Industrial Activities *(CD 1999/13/EC)*	1999	297
		Sulphur Dioxide, Nitrogen Dioxide & Nitrogen Oxides, Particulates & Lead in the Ambient Air *(CD 1999/30/EC)* (First Daughter Directive)	1999	297

Jurisdiction	Tier	Title	Date	Page No.
EUROPEAN	Council Directives	Information on the Fuel Consumption & CO_2 Emissions of New Cars *(CD 1999/94/EC)*	1999	297
		Noise Emitted by Equipment Used Outdoors *(Dir 2000/14/EC)*	2000	297
		Limit Values for Benzene & Carbon Monoxide in Ambient Air (Second Daughter Directive) *(Dir 2000/69/EC)*	2000	298
		National Emission Ceilings for Certain Atmospheric Pollutants *(Dir 2001/81/EC)*	2001	298
		Third Daughter Directive Relating to Ozone *(Dir 2002/3/EC)*	2002	298
		Establishment of Rules & Procedures with regard to the Introduction of Noise-related Operating Restrictions at Community Airports *(Dir 2002/30/EC)*	2002	299
		Assessment & Management of Environmental Noise *(Dir 2002/49/EC)*	2002	300
		Establishing a Scheme for Greenhouse Gas Emission Allowance Trading within the Community (amending Council Directive 96/61/EC) *(Dir 2003/87/EC)*	2003	300
		Establishing a Scheme for Greenhouse Gas Emission Allowance Trading within the Community, in respect of the Kyoto Protocol's Project Mechanisms (amending Dir 2003/87/EC) *(Dir 2004/101/EC)*	2004	300
	Council Decisions	Mechanism for Monitoring CO_2 & other Greenhouse Gas Emissions *(Dec 93/389/EEC Amended by Dec 99/296/EC)*	1993 1999	301
		Reciprocal exchange of Information from Networks & Individual Stations Measuring Ambient Air Quality *(C Dec 97/101/EC Amended by C Dec 2001/752/EC)*	1997 2001	301
		CO_2 Emissions from New Passenger Cars: Monitoring *(Dec No. 1753/2000/EC)*	2000	301
		Protocol on Heavy Metals	2001	301
		Kyoto Protocol on Climate Change *(Dec 2002/358/EC)*	2002	301
		The Conclusion of the Stockholm Convention on Persistent Organic Pollutants *(COM (2003) 331)*	2003	302
		Establishing Guidelines for the Monitoring & Reporting of Greenhouse Gas Emissions pursuant to Directive 2003/87/EC *(C (2004) 130)*	2004	302
		Limit Values for Certain Pollutants in Ambient Air *(C Dec 2004/224/EC)*	2004	302
		The Conclusion of the Protocol to the 1979 Convention on Long Range Transboundary Air Pollution on Persistent Organic Pollutants *(C Dec 2004/259/EC)*	2004	302
	Council Communications	Transport & CO_2 – Developing a Community Approach *(COM (98) 204)*	1998	303
		Climate Change – Towards an EU Post-Kyoto Strategy *(COM (98) 353)*	1998	303
		Phase-out of CFCs in Metered Dose Inhalers *(COM (1998) 603)*	1998	303

Jurisdiction	Tier	Title	Date	Page No.
EUROPEAN	Council Communications	Preparing for Implementation of the Kyoto Protocol *(COM (1999) 230)*	1999	303
		Policies & Measures to Reduce Greenhouse Gas Emissions: towards a European Climate Change Programme (ECCP) *(COM (2000) 88)*	2000	304
		Clean Air For Europe (CAFE) Programme *(COM (2001) 245)*	2001	304
		Implementation of the European Climate Change Programme (ECCP) *(COM (2001) 580)*	2001	304
		Towards a Thematic Strategy for Soil Protection *(COM (2002) 179)*	2002	304
		On the Road to Sustainable Production *(COM (2003) 354)*	2003	305
		Implementation of Criteria Listed in Annex III of Directive 2003/87/EC – Greenhouse Gas Emission Allowance Trading *(COM (2003) 830)*	2004	305
	Miscellaneous	Commission Green Paper on future noise policy *(COM (96) 540)*	1996	305
		Nitrogen Emissions from Civil Subsonic Jet Aeroplanes *(COM (97) 629)*	1997	306
		Guidance Report on Preliminary Assessment under EC Air Quality Directives	1998	306
		Wheeled Agricultural or Forestry Tractors: Pollutant Gases *(COM (1998) 444)*	1998	306
		Environmental Problems Caused by Heavy Goods Vehicles (HGVs) *(COM (98) 472)*	1998	307
		Registration & Operation within the Community of Certain Types of Civil Subsonic Jet Aeroplanes *(C Reg No. 925/1999)*	1999	307
		EU Focus on Clean Air	1999	307
		Guidance on Assessment under the EU Air Quality Directives – Final Draft	2000	307
		Green Paper on Greenhouse Gas Emissions Trading within the EU *(COM (2000) 87)*	2000	308
		Proposal for a Directive relating to the Assessment & Management of Environmental Noise *(COM (2000) 468 Amended by Reg No. 2038/2000; Reg No. 2039/2000)*	2000	308
		The Behaviour of PVC in Landfill	2000	308
		The Noise Policy of the European Union – Towards Improving the Urban Environment & Contributing to Global Sustainability	2000	308
		The Auto Oil II Programme – A Report from the Services of the European Commission	2000	309
		Substances which Damage the Ozone Layer *(Reg No. 2037/2000)*	2000	309
		Greenhouse Gas Emissions Trading Scheme *(COM (2001) 581)*	2001	309
		Economic Evaluation of Sectoral Emission Reduction Objectives for Climate Change Top-down Analysis of Greenhouse Gas Emission Reduction Possibilities in the EU	2001	310

Jurisdiction	Tier	Title	Date	Page No.
UK	Strategies & Reports	Dealing with Contaminated Land in England – Progress in 2002 with implementing the Part IIA regime	2002	317
		Implementation of Directive 96/61/EC Concerning Integrated Pollution Prevention & Control	2003	318
		Audit of UK Soil Research – Final Report	2003	318
		Use of Multi-Criteria Analysis in Air Quality Policy	2003	318
		Plans & Programmes under the 1st Air Quality Daughter Directive	2003	319
		Report to the European Commission on Plans & Programmes to meet Limit Values under the 1st Air Quality Daughter Directive (1999/30/EC)	2003	319
		The First Soil Action Plan for England: 2004–2006	2004	319
		The State of Soils in England & Wales	2004	320
		Expert Group Report on Nitrogen Dioxide – Nitrogen Dioxide in the United Kingdom	2004	320
		Performance Review 2003/4 of Local Authority Implementation of LAPC, LAPPC & LA-IPPC Regimes – Final Report	2004	321
		Valuation of Health Benefits Associated with Reductions in Air Pollution – Final Report	2004	321
	Consultation Papers	Proposed Regulations Transposing the EC Air Quality Framework Directive & First Daughter Directive – Consultation Paper	2001	322
		Draft Soil Strategy for England – Consultation Paper	2001	322
		Consultation Paper on Implementation of Council Directive 2000/76/EC on the Incineration of Waste	2002	322
		Second & Third Air Quality Daughter Directives: Consultation Paper	2002	322
		Plans & Programmes under the First Air Quality Daughter Directive: A consultation paper on the UK's Plans & Programmes to meet limit values set in EC Directive 1999/30/EC, the First Daughter Directive	2004	323
		EU Emissions Trading Scheme – Proposed Amendments to the UK National Allocation Plan Consultation Document	2004	323
	Guidance Notes & Circulars	Contaminated Land (England) Regulations 2000 & Statutory Guidance: Regulatory Impact Assessment (Final)	2000	323
		Contaminated Land: DETR Circular 2/2000	2000	323
		Integrated Pollution Prevention & Control a Practical Guide – Second Edition	2002	324
		Integrated Pollution Prevention & Control: A Practical guide – Third Edition	2004	325
NORTHERN IRELAND	Strategies & Reports	Air Quality Monitoring in Northern Ireland: 2000–2001	2002	326
		Implementation of the Environmental Noise Directive	2003	326
		Air Quality Monitoring in Northern Ireland: 2002	2004	326

Jurisdiction	Tier	Title	Date	Page No.
NORTHERN IRELAND	Consultation Papers	EC Directive 96/61 on Integrated Pollution Prevention & control – Arrangements for Transposition in NI (A Consultation Paper)	2001	327
		Consultation Document on a Proposed Bill for Local Air Quality Assessment & Management in Northern Ireland	2001	327
		The Environment (Northern Ireland) Order 2002 – Local Air Quality Management Draft Policy Guidance	2002	327
	Guidance Notes & Circulars	Pollution Prevention Guideline – PPG1: General Guide to Pollution Prevention	2000	328
		PPG5: Works in, Near or Liable to Affect Watercourses	2001	328
		PPG6: Working at Demolition & Construction Sites	2001	328
		PPG11: Industrial Sites	2001	328
		PPG15: Retail Premises	2001	329
		IPC Guidance Note 1: Introduction to the Order	2002	329
		Guidance on Drafting Codes of Practice for Minimising Noise in Northern Ireland	2003	329
		Integrated Pollution Prevention & Control (Northern Ireland) – A Practical Guide: Edition One	2003	330
		Integrated Pollution Prevention & Control (IPPC) – Guide for Applicants	2003	330
		EU Emissions Trading Scheme Guidance Note One	2003	331
		Depolluting End-of Life Vehicles – Guidance for Authorised Treatment Facilities	2004	331

8.5 Pollution Policy Tables

Table 8.5.1 International Level Pollution Policy.

Title	Venue	Purpose	Key Outcome(s)	Website
International Pollution Prevention Summit	Montreal, Canada (2000)	The Summit provided an opportunity for national and regional roundtables to explore how pollution prevention could be better understood and implemented as a result of roundtables working together and in partnership with other organisations and agencies.	Outcomes from the Summit included: ▪ action plans for fostering change, expanding education and awareness, financing and national-level policy development. ▪ the launch of the Global Information Network, a cross-regional program linking innovation, ideas, practitioners and decision-makers.	http://www.c2p2online.com/main.php3?doc_id=221&section=50
Convention on the Prevention of Marine Pollution by Dumping Waste & Other Matter (London Dumping Convention)	London, UK (1972)	To improve protection of the marine environment by encouraging States with a common interest in particular geographical areas to enter into appropriate agreements supplementary to this Convention. Superseded by the 1996 Protocol.	An agreement that Contracting Parties shall individually and collectively promote the effective control of all sources of pollution of the marine environment, and pledge themselves especially to take all practicable steps to prevent the pollution of the sea by the dumping of waste and other matter that is liable to create hazards to human health, to harm living resources and marine life, to damage amenities or to interfere with other legitimate uses of the sea. A comprehensive listing of substances which are prohibited from ocean disposal. Regulations for the control of incineration of wastes and other matter at sea.	http://www.londonconvention.org/lc1972.doc http://www.londonconvention.org/1996%20PROTOCOL.doc
International Convention for the Prevention of Pollution from Ships (MARPOL 73/78)	London, UK (1973)	The MARPOL Convention is the main international convention covering prevention of pollution of the marine environment by ships from operational or accidental causes. It is a combination of two treaties adopted in 1973 and 1978 respectively and has been updated by amendments through the years.	The Convention includes regulations aimed at preventing and minimising pollution from ships – both accidental pollution and that from routine operations – and currently includes six technical Annexes: ▪ Annex I – Regulations for the Prevention of Pollution by Oil ▪ Annex II – Regulations for the Control of Pollution by Noxious Liquid Substances in Bulk	http://www.imo.org/Conventions/mainframe.asp?topic_id=255

		The Convention covered pollution by oil, chemicals, harmful substances in packaged form, sewage and garbage. (modified by the Protocol of 1978 relating thereto (MARPOL 73/78).)	■ Annex III – Prevention of Pollution by Harmful Substances Carried by Sea in Packaged Form ■ Annex IV – Prevention of Pollution by Sewage from Ships (not yet in force) ■ Annex V – Prevention of Pollution by Garbage from Ships ■ Annex VI – Prevention of Air Pollution from Ships (adopted September 1997 – not yet in force)	
Convention on Long-Range Transboundary Air Pollution (UNECE)	Geneva, Switzerland (1979)	This Convention addressed some of the major environmental problems of the UNECE region. This has been achieved through a process of scientific collaboration and policy negotiation.	The Convention was the first international legally binding instrument to deal with problems of air pollution on a broad regional basis. Besides laying down the general principles of international co-operation for air pollution abatement, the Convention set up an institutional framework bringing together research and policy. The Convention on Long-range Transboundary Air Pollution entered into force in 1983. It has been extended by eight specific protocols which identify specific obligations or measures to be taken by Parties.	http://www.unece.org/env/lrtap/
Vienna Convention for the Protection of the Ozone Layer	Vienna, Austria (1985)	For the first time nations agreed in principle to tackle a global environmental problem before its effects were felt, or even scientifically proven.	Nations agreed to take 'appropriate measures . . . to protect human health and the environment against adverse effects resulting or likely to result from human activities which modify or are likely to modify the Ozone Layer', but the measures were unspecified. (See later Montreal Protocol 1987.)	http://www.unep.org/ozone/vienna.shtml
Convention on Oil Pollution Preparedness, Response & Co-operation (OPRC)	London, UK (1990)	Recognition of the serious threat posed to the marine environment by oil pollution incidents involving ships, seaports and oil handling facilities. The Convention was designed to facilitate international co-operation and mutual assistance in preparing for and responding to a major oil pollution incident and to encourage States to develop and maintain an adequate capability to deal with oil pollution emergencies.	It requires Governments to establish a national system for responding promptly and effectively to oil pollution incidents. This includes, as a basic minimum, the creation of a national contingency plan. In addition, seaports and oil handling facilities have to maintain Oil Pollution Emergency Plans (OPEPs) if there exists risks of oil pollution. Masters of ships have to report any observed event involving a discharge of oil at sea or the presence of oil in the sea.	http://www.imo.org/Conventions/mainframe.asp?topic_id=258&doc_id=682

Continued

Table 8.5.1 International Level Pollution Policy. *Continued*

Title	Venue	Purpose	Key Outcome(s)	Website
Convention on Environmental Impact Assessment in a Transboundary Context	Espoo, Finland (1991)	To enhance international co-operation in assessing environmental impact in particular in a transboundary context.	Stipulates the obligations of Parties to assess the environmental impact of certain activities at an early stage of planning. Lays down the general obligation of states to notify & consult each other on all major projects under consideration that are likely to have significant adverse environmental impact across boundaries. Appendices covering: ■ I – List of activities for which an EIA is required ■ II – Content of the EIA documentation ■ III – General criteria to assist in the determination of the environmental significance of activities not listed in Appendix I ■ IV – Inquiry procedure ■ V – Post project analysis ■ VI – Elements for bilateral & multi-lateral co-operation ■ VII – Arbitration.	http://www.unece.org/env/eia/eia.htm
UN Framework Climate Change Convention (UNFCCC)	Rio de Janeiro, Brazil (1992)	The ultimate objective of the Convention is 'to achieve stabilisation of atmospheric concentrations of greenhouse gases at levels that would prevent dangerous anthropogenic (human-induced) interference with the climate system . . .'. (See later Kyoto Protocol 1997)	Parties to the UNFCCC have a number of commitments including: ■ Submit a national inventory of emissions/removals of greenhouse gases (NGGI) ■ Implement national programmes to mitigate climate change & adapt to its impacts ■ Strengthen scientific & technical research & systematic observation related to the climate system, & promote the development & diffusion of relevant technologies ■ Promote education programs & public awareness about climate change & its likely effects ■ Periodically submit comprehensive national reports on activities to implement commitments under the convention.	http://unfccc.int/resource/docs/convkp/conveng.pdf

Convention on the Prior Informed Consents (PIC) Procedure for Certain Hazardous Chemicals & Pesticides in International Trade	Rotterdam, The Netherlands (1998)	The Convention establishes a first line of defence by giving importing countries the tools and information they need to identify potential hazards and exclude chemicals they cannot manage safely. If a country agrees to import chemicals, the Convention promotes their safe use through labelling standards, technical assistance, and other forms of support. It also ensures that exporters comply with the requirements.	The Convention requires that hazardous chemicals and pesticides that have been banned or severely restricted in at least two countries shall not be exported unless explicitly agreed by the importing country. It also includes pesticide formulations that are too dangerous to be used by farmers in developing countries. Countries are also obliged to stop national production of those hazardous compounds. The treaty covers a list of five industrial chemicals and 22 pesticides, including Aldrin, DDT, Dieldrin, HCH, Lindane, Mercury compounds, Polychlorinated Biphenyls (PCBs) and others. Many more are likely to be added in future. For developing countries, technical assistance will be promoted to help them develop their infrastructure and capacity for managing chemicals and pesticides safely.	http://www.fao.org/waicent/FaoInfo/Agricult/AGP/AGPP/Pesticid/PIC/Download/fin_e.pdf
Stockholm Convention on Persistent Organic Pollutants (POPs)	Stockholm, Sweden (2001)	The Stockholm Convention is a global legally binding treaty to protect human health and the environment from persistent organic pollutants (POPs). The Convention seeks the elimination or restriction of production & use of all intentionally produced POPs (i.e. industrial chemicals & pesticides). It also seeks the continuing minimisation, & where feasible, ultimate elimination of the releases of unintentionally produced POPs such as dioxins & furans.	General obligations (including developing an implementation plan, promoting awareness, reporting to COP (Conference of Parties) on measures implemented). Control provisions on intentionally produced POPs; unintentionally produced POPs; & stockpiles & wastes.	http://www.pops.int/documents/convtext/convtext_en.pdf
Montreal Protocol on Substances that Deplete the Ozone Layer	Montreal, Canada (1987)	To further the issues discussed at the Vienna Convention & where possible sets appropriate measures/targets.	The final agreement contains clauses to cover the special circumstances of several groups of countries, especially developing countries with low consumption rates who do not want the Protocol to hinder their development. However it is constructively flexible – it can be tightened or 'adjusted' as the scientific evidence is strengthened, without having to be completely renegotiated. It sets the 'elimination' of ozone-depleting substances as its 'final objective.' (Subsequently amended at London 1990, Copenhagen 1992, Vienna 1995, Montreal 1997 & Beijing 1999).	http://www.unep.org/ozone/pdf/Montreal-Protocol2000.pdf

Continued

Table 8.5.1 International Level Pollution Policy. *Continued*

Title	Venue	Purpose	Key Outcome(s)	Website
Sofia Protocol Concerning the Control of Emissions of Nitrogen Oxides	Sofia, Bulgaria (1988)	This Protocol required as a first step, to freeze emissions of nitrogen oxides or their transboundary fluxes. The general reference year is 1987 (with the exception of the United States that chose to relate its emission target to 1978). The second step to the NOx Protocol required the application of an effects-based approach. Applying the multi-pollutant, multi-effect critical load approach, a new instrument was to provide for further reduction of emissions of nitrogen compounds, including ammonia, & volatile organic compounds, in view of their contribution to photochemical pollution, acidification & eutrophication, & their effects on human health, the environment & materials, by addressing all significant emission sources.	Key features of the Protocol include: ▪ Control and/or reduce national annual emissions of nitrogen oxides by 31st December 1994 ▪ Apply national emissions standards to major new stationary/mobile sources ▪ Introduce pollution control measures for major existing stationary sources ▪ Co-operation between parties to establish (i) critical loads; (ii) reductions in national annual emissions of nitrogen oxides; (iii) measures & a timetable commencing no later than 1st January 1996 for achieving such reductions.	http://www.unece.org/env/lrtap/full%20text/1988.NOX.e.pdf
Geneva Protocol concerning the Control of Emissions of Volatile Organic Compounds	Geneva, Switzerland (1991)	This protocol saw parties determined to take effective action to control & reduce national annual emissions of VOCs or the transboundary fluxes of VOCs & the resulting secondary photochemical oxidant products, in particular by applying appropriate national or international emission standards to new mobile & new stationary sources & retrofitting existing major stationary sources, & also by limiting the content of components in products for industrial & domestic use that have the potential to emit VOCs.	This Protocol specifies three options for emission reduction targets that have to be chosen upon signature or upon ratification: ▪ 30% reduction in emissions of volatile organic compounds (VOCs) by 1999 using a year between 1984 and 1990 as a basis. ▪ The same reduction as for the above within a Tropospheric Ozone Management Area (TOMA) specified in Annex I to the Protocol and ensuring that by 1999 total national emissions do not exceed 1988 levels. ▪ Finally, where emissions in 1988 did not exceed certain specified levels, Parties may opt for a stabilisation at that level of emission by 1999.	http://www.unece.org/env/lrtap/full%20text/1991.VOC.e.pdf

Oslo Protocol on Further Reduction of Sulphur Emissions	Oslo, Norway (1994)	The 1994 Oslo Protocol on Further Reduction of Sulphur Emissions entered into force on 5 August 1998. An effects-based approach, the critical load concept, best available technology, energy savings, the application of economic instruments and other considerations was applied in the preparation of the Protocol. This has led to a differentiation of emission reduction obligations of Parties to the Protocol.	Key features of this protocol include: ▪ Control & reduce sulphur emissions, also to ensure, as far as possible, without entailing excessive costs, that depositions of oxidised sulphur compounds do not exceed the critical loads in Annex 1; ▪ Reduce & maintain annual sulphur emissions in accordance with the timing & levels specified in Annex 2; ▪ Make use of the most effective measures for the reduction of sulphur emissions including: – Measures to increase energy efficiency; – Measures to increase the use of renewable energy; – Measures to reduce the sulphur content of certain fuels & to encourage use of fuel with a low sulphur content; – Measures to apply best available control technologies not entailing excessive cost.	http://www.unece.org/env/lrtap/full%20text/1994.Sulphur.e.pdf
Kyoto Protocol to the UN Framework Convention on Climate Change (COP-3)	Kyoto, Japan (1997)	The Kyoto Protocol, negotiated by more than 160 nations in December 1997, aims to reduce net emissions of certain greenhouse gases (primarily CO_2).	Protocol commits industrialised nations to legally binding reductions in emissions of six 'greenhouse gases'. The developed countries commit themselves to reducing their collective emissions of six key greenhouse gases by at least 5%. Each of the participating developed countries must decide how to meet its respective reduction goal during a five-year period (2008–2012); but specific ground rules remain to be worked out at future negotiating sessions. The Protocol encourages governments to co-operate with one another, improve energy efficiency, reform the energy and transportation sectors, promote renewable forms of energy, phase out inappropriate fiscal measures and market imperfections, limit methane emissions from waste management and energy systems, and protect forests and other carbon 'sinks'.	http://unfccc.int/resource/docs/convkp/kpeng.pdf

Table 8.5.2 European Level Pollution Policy.

Policy Type	Programme Title	Policy Objective	Community Measure	Document Reference	Website
Council Directives	Motor Vehicles with Trailers – Permissible Sound Level	To implement Community regulations relating to vehicle noise based on complete harmonisation.	Council Directive 70/157/EEC on the approximation of the laws of the Member States relating to the permissible sound level & the exhaust system of motor vehicles (06/02/70) Amended by the following: ▪ Commission Directive 73/350/EEC (07/11/73) ▪ CD 77/212/EEC (08/03/77 ▪ Commission Directive 81/334/EEC (13/04/81) ▪ CD 84/372/EEC (03/07/84 ▪ CD 84/424/EEC (03/09/84 ▪ CD 87/354/EEC (25/06/87 ▪ Commission Directive 89/491/EEC (17/07/89) ▪ CD 92/97/EEC (10/11/92) ▪ Commission Directive 96/20/EC (27/03/96)	▪ Off Jnl L42, 23/02/70 ▪ Off Jnl L321, 22/11/73 ▪ Off Jnl L66, 12/03/77 ▪ Off Jnl L131, 18/05/81 ▪ Off Jnl L196, 26/07/84 ▪ Off Jnl L238, 06/09/84 ▪ Off Jnl L192, 11/07/87 ▪ Off Jnl L238, 15/08/89 ▪ Off Jnl L371, 19/12/92 ▪ Off Jnl L92, 13/04/96	http://europa.eu.int/scadplus/leg/en/lvb/l21045b.htm

Motor Vehicles with Trailers: Petrol & Diesel Engines	To establish limit values for emissions from petrol & diesel engine passenger cars & light commercial vehicles.	CD 70/220/EEC on the approximation of the laws of Member States relating to measures to be taken against air pollution by emission from motor vehicles (20/03/70) Amended by the following: ▪ CD 74/290/EEC (28/05/74) ▪ Com Dir 77/102/EEC (30/11/76) ▪ Com Dir 78/665/EEC (14/07/78) ▪ CD 83/351/EEC (16/06/83) ▪ CD 88/76/EEC (16/06/88) ▪ Com Dir 89/491/EEC (17/07/89) ▪ CD 89/458/EEC (18/07/89) ▪ CD 91/441/EEC (26/06/91) ▪ CD 93/59/EEC (28/06/93) ▪ CD 94/12/EC (23/03/94) ▪ Com Dir 96/44/EC (01/07/96) ▪ CD 96/69/EC (08/10/96) ▪ CD 98/69/EC (13/10/98) ▪ Com Dir 99/102/EC (15/12/99) ▪ CD 2001/1/EC (22/01/01)	▪ COM (95) 689 Final ▪ COM (2000) 626 Final ▪ Off Jnl L76, 06/04/70	http://europa.eu.int/scadplus/leg/en/lvb/l21047.htm
Two or Three Wheeled Motor Vehicles – Permissible Sound Level of Motorcycles	To harmonise national regulations on the sound level & exhaust system of motorcycles.	CD 78/1015/EEC on the approximation of the laws of the Member States on the permissible sound level & exhaust system of motorcycles (23/11/78) Amended by the following: ▪ CD 87/56/EEC (18/12/86) ▪ CD 89/235/EEC (13/03/89)	▪ Off Jnl L349, 13/12/78 ▪ Off Jnl L24, 27/01/87 ▪ Off Jnl L98, 11/04/89	http://europa.eu.int/scadplus/leg/en/lvb/l21045a.htm

Continued

Table 8.5.2 European Level Pollution Policy. *Continued*

Policy Type	Programme Title	Policy Objective	Community Measure	Document Reference	Website
Council Directives	Concentration of Lead in the Air	To monitor & limit the concentration of lead in the air.	CD 82/884/EEC on a limit value for lead in the air (03/12/82) Amended by: ■ CD 90/656/EEC (04/12/90 and ■ CD 91/692/EEC (23/12/91	■ Off Jnl L378, 31/12/82 ■ Off Jnl L353, 17/12/90 ■ Off Jnl L377, 31/12/91	http://europa.eu.int/scadplus/leg/en/lvb/l28034.htm
	Nitrogen Dioxide	To monitor & limit the cuantity of nitrogen dioxide in the atmosphere.	CD 85/203/EEC on air quality standards for nitrogen dioxide (07/03/83)	■ Off Jnl L87, 27/03/85	http://europa.eu.int/scadplus/leg/en/lvb/l28031b.htm
	Airborne Noise Emitted from Household Appliances	To provide the public with information on levels of noise emitted by household appliances, harmonising only those requirements necessary for measuring noise.	CD 86/594/EEC on airborne noise emitted by household appliances (01/12/86)	■ Off Jnl L344, 06/12/86	http://europa.eu.int/scadplus/leg/en/lvb/l21187a.htm
	Tower Cranes – Sound Levels	To consolidate into one directive all the technical provisions required to determine the sound levels of tower cranes.	CD 87/405/EEC amending Directive 84/534/EEC on the approximation of the laws of the Member States relating to the permissible sound power level of tower cranes (25/06/87)	■ Off Jnl L200, 08/08/87 ■ Off Jnl L184, 30/06/89	http://europa.eu.int/scadplus/leg/en/lvb/l21182.htm
	Motor Vehicles with Trailers: Emission of Gaseous Pollutants from Diesel Engines	To lay down limit values for emissions from diesel engines.	CD 88/77/EEC on the approximation of the laws of the Member States relating to the measures to be taken against the emission of gaseous pollutants from diesel engines for use in vehicles (03/12/87)	■ Off Jnl L36, 09/02/88	http://europa.eu.int/scadplus/leg/en/lvb/l21048.htm

		Amended by the following: ■ CD 91/542/EEC (01/10/91) ■ CD 96/1/EC (22/01/96) ■ CD 1999/96/EC (13/12/99)	■ Off Jnl L295, 25/10/91 ■ Off Jnl L40, 17/02/96 ■ Off Jnl L44, 16/02/00	
Pollution from Large Combustion Plants	To limit pollutant emissions from large combustion plants.	CD 88/609/EEC on the limitation of emissions of certain pollutants into the air from large combustion plants (24/11/88) Amended by CD 94/66/EC (15/12/94) Dir 2001/80/EC (23/10/01)	■ Off Jnl L336, 07/12/88 ■ Off Jnl L337, 24/12/94 ■ Off Jnl L309, 27/11/01	http://europa.eu.int/scadplus/leg/en/lvb/l28028.htm
Air Pollution by Ozone	To establish a procedure for the reciprocal exchange of information between Member States & the Commission in order to control & limit the effects of ozone pollution on human health & the environment.	CD 92/72/EEC on air pollution by ozone (21/09/92)	■ Off Jnl L297, 13/10/92	http://europa.eu.int/scadplus/leg/en/lvb/l28033a.htm
Sulphur Content of Certain Liquid Fuels	To reduce emissions of dioxide & sulphur resulting from the combustion of certain types of liquid fuels, & to limit the harmful effects of such pollution on human beings & the environment.	CD 93/12/EEC relating to the sulphur content of gas oil (22/03/93) Amended by the following: ■ Directive 98/70/EC (13/10/98) ■ CD 99/32/EC (26/04/99)	■ Off Jnl L74, 27/03/93 ■ Off Jnl L350, 28/12/98 ■ Off Jnl L121, 11/05/99	http://europa.eu.int/scadplus/leg/en/lvb/l21050.htm
Volatile Organic Compounds (VOCs) Resulting from the Storage of Petrol	To reduce losses due to evaporation of the petrol at all stages of the fuel storage & distribution chain.	CD 94/63/EC on the control of VOC emissions resulting from the storage of petrol & its distribution from terminals to service stations (20/12/94)	■ Off Jnl L365, 31/12/94	http://europa.eu.int/scadplus/leg/en/lvb/l28029a.htm

Continued

Table 8.5.2 European Level Pollution Policy. *Continued*

Policy Type	Programme Title	Policy Objective	Community Measure	Document Reference	Website
Council Directives	Integrated Pollution Prevention & Control: Directive IPPC	To prevent or minimise air, water & soil pollution by emissions from industrial installations in the Community, with a view to achieving a high level of environmental protection.	CD 96/61/EC concerning integrated pollution prevention & control (24/09/96)	▪ Off Jnl L257, 10/10/96	http://europa.eu.int/scadplus/leg/en/lvb/l28045.htm
	Management & Quality of Ambient Air	To establish the basic principles of a common strategy to define & set objectives for ambient air quality in order to avoid, prevent or reduce harmful effects on human health & the environment. To assess ambient air quality in the Member States & inform the public, notably by means of alert thresholds.	CD 96/62/EC on ambient air quality assessment & management (27/09/96)	▪ Off Jnl L296, 21/11/96	http://europa.eu.int/scadplus/leg/en/lvb/l28031a.htm
	Non-Road Mobile Machinery: Gaseous Pollutants	To reduce air pollution caused by engines of non-road mobile machinery by laying down acceptable emission standards for such engines at Community level.	Directive 97/68/EC on the approximation of the laws of the Member States relating to measures against the emission of gaseous & particulate pollutants from internal combustion engines to be installed in non-road mobile machinery (16/12/97)	▪ Off Jnl L59, 27/02/98	http://europa.eu.int/scadplus/leg/en/lvb/l21219.htm
			Amended by CD 2001/63/EC (17/10/01)	▪ Off Jnl L227, 23/08/01	
	Quality of Petrol & Diesel Fuels	To reduce pollution from car emissions by introducing new environmental specifications applicable to petrol & diesel fuels, & to ban leaded petrol from the market from the year 2000.	Directive 98/70/EC relating to the quality of petrol & diesel fuels amending CD 93/12/EEC (13/10/98)	▪ Off Jnl L350, 28/12/98	http://europa.eu.int/scadplus/leg/en/lvb/l28077.htm

Volatile Organic Compounds (VOCs) Resulting from Certain Industrial Activities	To prevent or reduce the direct & indirect effects of emissions of VOCs in the environment & on human health, by setting emission limits for such compounds & laying down operating conditions for industrial installations using organic solvents.	CD 1999/13/EC on the limitation of emissions of VOCs due to the use of organic solvents in certain activities & installations (11/03/99)	Off Jnl L85, 29/03/99	http://europa.eu.int/scadplus/leg/en/lvb/l28029b.htm
Sulphur Dioxide, Nitrogen Dioxide & Nitrogen Oxides, Particulates & Lead in the Ambient Air	To maintain or improve the quality of the ambient air by establishing limit values for the concentrations of sulphur dioxide, nitrogen dioxide & nitrogen oxides, particulates & lead, together with alert thresholds for concentrations of sulphur dioxide & nitrogen dioxide in the ambient air by evaluating those concentrations on the basis of common methods & criteria, & by bringing together suitable information on such concentrations in order to keep the public informed.	CD 1999/30/EC laying down limit values for sulphur dioxide, nitrogen dioxide & nitrogen oxides, particulates & lead in the ambient air (22/04/99)	Off Jnl L163, 29/06/99	http://europa.eu.int/scadplus/leg/en/lvb/l28098.htm
Information on the Fuel Consumption & CO_2 Emissions of New Cars	To provide potential purchasers of new passenger cars with useful information on the fuel consumption & CO_2 emissions of these vehicles so as to influence their choice in favour of those models that use the least fuel.	CD 1999/94/EC relating to the availability of consumer information on fuel economy & CO_2 emissions in respect of the marketing of new passenger cars (13/12/99)	Off Jnl L12, 18/12/99	http://europa.eu.int/scadplus/leg/en/lvb/l32034.htm
Noise Emitted by Equipment Used Outdoors	To control noise emitted by equipment used outdoors.	Directive 2000/14/EC of the European Parliament & the Council on the approximation of the laws of Member States relating to the noise emission in the environment by equipment for use outdoors (2000)	Off Jnl L162, 03/07/00	http://europa.eu.int/scadplus/leg/en/lvb/l28048.htm

Continued

Table 8.5.2 European Level Pollution Policy. *Continued*

Policy Type	Programme Title	Policy Objective	Community Measure	Document Reference	Website
Council Directives	Limit Values for Benzene & Carbon Monoxide in Ambient Air (Second Daughter Directive)	The objectives of this Directive shall be: ▪ to establish limit values for concentrations of benzene & carbon monoxide in ambient air intended to avoid, prevent or reduce harmful effects on human health & the environment as a whole; ▪ to assess concentrations of benzene & carbon monoxide in ambient air on the basis of common methods & criteria; ▪ to obtain adequate information on concentrations of benzene & carbon monoxide in ambient air & ensure that it is made available to the public; ▪ to maintain ambient air quality where it is good & improve it in other cases with respect to benzene & carbon monoxide.	Dir 2000/69/EC relating to limit values for benzene & carbon monoxide in ambient air (16/11/00)	▪ Off Jnl L313, 13/12/00	http://europa.eu.int/eur-lex/pri/en/oj/dat/2000/l_313/l_31320001213en00120021.pdf
	National Emission Ceilings for Certain Atmospheric Pollutants	To set national emission ceilings for pollutants causing acidification & eutrophication & for ozone precursors in order to provide fuller protection for the environment & human health against their adverse effects.	Dir 2001/81/EC of the European Parliament & the Council on national emission ceilings for certain atmospheric pollutants (2001)	▪ Off Jnl L309, 27/11/01	http://europa.eu.int/scadplus/leg/en/lvb/l28095.htm
	Third Daughter Directive Relating to Ozone	The purpose of this Directive is: ▪ to establish long-term objectives, target values, an alert threshold & an information threshold for concentrations of ozone in ambient air in the Community, designed to avoid, prevent or reduce harmful effects on human health & the environment as a whole;	▪ Dir 2002/3/EC relating to ozone (12/02/02)	▪ Off Jnl L67, 09/03/02	http://europa.eu.int/eur-lex/pri/en/oj/dat/2002/l_067/l_06720020309en00140030.pdf

	▪ to ensure that common methods & criteria are used to assess concentrations of ozone &, as appropriate, ozone precursors (oxides of nitrogen & volatile organic compounds) in ambient air in the Member States; ▪ to ensure that adequate information is obtained on ambient levels of ozone & that it is made available to the public; to ensure that, with respect to ozone, ambient air quality is maintained where it is good, & improved in other cases; ▪ to promote increased co-operation between the Member States, in reducing ozone levels, use of the potential of transboundary measures & agreement on such measures.			
Establishment of Rules & Procedures with regard to the Introduction of Noise-related Operating Restrictions at Community Airports	The objectives of this Directive are: ▪ to lay down rules for the Community to facilitate the introduction of operating restrictions in a consistent manner at airport level so as to limit or reduce the number of people significantly affected by the harmful effects of noise; ▪ to provide a framework which safeguards internal market requirements; ▪ to promote development of airport capacity in harmony with the environment; ▪ to facilitate the achievement of specific noise abatement objectives at the level of individual airports; ▪ to enable measures to be chosen from those available with the aim of achieving maximum environmental benefit in the most cost-effective manner.	Directive 2002/30/EC of the European Parliament & of the Council on the establishment of rules and procedures with regard to the introduction of noise-related operating restrictions at Community airports (26/03/02)	▪ Off Jnl L85/40, 28/03/02	http://europa.eu.int/eur-lex/pri/en/oj/dat/2002/l_085/l_08520020328en00400046.pdf

Continued

Table 8.5.2 European Level Pollution Policy. *Continued*

Policy Type	Programme Title	Policy Objective	Community Measure	Document Reference	Website
Council Directives	Assessment & Management of Environmental Noise	This Directive should *inter alia* provide a basis for developing & completing the existing set of Community measures concerning noise emitted by the major sources, in particular road & rail vehicles & infrastructure, aircraft, outdoor & industrial equipment & mobile machinery, & for developing additional measures, in the short, medium & long term.	Directive 2002/49/EC of the European Parliament & the Council relating to the assessment and management of environmental noise (25/06/02)	■ L189/12	http://europa.eu.int/eur-lex/pri/en/oj/dat/2002/l_189/l_18920020718en00120025.pdf
	Establishing a Scheme for Greenhouse Gas Emission Allowance Trading within the Community (amending Council Directive 96/61/EC)	This Directive establishes a scheme for greenhouse gas emission allowance trading within the Community in order to promote reductions of greenhouse gas emissions in a cost-effective & economically efficient manner.	Directive 2003/87/EE of the European Parliament & of the Council establishing a scheme for greenhouse gas emission allowance trading within the Community (amending Council Directive 96/61/EC) (13/10/03)	■ Off Jnl L275/32, 25/10/03	http://europa.eu.int/eur-lex/pri/en/oj/dat/2003/l_275/l_27520031025en00320046.pdf
	Establishing a Scheme for Greenhouse Gas Emission Allowance Trading within the Community, in respect of the Kyoto Protocol's Project Mechanisms (amending Dir 2003/87/EC)	Since the objective of the proposed action, namely the establishment of a link between the Kyoto project-based mechanisms and the Community scheme, cannot be sufficiently achieved by the Member States acting individually, & can therefore by reason of the scale & effects of this action be better achieved at Community level, the Community may adopt measures, in accordance with the principle of subsidiarity as set out in Article 5 of the Treaty. In accordance with the principle of proportionality, as set out in that Article, this Directive does not go beyond what is necessary in order to achieve that objective.	Directive 2004/101/EC of the European Parliament & of the Council amending Directive 2003/87/EC establishing a scheme for greenhouse gas emission allowance trading within the Community, in respect of the Kyoto Protocol's project mechanisms (27/10/04)	■ Off Jnl L338/18, 13/11/04	http://europa.eu.int/comm/environment/climat/emission/pdf/dir_2004_101_en.pdf

Council Decisions	Mechanism for Monitoring CO_2 & other Greenhouse Gas Emissions	To update the mechanism for monitoring greenhouse gas emissions established in the Community so as to be able to evaluate more accurately & more regularly the progress made in reducing emissions with a view to complying with the Community's commitments under the UN Framework Convention on Climate Change (UNFCCC) & under the Kyoto protocol.	Council Decision 93/389/EEC for a monitoring mechanism of Community CO_2 & other greenhouse gas emissions (24/06/93) Amended by Council Decision 99/296/EC (26/04/99)	Off Jnl L167, 09/07/93 Off Jnl L117, 05/05/99	http://europa.eu.int/scadplus/leg/en/lvb/l28044.htm
	Reciprocal exchange of Information From Networks & Individual Stations Measuring Ambient Air Quality	This Council Decision highlights the establishment of a reciprocal exchange mechanism for storing information and data from both networks and individual stations on ambient air quality measurements.	Council Decision 97/101/EC establishing a reciprocal exchange of information & data from networks & individual stations measuring ambient air quality (27/01/97) Amended by C Dec 2001/752/EC	Off Jnl L035, 05/02/97	http://europa.eu.int/smartapi/cgi/sga_doc?smartapi!celexapi!prod!CELEXnumdoc&lg=en&numdoc=31997D0101&model=guichett
	CO_2 Emissions from New Passenger Cars: Monitoring	To set up a mechanism to monitor CO_2 emissions from new passenger cars in order to check on the implementation of the strategy to reduce the CO_2 emissions they produce.	Decision No. 1753/2000/EC of the European Parliament & Council, establishing a scheme to monitor the average specific emissions of CO_2 from new passenger cars (22/06/00)	COM (2000) 615 Off Jnl L202, 10/08/00	http://europa.eu.int/scadplus/leg/en/lvb/l28055.htm
	Protocol on Heavy Metals	To reduce the emissions of heavy metals that are subject to long-range transboundary atmospheric transport & are likely to have adverse effects on human health & the environment.	Council Decision on the approval, on behalf of the European Community, of the protocol to the 1979 Convention on long-range transboundary air pollution on heavy metals (04/04/01)	Off Jnl L134, 17/05/01	http://europa.eu.int/scadplus/leg/en/lvb/l28035.htm
	Kyoto Protocol on Climate Change	To tackle climate change by means of international action to reduce the emissions of certain greenhouse gases responsible for global warming.	Council Decision 2002/358/EC concerning the approval, on behalf of the European Community, of the Kyoto Protocol to the UN Framework Convention on Climate Change & the joint fulfilment of commitments thereunder (25/04/02)	Off Jnl L130, 15/05/02	http://europa.eu.int/scadplus/leg/en/lvb/l28060.htm

Continued

Table 8.5.2 European Level Pollution Policy. *Continued*

Policy Type	Programme Title	Policy Objective	Community Measure	Document Reference	Website
Council Decisions	The Conclusion of the Stockholm Convention on Persistent Organic Pollutants	The Stockholm Convention on Persistent Organic Pollutants is hereby approved on behalf of the Community.	Council Decision concerning the conclusion, on behalf of the European Community, of the Stockholm Convention on Persistent Organic Pollutants (12/06/03)	▪ COM (2003) 331	http://europa.eu.int/eur-lex/en/com/pdf/2003/com2003_0331en01.pdf
	Establishing Guidelines for the Monitoring & Reporting of Greenhouse Gas Emissions pursuant to Directive 2003/87/EC	The guidelines contained in this Decision set out detailed criteria for the monitoring & reporting of greenhouse gas emissions resulting from the activities listed in Annex I of Directive 2003/87/EC of greenhouse gases specified in relation to those activities, based on the principles for monitoring & reporting set out in Annex IV of that Directive.	Commission Decision establishing guidelines for the monitoring & reporting of greenhouse gas emissions pursuant to Directive 2003/87/EC of the European Parliament & of the Council (29/01/04)	▪ C (2004) 130	http://europa.eu.int/comm/environment/climat/pdf/c2004_130_en.pdf
	Limit Values for Certain Pollutants in Ambient Air	When submitting the information on the plans or programmes referred to in Article 8(3) of Directive 96/62/EC, as required by Article 11(1)(a)(iii) thereof, with regard to the limit values established by Directives 1999/30/EC and 2000/69/EC, Member States shall present that information in accordance with the structure set out in the Annex to this Decision.	Commission Decision laying down arrangements for the submission of information on plans or programmes required under Council Directive 96/62/EC in relation to limit values for certain pollutants in ambient air (20/02/04)	▪ C Dec 2004/224/EC ▪ Off Jnl L68/27, 06/03/04	http://europa.eu.int/eur-lex/pri/en/oj/dat/2004/l_068/l_06820040306en00270033.pdf
	The Conclusion of the Protocol to the 1979 Convention on Long Range Transboundary Air Pollution on Persistent Organic Pollutants	The Protocol to the 1979 Convention on Long Range Transboundary Air Pollution on Persistent Organic Pollutants, is hereby approved on behalf of the Community.	Council Decision concerning the conclusion, on behalf of the European Community, of the Protocol to the 1979 Convention on Long Range Transboundary Air Pollution on Persistent Organic Pollutants (19/02/04)	▪ C Dec 2004/259/EC ▪ Off Jnl L81/35, 19/03/04	http://europa.eu.int/eur-lex/pri/en/oj/dat/2004/l_081/l_08120040319en00350036.pdf

Council Communications	Transport & CO_2	To protect the environment by reducing emissions of CO_2 from transport in the framework of the undertakings made at the Kyoto conference.	Communication from the Commission to the Council, European Parliament, Economic & Social Committee & the Committee of the Regions on transport & CO_2 – developing a community approach (1998)	■ COM (98) 204	http://europa.eu.int/scadplus/leg/en/lvb/l28049.htm
	Climate Change – Towards an EU Post-Kyoto Strategy	To develop a Community strategy for achieving the targets laid down in the Kyoto Protocol & the Buenos Aires Action Plan.	Communication from the Commission to the Council & European Parliament on climate change – towards an EU post-Kyoto strategy (03/06/98)	■ COM (98) 353	http://europa.eu.int/scadplus/leg/en/lvb/l28074.htm
			Communication from the commission on preparing for implementation of the Kyoto Protocol (19/05/99)	■ COM (1999) 230	
	Phase-out of CFCs in Metered Dose Inhalers	To work out a Community Strategy for the phase-out of Chlorofluorocarbons (CFCs) in metered dose inhalers.	Communication from the Commission to the European Parliament & Council on the phase-out of CFCs in metered dose inhalers (23/10/98)	■ COM (1998) 603 Final ■ Off Jnl C355, 20/11/98	http://europa.eu.int/scadplus/leg/en/lvb/l28073.htm
	Preparing for Implementation of the Kyoto Protocol	Key elements in such a preparation are compliance questions in a broad political & practical perspective. The political dimension deals with how the EC & its Member States can reach the 8% reduction of greenhouse gas emissions in the commitment period 2008–2012 compared to 1990 levels & be on track for further reductions after 2012. The practical dimension deals with how to establish a comprehensive monitoring system to accompany & follow up the implementation process.	Commission Communication to the Council and the Parliament on Preparing for Implementation of the Kyoto Protocol (19/05/99)	■ COM (1999) 230	http://europa.eu.int/comm/environment/docum/pdf/99230_en.pdf

Continued

Table 8.5.2 European Level Pollution Policy. *Continued*

Policy Type	Programme Title	Policy Objective	Community Measure	Document Reference	Website
Council Communications	Greenhouse Gas Emissions Trading & Climatic Change Programme	To launch a public debate with a view to implementing a possible Community policy on emissions trading before the Kyoto protocol is applied.	Green Paper on greenhouse gas emissions trading within the EU Communication from the Commission to the Council & European Parliament on EU policies & measures to reduce greenhouse gas emissions: towards a European climate change programme (ECCP) (2000)	▪ COM (2000) 87 ▪ COM (2000) 88	http://europa.eu.int/scadplus/leg/en/lvb/l28109.htm
	Clean Air For Europe (CAFE) Programme	To establish a long-term, integrated strategy to tackle air pollution & to protect against its effects on human health & the environment.	Commission Communication – The Clean Air For Europe (CAFE) Programme: towards a thematic strategy for air quality (04/05/01)	▪ COM (2001) 245	http://europa.eu.int/scadplus/leg/en/lvb/l28026.htm
	Implementation of the European Climate Change Programme (ECCP)	To combat climate change by means of various cross-cutting measures & measures in the fields of energy, industry & transport in implementing the ECCP.	Communication from the Commission on the implementation of the 1st phase of the European Climate Change Programme (23/10/01)	▪ COM (2001) 580	http://europa.eu.int/scadplus/leg/en/lvb/l28118.htm
	Towards a Thematic Strategy for Soil Protection	The purpose of this Communication is to build on the political commitment to soil protection in order that it be achieved more fully & systematically in coming years. As it is the first communication addressing soil protection, it is both descriptive & action orientated in order that a full picture of the complexity of the issue can be fully understood & it can serve as a basis for	Communication from the Commission to the Council, the European Parliament, the Economic & Social Committee & the Committee of the Regions on a thematic strategy for soil protection (16/4/02)	▪ COM (2002) 179	http://europa.eu.int/eur-lex/en/com/pdf/2002/com2002_0179en01.pdf

		future work. A distinction is made between soil, which is the subject of this communication & land use, which will be the subject of a separate communication addressing the territorial dimension to be published in 2003.			
	On the Road to Sustainable Production	This Communication is about Integrated Pollution Prevention & Control. In order to promote effective implementation of Council Directive 96/61/EC on Integrated Pollution Prevention & Control (the 'IPPC Directive'), the Commission reports on progress made so far in the Member States & Candidate Countries. The Sixth Community Environment Action Programme, adopted in July 2002, confirms that more effective implementation & enforcement of Community legislation on the environment is a priority.	Communication from the Commission to the Council, the European Parliament, the European Economic & Social Committee & the Committee of the Regions – progress in implementing Council Directive 96/61/EC concerning Integrated Pollution Prevention & Control (19/06/03)	▪ COM (2003) 354	http://europa.eu.int/eur-lex/en/com/cnc/2003/com2003_0354en01.pdf
	The Implementation of the Criteria Listed in Annex III to Directive 2003/87/EC Establishing a Scheme for Greenhouse Gas Emission Allowance Trading within the Community.	The purpose of this guidance document is three-fold: ▪ First, to assist Member States in drawing up their national allocation plans, by indicating the scope of interpretation of the Annex III criteria that the Commission deems acceptable; ▪ Second, to support the Commission assessment of notified national allocation plans, pursuant to Article 9(3); ▪ Third, to describe the circumstances under which *force majeure* is demonstrated.	Communication from the Commission on guidance to assist Member States in the implementation of the criteria listed in Annex III to Directive 2003/87/EC establishing a scheme for greenhouse gas emission allowance trading within the Community (07/01/04)	▪ COM (2003) 830	http://europa.eu.int/eur-lex/en/com/cnc/2003/com2003_0830en01.pdf
Miscellaneous	Action against Noise – Green Paper	To launch a debate on the Community's future noise policy.	Commission Green Paper on future noise policy (04/11/96)	▪ COM (96) 540	http://europa.eu.int/scadplus/leg/en/lvb/l21224.htm

Continued

Table 8.5.2 European Level Pollution Policy. *Continued*

Policy Type	Programme Title	Policy Objective	Community Measure	Document Reference	Website
Miscellaneous	Nitrogen Emissions from Civil Subsonic Jet Aeroplanes	To reduce atmospheric pollution by restricting the level of nitrogen oxide emitted by civil subsonic jet aeroplanes.	Proposal for a CD on the limitation of the emission of oxides of nitrogen from Civil Subsonic Aeroplanes (1997)	▪ COM (97) 629	http://europa.eu.int/scadplus/leg/en/lvb/l28052.htm
	Guidance Report on Preliminary Assessment under EC Air Quality Directives	In this report, guidance is provided on undertaking an assessment of air pollution levels as required by Article 5. It is recommended to use information from three main assessment methods: measurements, emission inventories, & modelling. Information on measurement methods concentrates on indicative measurements for the case where data from representative monitoring are not available or incomplete. For emissions, information is provided on the CORINAIR methodology; some information is also provided for some pollutants currently not covered in CORINAIR, such as PM10. Guidance is also provided on selecting & using models for the calculation of air pollution levels from the emissions, & comparing the results with measurements.	Technical Report No. 11 European Environment Agency (Jan 1998)	▪ EEA Technical Report No. 11	http://reports.eea.eu.int/TEC11a/en/tech11.pdf
	Wheeled Agricultural or Forestry Tractors: Pollutant Gases	To reduce the atmospheric pollution caused by agricultural or forestry tractor engines by laying down, at Community level, standards for acceptable emissions that apply to those engines.	Proposal for a Parliament & Council Directive on action to be taken against emissions of pollutant gases & pollutant particles by engines intended to power agricultural or forestry tractors, & amending CD 74/150/EEC (1998)	▪ COM (98) 472 ▪ Off Jnl C303, 02/10/98	http://europa.eu.int/scadplus/leg/en/lvb/l28070.htm

Environmental Problems Caused by Heavy Goods Vehicles (HGVs)	To give an overview of existing & prospective policies which can form the basis of an integrated policy on the environment problems caused by HGVs.	Commission working paper – Towards a framework for the solution of the environmental problems caused by traffic of HGVs (14/07/98)	▪ COM (1998) 444	http://europa.eu.int/scadplus/leg/en/lvb/l28063.htm
Subsonic Civil Jet Aircraft	To prevent worsening noise pollution within the Community caused by older types of subsonic civil jet aircraft which have been recertified, & other environmentally undesirable effects.	Council Regulation (EC) No. 925/1999 on the registration & operation within the Community of certain types of civil subsonic jet aeroplanes which have been modified & recertified as meeting the standards of Volume 1, Part II, Chapter 3 of Annex 16 to the Convention on International Civil Aviation (3rd Edition, July 1993) (1999)	▪ Council Regulation (EC) No. 925/1999 ▪ Off Jnl L115, 04/05/99	http://europa.eu.int/scadplus/leg/en/lvb/l28068.htm
EU Focus on Clean Air	The European air strategy was addressed in this brochure. It was intended to inform local & regional players, NGOs, policymakers at all levels, social partners & consumers, as well as citizens. It was also hoped the information would give ideas on how to contribute towards resolving environmental problems.	DG Environment Brochure on air quality (1999)	▪ DG XI Brochure	http://europa.eu.int/comm/environment/eufocus/clean_air.pdf
Guidance on Assessment under the EU Air Quality Directives – Final Draft	This report gives technical guidance on how to assess air quality under the EU air quality directives, in particular the Framework Directive 96/62/EC & the First Daughter Directive 1999/30/EC. The approach is to clarify which assessment outputs the directives require Member States to compile & report & to present possible assessment methods to generate this output. All reporting items are individually discussed. Methods are presented to designate zones, to measure concentrations, to determine the spatial concentration distribution in zones & to analyse the causes of air pollution problems. An important place is given to practical examples of assessment methods that are already in use in Member States.	European Commission Report (2000)	▪ EC Report	http://europa.eu.int/comm/environment/air/pdf/guidanceunderairquality.pdf

Continued

Table 8.5.2 European Level Pollution Policy. *Continued*

Policy Type	Programme Title	Policy Objective	Community Measure	Document Reference	Website
Miscellaneous	Greenhouse Gas Emissions Trading & Climatic Change Programme	To launch a public debate with a view to implementing a possible Community policy on emissions trading before the Kyoto protocol is applied.	Green Paper on greenhouse gas emissions trading within the EU (2000) Communication from the Commission to the Council & European Parliament on EU policies & measures to reduce greenhouse gas emissions: towards a European climate change programme (ECCP)	▪ COM (2000) 87 ▪ COM (2000) 88	http://europa.eu.int/scadplus/leg/en/lvb/l28109.htm
	Assessment & Management of Environmental Noise	To establish a common approach to the management of environmental noise, with a view to protecting the public against the harmful effects of exposure to noise in the domestic environment. This approach is based on assessments using common methods, on public information & on common action at European level.	Proposal for a Directive of the European Parliament & the Council relating to the assessment & management of environmental noise (2000)	▪ COM (2000) 468	http://europa.eu.int/scadplus/leg/en/lvb/l21180.htm
	The Behaviour of PVC in Landfill	This is the Final Report of the Study in the 'Behaviour of PVC in Landfills'. The study was carried out between October 1998 and September 1999 with the objective of evaluating the environmental impact of PVC waste after final disposal in landfills.	European Commission DG XI Report (2000)	▪ DG XI Report	http://europa.eu.int/comm/environment/waste/studies/pvc/landfill.pdf
	The Noise Policy of the European Union – Towards Improving the Urban Environment & Contributing to Global Sustainability	Reports showed that there are many ways, many different times, & different measuring places & methods to get to the noise value in decibels. This booklet was designed as a brief overview of the position achieved as the Noise Team prepared the Commission's Proposal in spring 2000. It was neither final nor authoritative, but was regarded as a useful little guide.	DG Environment Booklet (2000)	▪ DG XI Booklet	http://europa.eu.int/comm/environment/noise/pdf/noisebrochure.pdf

The Auto Oil II Programme – A Report from the Services of the European commission	The scope of AOP II included: ▪ the main 'conventional' pollutants & effects on CO_2 emissions; ▪ the period 1990–2020 but with the focus on introducing measures in 2005 which might help meet air quality objectives by 2010; ▪ the EU15 but with a focus on ten 'auto-oil' cities and their nine host countries; ▪ air quality modelling at regional, urban & local scale; ▪ a detailed assessment of the full range of potential road transport measures, including 'nontechnical' measures; ▪ comparison with other sources than road transport; & ▪ an evaluation of the benefits of reducing emissions.	DG Environment Report (2000)	▪ DG XI Report	http://europa.eu.int/comm/environment/autooil/auto-oil_en.pdf
Substances which Damage the Ozone Layer	To limit & monitor the Community production, marketing & use of substances which deplete the ozone layer & exports of those substances to third countries, in order to protect human health & the environment.	Regulation (EC) No. 2037/2000 of the European Parliament & the Council on substances that deplete the ozone layer (29/06/00). Amended by Reg (EC) No. 2038/2000 (28/09/00) and Reg (EC) No. 2039/2000 (28/09/00).	▪ Off Jnl L244, 29/09/00	http://europa.eu.int/scadplus/leg/en/lvb/l28064.htm
Greenhouse Gas Emissions Trading Scheme	To establish a greenhouse gas emissions trading scheme for the cost effective reduction of such emissions in the Community.	Proposal for a Directive of the European Parliament & the Council, establishing a scheme for greenhouse gas emission allowance trading within the Community & amending CD 96/61/EC (2001)	▪ COM (2001) 581 ▪ Off Jnl C75E, 26/03/02	http://europa.eu.int/scadplus/leg/en/lvb/l28012.htm

Continued

Table 8.5.2 European Level Pollution Policy. *Continued*

Policy Type	Programme Title	Policy Objective	Community Measure	Document Reference	Website
Miscellaneous	Economic Evaluation of Sectoral Emission Reduction Objectives for Climate Change Top-down Analysis of Greenhouse Gas Emission Reduction Possibilities in the EU	This report is a contribution to the study 'Economic Evaluation of Sectoral Emission Reduction Objectives for Climate Change' launched by the Environment DG of the European Commission in 1999. The analysis presented in this report is based on the energy market system model developed & maintained at the National Technical University of Athens. However, the analysis has been complemented by the results of another analysis of non-CO_2 greenhouse gases.	A study for the DG Environment (Mar 2001)	▪ DG Env Report	http://europa.eu.int/comm/environment/enveco/climate_change/top_down_analysis.pdf
	A Billion Euro Question: 'How Much Should We Pay for Noise Control, & How Much is it Worth?'	In conjunction with Inter-noise 2001 congress, the Directorate General Environment from the European Commission organised a half-day workshop on cost-benefit analysis (CBA) in noise policy. The workshop consisted of three practical presentations made by specialists in the field of cost-benefit analysis. Each presentation was followed up by two commentators who raised critical questions & the audience joined to the discussion. This report gives the conclusions & recommendations of the workshop as well as the main points made during the discussion.	Workshop on cost–benefit analysis in noise policy Inter-noise 2001 – The Hague, Netherlands (29/08/01)	▪ Inter-noise Workshop	http://europa.eu.int/comm/environment/noise/pdf/internoise_workshop.pdf
	Overview of Methods & Results of the Preliminary Assessment of	The purpose of the preliminary assessment is to prepare an initial overview of the concentration levels, so as to be able to divide the country into 'zones' & to set up an appropriate air quality assessment (i.e.	European Commission DG Environment Unit C1 'Air & Noise' Report (May 2002)	▪ DG Env Report	http://europa.eu.int/comm/environment/air/pdf/reportpreliminaryasses.pdf

Air Quality in Europe under Directives 96/62/EC & 1999/30/EC	monitoring & modelling) system for the regular assessment under the Daughter Directive. Zones are the basic territorial units for air quality assessment & management under the Framework Directive. The assessment requirements, which are to be applied per zone, depend on whether – anywhere in the zone – levels exceed the Upper Assessment Threshold (UAT) or Lower Assessment Threshold (LAT).			
Monitoring Mechanism of Community Greenhouse Gas Emissions	This is the third progress report under Council Decision 93/389/EEC as amended by Decision 99/296/EC for a monitoring mechanism of Community CO_2 & other greenhouse gas emissions. It assesses the actual & projected progress of Member States & the Community towards fulfilling their greenhouse gas emission commitments under the United Nations Framework Convention on Climate Change (UNFCCC) & the Kyoto Protocol. The Convention was aimed at returning greenhouse gas emissions to their 1990 level by 2000, while the Kyoto Protocol requires the EC to reduce greenhouse gas emission by 8% 2008–2012.	Report from the Commission under Council Decision 93/389/EEC as amended by Decision 99/296/EC for a monitoring mechanism of Community greenhouse gas emissions (09/12/02)	COM (2002) 702	http://europa.eu.int/eur-lex/en/com/rpt/2002/com2002_0702en01.pdf
Proposal for a Decision for a Monitoring Mechanism of Community Greenhouse Gas Emissions & the Implementation of the Kyoto Protocol	This proposal for a Decision regarding a monitoring mechanism of Community greenhouse gas emissions & the implementation of the Kyoto Protocol will replace Council Decision 93/389/EEC for a monitoring mechanism of Community CO_2 and other greenhouse gas emissions, which established a mechanism for monitoring anthropogenic greenhouse gas emissions & evaluating progress towards meeting commitments in respect of these emissions.	Proposal for a Decision of the European Parliament & of the Council for a monitoring mechanism of Community greenhouse gas emissions & the implementation of the Kyoto Protocol (05/02/03)	COM (2003) 51	http://europa.eu.int/eur-lex/en/com/pdf/2003/com2003_0051en01.pdf

Continued

Table 8.5.2 European Level Pollution Policy. *Continued*

Policy Type	Programme Title	Policy Objective	Community Measure	Document Reference	Website
Miscellaneous	Proposal for a Regulation on Persistent Organic Pollutants (amending Directives 79/117/EEC & 96/59/EC)	The purpose of the proposed Regulation was to give effect to the main provisions of the Convention and the Protocol which were not then sufficiently covered by Community legislation. Separate proposals for Council Decisions concerning the Community's conclusions on these two international agreements were submitted in parallel to this proposal.	Proposal for a Regulation of the European Parliament & of the Council on persistent organic pollutants (amending Directives 79/117/EEC and 96/59/EC) (12/06/03)	▪ COM (2003) 333	http://europa.eu.int/eur-lex/en/com/pdf/2003/com2003_0333en01.pdf
	Proposal for a Directive on Arsenic, Cadmium, Mercury, Nickel & Polycyclic Aromatic Hydro Carbons in Ambient Air	The proposed Directive relates to Annex I of the Air Quality Framework Directive, which lists atmospheric pollutants to be taken into consideration in the assessment & management of ambient air quality. The proposal introduced Community legislation on arsenic, cadmium, mercury, nickel & polycyclic aromatic hydrocarbons in fulfillment of obligations under Directive 96/62/EC.	Proposal for a Directive of the European Parliament & of the Council relating to arsenic, cadmium, mercury, nickel & polycyclic aromatic hydrocarbons in ambient air (16/07/03)	▪ COM (2003) 423	http://europa.eu.int/eur-lex/en/com/pdf/2003/com2003_0423en01.pdf
	Proposal for a Directive Amending the Directive Establishing a Scheme for Greenhouse Gas Emission Allowance Trading in the Community, in Respect of the Kyoto Protocol Project Mechanisms	This proposal will boost Joint Implementaction and the Clean Development Mechanism by providing additional incentives for business to engage in these mechanisms. It thereby promotes technology transfer to industrialised countries, for example Russia, & to developing countries while reducing the costs of meeting commitments under the EC emissions allowances trading scheme. The proposal sends a strong signal to other Parties to the Kyoto Protocol that the Community is fully committed to the Kyoto flexible mechanisms & the advantages that they entail for both the Community & other parties. At the same time, industrialised	Proposal for a Directive of the European Parliament & of the Council amending the Directive establishing a scheme for greenhouse gas emission allowance trading in the community, in respect of the Kyoto Protocol project mechanisms (23/07/03)	▪ COM (2003) 403	http://europa.eu.int/eur-lex/en/com/pdf/2003/com2003_0403en01.pdf

	countries have a responsibility to reduce their emissions of greenhouse gases through domestic measures, given their historic levels of emissions & current higher *per capita* emissions than developing countries.			
Proposal for a Regulation on Certain Fluorinated Greenhouse Gases	This proposal for a new EC Regulation on fluorinated gases is a key element of the first phase of the European Climate Change Programme. It will put in place a legislative framework to reduce emissions of hydrofluorocarbons (HFCs), perfluorocarbons (PFCs) & sulphur hexafluoride, which are powerful greenhouse gases covered by the Kyoto Protocol. The proposal includes provisions on the containment, reporting and marketing & use of fluorinated gases.	Proposal for a regulation of the European Parliament & of the Council on certain fluorinated greenhouse gases (11/08/03)	COM (2003) 492	http://europa.eu.int/eur-lex/en/com/pdf/2003/com2003_0492en01.pdf
A Monitoring Mechanism of Community Greenhouse Gas Emissions	This is the fourth progress report under Council Decision 93/389/EEC as amended by Decision 99/296/EC for a monitoring mechanism of Community carbon dioxide (CO_2) & other greenhouse gas emissions. It assesses the actual & projected progress of Member States & the Community towards fulfilling their greenhouse gas (GHC) emission commitments under the United Nations Framework Convention on Climate Change (UNFCCC) & the Kyoto Protocol.	Report from the Commission under Council Decision 93/389/EEC as amended by Dec 99/296/EC for a monitoring mechanism of Community Greenhouse Gas emissions (28/11/03)	COM (2003) 735	http://europa.eu.int/eur-lex/en/com/rpt/2003/com2003_0735en01.pdf
Demonstration of Equivalence of Ambient Air Monitoring Methods	This document describes the principles and methodologies for the demonstration of the equivalence of alternative methods to the CEN (Committee for Air Quality Measurement) Standard Methods. It is intended to be used by laboratories nominated by National Competent Authorities (see Directive 96/62/EC [1]) to perform the tests relevant to the demonstration of method equivalence.	Report by an EC Working group on Guidance for the Demonstration of Equivalence (Jan 2004)	EC Working Group Report	http://europa.eu.int/comm/environment/air/cafe/pdf/equivalence_report_final.pdf

Table 8.5.3 UK Level Pollution Policy.

Policy Type	Document Title	Objective	Contents	Website
Strategies & Reports	Development of an Overall Indicator of Air Pollution Concentrations (23/11/98)	This report presented the findings of a study to produce an overall indicator of air pollution. This was intended for inclusion in the proposed set of Headline Indicators which were being developed to stimulate public interest in the Government's Sustainable Development Strategy.	Introduction; Preface; List of table/figures; Executive summary; Chapter 1: Introduction; Chapter 2: Indicator format; Chapter 3: Data inclusion criteria; Chapter 4: Annual indicator; Chapter 5: Bi-annual & quarterly indicator; Chapter 6: Air pollution headline indicator; Chapter 7: References	http://www.defra.gov.uk/environment/airquality/airpollconc/index.htm
	Air Quality Strategy for England, Scotland, Wales & N. Ireland (19/01/00)	This Strategy describes plans of government and devolved administrations to improve and protect ambient air quality in the UK in the medium term. The proposals aimed to protect people's health and the environment without imposing unacceptable economic or social costs. They formed an essential part of the Government's strategy for sustainable development.	Executive summary; Chapter 1: Introduction; Chapter 2: Setting the scene; Chapter 3: Legislative & policy framework; Chapter 4: Air quality standards & objectives; Chapter 5: Delivering cleaner air; Chapter 6: Next steps; Technical annexes; Glossary	http://www.defra.gov.uk/environment/airquality/strategy/index.htm
	Risk Assessment Method for Local Air Pollution Control (Nov 2000)	This report provides the results of stage 1 of a study conducted by Risk Policy Analysts Ltd for the Air & Environment Quality Division of the DETR, in partnership with the National Assembly of Wales & the DTI. The study is concerned with the development of a risk assessment method for use by local authority inspectors under the local air pollution control regime (LAPC).	Executive summary; Introduction; The need for & purpose of a risk assessment method; Other risk assessment schemes; Development of risk assessment methods; Linking risk assessment to inspection planning & changing; Appraisal of costs & benefits; Conclusions & recommendations; References & bibliography; Annexes	http://wwww.defra.gov.uk/environment/airquality/riskam/03.htm
	Feasibility Study – Pollution Inventory (Jan 2001)	The principal objective of this study was to assess the feasibility of extending the Pollution Inventory to include emissions from Part B processes. The study covers three key issues & the report has been divided into three main sectors accordingly: ▪ Evaluation of the substances to be reported on; ▪ Methods of data collection; & ▪ Methods of estimating emissions.	Introduction; Methodology; Range of substances; Methods of data collection & transfer; Methods of estimating emissions; Potential cost & resource implications; key conclusions & recommendations; Appendices	http://www.defra.gov.uk/environment/airquality/pollinven/study/index.htm

Contaminated Land Inspection Strategies – Technical Advice for Local Authorities (May 2001)	This advice note sets out a procedure for the formulation of local authority inspection strategies, and is intended to assist local authorities in fulfilling their statutory obligations & complying with the underlying principles of the Part IIA regime. It is not intended to be mandatory, but provides advice to local authorities on good practice in developing & writing an inspection strategy.	Foreword; Part A: Introduction & overview; Part B: Procedure for developing an inspection strategy; Part C: Suggested outline for a typical Local Authority strategy; Annex 1: Environment Agency information requirements for the state of contaminated land report; Annex 2: Primary Environment Agency contacts for liaison with LAs on contaminated land issues	http://www.defra.gov.uk/environment/landliability/pdf/la-inspect.pdf
Air Quality Strategy: Particles, Benzene, Carbon Monoxide & Polycyclic Aromatic Hydrocarbons (17/09/01)	The chapters of this consultation document explain the changes that the government & the devolved administrations propose to make to the Strategy's objectives to take account of the latest health evidence. The proposals also seek to set a longer-term focus for the Strategy to reflect recent developments at the European Union level & to influence the development of wider policies that impact on air quality.	Executive Summary; Chapter 1: Introduction; Chapter 2: Principles & approaches; Chapter 3: Proposals for air quality objectives for particles; Chapter 4: Proposals for air quality objectives for benzene; Chapter 5: Proposals for air quality objectives for carbon monoxide; Chapter 6: Proposals for air quality objectives for polycyclic aromatic hydrocarbons; Technical annexes	http://www.defra.gov.uk/environment/consult/airqual01/contents.htm
Assessment of Risks to Human Health from Land Contamination: An Overview of the Development of Soil Guideline Values and Related Research (March 2002)	This report, CLR7, is an introduction to those that follow (CLR8–10). It gives an overview of the development of Soil Guideline Values and related material relevant to the assessment of risks to human health from land contamination.	Introduction; Legal framework; Development of soil guideline values & their use in the assessment of risks to human health from land contamination; Approaches to soil sampling when using soil guideline values in different regulatory contexts; Summary of reports CLR8, 9 & 10; References	http://www.defra.gov.uk/environment/landliability/pdf/clr7.pdf
Priority Contaminants for the Assessment of Land (March 2002)	This report presents a selection of contaminants that may be relevant for the assessment of contaminated land because they are likely to be found on a large number of industrial sites in the UK and have the potential to affect human health and the environment. The primary purpose of the selection has been to provide DEFRA with a guide to the substances it should cover in its research work on contaminated land. The report is one of a series published by DEFRA & the Environment Agency that is relevant to the assessment of the risks to human health arising from long-term exposure to soil contamination.	Introduction; Potential contaminants for the assessment of industrial land; Use of the selection of potential contaminants for the on-going research programme; Appendix A: Risk posed by contaminants on land; Appendix B: Review of individual contaminants not selected; Appendix C: Naming of chemical substances; References	http://www.defra.gov.uk/environment/landliability/pdf/clr8.pdf

Continued

Table 8.5.3 UK Level Pollution Policy. *Continued*

Policy Type	Document Title	Objective	Contents	Website
Strategies & Reports	Contaminants in soils: Collation of Toxicological Data and Intake Values for Humans (March 2002)	This report describes a framework for the collation of toxicological data to support the derivation of soil contaminant intakes that are protective of human health. It has been developed by DEFRA & the Environment Agency & involves assessing the potential harm to human health from contaminants in soil & using this information for deriving health criteria. Tolerable daily intakes (TDIs) & minimal risk levels (Index Doses) are derived for threshold & non-threshold contaminants, respectively. The latter refers to chemicals for which a threshold for health effects cannot be assumed, such as genotoxic carcinogens & mutagens. TDIs and Index Doses are derived from expert judgement of all the relevant data on the toxicological properties of the chemical.	Introduction; Basic concepts & definitions; Technical framework for delivering soil contaminant intakes for the protection of human health; Dealing with mixtures of contaminants; Appendix A: Overview of International approaches used for the purposes of deriving health criteria for the protection of human health; References; Abbreviations; Glossary	http://www.defra.gov.uk/environment/landliability/pdf/clr9.pdf
	Valuation of Air Pollution Effects on Ecosystems: A Scoping Study (08/04/02)	The aim of this paper is to assess current prospects for valuing ecosystem recovery in the context of air pollution & to give guidance regarding future research needs, in terms of economic valuation & scientific understanding. Specific objectives include: ▪ to review the literature on ecosystem valuation in the context of air pollution ▪ to describe the main challenges confronting valuation of ecosystem recovery ▪ to assess future research needs & establish priorities.	Executive summary; Introduction; Chapter 1: Ecosystem valuation; Chapter 2: Review of ecosystem valuation literature; Chapter 3: Ecosystem valuation – the challenges; Chapter 4: Valuing recovery in the freshwater ecosystem; Chapter 5: General conclusions & recommendations; Bibliography; Appendix 1, Appendix 2	http://www.defra.gov.uk/environment/airquality/ecosystem/pdf/airpolln_ecovalue.pdf

Audit of Contaminated Land Research in the UK – Final Report (April 2002)	An audit of contaminated land research in the UK with the aim of: ▪ Establishing if contaminated land research is meeting the requirements of the stakeholders; ▪ Identifying how co-ordination between the various research programmes can be made more effective; ▪ Identifying how liaison & communication of research between stakeholders can be improved.	Executive summary; Introduction; Aims & objectives of audit; Work programme; Main UK research programmes; Industry view of UK research; Industry needs; United States research; European Union research; Discussion; Summary & conclusions; References; Appendices.	http://www.defra.gov.uk/environment/landliability/pdf/contamland_audit.pdf
National Strategy to Combat Acidification, Eutrophication & Ground-Level Ozone (August 2002)	This paper summarises the current policies and proposed measures that will serve to reduce acidification, eutrophication and ground level ozone and meet the UK's obligations under the National Emission Ceilings Directive (NECD) and the Gothenburg Protocol. The paper also contains a copy of draft regulations for transposing the NECD into domestic legislation. The proposed course of legislative action is through UK Regulations and is being taken forward for DEFRA on behalf of and in consultation with the devolved administrations.	Section 1: National emissions ceilings; Section 2: Draft national emissions ceilings regulations; Section 3: Gothenburg protocol emission limit values; Section 4: REIA (Regulatory and Environmental Impact Assessment) for the National emission ceilings directive; Section 5: REIA for the Gothenburg protocol; Annex 1: EC National emission ceilings directive Annex 2: UNECE (United Nations Economic Commission for Europe) Gothenburg protocol	http://www.defra.gov.uk/environment/consult/air-natstrat/pdf/airqual_natstrat0802.pdf
Dealing with Contaminated Land in England – Progress in 2002 with implementing the Part IIA regime (Sept 2002)	This report fulfils a statutory duty to publish a report on the state of contaminated land in England (as defined by section 78U, Part IIA, EPA 1990). The report: ▪ Describes progress in implementing the Part IIA regime, including the development of Local Authority strategies for inspecting their area; ▪ Gives the First report of a number of sites determined as contaminated land (as of 31st March 2002); ▪ Presents some of the sources of national data that may be used to assess the nature & extent of the wider legacy of land affected by contamination.	Chapter 1: Introduction; Chapter 2: Background & legislation; Chapter 3: Preparation for Part IIA; Chapter 4: LA strategies for inspecting contaminated land; Chapter 5: Progress in identifying contaminated land under Part IIA; Chapter 6: Other sources of information about land that may be affected by contamination; Chapter 7: The way forward; Annex 1: LA liaison groups for contaminated land; Annex 2: Environment Agency research outputs on land affected by contamination; Annex 3: Production of LA inspection strategies July 2002; Annex 4: Analysis of historic land uses; Annex 5: Notes to accompany figures in section 6; References	http://www.environment-agency.gov.uk/commondata/105385/dealing_with_contaminated_land_i

Continued

Table 8.5.3 UK Level Pollution Policy. *Continued*

Policy Type	Document Title	Objective	Contents	Website
Strategies & Reports	Implementation of Directive 96/61/EC Concerning Integrated Pollution Prevention & Control (Oct 2003)	Report required by the European Commission under the terms of article 16(3) of directive 96/61/EC. Commission Decision 1999/391/EC prescribes a questionnaire for use as a basis for drawing up the required report. This UK report reproduces each of the questions, under 16 main headings, & presents the UK response to each in turn. This report has been assembled with input from UK government departments, devolved administrations & regulatory bodies in England & Wales, Scotland, Northern Ireland & Gibraltar. It was finalised on 17 September 2003 for submission to the Commission by the 30 September deadline.	General description; Coverage of installations; Basic operation obligations; Existing installations; Permit applications; Co-ordination of the permitting procedure & conditions; Permit conditions; General binding rules; Environmental quality standards; Development in best available techniques; Changes to installations; Reconsideration & updating of permit conditions; Compliance with permit conditions; Information & participation of the public; Transboundary co-operation; Relationship with other Community instruments.	http://www.defra.gov.uk/environment/ppc/ippc-implement.pdf
	Audit of UK Soil Research – Final Report (Oct 2003)	The aim of this audit commissioned by DEFRA & SNIFFER (Scotland and Northern Ireland Forum for Environmental Research) was to review programmes of UK Soil Research from 1995 to the present, to recommend future directions for soil research & to identify improvements in the management of soil research so that funders, providers & end users could maximise the value of the research. Carrying out a soil research audit was one of the proposed actions in the draft Soil Strategy for England.	Executive summary; Introduction; Scope & classification of UK soil research; Collecting & presenting the data; Profile of the UK soil research programme; Assessment of the UK soil research programme; Meeting user needs; International comparisons; Future directions; Conclusions & recommendations; Acknowledgements; References.	http://www.defra.gov.uk/environment/landliability/soil/pdf/audit-report.pdf
	Use of Multi-Criteria Analysis in Air Quality Policy (Nov 2003)	This project was jointly funded by DEFRA and the Environment Agency to explore the use of multi-criteria decision analysis (MCDA) to inform the UK Government's air quality strategy (AQS) objectives. Specifically, these objectives were to be met:	Executive summary; Background; Multi-criteria decision analysis & cost–benefit analysis compared; Method; Ecosystems model; Air quality model; Conclusion; Glossary	http://www.defra.gov.uk/environment/airquality/mcda/mcda-report.pdf

	■ To assess how MCDA techniques compare with currently employed CBA (cost–benefit analysis) techniques employed by DEFRA. ■ To demonstrate the benefits of using MCDA techniques as a supplement to existing processes. ■ To identify areas of AQS work where MCDA techniques can most appropriately be used.		
Plans & Programmes under the First Air Quality Daughter Directive (Nov 2003)	This consultation paper seeks views on the UK government's and devolved administrations' report to the European Commission on plans and programmes to meet the ambient air quality limit values for pollutants set under the First Air Quality Daughter Directive (1999/30/EC), produced in accordance with requirements under the Air Quality Framework Directive (1996/62/EC).	Summary; Policy framework; The air quality framework directive (1996/62/EC); The First Daughter Directive (1999/30/EC); UK annual report for 2001; Report on plans & programmes; Annex A – limit values in Directive 1999/30/EC	http://www.defra.gov.uk/corporate/consult/air-1daughter/consultation.pdf
Report to the European Commission on Plans & Programmes to meet Limit Values under the First Air Quality Daughter Directive (1999/30/EC) (Nov 2003)	This report summarises the UK approach to plans and programmes in place to deliver the limit values set under the First Air Quality Daughter Directive (1999/30/EC). The report is intended to fulfil obligations set out in Articles 8 and 11 of Directive 1996/62/EC1 as supplemented by Commission Decision.	Exceedances reported under Directive 1999/30/EC – UK annual report for 2001; UK plans & programmes; National & European measures contributing to delivery of limit values in the UK; Local air quality management; Future measures; Section 2: Forms; Annexes	http://www.defra.gov.uk/corporate/consult/air-1daughter/draft-report.pdf
The First Soil Action Plan for England: 2004–2006 (March 2004)	The plan sets out an ambitious programme of work for the next three years, to help move towards a clearly stated vision for the nation's soils. The actions are often only the first, important step in the process. The aim for this first plan is to achieve as much as possible by properly embedding soils into ongoing work; to gather the evidence; & to build consensus and partnerships with others in Government & outside to provide the foundation for future action.	Foreword; Summary; Introduction; Our vision & aims for England's soils; The structure & aims of this plan; Core actions to achieve our aims; Overarching soil protection issues; Soils for agriculture & forestry; Interactions between soils, air & water; Soils & biodiversity; Soils, the landscape & cultural heritage; Soils in minerals extraction, construction & the built environment; Measuring & reporting progress; Annexes	http://www.defra.gov.uk/environment/land/soil/pdf/soilactionplan.pdf

Continued

Table 8.5.3 UK Level Pollution Policy. *Continued*

Policy Type	Document Title	Objective	Contents	Website
Strategies & Reports	The state of soils in England & Wales (March 2004)	The state of soils in England & Wales is a report that summarises current knowledge about the condition of soils. Healthy soils are vital to a sustainable environment, but there are increasing signs that their condition has been neglected. As it cannot be assumed that soil loss & damage will be recoverable, it is essential that soils are managed sustainably to keep them healthy for future generations. The report covers the main functions of soil & highlights the issues that need to be addressed. It looks at the quality of soil, the pressures acting on it & the impacts that these pressures are having on it.	Foreword; Summary; Introduction; The interdependence of soil, water & air; Soil, wildlife & biodiversity; Producing food & fibre; Providing raw materials; The built environment; Cultural heritage; References	http://www.environment-agency.gov.uk/commondata/acrobat/stateofsoils_775492.pdf
	Expert Group Report on Nitrogen Dioxide – Nitrogen Dioxide in the United Kingdom (April 2004)	This report provides a detailed analysis of nitrogen dioxide in the UK. It addresses: ▪ Emissions sources & their past & projected trends. ▪ Meteorological & chemical processes that affect ambient nitrogen dioxide concentrations. ▪ Measurement methods & networks. ▪ Modelling & mapping methods & their coupling with measurements at monitoring sites to predict future concentrations of nitrogen dioxide. ▪ Analysis of monitoring data & projections of future concentrations & likely exceedances of air quality objectives & limit values.	Executive summary; Introduction; NO_x emissions & emission inventories; Meteorological & chemical processes; Measurement methods & UK monitoring networks for NO_2; Modelling; NO_2 concentrations & recent trends; Projected NO_2 concentrations in 2005 & 2010; Conclusions; Research recommendations; Appendices.	http://www.defra.gov.uk/environment/airquality/aqeg/nitrogen-dioxide/index.htm
	Performance Review 2003/4 of Local Authority Implementation of LAPC, LAPPC &	This report presents the findings of Atkins review of Local Authority Implementation of the Local Air Pollution Control Regime, undertaken in January & February 2004. The aims & objectives of the review involved collecting a mixture of	Executive Summary; Introduction; Methodology; Questionnaire Responses & Analysis; Characteristic Identification; Conclusions & Recommendations; Appendices	http://www.defra.gov.uk/environment/airquality/lapc/performreview0304/pdf/report.pdf

	LA-IPPC Regimes – Final Report (April 2004)	quantitative & qualitative information from each participating authority. The data was then to be cross-referenced to produce a picture of how different types of authority are implementing pollution control & to identify examples of best practice that other authorities may use to improve their performance. The review also had to provide recommendations as to the adequacy or otherwise of the current levels of fees & charges, with supporting evidence.		
	Valuation of Health Benefits Associated with Reductions in Air Pollution – Final Report (May 2004)	A key recommendation of the EAHEAP (Economic Appraisal of Health Effects of Air Pollution) & the subsequent 1999 ICCB (Inter-departmental Group on Costs and Benefits) report was to undertake empirical studies of WTP (Willingness to Pay) for health benefits associated with reductions in air pollution. As a result, this study was commissioned with the aim of generating a range of monetary values for various key mortality & morbidity benefits. The study focused on two types of mortality effects – chronic (loss of life expectancy in normal health) & acute (loss of life expectancy when elderly & in poor health); & two types of morbidity effects – a hospital admission with breathing difficulties & breathing discomfort on 2 or 3 days every year.	Executive summary; Background; Methodology; Results; Follow-up interviews; Policy implications; Concluding remarks; Annexes	http://www.defra.gov.uk/environment/airquality/airpoll_health/airpollution_reduction.pdf
Consultation Papers	Proposed Regulations Transposing the EC Air Quality Framework Directive & First Daughter Directive – Consultation Paper (05/03/01)	This consultation paper seeks views on the proposed Air Quality Limit Values (England) Regulations 2001. The Regulations have the effect of transposing two European Community (EC) Directives, 96/62/EC and 99/30/EC, relating to the setting of limit values for air pollutants. Similar consultation papers covering Scotland, Wales and Northern Ireland will be published by each individual devolved administration.	Part I – Introduction & summary; Part II – Provisions of Directives 96/62/EC & 99/30/EC; Part III – Implementation proposals; Part IV – Regulatory impact assessment; Part V – Proposed air quality limit values (England) regulations 2001; Annex A: Directive 96/62/EC; Annex B: Directive 99/30/EC; Annex C: Delineation of zones & agglomerations map	http://www.defra.gov.uk/environment/consult/aqframe/index.htm

Continued

Table 8.5.3 UK Level Pollution Policy. *Continued*

Policy Type	Document Title	Objective	Contents	Website
Consultation Papers	Draft Soil Strategy for England – Consultation Paper (26/11/01)	This draft soil strategy sets out the government's overall approach for the sustainable use and protection of soil. It is being published for consultation so that all stakeholders have the opportunity to provide their comments, views and suggestions on what the final strategy should say.	Introduction; Chapter 1: The overall strategy; Chapter 2: The extent of soil; Chapter 3: The diversity of soil; Chapter 4: The quality of soil; Chapter 5: Monitoring & Research; Chapter 6: Redeveloping the strategic approach to soil; References; Code of practice on consultation – the consultation criteria	http://www.defra.gov.uk/environment/consult/dss/index.htm
	Consultation Paper on Implementation of Council Directive 2000/76/EC on the Incineration of Waste (June 2002)	This consultation paper addresses how the Government intends to implement EC Directive 2000/76/EC on the Incineration of Waste (WID) and invites views where there appear to be options for national controls. It is aimed at operators of incineration and co-incineration plant, regulators, waste producers and waste managers, all of whom will be affected by the implementation of WID. This document is also aimed at waste and environmental interest groups who have an interest in environmentally sound regulation.	Introduction; Part I – Chapter 1: The regulatory framework; Chapter 2: The scope of the Directive; Chapter 3: Issues arising from the Directive; Part II – Explanatory notes; Draft regulations & directions; Annex 1: Regulatory impact assessment; Annex 2: List of questions contained in the paper; Annex 3: Directive 2000/76/EC on the incineration of waste; Annex 4: Annexes IIA & IIB of the Waste Framework Directive 76/442/EEC	http://www.defra.gov.uk/environment/consult/wasteincin/pdf/waste_incineration1.pdf
	Second & Third Air Quality Daughter Directives: Consultation Paper (Aug 2002)	This consultation paper seeks views on the proposed Air Quality limit values (Amendment) Regulations 2002 & Air Quality limit values (Amendment) Regulations 2003 which have the effect of transposing the EC Directive 2000/69/EC (the Second Daughter Directive) which sets health based limit values for carbon monoxide & benzene, & Directive 2002/3/EC (Third Daughter Directive) which sets target values & long-term objectives for ozone respectively. Similar papers covering Scotland, Wales & NI will be published by the devolved administrations.	Summary; Section 1: Background to Second Daughter Directive 2000/69/EC; Section 2: Second Daughter Directive; Section 3: Background to Third Daughter Directive 2002/3/EC; Section 4: Third Daughter Directive; Section 5: Regulatory Impact Assessment for the Second Daughter Directive; Section 6: Regulatory Impact Assessment for the Third Daughter Directive	http://ww.defra.gov.uk/environment/consult/air-23daughter/pdf/airqual_2-3-daughter.pdf

	Plans & Programmes under the First Air Quality Daughter Directive: A consultation paper on the UK's Plans & Programmes to meet limit values set in EC Directive 1999/30/EC, the First Daughter Directive (Sept 2004)	This consultation paper seeks views on report of the UK government & devolved administrations to the European Commission on plans & programmes to meet the ambient air quality limit values for pollutants set under the First Air Quality Daughter Directive (1999/30/EC), produced in accordance with requirements under the Air Quality Framework Directive (1996/62/EC).	Summary; Policy framework; The Air Quality Framework Directive (1996/62/EC); The First Daughter Directive (1999/30/EC); UK annual report for 2002; Report on plans and programmes; Annex A – Limit Values in Directive 1999/30/EC; Annex B – Map of UK zones and agglomerations under Directive 1999/30/EC 10	http://www.defra.gov.uk/corporate/consult/air-1daughter2004/consultation.pdf
	EU Emissions Trading Scheme – Proposed Amendments to the UK National Allocation Plan Consultation Document (Nov 2004)	The purpose of this consultation is to seek public comments on proposed amendments to the UK National Allocation Plan. The National Allocation Plan (NAP) sets out the total quantity of allowances which the UK intends to allocate to installations covered by the EU Emissions Trading Scheme & how it intends to allocate those allowances to installations.	Process for Responding; Introduction; The EU Emissions Trading Scheme; Development of the UK National Allocation Plan; Proposed amendment to the NAP; Proposal; Total quantity of allowances to be issued for Phase 1; Allocation methodology for Power Stations; Methodology for calculating new entrant reserve; Rationale; Questions for stakeholders	http://www.defra.gov.uk/corporate/consult/euetsnap-amend/consult.pdf
Guidance Notes	Contaminated Land (England) Regulations 2000 & Statutory Guidance: Regulatory Impact Assessment (Final) (16/02/00)	This final regulatory impact assessment (RIA) deals with the regulations and statutory guidance which form key elements in the implementation of a new regulatory regime for the legacy of contaminated land. The main regime is set out in Part IIA of the Environmental Protection Act 1990, and has already been appraised – a compliance cost assessment (CCA) was prepared when the Environment Bill containing Part IIA was originally introduced into Parliament in 1994.	Regulatory impact assessment (final); Annex A: The new contaminated land regime; Annex B: The risk imposed by contaminated land; Annex C: The options; Annex D: Examples of remediation costs & small businesses	http://www.defra.gov.uk/environment/landliability/ria/index.htm
	Contaminated Land: DETR Circular 2/2000 (20/03/00)	Contaminated land is an archetypal example of our failure in the past to move towards sustainable development. This circular has two functions: first it promulgates the statutory guidance which is an essential part of the new regime; secondly, it sets out the way in which the new regime is expected to work, by providing a summary of government policy in this field, a description of the new regime, a guide to the Regulations & a note on the saving provision in the Commencement Order.	Circular; Annex 1: A statement of government policy; Annex 2: A description of the new regime for contaminated land; Annex 3: Statutory guidance; Annex 4: A guide to the contaminated land (England) Regulations 2000; Annex 5: A guide to the Environment Act 1995 (Commencement No. 16 & saving provision) (England) Order 2000; Annex 6: Glossary of terms	http://www.defra.gov.uk/environment/landliability/circ2-2000/index.htm

Continued

Table 8.5.3 UK Level Pollution Policy. *Continued*

Policy Type	Document Title	Objective	Contents	Website
Guidance Notes	Integrated Pollution Prevention & Control a Practical Guide – Second Edition (June 2002)	The system of Integrated Pollution Prevention and Control (IPPC) applies an integrated environmental approach to the regulation of certain industrial activities. This means that emissions to air, water (including discharges to sewer) & land, plus a range of other environmental effects, must be considered together. It also means that regulators must set permit conditions so as to achieve a high level of protection for the environment as a whole. These conditions are based on the use of the Best Available Techniques. (BAT), which balance the costs to the operator against the benefits to the environment. This Guide is to help those operating or regulating activities prescribed under the terms of the Pollution Prevention & Control (PPC) Regulations 2000. It describes the main provisions of IPPC and sets out the views of the Secretaries of State for the Department for Environment, Food & Rural Affairs (DEFRA) & for Wales (the Secretary of State) on how IPPC should be applied & how particular terms should be interpreted.	Chapter 1: Introduction; Chapter 2: Overview of the regulatory process; Chapter 3: IPPC activities, installations, operators & non-IPPC landfills; Chapter 4: Timeframe for obtaining IPPC permits; Chapter 5: Permit applications; Chapter 6: Consultation on permit applications; Chapter 7: Determination of permit applications; Chapter 8: Management systems & operator competence; Chapter 9: Required standards & best available techniques (BAT); Chapter 10: Environmental quality standards; Chapter 11: Changes to installations that have already been permitted; Chapter 12: Permit transfers; Chapter 13: Permit reviews; Chapter 14: Site assessment & restoration; Chapter 15: Special consideration for activities involving waste; Chapter 16: Checking & enforcing compliance; Chapter 17: Public registers & information; Chapter 18: Charging; Chapter 19: Appeals; Chapter 20: Connections with other legislation; Annex 1: Other information; Annex 2: Definitions;	http://www.defra.gov.uk/environment/ppc/ippcguide/pdf/ippcguide_ed2.pdf

			Annex 3: Statutory consultees; Annex 4: Outline of the main IPPC activities & the transition timetable for existing installations; Annex 5: Content of applications; Annex 6: Indicative list of pollutants; Annex 7: Factors to be considered in determining BAT; Annex 8: EC environmental quality standards relevant to IPPC.	
	Integrated Pollution Prevention & Control: A Practical guide – Third Edition (Feb 2004)	This Guide is to help those operating or regulating activities prescribed under the terms of the Pollution Prevention and Control (England & Wales) Regulations 2000 (as amended). It describes the main provisions & sets out the views of the Secretaries of State for the Department for Environment, Food & Rural Affairs (DEFRA) & for Wales on how they should be applied & how particular terms should be interpreted.	Introduction; Overview of the regulatory process; IPPC activities, installations & operators; Timeframe for obtaining IPPC permits; Permit applications; Consultation on permit applications; Determination of permit applications; Management systems & operator competence; Required standards & best available techniques (BAT); Environmental quality standards; Changes to installations that have already been permitted; Permit transfers; Permit reviews; Site assessment, protection, monitoring & restoration; Special considerations for activities involving waste; Checking & enforcing compliance; Public registers & information; Charging; Appeals; Connections with other legislation; Annexes	http://www.defra.gov.uk/environment/ppc/ippcguide/pdf/ippcguide_ed3.pdf

Table 8.5.4 Northern Ireland Level Pollution Policy.

Policy Type	Document Title	Objective	Contents	Website
Strategies & Reports	Air Quality Monitoring in Northern Ireland: 2000–2001 (Nov 2002)	This report on air quality monitoring in Northern Ireland combines data from large-scale networks such as the Automatic Urban Network & Non-Automatic Networks, with data from the considerable number of other monitoring sites operated by Northern Ireland's District Councils. The period covered is calendar years 2000 & 2001. The report follows on from the series of six previous reports produced by the Chief Environmental Health Officers Group (CEHOG) for Northern Ireland.	Foreword; Executive summary; Introduction; Air quality development in 2000 & 2001; Carbon monoxide; Nitrogen dioxide; Sulphur dioxide; Particulate matter; Ozone; Hydrocarbons; Metals; Conclusions; References; Appendices.	http://www.ehsni.gov.uk/pubs/publications/AQMNIreport_web.pdf
	Implementation of the Environmental Noise Directive (May 2003)	EC Directive 2002/49/EC relating to the assessment and management of environmental noise was to be transposed into Northern Ireland statute by 18 July 2004. The Department of the Environment developed the necessary legislative provisions to effect this transposition. In the development of the necessary provisions, the Department explored the areas of the Directive affording a degree of discretion in the implementation methods used.	About this document; The provisions of the directive; Implementation of the directive in NI; Annex: The Environmental Noise Directive	http://www.ehsni.gov.uk/pubs/publications/Implementation_of_the_Environmental_Noise_Directive.pdf
	Air Quality Monitoring in Northern Ireland: 2002 (Jan 2004)	This report presents a summary of air quality in Northern Ireland over the calendar year 2002. It is intended to bring together in one report results from all the District Councils who carried out air quality measurements in Northern Ireland over this period, both as part of larger monitoring networks & for other purposes. The report aims to provide information on the main pollutants of concern, details of the air quality monitoring undertaken in 2002, & a summary of results for each pollutant.	Foreword; Executive summary; Introduction; Air quality development in 2000 & 2001; Carbon monoxide; Nitrogen dioxide; Sulphur dioxide; Particulate matter; Ozone; Hydrocarbons; Metals; Conclusions; References; Appendices.	http://www.ehsni.gov.uk/pubs/publications/AQReport_web.pdf

Consultation Papers	EC Directive 96/61 on Integrated Pollution Prevention & control – Arrangements for Transposition in NI (A Consultation Paper) (June 2001)	This consultation paper invites views on proposals for primary legislation, the main purpose of which will be to provide the statutory framework necessary to transpose the requirements of EC Directive 96/61 on integrated pollution prevention & control. Detailed implementation of the new regime will be by regulations. A copy of the Directive is at Annex A.	Part 1: Introduction – The industrial pollution control (NI) order 1997; Part 2: The Directive; Part 3: Timeframe for obtaining IPPC permits; Part 4: Enforcement arrangements; Part 5: IPPC system – regulatory issues; Part 6: Miscellaneous; Part 7: Summary of recommendations; Part 8: comments; App A: CD 96/61/EC concerning IPPC; App B: Provisional estimate of the number of installations in NI covered by the Directive; App C: Meaning of installation	http://www.ehsni.gov.uk/pubs/publications/EC_Directive_96_61_Integrated_Pollution_Blue.pdf
	Consultation Document on a Proposed Bill for Local Air Quality Assessment & Management in Northern Ireland (Oct 2001)	This consultation document sets out proposals for the future legislative framework & policy for ambient air quality assessment & management in Northern Ireland. It seeks views on the Department's proposals for legislation to provide a framework for the assessment & management of air quality in Northern Ireland.	Introduction; The UK air quality strategy; The EC air quality framework directive; NI air quality; Proposed legislation – local air quality management bill; Appendices	http://www.ehsni.gov.uk/pubs/publications/LAQM_consultation_paper_oct.pdf
	The Environment (Northern Ireland) Order 2002 – Local Air Quality Management Draft Policy Guidance (Mar 2002)	This consultation paper seeks views on proposed Statutory Guidance issued under Article 16 of the Environment (Northern Ireland) Order 2002 for Local Air Quality Management (LAQM). District councils & other relevant authorities in Northern Ireland are required to have regard to this guidance when carrying out their LAQM duties under the Environment (Northern Ireland) Order 2002. This guidance document represents major aspects of policy in regards to ambient air quality management. It includes policy guidelines on air quality reviews & assessments, air quality action planning, transport planning & land use planning.	Introduction; Role of district councils; Air quality management areas; Air quality action plans; Consultation & liaison; Developing local & regional air quality strategies; Air quality & transport; National context; Regional context; Local toolkit measures; Other local measures; Air quality & landuse planning; Appendices.	http://www.ehsni.gov.uk/pubs/publications/POLICYGUIDANCE_12.pdf

Continued

Table 8.5.4 Northern Ireland Level Pollution Policy *Continued*

Policy Type	Document Title	Objective	Contents	Website
Guidance Notes	Pollution Prevention Guideline – PPG1: General Guide to Pollution Prevention (Aug 2000)	The series of Pollution Prevention Guidance notes (known as PPGs), of which this is the first, provide practical advice on avoiding causing pollution, minimising waste & complying with the requirements of the law. Often the necessary measures cost little, especially if thought about early on, for example at the design stage, & can save money. In contrast, the fines for failing to comply with the relevant regulations or the costs of cleaning up pollution (which are recovered from the polluter wherever possible) can be very high.	Introduction; Availability; Legal framework; Site drainage; Waste storage & disposal; Deliveries & security; Oil storage & pipelines; Contingency plans; Construction & demolition; Agriculture; Groundwater pollution; References.	http://www.ehsni.gov.uk/pubs/publications/PPG01.pdf
	PPG5: Works in, Near or Liable to Affect Watercourses (April 2001)	These guidelines have been drawn up to assist all those who may have cause to work in or near watercourses. Compliance with this guidance should minimise the risk of pollution occurring.	Legal framework; Introduction; General precautions; Silt; Concrete & cement; Oil & chemicals; Bridge cleaning & repainting; Herbicide use; Emergencies; References	http://www.ehsni.gov.uk/pubs/publications/PPG05.pdf
	PPG6: Working at Demolition & Construction Sites (April 2001)	These guidelines are intended to assist those in the construction & demolition industry with responsibility for managing the environmental impact of their activities. Compliance with these should minimise the effect of the work on the environment.	Legal framework; Introduction; Planning & preparation; Site drainage; Deliveries; Storage; Waste management; Silt; Refuelling; Concrete; Emergencies; References	http://www.ehsni.gov.uk/pubs/publications/PPG06.pdf
	PPG11: Industrial Sites (April 2001)	These guidelines are intended to provide some information to industrial users on the potential impact of their practices on the surrounding environment.	Legal framework; Introduction; Site drainage; Sewage & waste water disposal; Surface water drainage; Delivery & handling of materials; Storage facilities; Waste management; Routine inspection & maintenance; Cleaning & degreasing; Demolition & construction; Staff training; Emergency plans; Security; References	http://www.ehsni.gov.uk/pubs/publications/PPG11.pdf

PPG15: Retail Premises (April 2001)	This document has been drawn up to assist those who design & operate major retail outlets & similar sites, such as distribution warehouses. Compliance with these guidelines should minimise the risk of pollution occurring.	Introduction; Site drainage; Sewage & waste water disposal; Surface water drainage; Waste management; Oil storage; References	http://www.ehsni.gov.uk/pubs/publications/PPG15.pdf
IPC Guidance Note 1: Introduction to the Order (Sept 2002)	The Industrial Pollution Control (NI) Order 1997 (the Order), which came into force on 2 March 1998, enacts proposals for enhanced industrial pollution control in NI which were contained in a series of consultation papers published by the DOE from 1987–1997. The first & second of these, published in 1987 & 1993 respectively, reviewed the system of air pollution control & outlined proposals to implement the EC Air Framework Directive (84/360/EEC) dealing with air pollution from industrial plant. A further consultation paper was published in 1995 proposing the introduction of a system of Integrated Pollution Control (IPC) for NI. The guidance contained in this Note falls into two categories: ▪ Guidance on the interpretation of key terms used in the legislation, notably: BATNEEC, substantial change, existing processes, capacity to cause harm, & commercially confidential information; ▪ Guidance on the procedural aspects of the legislation including an explanation of the variation & appeals procedures, enforcement options, the interface with other legislation, the charging scheme, & the procedures for handling mobile plant.	Introduction; Main features of the system of local air pollution control; The Industrial Pollution Control (Prescribed Processes & Substances) Regulations (NI) 1998; Timetable for implementation; Best available techniques not entailing excessive costs – BATNEEC; Variation notice procedures; Meaning of substantial change; Handling commercially confidential information; Procedures for dealing with mobile plant; Interface with DED health & safety division; Charging scheme; Enforcement; Transfer of authorisations; Delegation of decisions; Process guidance notes – application to Public Health (Ireland) Act 1878; Directions under article 5(5) of the order; Articles 31–33 of the industrial pollution control (NI) order 1997; Appendices	http://www.ehsni.gov.uk/pubs/publications/EPDGG1.pdf
Guidance on Drafting Codes of Practice for Minimising Noise in Northern Ireland (May 2003)	To ensure that all Codes of Practice on noise are considered fairly, the Department of the Environment has prepared a guidance document to assist authors wishing to produce Codes. The guidance sets out considerations to be taken into account & recommends the procedures that should be followed if a Code is to be submitted for approval & command the confidence of the courts & others applying it.	About this guidance; Guidance on drafting codes of practice; Appendix A: List of approved codes of practice	http://www.ehsni.gov.uk/pubs/publications/guidance_drafting_codes_practice_minimising_noise.pdf

Continued

Table 8.5.4 Northern Ireland Level Pollution Policy *Continued*

Policy Type	Document Title	Objective	Contents	Website
Guidance Notes	Integrated Pollution Prevention & Control (Northern Ireland) – A Practical Guide Edition One (May 2003)	This guide is to help those operating or regulating activities prescribed under the terms of the Pollution Prevention & Control (PPC) Regulations Northern Ireland 2003. It describes the main provisions of IPPC & sets out the views of the Department of the Environment on how IPPC should be applied & how particular terms should be interpreted. The Guide explains the main legal provisions of IPPC, but the precise requirements can be only determined by referring directly to the law itself.	Introduction; Overview of the regulatory process; IPPC activities, installations & operators; Timeframe for obtaining IPPC permits; Permit applications; Consultation on permit applications; Determination of permit applications; Management systems & operator competence; Required standards & best available techniques (BAT); Environmental quality standards; Changes to installations that have already been permitted; Permit transfers; Permit reviews; Site assessment & restoration; Special consideration for activities involving waste; Checking & enforcing compliance; Public registers & information; Charging; Appeals; Connections with other legislation; Annexes	http://www.ehsni.gov.uk/pubs/publications/IPPC_Practical_Guide.pdf
	Integrated Pollution Prevention & Control (IPPC) – Guide for Applicants (July 2003)	This Guide is intended to help those regulated under Part A of the Integrated Pollution Prevention and Control (IPPC) regime. It provides a brief introduction to IPPC and explains where more detailed information can be found. It then gives advice on how to prepare each of the four main types of applications that may be required.	Introduction; Getting a new IPPC permit; After a permit has been issued; Making an application; Explanatory notes on Part A of the application form; Explanatory notes on Part B of the application form; Explanatory notes on Part C of the application form; Explanatory notes on Part D of the application form; Explanatory notes on Part E of the application form; Explanatory notes on Part F of the application form; Submitting your application; Glossary; Annexes	http://www.ehsni.gov.uk/pubs/publications/SiteReportGuidanceRLLPWordLogoTNR2.pdf

	EU Emissions Trading Scheme Guidance Note One (Dec 2003)	This guidance was drawn up on the basis of the draft EU Greenhouse Gas Emissions Trading Scheme Regulations 2003 and was therefore subject to change pending further work on the regulations.	How do I know if I am included? Who should apply for a permit? How should I define the scope of the installation? When should capacities be aggregated to determine whether a schedule 1 activity is being carried out? How do I define a directly associated activity? Examples of the directly associated activity rule; If I had a directly associated activity on site, do I need to account for the emissions from that activity? I have stand-by generation or boiler capacity on site – should I include them in the aggregation? What is & isn't included in the definition of 'combustion installation'? What happens if there is a CHP plant on site that supplies the process with electricity, but is operated by a different organisation? How do I find out what the rated thermal input of my combustion installation is? Further examples; Checklist; Annexes	http://www.ehsni.gov.uk/pubs/publications/eu-ets-guidance01.pdf
	Depolluting End-of Life Vehicles – Guidance for Authorised Treatment Facilities (Jan 2004)	This is guidance on how to depollute an ELV to achieve a depolluted ELV classed as non-hazardous. The particular guidance methods are not mandatory (provided the same levels of depollution are demonstrated). Non-destructive methods of removing hazardous components may be carried out, provided that component integrity is not compromised.	Introduction; Example of the depollution process; Preliminary activities; Removal of fluids & other items; Removal or deployment of air bags; End of depollution procedure	http://www.ehsni.gov.uk/pubs/publications/DepollutingELVs.pdf

9 Social Issues

9.1 Social Issues and the Built Environment

Social issues have been at the heart and cause of unrest, inequality and concern throughout the ages as people have attempted to eke out a living, seek a better quality of life or achieve an enhanced status in the company of and in opposition to their fellow men. This tension has been expressed dramatically in warfare as societies have been decimated as a result of greed, perceived inequality or the pursuit of social justice. However, the social condition of society remains in a fluctuating and sometimes volatile state due to opportunities and threats which challenge the consolidation of all types and sizes of communities.

As people live, work and play in the built environment, its social fabric and structure can determine the social condition of its inhabitants. For example, a low income community which lives in poorly heated and ventilated properties is more likely to have children who suffer from asthma and poor health, while those in high quality homes on high wages have the opportunity to develop good health, leading to high school attendance, better performance and subsequently full employment. This demonstrates the complex interactions between the social status of people and various elements of the built environment, with a range of implications for those who can access better living and working environments, and this is further complicated when considered against the backdrop of a growing global population.

The UN Conference on Human Settlements (1976) first addressed these issues of physical and spatial organisation of human life on this planet, and considered the national and international actions needed to accommodate the growing population in urban and rural communities. The Vancouver Declaration was agreed which established the concept of 'human settlements consisting of several elements previously considered separately from one another i.e. housing, building, planning and their relationship with other activities such as environmental change and national/international development'.

This theme was re-visited at the World Conference on Human Rights in Austria (1993), at which the Vienna Declaration advocated a common plan for the strengthening of human rights work around the world. It also examined the link between development, democracy and economic, social, cultural, civil and political rights. Furthermore it evaluated the effectiveness of United Nations methods and mechanisms with the aim of recommending ways to ensure adequate financial and other resources for all United Nations human rights activities.

These high level matters are balanced by the practical needs of society as it strives for progress. At the most fundamental level, society requires a 'basket of basic human needs' such as

- housing and shelter;
- clothing and warmth;
- food and nutrition;
- healthcare and life expectancy;

- education and employment;
- energy for survival;
- mobility and movement; and
- culture and leisure.

As society progresses and this basic basket swells, the opportunity for a more creative life arises through wealth creation and knowledge attainment. This defines a society's social status in practical and cultural terms and from both an individual and community perspective. There remains a need for transparent and established governance structures to both facilitate and influence progress within a democracy in which a social arrangement ensures that power is vested in the people collectively or in officers appointed or elected to administer powers. This raises questions of how democracy works, how do elected representatives stay in touch, how many people within a community make their views known, and how does it all work in practice – with many of these issues resolved through the dynamic processes of social inclusion and community engagement.

Non-governmental organisations (NGOs) have a key role in representing and developing the needs and solutions for all communities, especially those which are beleaguered due to low incomes, inequalities and social conditions. The World Bank defines NGOs as 'private organisations that pursue activities to relieve suffering, promote the interests of the poor, protect the environment, provide basic social services, or undertake community development'. In wider usage, the term NGO can be applied to any non-profit organisation which is independent from government. NGOs are typically value-based organisations which depend, in whole or in part, on charitable donations and voluntary service, where the principles of altruism and voluntarism remain key defining characteristics. Research shows that NGOs tend to focus on one of three key issues – environmentally sustainable development, human rights, or women in development. Many of these NGOs also address broader or related issues such as poverty, hunger and humanitarian aid.

Industry and business are driven and guided by the concept of Corporate Social Responsibility in responding to the range of issues and practices which describe their interaction with the built environment, while individuals and their respective community groups respond to a different set of core values. An individual will be guided by a social conscience of environmental stewardship connecting the use and preservation of natural materials with the need to protect the environment. However corporate businesses are more likely to be influenced by a desire to manage resources and maximise profit margins in this cross-cutting consideration. Similarly, the issue of how natural resources are prudently used while considering economic and employment matters is guided by a desire to achieve good value for money for both community and business groups. The latter will also wish to adopt some form of environmental management system to ensure consistency and stability over a longer period.

9.2 Key Elements of Social Issues

The elements which form the social structure of society are drawn from several areas of activity as they impinge on people's quality of life. These social elements are best understood or defined by the way in which a society provides the broad sections within the 'basket of basic human needs' referred to above. Understanding the interaction between the conditions which deliver or prevent the basket of basic human needs existing is critical for society to prevail and develop its pursuit of a better quality of life.

Access to the continuum of basic education, skills development and employment opportunities for all people is essential for the growth of a community and this is particularly true for children. Exposure to the fundamentals of reading, writing and numeracy education provides the basis for the development of physical, motor and social skills, creating the potential for gaining employment and aligning with the concept of lifelong learning. The ability of individuals and society to develop creative and innovative minds is complex, but this can be enhanced by a basic education and exposure to different ethnic backgrounds, experiences and professions.

Social issues are also influenced by conditions of poverty and deprivation. Social deprivation implies the absence of a complex mix of aspects including education, employment, income, health, housing and social conditions. Deeper understanding involves both quantitative and qualitative assessments, encompassing all areas of social sciences. Social deprivation is arguably best judged by its effect on areas of behaviour such as smoking, suicide, accidents, low educational attainment, mental health problems (including alcohol and drug abuse), the average age of first pregnancy, and the numbers of elderly people in an area. Poverty plays a key role in causing the conditions for social deprivation to prevail.

Health is strongly influenced by social conditions. Even in the most affluent nations, people who are less well off have substantially shorter life expectancies and more illnesses than the rich. These differences in health quality are an important social injustice and draw scientific attention to the social determinants of health within society. Health is sensitive to the social environment, with evidence that a large range of factors play simultaneous and inter-related roles in the health of individuals and communities. The social determinants of health include:

- poor social conditions;
- stressful and worrying circumstances;
- conditions of hardship, resentment and poverty;
- workplace stress;
- job security or high levels of unemployment;
- levels of addiction to alcohol, drugs and tobacco;
- access to and availability of healthy food;
- friendship, good social relations and strong supportive networks;
- a good start in life for babies and support for mothers; and
- access to public transport facilitating more walking and cycling.

For many years, gender and ethnic issues have plagued and dictated the social conditions prevalent in communities. The role of women has changed with legislation, and later in practice, and attitudes to ethnic minorities, immigrants and those of differing religious interests have evolved. There have been strident moves strengthening the voice of women in environmental decision-making and in policy areas such as water, human settlements and the environment. Similarly, ethnic groups are being protected by laws that provide equality of opportunity and fair employment. However both gender and ethnic prejudices run deep in many societies and these remain as matters which strongly shape communities.

Access to culture, arts and leisure facilities are viewed as a measure of social progress, and while these elements are more obviously prevalent in the developed world with a specialised entertainment industry, strongly felt cultural customs and leisure activity are supported by all societies. The development of music, drama, painting and all forms of performing arts allows a society to project itself and appreciate the talents and therapeutic value of these cultural pursuits. Similarly, access to leisure, recreation and sport enables individuals and teams to improve their health, and develop attributes such as self-worth and skilful performances to help achieve personal goals. There is equal merit in the enjoyment gained from study and interaction with historical and heritage sites or artefacts, performance of traditional ceremonies and reverence for the past. These are all dynamic contributors to the social well-being of individuals and communities.

The sustainability of communities is determined by a combination of many of these elements, but can be hampered by anti-social activities such as high levels of crime and lack of safety for people and property. Crime and safety issues are indicators of social governance, but are also driven by political and cultural influences when pressure groups are permitted to influence the structure and behaviour of society. Good role models and appropriate legal support mechanisms can assist in establishing stable social conditions, while behavioural change arises from reason and social conscience, with faith groups having the opportunity to help initiate change.

9.3 Sustainability Challenges for Social Issues

Social issues are implicit with the 'social progress which recognises the needs of everyone' objective of sustainable development, but also have an impact at a higher level and relate to the other objectives. The wider links with the objectives of sustainable development are shown below:

- *Effective protection of the environment* through the provision of structures for environmental justice, in which NGOs can take advocacy roles while acting in the capacity of lobbyists or pressure groups, advisory bodies and negotiators. This attention to environmental protection can be carried out by practices such as water conservation, waste minimisation, energy usage reduction and support for transport integration while contributing simultaneously to social inclusion.

- *Social progress which recognises the needs of everyone* is at the heart of the eradication of social exclusion and contributes to the aspiration of total social inclusion. The voice of all sections of society must be heard and understood for progress to be achieved, and it must address all the elements of the basket of life's essentials, delivering stable and sustainable communities and providing a good quality of life for all. Advocacy and leadership are congruent with this comprehensive and practical approach.
- *Prudent use of natural resources* can be achieved through a strong sense of environmental awareness and stewardship and by a series of practical and beneficial actions at a micro-level. Societies can make a range of small actions which contribute to a much greater impact when combined with other similar activities. Actions such as water usage reduction, improved household plumbing, collection of rainwater to assist with watering of gardens, use of low energy bulbs, greater use of natural light in homes and offices, campaigns for turning off of TVs and computers during periods of non-usage, and control of vehicle speeds to minimise fuel consumption highlight some of the small actions by which communities can contribute to the preservation of natural resources.
- *Maintenance of high and stable levels of economic growth and employment* can be secured at community and business levels through generating a full understanding of the need to educate and provide skills for all persons. The development of knowledge, opportunities to access a range of training services and a structure which gives equality to all will facilitate business and cultural progress. However society needs to recognise human and cultural diversity, as not all members of society will have the same aspirations.

The dynamics and structure of the social conditions in a community influences and dictates its sustainability in the widest sense. This results in a number of challenges that need to be faced, now and in the future, within the realm of social inclusion:

- *Develop sustainable communities within all urban and rural developments* by promoting community governance and integration which adopts a wide sustainability vision and facilitates good practice and community ownership of systems and processes.
- *Establish equality across all legislation and strategies at national and local levels*, by addressing gender issues through changing women's roles and eradicating oppression, by protecting and advising racial and ethnic minorities and immigrants, by removing religious prejudice and discrimination, and reducing homelessness by considering the social stigma, psychological impact and underlying causes.
- *Legislate for the human rights of all spectrums of society*, by providing social justice, regardless of gender, religious persuasion or ethnic background, by recognising the specific needs of the elderly, and issues of ageism, pension and benefit dependence, and by ensuring that the requirements of an Equality Impact Assessment (EQIA) are fully delivered.
- *Facilitate community cohesion* by eliminating poverty in developed and developing nations, by seeking community engagement through the empowerment of community members, and by encouraging capacity building, citizenship, identity, community ownership and cohesion.
- *Secure good quality community infrastructure, investment and networks* through the development of several levels of partnerships, engaging with public, private, community and voluntary agencies, and through adopting the principle and practice of corporate social responsibility by local businesses to help build public trust.
- *Develop neighbourhood renewal* by supporting sustainable human settlements, decent housing, economic prosperity, safe communities, quality education, good healthcare and equality, by reviewing and responding to the implications of population growth, by improving transport accessibility and personal movement for all levels of physical ability, by reaching out to isolated or detached communities, and by supporting rural revival and regeneration to ensure that rural communities do not die out.

- *Support and finance improved health* through better healthcare to cater for all ages, by facilitating 'active ageing', supported by investment in drugs and pensions, and by advising on appropriate diets and nutrition to reduce issues such as obesity, diabetes and heart disease while warning of the consequences of poor diet.
- *Eliminate environmental threats to human life* by providing for and monitoring genetically modified food production and its safety, by preventing pollution, by minimising communicable diseases, and by advising on and planning for chemical threats.
- *Eradicate health inequalities* by specifically targeting disadvantaged areas which suffer from environmental problems caused by pollution, poor hygiene and inadequate services, by increasing access to food to eliminate hunger and malnutrition in third world countries and by supporting active aid missions and education of local communities.
- *Reduce crime levels* through support for urban safety and crime prevention, removal of anti-social behaviour due to alcohol and drug dependence, and a concerted effort to change the mindset and culture of young people.
- *Increase employment levels* by removing employment barriers, by improving education, by developing persuasive links between education and employment, by encouraging national full employment policies, and by eliminating all forms of discrimination.
- *Encourage a sense of globalisation* by increasing an understanding of global markets, their competitiveness, and the issues associated with local businesses falling victim to price wars with larger conglomerates, and by developing entrepreneurship in capturing new innovative ideas and pushing the boundaries of sustainable technological advancement.
- *Develop cultural diversity* by increasing multi-cultural appreciation, by unlocking creativity, cultural development and promotion, by investing in cultural development to help preserve cultural heritage for future generations, and by ensuring local traditions are maintained.

9.4 Social Issues Theme Index

Jurisdiction	Tier	Title	Date	Page No.
INTERNATIONAL	Conferences	UN Conference on the Human Environment	1972	342
		First UN Conference on Human Settlements (Habitat 1 – Vancouver)	1976	342
		World Conference on Human Rights	1993	343
		International Conference on Population & Development (ICPD)	1993	343
		The World Summit for Social Development (WSSD)	1995	343
		The Fourth World Conference on Women (FWCW)	1995	343
		World Food Summit (WFS)	1996	344
		Second UN Conference on Human Settlements (Habitat II – Istanbul)	1996	344
		International Meeting on Urban Poverty	1996	344
		International Conference on Urban Poverty	1997	345
		Third International Conference on Social Integration & Security for the Urban Poor – Towards Cities for All	1999	345
		Fourth International Conference on Productive & Inclusive Cities – Towards Cities for All	2001	345
		UNEP's Global Women's Assembly on Environment (WAVE)	2004	346

Jurisdiction	Tier	Title	Date	Page No.
INTERNATIONAL	Conventions	Convention on Access to Information, Public Participation in Decision-making & Access to Justice in Environmental Matters	1998	346
EUROPEAN	Council Directives	The Freedom of Access to Information in the Environment *(CD 90/313/EC)*	1990	347
		Providing for Public Participation in Respect of the Drawing up of Certain Plans & Programmes Relating to the Environment *(Dir 2003/35/EC amending CD 85/337/EEC & CD 96/61/EC)*	2003	347
	Council Communications	Cohesion Policy & the Environment *(COM (1995) 509)*	1995	347
		Building an Inclusive Europe *(COM (2000) 79)*	2000	348
		Acting Locally for Employment a Local Dimension for the European Employment Strategy *(COM (2000) 196)*	2000	348
		The Health Strategy of the European Community *(COM (2000) 285)*	2000	349
		Towards a Community Framework Strategy on Gender Equality (2001–2005) *(COM (2000) 335)*	2000	349
		Community Initiative: EQUAL *(COM (2000) 853)*	2000	349
		Social Policy Agenda *(COM (2000) 379)*	2000	350
		Supporting National Strategies for Safe & Sustainable Pensions through an Integrated Approach *(COM (2001) 362)*	2001	350
		Strengthening the local dimension of the European Employment Strategy *(COM (2001) 629)*	2001	350
		First progress report on economic & social cohesion *(COM (2002) 46)*	2002	351
		Consumer Policy Strategy 2002–2006 *(COM (2002) 208)*	2002	351
		The future of the European Employment Strategy (EES) – A strategy for full employment & better jobs for all *(COM (2003) 6)*	2003	351
		Scoreboard on Implementing The Social Policy Agenda *(COM (2003) 57)*	2003	352
		Immigration, Integration & Employment *(COM (2003) 336)*	2003	352
		A European Environment & Health Strategy *(COM (2003) 338)*	2003	353
		Improving Quality in Work: A Review of Recent Progress *(COM (2003) 728)*	2003	353
		Joint Report on Social Exclusion *(COM (2003) 773)*	2003	354
		Modernising Social Protection for the Development of High-quality, Accessible & Sustainable Health Care & Long-term Care: Support for the National Strategies Using the 'Open Method of Co-ordination'	2004	354
		The European Environment & Health Action Plan 2004–2010 *(COM (2004) 416)*	2004	355

Jurisdiction	Tier	Title	Date	Page No.
EUROPEAN	Council Decisions	Setting up a Network for the Epidemiological Surveillance & Control of Communicable Diseases in the Community *(Dec No. 2119/98/EC)*	1998	355
		Establishing the Culture 2000 Programme *(Dec No. 508/2000/EC)*	2000	355
		Establishing a Programme of Community Action to Encourage Co-operation between Member States to Combat Social Exclusion *(Dec No. 50/2002/EC)*	2002	356
	Miscellaneous	Access to Information, Public Participation & Access to Justice in Environmental Matters *(COM (98) 344)*	1998	356
		White Paper on Food Safety *(COM (1999) 719)*	1999	357
		The Lisbon European Council – An Agenda of Economic & Social Renewal for Europe *(COM (2000) 7)*	2000	357
		Charter of Fundamental Rights of the European Union	2000	357
		Second report on economic & social cohesion *(COM (2001) 24)*	2001	358
		White Paper Strategy for a future Chemicals Policy *(COM (2001) 88)*	2001	358
		Promoting a European Framework for Corporate Social Responsibility *(COM (2001) 366)*	2001	358
		Report on Indicators in the Field of Poverty & Social Exclusion	2001	359
		Increasing Labour-Force Participation & Promoting Active Ageing	2002	359
		Adequate & Sustainable Pensions	2003	360
		The Impact of the New Chemicals Policy on Health & the Environment	2003	360
		Recommendations on Health & EU Social Policy	2003	360
		Report on Equality Between Women & Men *(COM (2004) 115)*	2004	361
		Joint Report by the Commission & the Council on Social Inclusion	2004	361
		Enabling Good Health For All: A Reflection Process for a new EU Health Strategy	2004	361
UK	Strategies & Reports	Eliminating World Poverty: A Challenge for the 21st Century – White Paper on International Development	1997	362
		Improving Shopping Access for People Living in Deprived Neighbourhoods	1999	362
		Jobs for All – National Strategy for Neighbourhood Renewal	1999	363
		Social Capital & Health	1999	363
		Eliminating World Poverty: Making Globalisation Work for the Poor – White Paper on International Development	2000	364
		A New Commitment to Neighbourhood Renewal: A National Strategy Action Plan	2001	364

Jurisdiction	Tier	Title	Date	Page No.
UK	Strategies & Reports	National Strategy for Neighbourhood Renewal – Policy Action Team Audit	2001	365
		Preventing Social Exclusion	2001	365
		Neighbourhood Renewal: Increasing Employment Amongst Deprived Areas & Groups – Implementation Strategy	2001	365
		Equality in Practice – Report of a Survey on the Adoption & Implementation by English Councils of *Racial Equality Means Quality*, the CRE Standard for Local Government	2001	366
		Secure Borders, Safe Haven – Integration with Diversity in Modern Britain	2002	366
		Making the Connections: Transport & Social Exclusion	2002	366
		Vision for Sustainable Regeneration – Environment & Poverty: The Missing Link?	2002	367
		Active Communities: Initial Findings from the 2001 Home Office Citizenship Survey	2002	367
		Making the Connections: Transport & Social Exclusion – Final Report	2003	367
		Ensuring the Quality of Community Investment – A Report by People for Action	2003	368
		Realising the Potential – Recognising Residents' Achievement in Neighbourhood Renewal	2003	368
		Changing Neighbourhoods, Changing Lives – The vision for neighbourhood renewal	2003	369
		Changing World, Changing Lives – Women in the UK since 1999	2003	369
		Community Cohesion Pathfinder Programme: The first six months	2003	370
		Community Cohesion Advice for Those Designing, Developing & Delivering Area Based Initiatives	2003	370
		Building Civic Renewal – Review Findings from the Civil Renewal Unit	2003	370
		Achieving Rural Revival – Report of the LGA Inquiry into Rural Revival	2003	370
		Tackling Health Inequalities – A Programme for Action	2003	371
		Research Report 5: Business-led Regeneration of Deprived Areas – A Review of the Evidence Base	2003	371
		Creating Safer & Stronger Communities – The Contribution of Cultural Services	2004	371
		Simplicity, Security & Choice: Informed Choices for Working & Saving – Pensions White Paper	2004	372
		Understanding & Engaging Deprived Communities	2004	372
		Skills For Sustainable Communities – The Egan Review	2004	372
		Fairness for All: A New Commission for Equality & Human Rights – White Paper	2004	372

Jurisdiction	Tier	Title	Date	Page No.
UK	Strategies & Reports	Choosing Health? Choosing a Better Diet A Consultation on Priorities for a Food & Health Action Plan	2004	373
		Transformation & Sustainability – Future Support, Management & Monitoring of the New Deal for Communities Programme	2004	373
		Crime in England & Wales 2003/2004	2004	373
		The English Indices of Deprivation 2004 (revised)	2004	374
		Get a Public Life: Evaluation of the 2002 National Outreach Campaign	2004	374
		Sustainable Solutions to Anti-social Behaviour: Local Government's Joined-up Approaches to Tackling Anti-social Behaviour	2004	374
		Releasing the Potential for the Public's Health	2004	375
		Breaking the Cycle: Taking Stock of Progress & Priorities for the Future – A Report by the Social Exclusion Unit	2004	375
		What Works in Community Involvement in Area-based Initiatives? A Systematic Review of the Literature	2004	375
		Choosing Health: Making Healthier Choices Easier – White Paper	2004	376
		Towards Self-Governing Communities: The Role of Local Government in Civil Renewal	2004	376
	Consultation Papers	Analysis of Consultation Responses on the Review of Community Participation	2003	376
	Guidance Notes & Circulars	Factsheet No.11: Employment & Neighbourhood Renewal	2002	377
		Guidance on Community Cohesion	2002	377
		Community Cohesion Our Responsibility	2002	377
		Health & Neighbourhood Renewal Guidance from the Department of Health & the Neighbourhood Renewal Unit	2002	378
		How to Make Your Neighbourhood A Better Place to Live – Sustainable Communities Programme handbook	2004	378
		Building Community Cohesion into Area Based Initiatives: A guide for residents and practitioners	2004	379
		Community Cohesion – An Action Guide: Guidance for Local Authorities	2004	379
NORTHERN IRELAND	Strategies & Reports	Community Safety Auditing Sound Options – Hard facts Guidance Paper III: A Model for Reviewing Neighbourhood Levels of Crime, Anti-social Behaviour & the Concerns of Local People	2000	380
		Unlocking Creativity – A Strategy for Development	2001	380
		Face-to-Face: A Vision for Arts & Culture in Northern Ireland	2001	380
		Unlocking Creativity – Making it Happen	2001	381
		New TSN Action Plan – Department of Culture, Arts & Leisure	2002	381

Jurisdiction	Tier	Title	Date	Page No.
NORTHERN IRELAND	Strategies & Reports	Weak Community Infrastructure Report	2002	381
		Report on Research into Evaluating Community-Based & Voluntary Activity in NI	2003	382
		Creating a Safer Northern Ireland through Partnership: A Strategy Document	2003	382
		Young People, Unemployment Duration & the New Deal in Northern Ireland	2003	383
		Entrepreneurship & Education Action Plan	2003	383
		People & Place – A Strategy for Neighbourhood Renewal	2003	383
		Think, Create, Innovate – The Regional Innovation Strategy for Northern Ireland	2003	384
		Position Paper Policy on Homelessness in the UK & Republic of Ireland	2003	384
		Unlocking Creativity – A Creative Region	2004	384
		People & Place – Neighbourhood Renewal in Belfast Draft Implementation Plan	2004	385
		Equality Impact Assessment (EQIA) Urban Regeneration – Neighbourhood Renewal Implementation Plan For Belfast	2004	385
		People & Place – Neighbourhood Renewal in Regional Towns & Cities Draft Implementation Plan	2004	385
		Evaluation of Partners for Change – Final Report	2004	386
		People & Place: Neighbourhood Renewal in Derry/Londonderry Implementation Plan – Discussion Paper	2004	386
		Ending Fuel Poverty: A Strategy for Northern Ireland	2004	386
		First Review & Progress Report of the Taskforce on Employability & Long-Term Unemployed	2004	387
	Guidance Notes	Guidance to District Councils on the Development of Local Cultural Strategies	2001	387

9.5 Social Issues Policy Tables

Table 9.5.1 International Level Social Issues Policy.

Title	Venue	Purpose	Key Outcome(s)	Website
UN Conference on the Human Environment	Stockholm, Sweden (1972)	First major modern international gathering focusing on human activities in relation to the environment.	Heralded the beginning of environmental awareness in the international community. The Stockholm Declaration a non-binding statement of principles, 'to inspire and guide the peoples of the world in the preservation and enhancement of the human environment'. The creation of the United Nations Environmental Programme (UNEP).	http://www.unep.org/Documents/Default.asp?DocumentID=97
UN Conference on Human Settlements (Habitat I)	Vancouver, Canada (1976)	In 1976 in Vancouver, the United Nations held its first conference on the issue of physical & spatial organisation of human life on this planet, & on the national & international actions needed to accommodate the growing of population in urban & rural communities. This conference established the concept of human settlements to consist of several elements that had been previously considered separately from one another – housing, building, planning & the relationship of these & such other activities as environmental change & national & international development.	Key outcomes included the Vancouver declaration on human settlements which highlighted 19 general principles & 10 guidelines for action.	http://www.unchs.org/declarations/The_Vancouver_Declaration.pdf

World Conference on Human Rights	Vienna, Austria (1993)	Development of a common plan for the strengthening of human rights work around the world. Also the examination of the link between development, democracy and economic, social, cultural, civil and political rights, and the evaluation of the effectiveness of United Nations methods and mechanisms with the aim of recommending ways to ensure adequate financial and other resources for United Nations human rights activities.	The Vienna Declaration and Programme of Action.	http://www.unhchr.ch/html/menu5/wchr.htm
International Conference on Population & Development (ICPD)	Cairo, Egypt (1994)	To consider the broad issues of and interrelationships between population, sustained economic growth and sustainable development, and advances in the education, economic status and empowerment of women.	The present Programme of Action recommends to the international community a set of important population and development objectives, as well as qualitative and quantitative goals that are mutually supportive and of critical importance to these objectives. Among these objectives and goals are: sustained economic growth in the context of sustainable development; education, especially for girls; gender equity and equality; infant, child and maternal mortality reduction; and the provision of universal access to reproductive health services, including family planning and sexual health.	http://www.unhchr.ch/html/menu5/popdev94.htm
The World Summit for Social Development (WSSD)	Copenhagen, Denmark (1995)	The Social Summit was the largest gathering ever of world leaders at that time. It pledged to make the conquest of poverty, the goal of full employment and the fostering of social integration overriding objectives of development.	Copenhagen Declaration on Social Development WSSD Programme of Action Ten Commitments of Conference	http://www.un.org/esa/socdev/wssd/
The Fourth World Conference on Women (FWCW)	Beijing, China (1995)	To advance the goals of equality, development and peace for all women everywhere in the interest of all humanity.	The Beijing Declaration and Platform for Action (an agenda for women's empowerment).	http://www.un.org/womenwatch/confer/beijing/reports/plateng.htm

Continued

Table 9.5.1 International Level Social Issues Policy. *Continued*

Title	Venue	Purpose	Key Outcome(s)	Website
World Food Summit (WFS)	Rome, Italy (1996)	To show commitment to achieving food security for all and to an ongoing effort to eradicate hunger in all countries, with an immediate view to reducing the number of undernourished people to half their present level no later than 2015. The common objective of food security, at the individual, household, national, regional and global levels.	The Rome Declaration on World Food Security. World Food Summit Plan of Action. Seven Commitments.	http://www.fao.org/docrep/003/w3613e/w3613e00.htm
Second UN Conference on Human Settlements (Habitat II)	Istanbul, Turkey (1996)	The two major themes of the Conference were – adequate shelter for all and sustainable human settlements development in an urbanising world.	The Habitat Agenda – a commitment to a political, economic, environmental, ethical and spiritual vision of human settlements based on the principles of equality, solidarity, partnership, human dignity, respect and co-operation. The Istanbul Declaration on Human Settlements.	http://www.unchs.org/unchs/english/hagenda/index.htm
International Meeting on Urban Poverty	Recife, Brazil (1996)	Participants carried out an intensive programme of discussion in sectoral groups (focusing on employment & the urban informal sector, access to land & basic services, & social integration), in groups working on poverty reduction issues at different levels (community, municipal, & national), as well as in plenary sessions, addressing general cross-cutting problems & operational proposals. The Recife Meeting discussed strategies for urban poverty reduction to provide recommendations to the City Summit (Habitat II), to community organisations & to institutions at the local, national & global levels.	Recife Declaration – six key areas are covered in the Declaration: ▪ Globalisation & the challenge of urban poverty; ▪ Defining poverty: Understanding unity & diversity; ▪ Transforming public & private action: forging new relationships with the poor; ▪ Towards enabling policies for cities: investing in the poor; ▪ Organising for action: articulating public & private actors; & ▪ The future of our cities: our common future.	http://www.unhabitat.org/programmes/ifup/rde.asp

International Conference on Urban Poverty	Florence, Italy (1997)	The objectives of the Florence Conference were two-fold: ▪ to formally launch the Forum & to put in place mechanisms for it to function; ▪ to bring together representatives from the North & South to deliberate on the theme: Governance & Participation – practical approaches to urban poverty reduction.	The key thematic areas covered at the conference included: ▪ Governance & participation; ▪ From urban violence & exclusion to safer cities; ▪ Shelter, employment & the informal city; ▪ Access to affordable transport for the urban poor; ▪ Urban children.	http://www.unhabitat.org/programmes/ifup/fc.asp
Third International Conference on Social Integration & Security for the Urban Poor – Towards Cities for All	Nairobi, Kenya (1999)	The main objectives of the Conference were to examine the relationships between social integration, poverty alleviation & human security in cities, to identify the main constraints & obstacles for their achievement, to learn from current practices designed to achieve these objectives, to share the lessons from those practices, & to promote & facilitate action plans seeking to eradicate poverty & increase social integration & security in cities.	The conference output was based around four separate workshops: ▪ Informal cities & security of tenure; ▪ Urban safety for the poor & crime prevention; ▪ Safety in urban transport; & ▪ Safety of children & youth in urban areas.	http://www.unhabitat.org/programmes/ifup/ifupen.asp http://www.unchs.org/programmes/ifup/DOWNLOAD/dpu.doc
Fourth International Conference on Productive & Inclusive Cities – Towards Cities for All	Marrakech, Morocco (2001)	The objectives of the conference were as follows: ▪ To present & discuss the achievements since the Nairobi Conference; ▪ To draw the lessons from activities undertaken under the two thematic issues & in each region; ▪ To prepare the creation of the Urban Forum.	The output of the conference was organised around two key areas: ▪ Inclusive cities; & ▪ Productive cities. There were also a series of recommendations developed on: ▪ Competitiveness & equity; ▪ Sustainability & profit; & ▪ Exclusion against peace & development.	http://www.unhabitat.org/programmes/ifup/conf/morocco1.htm

Continued

Table 9.5.1 International Level Social Issues Policy. *Continued*

Title	Venue	Purpose	Key Outcome(s)	Website
UNEP's Global Women's Assembly on Environment (WAVE)	Nairobi, Kenya (2004)	The first Global Women's Assembly on Environment: Women as the Voice for the Environment (WAVE) was held at the headquarters of the United Nations Environment Programme (UNEP), Nairobi, Kenya, & was attended by some 150 participants from 65 countries. The aim of the meeting was the strengthening women's voice in environmental decision-making, implementation of the Beijing Platform for Action & the goals contained in the Millennium Declaration, & policy issues including those on water, human settlements & the environment.	The Assembly's aim was to put women's issues at the centre of the global environmental effort to deliver the United Nations Millennium Development Goals and the World Summit on Sustainable Development's Plan of Implementation. A manifesto & action plan recommendations were also presented.	http://www.unep.org/dpdl/cso/wave/Official.asp http://www.unep.org/DPDL/Civil_Society/PDF_docs/UNEP-WAVE_Outcome_Document_English.pdf
Convention on Access to Information, Public Participation in Decision-making & Access to Justice in Environmental Matters	Aarhus, Denmark (1998)	In order to contribute to the protection of the right of every person of present and future generations to live in an environment adequate to his or her health and well-being, each Party shall guarantee the rights of access to information, public participation in decision-making, and access to justice in environmental matters in accordance with the provisions of this Convention.	The Convention Treaty contains various articles & Annexes governing: ▪ Access to environmental information ▪ Collection & dissemination of environmental information ▪ Public participation in decisions on specific activities ▪ Public participation concerning plans, programmes & policies relating to the environment ▪ Public participation during the preparation of executive regulations &/or generally applicable legally binding normative instruments ▪ Access to justice ▪ Annex I – list of activities potentially harmful to the environment ▪ Annex II – arbitration in event of a dispute.	http://www.unece.org/env/pp/documents/cep43e.pdf

Table 9.5.2 European Level Social Issues Policy.

Policy Type	Programme Title	Policy Objective	Community Measure	Document Reference	Website
Council Directives	Freedom of Access to Information in the Environment	To ensure freedom of access to, & dissemination of information on the environment held by public authorities & to set out the basic terms & conditions under which such information should be available.	Council Directive 90/313/EC on the freedom of access to information in the environment (07/06/90)	▪ COM (2000) 400 ▪ Off Jnl L158, 26/06/90	http://europa.eu.int/scadplus/leg/en/lvb/l28091.htm
	Providing for Public Participation in Respect of the Drawing up of Certain Plans & Programmes Relating to the Environment (amending Council Directives 85/337/EEC & 96/61/EC)	The objective of this Directive is to contribute to the implementation of the obligations arising under the Aarhus Convention, in particular by providing for public participation in respect of the drawing up of certain plans & programmes relating to the environment; improving the public participation & providing for provisions on access to justice within Council Directives 85/337/EEC & 96/61/EC.	Directive 2003/35/EC of the European Parliament & of the Council providing for public participation in respect of the drawing up of certain plans & programmes relating to the environment – amending Council Directives 85/337/EEC & 96/61/EC (26/05/03)	▪ Off Jnl L156/17 25/06/03	http://europa.eu.int/eur-lex/pri/en/oj/dat/2003/l_156/l_15620030625en00170024.pdf
Council Communication	Cohesion Policy & the Environment	This Communication examines the developing relationship between Cohesion & Environmental Policies & illustrates options for achieving greater synergy between them during the implementation of Structural Funds programmes & Cohesion Fund projects.	Communication from the Commission to the Council, the European Parliament, the Economic & Social Committee & the Committee of the Regions – Cohesion Policy & the Environment (22/11/95)	▪ COM (1995) 509	http://europa.eu.int/comm/regional_policy/sources/docoffic/official/communic/pdf/envir/envir_en.pdf

Continued

Table 9.5.2 European Level Social Issues Policy. *Continued*

Policy Type	Programme Title	Policy Objective	Community Measure	Document Reference	Website
Council Communication	Building an Inclusive Europe	This initiative proposed by the European Commission expresses the ambition & commitment of the Union to promoting more inclusive economies & societies thereby responding to the strong expectations which have been expressed by the Parliament, NGOs, social partners, local authorities & confirmed by the Member States. This initiative would use the significant new Treaty provisions agreed in Amsterdam to promote co-operation with & between Member States, set up specific measures to support their action & complement existing Community policies which contribute (directly or indirectly) to social cohesion. It would promote partnership between all relevant stakeholders by strengthening their capacity to progress towards an inclusive Europe.	Communication from the Commission – Building an Inclusive Europe (01/03/00)	▪ COM (2000) 79	http://europa.eu.int/eur-lex/en/com/cnc/2000/com2000_0079en02.pdf
	Acting Locally for Employment: a Local Dimension for the European Employment Strategy	The purpose of this Communication is: ▪ to summarise developments which have led Member States & the EU to turn increasingly to the local level for job growth; ▪ to consider how the various players at the local level – including local authorities, businesses, the 'Third system'/social economy or voluntary sector, social partners, public employment services – but also those at higher levels whose actions influence what can be done locally – states & regions, even the EU itself in the areas where it has competence – can help the job creation/job maintenance process at local level; &	Communication from the Commission to the Council, the European Parliament, the Economic & Social Committee & the Committee of the Regions – Acting Locally for Employment: a Local Dimension for the European Employment Strategy (07/04/00)	▪ COM (2000) 196	http://europa.eu.int/eur-lex/en/com/cnc/2000/com2000_0196en01.pdf

	■ to raise a number of questions for debate over the six month period from April to Oct 2000.			
The Health Strategy of the European Community	This Communication sets out the Community's broad health strategy – how it is working to achieve a coherent & effective approach to health issues across all the different policy areas. A key element of this is a new public health framework which includes the attached proposal for a decision of the European Parliament & of the Council for a programme of Community action in the field of public health.	Communication from the Commission to the Council, the European Parliament, the Economic & Social Committee & the Committee of the Regions on the health strategy of the European Community (16/05/00)	■ COM (2000) 285	http://europa.eu.int/eur-lex/en/com/pdf/2000/en_500PC0285.pdf
Towards a Community Framework Strategy on Gender Equality (2001–2005)	This Communication sets out the Commission's ideas towards a strategy on gender equality over the next five years. Its purpose is to establish a framework for action within which all Community activities can contribute to attain the goal of eliminating inequalities & promoting equality between women & men, as set out in Article 3(2) of the Treaty.	Communication from the Commission to the Council, the European Parliament, the Economic & Social Committee & the Committee of the Regions – Towards a Community Framework Strategy on Gender Equality (2001–2005) (07/06/00)	■ COM (2000) 335	http://europa.eu.int/comm/employment_social/equ_opp/strategy/com2000_335_en.pdf
Community Initiative: EQUAL	To promote new ways of combating all forms of discrimination and inequalities in the labour market on the basis of transnational co-operation and to facilitate the social and occupational integration of asylum seekers.	Communication from the Commission to the Member States establishing the guidelines for the Community initiative EQUAL concerning transnational co-operation to promote new means of combating discrimination & inequalities in connection with the labour market (14/4/00)	■ COM (2000) 853 ■ Off Jnl C127 (5/5/00)	http://europa.eu.int/scadplus/leg/en/cha/c10237.htm

Continued

Table 9.5.2 European Level Social Issues Policy. *Continued*

Policy Type	Programme Title	Policy Objective	Community Measure	Document Reference	Website
Council Communication	Social Policy Agenda	This Social Policy Agenda forms part of the integrated European approach towards achieving the economic & social renewal outlined at Lisbon. Specifically, it seeks to ensure the positive & dynamic interaction of economic, employment & social policy, & to forge a political agreement which mobilises all key actors to work jointly towards the new strategic goal.	Communication from the commission to the council, the European Parliament, the Economic & Social Committee & the Committee of the Regions on the social policy agenda (28/06/00)	▪ COM (2000) 379	http://europa.eu.int/comm/employment_social/social_policy_agenda/com379_en.pdf
	Supporting National Strategies for Safe & Sustainable Pensions through an Integrated Approach	The present communication responds to the need for clear and integrated strategies to cope with the challenges for pension systems of an ageing society.	Communication from the Commission to the Council, the European Parliament & the Economic & Social Committee – Supporting national strategies for safe and sustainable pensions through an integrated approach (03/07/01)	▪ COM (2001) 362	http://europa.eu.int/comm/employment_social/soc-prot/social/com362/com362-2001_en.pdf
	Strengthening the Local Dimension of the European Employment Strategy	This Communication draws on past experience, on the lessons of existing policies & initiatives including the White Paper on Governance, & on the results of the Acting Locally consultation process. The communication suggests ways to help local actors play their full part in the European Employment Strategy in co-operation with national authorities & Community institutions, pooling their resources, & establishing local employment strategies where appropriate, thus eventually increasing the employment rate in Europe & providing stable, quality jobs. Candidate countries should be fully involved in this process in view of enlargement.	Communication from the Commission to the Council, the European Parliament, the Economic & Social Committee & the Committee of the Region – Strengthening the local dimension of the European Employment Strategy (06/11/01)	▪ COM (2001) 629	http://europa.eu.int/eur-lex/en/com/cnc/2001/com2001_0629en01.pdf

First Progress Report on Economic & Social Cohesion	This progress report on cohesion has two principal objectives: ▪ to update the analysis of economic and social cohesion presented in the Second cohesion report published in January 2001, including for the first time, an analysis of disparities in a Europe of 25 in the light of the enlargement to include the 10 new Member States which, according to the Laeken European Council, would be ready to join the Union in 2004 if negotiations continue at the current rhythm; ▪ to outline the state of the debate on future cohesion policy for the period after 2006, which began with the publication of the Second report & ▪ to prepare the next steps.	Commission Communication – First progress report on economic & social cohesion (30/1/02)	▪ COM (2002) 46	http://europa.eu.int/comm/regional_policy/sources/docoffic/official/reports/pdf/interim1/com_2002_046_en_acte.pdf
Consumer Policy Strategy 2002–2006	Consumer policy in this communication covers safety, economic & legal issues relevant to consumers in the market place, consumer information & education, the promotion of consumer organisations & their contribution with other stakeholders to consumer policy development.	Communication from the Commission to the European Parliament, the Council, the Economic & Social Committee & the Committee of the Regions Consumer policy strategy 2002–2006 (COM (2002) 208 final) (08/06/02)	▪ Off Jnl C137/2	http://europa.eu.int/eur-lex/pri/en/oj/dat/2002/c_137/c_13720020608en00020023.pdf
The future of the European Employment Strategy (EES) – A strategy for full employment & better jobs for all	The purpose of this Communication is to present an outline for the revised strategy complemented by examples of existing concrete objectives & targets & considerations & suggestions for possible new targets. This will offer the basis for an open discussion with all interested parties, especially the Member States, the European Parliament, representative organisations of the Social Partners & the civil society. In line with the new & streamlined approach for the overall Lisbon agenda follow-up, the Commission will make a formal proposal of Employment Guidelines & Recommendations, in the light of this debate & of the general political guidance issued by the Spring European Council.	Communication from the Commission to the Council, the European parliament, the Economic & Social Committee & the Committee of the Regions – The future of the European Employment Strategy (EES) 'A strategy for full employment and better jobs for all' (14/01/03)	▪ COM (2003) 6	http://europa.eu.int/eur-lex/en/com/pdf/2003/com2003_0006en01.pdf

Continued

Table 9.5.2 European Level Social Issues Policy. *Continued*

Policy Type	Programme Title	Policy Objective	Community Measure	Document Reference	Website
Council Communication	Scoreboard on Implementing The Social Policy Agenda	This is the third scoreboard the Commission presents to report on the implementation of the social policy agenda. This edition focuses essentially on the main achievements of the year. The presentation of an annual scoreboard responds to the invitation made both by the European Parliament & the European Council to keep track of the achievements & to verify the commitment & contributions from the different actors in executing this agenda. Its objective is not to provide any ranking of Member States' performance, but rather to monitor how the agenda is transformed into policy measures & concrete action. The Scoreboard complements the annual synthesis report to the Spring European Council with regards to the actions which are part of the Social Policy Agenda.	Communication from the commission to the council, the European Parliament, the European Economic & Social Committee & the Committee of the Regions – Scoreboard on Implementing the Social Policy Agenda (06/02/03)	■ COM (2003) 57	http://europa.eu.int/eur-lex/en/com/cnc/2003/com2003_0057en01.pdf
	Immigration, Integration & Employment	This communication: ■ responds to the Tampere conclusions by reviewing current practice & experience with integration policy at national & EU level; ■ examines the role of immigration in relation to the Lisbon objectives in the context of demographic ageing; & ■ outlines, on this basis, policy orientations & priorities, including actions at EU level, to promote the integration of immigrants.	Communication from the Commission to the Council, the European Parliament, the European Economic & Social Committee & the Committee of the Regions – on immigration, integration & employment (03/06/03)	■ COM (2003) 336	http://europa.eu.int/eur-lex/en/com/cnc/2003/com2003_0336en01.pdf

A European Environment & Health Strategy	The proposed strategy aims to achieve a better understanding of the environmental threats to human health to identify the disease burden caused by environmental factors in the EU & to plan policy responses to the challenges that emerge. The ultimate objectives of the proposed strategy are: ▪ to reduce the disease burden caused by environmental factors in the EU; ▪ to identify & to prevent new health threats caused by environmental factors; and ▪ to strengthen EU capacity for policymaking in this area. The main thrust of the strategy is to fill the knowledge gap on the link between environment & health, in a first phase focusing on a number of priority adverse health effects.	Communication from the Commission to the Council, the European Parliament & the European Economic & Social Committee – A European Environment & Health Strategy (11/06/03)	▪ COM (2003) 338	http://europa.eu.int/smartapi/cgi/sga_doc?smartapi!celexapi!prod!CELEXnumdoc&lg=en&numdoc=52003DC0338&model=guichett
Improving Quality in Work: A Review of Recent Progress	There is a positive link between employment growth & quality in work. Sustainable employment growth needs all of the following: an improved access to employment in general; an improved balance between flexibility & security & real opportunities for upward occupational mobility for those in low quality jobs, without making access to the labour market more difficult for those on the margins; a reduction of long-term unemployment without creating a situation based on repetitive spells in unemployment & low quality employment; & improved upward quality dynamics, helping people in low quality jobs & at high risk of unemployment to gain employment stability & to improve employability.	Communication from the Commission to the Council, the European Parliament, the European Economic & Social Committee & the Committee of the Regions – Improving quality in work: a review of recent progress (26/11/03)	▪ COM (2003) 728	http://europa.eu.int/comm/employment_social/employment_strategy/pdf/comm_en.pdf

Continued

Table 9.5.2 European Level Social Issues Policy. *Continued*

Policy Type	Programme Title	Policy Objective	Community Measure	Document Reference	Website
Council Communication	A Joint Report on Social Exclusion	The present report identifies key trends & challenges across the Union & good practice & innovative approaches of common interest. It will form the basis for the Joint Council/Commission Report on social inclusion to be presented to the Spring European Council of 2004. It assesses progress made towards meeting the Lisbon objective for poverty reduction & maps out the policy agenda for the future EU social inclusion process in a context which is set to change radically. The process will, from 2004, have to take account of the accession of 10 new Member States where the nature & extent of poverty & social exclusion are often sharply different. From 2006 onwards, it will form part of the new streamlined structure for EU-level policy co-ordination on social protection as proposed by the Commission in May 2003 & endorsed by the Council in October 2003.	Communication from the Commission to the Council, the European Parliament, the European Economic & Social Committee & the Committee of the Regions on social exclusion summarising the results of the examination of the Nations Action Plans for social inclusion (2003–05) (12/12/03)	■ COM (2003) 773	http://europa.eu.int/comm/employment_social/soc-prot/soc-incl/com_2003_773_jir_en.pdf
	Modernising Social Protection for the Development of High-quality, Accessible & Sustainable Health Care & Long-term Care: Support for the National Strategies Using the 'Open Method of Co-ordination'	The aim of this Communication is to define a common framework to support Member States in the reform & development of health care & long-term care, borne by the social protection system, using the 'open method of co-ordination'. A Resolution adopted by the European Parliament called for greater co-operation on health & long-term care & called on the Commission to present relevant proposals in the spring of 2004, allowing the Council to apply the 'open method of coordination' in this field & to adopt common objectives.	Communication from the Commission to the Council, the European Parliament, the European Economic & Social Committee & the Committee of the Regions – Modernising social protection for the development of high-quality, accessible & sustainable health care & long-term care: support for the national strategies using the 'open method of coordination' (20/04/04)	■ COM (2004) 304	http://europa.eu.int/comm/employment_social/soc-prot/healthcare/com_04_304_en.pdf

	The European Environment & Health Action Plan 2004–2010	This Action Plan covers the period of the first cycle & has been developed in close co-operation with experts from the Member States & representatives of the main stakeholders, including from the new Member States who have been involved since the beginning of this process (see Annex II for details on the consultation process). It has three main themes: ▪ improving the information chain to understand the links between sources of pollution & health effects (Actions 1–4); ▪ filling the knowledge gap by strengthening research & addressing the emerging issues on environment & health (Actions 5–8); ▪ reviewing policies & improving communication (Actions 9–13).	Communication from the Commission to the Council, the European Parliament, the European Economic & Social Committee – 'The European Environment & Health Action Plan 2004–2010' (09/06/04)	COM (2004) 416	http://europa.eu.int/comm/environment/health/pdf/com2004416.pdf
Council Decisions	Setting up a Network for the Epidemiological Surveillance & Control of Communicable Diseases in the Community	The objective of this Decision is to set up a network at Community level to promote co-operation & co-ordination between the Member States, with the assistance of the Commission, with a view to improving the prevention & control, in the Community, of the categories of communicable diseases specified in the Annex. This network shall be used for: ▪ the epidemiological surveillance of these diseases, & ▪ an early warning & response system for the prevention & control of diseases.	Decision No 2119/98/EC of the European Parliament & of the Council – Setting up a Network for the Epidemiological Surveillance and Control of Communicable Diseases in the Community (24/09/98)	Off Jnl L268/1 03/10/98	http://europa.eu.int/eur-lex/pri/en/oj/dat/1998/l_268/l_26819981003en00010006.pdf
	Establishing the Culture 2000 Programme	The Culture 2000 programme shall contribute to the promotion of a cultural area common to the European peoples. In this context, it shall support co-operation between creative artists, cultural operators, private & public promoters, the activities of the cultural networks, & other partners as well as the cultural institutions of the Member States & of the other participant States.	Decision No 508/2000/EC of the European Parliament & of the Council – Establishing the Culture 2000 Programme (14/02/00)	Off Jnl L63/1 10/03/00	http://europa.eu.int/comm/culture/eac/sources_info/pdf-word/decision_en.pdf

Continued

Table 9.5.2 European Level Social Issues Policy. *Continued*

Policy Type	Programme Title	Policy Objective	Community Measure	Document Reference	Website
Council Decisions	Establishing a Programme of Community Action to Encourage Co-operation Between Member States to Combat Social Exclusion	The programme shall support co-operation which enables the Community & the Member States to enhance the effectiveness & efficiency of policies to combat social exclusion by: ▪ improving the understanding of social exclusion & poverty with the help in particular of comparable indicators; ▪ organising exchanges on policies which are implemented & promoting mutual learning *inter alia* in the context of national action plans, with the help in particular of comparable indicators; ▪ developing the capacity of actors to address social exclusion & poverty effectively, & to promote innovative approaches, in particular through networking at European level, & by promoting dialogue with all those involved including at national & regional level.	Decision No 50/2002/EC of the European Parliament & of the Council establishing a programme of Community action to encourage co-operation between Member States to combat social exclusion (12/01/02)	▪ Off Jnl L10/1	http://europa.eu.int/comm/employment_social/news/2002/jan/03650_en.pdf
Miscellaneous	Access to Information, Public Participation & Access to Justice in Environmental Matters	To raise public awareness of environmental concerns by promoting access to information & participation in the decision-making process.	Proposal for a Council Decision on the signature by the European Community of the UN/ECE (United Nations Economic Commission for Europe) Convention on access to information, public participation & access to justice in environmental matters (Aarhus Convention) (1998)	▪ COM (98) 344	http://europa.eu.int/scadplus/leg/en/lvb/l28056.htm

White Paper on Food Safety	Assuring that the EU has the highest standards of food safety is a key policy priority for the Commission. This White Paper reflected this priority. A radical new approach was proposed driven by the need to guarantee a high level of food safety. This White Paper on Food Safety outlined a comprehensive range of actions needed to complement & modernise existing EU food legislation, to make it more coherent, understandable & flexible, to promote better enforcement of that legislation, & to provide greater transparency to consumers. This was to provide the response to the conclusions of the Helsinki European Council in Dec 1999.	EC White Paper On Food Safety (12/01/00)	COM (1999) 719	http://europa.eu.int/comm/dgs/health_consumer/library/pub/pub06_en.pdf
The Lisbon European Council – An Agenda of Economic & Social Renewal for Europe	This document highlights that the way forward for the Union is to confront challenges head on, building on the stability achieved & taking advantage of the favourable economic outlook to ensure sustained economic growth, full employment & social cohesion. This requires a new, positive vision of the future & a series of policies to deliver it. This is a complex task. It requires a fully integrated & operational approach linking economic, social & political objectives with dates & deadlines for achieving them. The core of this strategy is to continue & accelerate the structural changes already underway.	Contribution of the European Commission to the Special European Council in Lisbon 23/24th March 2000 (2/02/00)	COM (2000) 7	http://europa.eu.int/comm/lisbon_strategy/pdf/lisbon_en.pdf
Charter of Fundamental Rights of the European Union	This Charter reaffirms, with due regard for the powers & tasks of the Community and the Union & the principle of subsidiarity, the rights as they result, in particular, from the constitutional traditions & international obligations common to the Member States, the Treaty on European Union, the Community Treaties, the European Convention for the Protection of Human Rights & Fundamental Freedoms, the Social Charters adopted by the Community & by the Council of Europe & the case-law of the Court of Justice of the European Communities & of the European Court of Human Rights.	Charter of Fundamental Rights of the European Union (18/12/00)	Off Jnl C364/1	http://www.europarl.eu.int/charter/pdf/text_en.pdf

Continued

Table 9.5.2 European Level Social Issues Policy. *Continued*

Policy Type	Programme Title	Policy Objective	Community Measure	Document Reference	Website
Miscellaneous	Second Report on Economic & Social Cohesion	The Commission has chosen to present the Second Report on Economic & Social Cohesion at the beginning of 2001, which falls immediately after the first phase of the implementation of the reform of the Structural Funds, & after certain key decisions have been taken as regards financial allocations & geographical eligibility for support. It is, therefore, already possible at this stage to make a broad *ex ante* assessment of the possible impact of the reform. The second Report also contains an updating of the regional analysis contained in the Sixth & last Periodic Report on the situation & development of regions published in 1999. Such an updating is more necessary than before since the Cohesion Reports replace the Periodic Reports which the Commission has published since the beginning of the 1980s.	'Unity, solidarity and diversity for Europe, its people and its territory' – the Second Report on Economic and Social Cohesion (31/01/01)	▪ COM (2001) 24	http://europa.eu.int/comm/regional_policy/sources/docoffic/official/reports/contentpdf_en.htm
	White Paper Strategy for a future Chemicals Policy	This White Paper presents Commission proposals for a strategy on future chemicals policy in the Community with the overriding goal of sustainable development. Chemicals bring about benefits on which modern society is entirely dependent, for example, in food production, medicines, textiles, cars etc. They also make a vital contribution to the economic and social wellbeing of citizens in terms of trade and employment.	EC White Paper Strategy for a future Chemicals Policy (27/02/01)	▪ COM (2001) 88	http://europa.eu.int/comm/environment/chemicals/pdf/0188_en.pdf
	Promoting a European Framework for Corporate Social Responsibility	This Green Paper aims to launch a wide debate on how the European Union could promote corporate social responsibility at both the European & international level, in particular on how to make the most of	Green Paper on promoting a European Framework for Corporate Social Responsibility (18/07/01)	▪ COM (2001) 366	http://europa.eu.int/comm/employment_social/social_policy_agenda/csr_communication.pdf

	existing experiences, to encourage the development of innovative practices, to bring greater transparency & to increase the reliability of evaluation & validation. It suggests an approach based on the deepening of partnerships in which all actors have an active role to play.			
Report on Indicators in the Field of Poverty & Social Exclusion	Following the mandate from the Lisbon European Council, the Member States & the Commission have sought to develop common approaches & compatibility in regard to indicators. The work has been carried out by the Social Protection Committee & its technical subgroup on Indicators that started meeting in February 2001. In particular, the sub-group was concerned with improving indicators in the field of poverty & social exclusion. This follows from the political agreement reached at the European Council in Nice, defining appropriate objectives in the fight against poverty & social exclusion, & inviting Member States & the Commission to seek to develop commonly agreed indicators.	Report of the Social Protection Committee on Indicators in the Field of Poverty & Social Exclusion (Oct 2001)	▪ Social Protection Committee Report	http://europa.eu.int/comm/employment_social/news/2002/jan/report_ind_en.pdf
Increasing Labour-Force Participation & Promoting Active Ageing	This report focuses on action that is necessary to raise labour force participation through improvements in the functioning of labour markets. An in-depth analysis & the formulation of policy responses in relation to the macro-economic context & product & services market as well as in a range of adjacent areas would go beyond the scope of this report. Such issues are covered explicitly in documents that will be part of the overall package prepared for the Barcelona European Council on economic & social affairs in Spring 2002.	Joint Report from the Commission & the Council: Report requested by the Stockholm European Council 'Increasing Labour-Force Participation & Promoting Active Ageing' (08/03/02)	▪ Report requested by the Stockholm European Council	http://register.consilium.eu.int/pdf/en/02/st06/06707en2.pdf

Continued

Table 9.5.2 European Level Social Issues Policy. *Continued*

Policy Type	Programme Title	Policy Objective	Community Measure	Document Reference	Website
Miscellaneous	Adequate & Sustainable Pensions	The financial sustainability of pension systems is a necessary precondition for an adequate provision of pensions in the future & for this reason alone should be high on the priority lists of all Member States. Failure to ensure the financial sustainability of pensions systems in the long term will seriously jeopardise Member States' efforts to maintain or even raise the adequacy of their public pension systems & would have other adverse economic consequences as well. Public pension spending is projected to rise substantially in most Member States as a share of GDP over the coming decades if policies remain unchanged. Over the past few years, Member States have recognised the urgency of making pension systems financially stable in view of the limited window of opportunity that exists before the ageing population takes effect.	Joint report by the Commission & the Council on adequate & sustainable pensions (10/03/03)	■ Joint Commission & Council Report	http://europa.eu.int/comm/employment_social/soc-prot/pensions/2003jpr_en.pdf
	The Impact of the New Chemicals Policy on Health & the Environment	This report sets out the findings of a study that examines the potential impacts of Reach (Registration, Evaluation, Authorisation of Chemicals), in terms of the types of environmental & wider public health benefits that it may help to achieve. The aim of the study was to illustrate how a proactive approach towards chemicals legislation, i.e. the Reach system, may improve the environment, & public health in particular, by preventing the accumulation of potential pollutants until their effects are well known.	Report prepared for the European Commission Environment Directorate-General (June 2003)	■ DG Env Report	http://europa.eu.int/comm/environment/chemicals/pdf/envhlthimpact.pdf
	Recommendations on Health & EU Social Policy	Policy paper on the issue of health & social policy drafted by the EU Health Policy Forum which conducted a session on health & social policy. Members of the Forum, the European	Health Policy Forum Report – Recommendations on Health & EU Social Policy (Dec 2003)	■ Health Policy Forum Report	http://europa.eu.int/comm/health/ph_overview/health_forum/hforum_soc_en.pdf

	Public Health Alliance, EuroHealthNet, & the European Federation of Public Service Unions, delivered presentations outlining links between health & social policy.			
Report on Equality Between Women & Men	This report gives an overview of progress made so far regarding the status of women & men in main policy areas & addresses key challenges for the further promotion of equality between women & men.	Report from the Commission to the Council, the European Parliament, the European Economic & Social Committee & the Committee of the Regions – Report on Equality Between Women & Men (19/02/04)	COM (2004) 115	http://europa.eu.int/comm/employment_social/equ_opp/com_04_115_en.pdf
Joint Report by the Commission & the Council on Social Inclusion	This report, on the basis of an analysis of the NAPs Inclusion (National Action Plans against poverty and social exclusion), identifies key trends & challenges across the Union & good practice & innovative approaches of common interest. It assesses progress made towards meeting the Lisbon objective for poverty eradication & maps out the policy agenda for the future EU social inclusion process in a context which is set to change radically. The process will, from 2004, have to take account of the accession of 10 new Member States where the nature & extent of poverty & social exclusion are often sharply different.	Council of the European Union – Joint report by the Commission & the Council on social inclusion (05/03/04)	Commission & Council Joint Report	ttp://europa.eu.int/comm/employment_social/soc-prot/soc-incl/final_joint_inclusion_report_2003_en.pdf
Enabling Good Health For All: A Reflection Process for a new EU Health Strategy	Good health is a state of physical & mental well-being necessary to live a meaningful, pleasant & productive life. Good health is also an integral part of thriving modern societies, a cornerstone of well performing economies, & a shared principle of European democracies. Achieving good health for all means not just reacting to ill-health, but proactively promoting health, preventing diseases & helping people make healthy choices. This paper launches a reflection process to help define the future EU Health strategy.	Enabling Good Health for all: A Reflection Process for a new EU Health Strategy – European Commissioner for Health & Consumer Protection (15/07/04)	European Commissioner Report	http://europa.eu.int/comm/health/ph_overview/Documents/byrne_reflection_en.pdf

Table 9.5.3 UK Level Social Issues Policy.

Policy Type	Document Title	Objective	Contents	Website
Strategies & Reports	Eliminating World Poverty: A Challenge for the 21st Century – White Paper on International Development (Nov 1997)	This White Paper set out the Government's policies to achieve the sustainable development of this planet. It was first, & most importantly, about the single greatest challenge which the world faces – eliminating poverty. It was about ensuring that the poorest people in the world benefit in the move towards a new global society and creating partnerships with developing countries & their peoples, on the basis of specific & achievable targets, to bring that about.	Summary; The Challenge of Development; Building Partnerships; Consistency of Policies; Building Support for Development; Abbreviations	http://www.csr.gov.uk/pdf/e2eliminatingpoverty1.pdf
	Improving Shopping Access for People Living in Deprived Neighbourhoods (Nov 1999)	The report is built around two main principles considered critical to improving shopping access in deprived neighbourhoods: ▪ The need for locally-developed strategies, with the local community playing a full part in the process & working together with local retailers; & ▪ The need to produce solutions that are commercially viable in the long-term, building on good practice.	Executive summary; Proposed action points & recommendations; Introduction; Approach; Definition & problem analysis; Reversing the trend; The goal; The key elements to achieving our goal; A local retail strategy; Proactive planning & regeneration; Crime reduction & fear of crime; Improving business support for small retailers; Easing business burdens on small retailers; Next steps; Monitoring the impact of improving shopping access; Research; Annex 1: PAT (Project Action Team) 13 membership; Annex 2: Mapping groups membership; Annex 3: Community based & community-run initiatives; Annex 4: Private sector support – case studies; Annex 5: Community development & involvement; Annex 6: Coventry City Council area co-ordination initiative; Annex 7: Business in the community; Annex 8: What skills & knowledge do small retail businesses need? Annex 9: Public sector support for small businesses; Annex 10: Private sector	http://www.dh.gov.uk/assetRoot/04/04/29/30/04042930.pdf

		support; Annex 11: The national micro retail organisation; Annex 12: PAT 13 response to consultation on the proposals for a levy scheme for the Food Standards Agency; Annex 13: PAT 13 response to consultation on the small business service	
Jobs for All – National Strategy for Neighbourhood Renewal (1999)	The Project Action Team identified a number of key issues on which it wanted to gather evidence & form views. It was decided that the practical way forward was to form sub-teams which would look at each issue in depth and report back to the main PAT with emerging conclusions and recommendations. The sub-teams focused on: ▪ coherence & targeting of funding; ▪ what keeps people out of jobs & what helps them in deprived neighbourhoods & among people from ethnic minority backgrounds; ▪ empowering local communities supporting jobless people; ▪ partnerships with employers; ▪ financial incentives & disincentives to entering the labour market.	Executive summary; Chapter 1 – Introduction; Chapter 2 – Setting our sights; Chapter 3 – Supporting jobless people; Chapter 4 – Tackling racial discrimination; Chapter 5 – Building bridges: Partnerships with employers; Chapter 6 – Making work pay; Chapter 7 – Delivering the goods; Summary of recommendations/action plan	http://www.dwp.gov.uk/jfa/jfa.pdf
Social Capital & Health (1999)	This report presents detailed findings from a qualitative study of social capital & health carried out in England. It highlights a number of important issues for contemporary health promotion, in particular the importance of informal networks to connect people to each other & improve access to health-related information & services.	Executive summary; Introduction; The relevance of social capital to health promotion; Social capital in local communities in England: The Luton pilot study; Setting the context: perceptions of social capital in Farley & Sundon Park; Trust & local identity: the quality of community relationships in the past & present; Civic engagement 1: informal networks of friends; Civic engagement 2: Voluntary networks linked to leisure, hobbies & personal development; Civic engagement 3: local activist networks & perceived citizen power; Conclusions; References; Appendices	http://www.hda-online.org.uk/downloads/pdfs/socialcapital_health.pdf

Continued

Table 9.5.3 UK Level Social Issues Policy. *Continued*

Policy Type	Document Title	Objective	Contents	Website
Strategies & Reports	Eliminating World Poverty: Making Globalisation Work for the Poor – White Paper on International Development (Dec 2000)	Collective work is needed to tackle the problems of conflict & corruption, boost investment in education & health, spread the benefits of technology & research, strengthen the international financial system, reduce barriers to trade, tackle environmental problems & make development assistance more effective. This White Paper set out the UK Government's policies in all these areas. It reflected a commitment to work across all parts of government in order to help eliminate world poverty, & to co-operate with other governments & international institutions as part of a broader international effort. It is consistent with the determination to tackle poverty & social exclusion in the UK.	Key Policy Commitments; The Challenge of Globalisation; Promoting Effective Governments & Efficient Markets; Investing in People, Sharing Skills & Knowledge; Harnessing Private Finance; Capturing Gains from Trade; Tackling Global Environmental Problems; Using Development Assistance More Effectively; Strengthening the International System	http://www.csr.gov.uk/pdf/e2eliminatingpoverty2.pdf
	A New Commitment to Neighbourhood Renewal: A National Strategy Action Plan (Jan 2001)	This work has produced agreement on the vision that, within 10 to 20 years, no-one should be seriously disadvantaged by where they live. People on low incomes should not have to suffer conditions and services that are failing, & so different from what the rest of the population receives.	Foreword; Summary; Chapter 1: Analysis; Chapter 2: Success & good practice; Chapter 3: Strategy; Chapter 4: New policies, funding & targets; Chapter 5: Joining it up locally; Chapter 6: National & regional support; Chapter 7: Government commitments to neighbourhood renewal; Annex A: Main messages from the SEU's (Social Exclusion Unit) consultation; Annex B: Background to the action plan; Annex C: What different sectors can contribute?; Annex D: The neighbourhood renewal fund – published draft guidance; Annex E: Background – tracking changes from the consultation document; Annex F: Examples of good practice; Annex G: Checklist for developing local neighbourhood renewal strategies – steps in developing strategies; Further information; Abbreviations & notes	http://www.renewal.net/Documents/RNET/Policy%20Guidance/Newcommitmentneighbourhood.pdf

National Strategy for Neighbourhood Renewal – Policy Action Team Audit (Jan 2001)	The SEU published a report in 1998 setting out the need for a National Strategy for Neighbourhood Renewal – to be an agreed response to the problems of deprived neighbourhoods. The report proposed that one of the building blocks of the National Strategy should be 18 cross-cutting Policy Action Teams (PATs), set up to take forward an intensive programme of policy development. The Teams were made up of civil servants & outside experts from various backgrounds, with experience of working/living in deprived neighbourhoods.	Foreword; Overview; PAT1 Jobs; PAT2 Skills; PAT3 Business; PAT4 Neighbourhood management; PAT5 Housing management; PAT6 Neighbourhood wardens; PAT7 Unpopular housing; PAT8 Anti-social behaviour; PAT9 Community self-help; PAT10 Arts & sport; PAT11 Schools plus; PAT12 Young people; PAT13 Shops; PAT14 Financial services; PAT15 Information technology; PAT16 Learning lessons; PAT17 Joining it up locally; PAT18 Better information; Annex A: PAT remits; Annex B: PAT members	http://www.socialexclusionunit.gov.uk/downloaddoc.asp?id=57
Preventing Social Exclusion (March 2001)	Social exclusion is a relatively new term in British policy debate. It includes poverty and low income, but is broader and addresses some of the wider causes and consequences of poverty. The most important characteristic of social exclusion is that these problems are linked and mutually reinforcing, and can combine to create a complex and fast-moving vicious circle. Only when this process is properly understood and addressed will policies really be effective.	Foreword; Summary; Chapter 1: Social exclusion & why it matters; Chapter 2: Costs; Chapter 3: Causes; Chapter 4: A new approach; Chapter 5: Delivery; Chapter 6: What next?; Annex A: The social exclusion unit; Annex B: Truancy & school exclusion; Annex C: Rough sleepers; Annex D: Teenage pregnancy; Annex E: Bridging the gap; Annex F: National strategy for neighbourhood renewal; Notes	http://www.socialexclusionunit.gov.uk/downloaddoc.asp?id=232
Neighbourhood Renewal: Increasing Employment Amongst Deprived Areas & Groups – Implementation Strategy (Dec 2001)	This document sets out the Department for Work & Pensions' strategy for implementing its targets in support of the National Strategy for Neighbourhood Renewal. This implementation strategy describes how both mainstream & targeted activities will together address these additional barriers & help deliver the National Strategy's objective of narrowing the gap between the most deprived neighbourhoods & the rest of the country.	Introduction; Baseline Assumptions; Geographical Coverage; Programmes & Resources; Delivery Mechanisms & Partnerships; Risk & Performance Management; Annex A Strategy For Increasing Employment Rates for Lone Parents; Annex B Strategy For Increasing Employment Rates for People From Ethnic Minority Backgrounds; Annex C Strategy For Increasing Employment Rates for People over 50; Annex D Strategy For Increasing Employment Rates for Disabled People; Annex E Strategy For Increasing Employment Rates for 30 Local Authorities with the poorest initial Labour Market position; Annex F References	http://www.dwp.gov.uk/publications/dwp/2001/nsnr/nsnr.pdf

Continued

Table 9.5.3 UK Level Social Issues Policy. *Continued*

Policy Type	Document Title	Objective	Contents	Website
Strategies & Reports	Equality in Practice – Report of a Survey on the Adoption & Implementation by English Councils of *Racial Equality Means Quality*, the CRE Standard for Local Government (2001)	The report described the legislative & administrative framework within which local authorities were expected to operate & showed how the CRE (Commission for Racial Equality) Standard could help them meet their various requirements, such as the best value performance indicators set by the DETR; the duty they have under the Local Government Act 2000 to prepare community strategies for the development of their areas; the new public duty which the Race Relations (Amendment) Act 2000 placed on them; & the legal & positive obligations placed on them by the Human Rights Act 1998. The case for quality management as a means of ensuring that racial equality penetrates all aspects of employment & service delivery, underpins the CRE Standard itself & has been examined more closely in earlier studies, such as *Auditing for Equality* & *Measuring Up*.	Foreword; Preface; Introduction; The legislative and administrative framework; The survey findings; Conclusions and recommendations; Appendix 1 Methodology; Appendix 2 Bibliography and useful websites; Appendix 3 Sample questionnaire	http://www.cre.gov.uk/pdfs/remqsurv.pdf
	Secure Borders, Safe Haven – Integration with Diversity in Modern Britain (Feb 2002)	This White Paper set out the key challenges faced in developing a nationality, immigration & asylum policy & set out in detail the measures taken to produce a coherent strategy. A policy that recognises the need to change & develop. A nationality, immigration & asylum policy that secures the sustainable growth & social inclusion that are an essential part of the core principles & the Department's delivery targets.	Executive summary; Consultation; Chapter 1: Introduction & Overview; Chapter 2: Citizenship & Nationality; Chapter 3: Working in the UK; Chapter 4: Asylum; Chapter 5: Tackling Fraud – People Trafficking, Illegal Entry & Illegal Working; Chapter 6: Border Controls; Chapter 7: Marriage/Family Visits & War Criminals; Chapter 8: Conclusion; Annexes	http://www.official-documents.co.uk/document/cm53/5387/cm5387.pdf
	Making the Connections: Transport & Social Exclusion (May 2002)	This report contains an analysis of the problem of poor transport & highlights the social cost of this for different people and areas. The extent to which transport is a barrier to participation has not been the subject of extensive quantitative research. This report therefore makes use of considerable qualitative data.	Summary; Introduction; Chapter 1: The problem; Chapter 2: Why does it happen?; Chapter 3: What can be done?; Chapter 4: Barriers to success; Chapter 5: Towards an inclusive system	http://www.socialexclusionunit.gov.uk/downloaddoc.asp?id=70

Vision for Sustainable Regeneration – Environment & Poverty: The Missing Link? (Oct 2002)	This paper set out a vision for sustainable regeneration, based on a review of existing literature, policy development & initiatives. In doing so it argued for a new approach to regeneration which not only considers social & economic inequalities within society, but also environmental inequalities & the link between quality of the local environment & poverty. This vision is developed further through a study of the lessons & barriers to delivering sustainable regeneration within the UK, as well as identifying good practice examples of where barriers have been overcome.	Introduction; Environmental justice – the missing piece of the jigsaw; Do environmental injustices exist? Catching the wave: Growing political commitment; Making the links; Delivering the shared agenda – principles for sustainable regeneration; Next steps 'regeneration uncovered' – barriers & good practice in sustainable regeneration; Appendix 1: Background on the Sustainable Development Commission	http://www.sd-commission.org.uk/news/download_pdf.php?attach_id=A4R1STM-NEZ5D7E-35TL5KY-5C1KV5E
Active Communities: Initial Findings from the 2001 Home Office Citizenship Survey (2002)	This short report sets out some initial findings about active communities from the 2001 Home Office Citizenship Survey. Other reports to be published in due course will deal in greater detail with this & other related subjects (citizenship; identity, religion & language; racial prejudice and discrimination; & family networks and parenting).	Introduction; People & neighbourhoods; How do people feel about their neighbourhoods? How many other people in their neighbourhoods do they know? How often do people socialise with friends & neighbours? How much do people trust their neighbours? How many people are prepared to intervene for the common good? Active participation in communities? How many people participate in civic affairs & what do they do? How many people participate socially in groups & what do they do? How many people volunteer through employer supported volunteering schemes? Does people's participation in community activities vary according to the relative deprivation of the areas in which they live? Conclusions; Methodology	http://www.homeoffice.gov.uk/docs/infindin.pdf
Making the Connections: Transport & Social Exclusion – Final Report (Feb 2003)	This report examines the links between social exclusion, transport & the location of services. It is particularly focused on access to those opportunities that have the most impact on life-chances, such as work, learning & healthcare.	Foreword; Summary; The problem; The strategy; Solutions to the problem; Accessibility planning: a new approach; Improving transport services; Landuse planning; Safer streets; Access to work; Access to learning; Access to healthcare; Access to healthy affordable food; The action plan; Annexes	http://www.socialexclusionunit.gov.uk/downloaddoc.asp?id=229

Continued

Table 9.5.3 UK Level Social Issues Policy. *Continued*

Policy Type	Document Title	Objective	Contents	Website
Strategies & Reports	Ensuring the Quality of Community Investment – A Report by People for Action (March 2003)	This report presents a comparison of various approaches to accrediting & raising the standard of community investment undertaken by voluntary & community organisations; public sector bodies & private enterprises. The research was conducted by People for Action which is the national network of housing & regeneration organisations committed to putting power & influence into the hands of local people. The research focused on three approaches: Charter Mark; Investors in Communities; & the Community Mark. The research also considered a number of quality management approaches like Investors in People, the Business Excellence Model & PQASSO (Practical Quality Assurance System for Small Organizations) & sector-specific schemes such as Quality Mark (aimed at advice services) & the Foyer Accreditation Scheme (aimed at foyers). In total, the research looked into 15 approaches, models, schemes or areas of work.	Summary & draft recommendations; Brief; Context; Description of the schemes; Index of the schemes & approaches described in this report; Investors in communities; Charter mark; Community mark; PACE (Practical Award in Community Excellence); Prove it! Social audit; Comparative analysis; Assessment & accreditation; Equal opportunities; Soft outcome focus; End-user focus; Performance measurement & improvement; Comprehensiveness; Generic or sectoral? Complexity; End-user involvement; Technical innovation; Common good between schemes; Overview; Recommendations; Appendices	http://www.homeoffice.gov.uk/docs2/qualcomminvest2003.pdf
	Realising the Potential – Recognising Residents' Achievement in Neighbourhood Renewal (May 2003)	The report consists of three main sections: ▪ *Context*: Outlines the policy background to the project and identifies the key developments with which it connects, both nationally and regionally. ▪ *Findings*: Reports on effective practice, mainly within the region, which provides some building blocks for the future, and summarises the outcomes of focus groups involving residents and regeneration professionals. ▪ *Way Forward*: Suggests an approach to meeting residents' learning needs consistent with the Learning Curve and proposes the main elements of an action plan.	Introduction; Summary; Context; Findings; Way forward; Appendices; Learning curve framework; Focus group outcomes; Interview schedule; Acronyms deciphered; References	http://www.neighbourhood.gov.uk/formatteddoc.asp?id=469

Changing Neighbourhoods, Changing Lives – The vision for neighbourhood renewal (June 2003)	The Government set out an ambitious vision for narrowing the gap between deprived neighbourhoods and the rest of the country so that, within 10 to 20 years, no-one should be seriously disadvantaged by where they live. The aim is to deliver economic prosperity, safe communities, high quality education, decent housing and better health to the poorest parts of the country. Equality for black and minority ethnic groups is a cross-cutting theme, underpinning all of these objectives.	The challenge we face; Why it matters? A strategy for change; Better spending; Better partnerships; Better learning; Better information	http://www.neighbourhood.gov.uk/formatteddoc.asp?id=91
Changing World, Changing Lives – Women in the UK since 1999 (July 2003)	This fifth report on the implementation of the Convention illustrates how the process has been taken forward over the last four years, with the needs of women increasingly being placed at the heart of policymaking. The UK Government has continued to make progress in the areas of childcare, education, employment rights, health, civil partnership & reducing domestic violence. In doing so, it has underlined a commitment to strengthening the foundations of society by eliminating discrimination & providing women with the same opportunities & services as men.	Foreword; Introduction; Employment; Childcare; Political & public life; Education; Health; Violence against women; Gender mainstreaming; Gender equality in the law (since 1999)	http://www.womenandequalityunit.gov.uk/research/CEDAW_report5.pdf
Community Cohesion Pathfinder Programme: The first six months (Oct 2003)	This report is structured around the issues and challenges faced by the Pathfinders in the early stages of their programmes. These are: ▪ Engagement with leaders ▪ Working with the voluntary & community sector ▪ Communication ▪ Sustainability Each section provides both a condensed version of the main learning experiences & practical examples of how the Pathfinders & Shadows have translated the broad principles of community cohesion into pragmatic & workable community based programmes. The report concludes with a brief description of issues identified as areas of potential future learning.	The pathfinder programme; Engaging with leaders; Engaging with the voluntary sector & local communities; Communicating with a wider audience; Sustainability; Conclusion	http://www.homeoffice.gov.uk/docs3/comcohesion_pathfinderprog0303.pdf

Continued

Table 9.5.3 UK Level Social Issues Policy. *Continued*

Policy Type	Document Title	Objective	Contents	Website
Strategies & Reports	Community Cohesion Advice for Those Designing, Developing & Delivering Area Based Initiatives (ABIs) (Dec 2003)	This document provides advice on how best to address cohesion within both new & existing ABIs, to ensure that relations between recipient communities & their neighbours are not damaged. It is intended for Government departments, Government Offices for the Regions, Local Strategic Partnerships (LSPs), Regional Development Agencies (RDAs), Local Authorities, & the many statutory & non-statutory organisations involved in the development, delivery & monitoring of ABIs & regeneration programmes. It will also be of interest to community representatives involved in regeneration activity.	Executive summary; Delivering need; Flexibility & area focus; Relationships between communities; Consultation & community capacity building; Communication; Leadership & accountability; Mainstreaming community cohesion within ABIs; Conclusion	http://www.homeoffice.gov.uk/docs2/abi_advice.pdf
	Building Civic Renewal – Review Findings from the Civil Renewal Unit (Dec 2003)	This report does two things. Firstly, it offers proposals as to how community groups & community capacity building can be better supported at community or neighbourhood level. It complements the consultation document *Building Civic Renewal: Government support for community capacity building & proposals for change*, which posed a set of questions & asked for examples of good practice that could be learnt from & promoted more widely. Secondly, it sets out actions that Government intends to take in the light of the review on which this report is based, to improve its own policies & practices on community engagement & those of other public bodies.	Foreword; Executive summary; What's the problem? Key definitions, values & principles; Outcomes of community capacity building; Community capacity building practice; Getting it right locally; Getting the Government's act together; Appendices	http://www.homeoffice.gov.uk/docs2/civilrenewalrevfindings.pdf
	Achieving Rural Revival – Report of the LGA Inquiry into Rural Revival (2003)	The Rural white paper published at the end of 2000, led the Executive to undertake a Rural Revival Inquiry to focus on the role local government can play. This report outlines the findings and conclusions of the inquiry & makes recommendations for a series of actions for	Introduction; Executive summary; Rural services & social well-being; The rural voluntary & community sector; Economic revival; Land management & a high quality environment; Governance in rural areas; Appendices	http://www.lga.gov.uk/Documents/Publication/achievingruralrevival.pdf

	Government & national organisations, regional bodies, & local authorities to achieve & support positive & sustainable improvements in rural life.		
Tackling Health Inequalities – A Programme for Action (2003)	This Programme for Action sets out plans to tackle health inequalities over the next three years. It establishes the foundations required to achieve the challenging national target for 2010 to reduce the gap in infant mortality across social groups, and raise life expectancy in the most disadvantaged areas faster than elsewhere.	Foreword; Executive Summary; Chapter 1 – Introduction; Chapter 2 – Strategy; Chapter 3 – Progress to date; Chapter 4 – Delivering change; Chapter 5 – Roles & responsibilities; Chapter 6 – Performance management & inspection, measuring progress & managing risk; Annexes	http://www.renewal.net/Documents/RNET/Policy%20Guidance/Tacklinghealthprogramme.pdf
Research Report 5: Business-led Regeneration of Deprived Areas – A Review of the Evidence Base (2003)	This report summarises the available evidence relating to effective business-led regeneration in deprived areas. This study therefore aims to carry out a systematic review of the evidence relating to business-led regeneration in deprived areas in order to identify the lessons on how to most effectively encourage & stimulate business development as part of the regeneration process. As well as summarising what is known about the impact of different types of enterprise on deprived areas, it also involves looking at the evidence related to the effectiveness of different kinds of policies which, directly or indirectly, have had an influence on business development in deprived areas.	Executive Summary; Introduction; What is the evidence for successful business-led regeneration in deprived areas? What is the evidence relating to the impact of business support policy in deprived areas? What has been the impact of area based initiatives on business-led regeneration in deprived areas? What lessons can be learnt from the experience of policies which have directly or indirectly affected business-led regeneration in deprived areas? Where are the gaps in the current knowledge base & what further research may be needed?	http://www.neighbourhood.gov.uk/formatteddoc.asp?id=356
Creating Safer & Stronger Communities – The Contribution of Cultural Services (Jan 2004)	This paper focuses on crime diversion activities for young people. Crime & anti-social behaviour is a key issue facing most communities. The LGA (Local Government Association) believes that local government has an important role to play in tackling crime & anti-social behaviour, particularly in relation to providing community leadership, & the core services it delivers which support the preventative, enforcement & rehabilitation elements of a community safety strategy.	Executive summary; Cultural services & crime diversion; Risk factors associated with offending behaviour; What works – where's the evidence? What works? – examples of local projects; Cultural services: promoting community cohesion; Cultural services: rewarding bad behaviour; Helping young people to be part of the solution; Cultural services & work with young offenders; Conclusion	http://www.lga.gov.uk/Documents/Publication/creatings.pdf

Continued

Table 9.5.3 UK Level Social Issues Policy. *Continued*

Policy Type	Document Title	Objective	Contents	Website
Strategies & Reports	Simplicity, Security & Choice: Informed Choices for Working & Saving – Pensions White Paper (Feb 2004)	This paper sets out the action the Government is taking to allow individuals to take control of their retirement planning. In doing so, it fits within the Government's overarching pensions strategy. It follows on from the steps being taken to reform occupational pensions as announced in June 2003, as well as the programme of initiatives to do away with outdated, inflexible approaches to work in later life.	Foreword; Summary; Chapter 1: Introduction; Chapter 2: Making the most of pension provision; Chapter 3: Raising awareness & education; Chapter 4: Giving people the right information; Conclusion; Annexes	http://www.dsdni.gov.uk/publications/documents/White%20Paper.pdf
	Understanding & Engaging Deprived Communities (April 2004)	This study was designed to enhance understanding of deprived, high crime communities by exploring perceptions of the local area of different groups living & working in the community: children, parents, residents, local politicians & leaders, professionals & service providers. Four of the 24 On Track areas were selected as case study examples of a range of urban & rural contexts.	Acknowledgements; Executive summary; 1. Introduction; 2. Literature review; 3. Research methodology; 4. The On Track neighbourhoods; 5. Communities within community; 6. Crime, deprivation & community; 7. Findings; References; Appendix: Children's drawings	http://www.homeoffice.gov.uk/rds/pdfs04/rdsolr0704.pdf
	Skills For Sustainable Communities – The Egan Review (April 2004)	The approach to this Review was to ask a series of questions: what sort of communities are being created, & how should achievement be measured; what are the most effective delivery processes, & who is responsible for them; & finally what skills are needed to make the processes work effectively, & how is the gap bridged between current skills base & the skills wanted?	Executive summary; Introduction; Summary of recommendations; The common goal identifying & measuring sustainable communities; Responsibilities & processes for delivery; The right skills for the job; The way forward; Annexes	http://www.odpm.gov.uk/stellent/groups/odpm_urbanpolicy/documents/downloadable/odpm_urbpol_028300.pdf
	Fairness for All: A New Commission for Equality & Human Rights – White Paper (May 2004)	The purpose of this White Paper is to set out the Government's proposals for the Commission for Equality & Human Rights (CEHR) in detail, including its role, duties & powers, & to outline the way in which it will deliver to its key stakeholders. It also sets out the anticipated next steps towards establishment.	Foreword; The White Paper; Vision; Engaging & working with stakeholders; The functions of the CEHR; Tools to promote change; The governance of the CEHR; Promoting good relations; Supporting key customers; Regional arrangements of the CEHR; The Scottish & Welsh dimensions; Preparing the way; The road to implementation; Appendices	http://www.womenandequalityunit.gov.uk/equality/project/cehr_white_paper.pdf

Choosing Health? Choosing a Better Diet A Consultation on Priorities for a Food & Health Action Plan (May 2004)	This paper presents an important opportunity to identify priorities for action & clarify roles & responsibilities for improving diet & nutrition, within the context of the overall health improvement debate set by *Choosing Health?* The *Choosing a Better Diet* consultation presented an opportunity for a wider group of people to inform the development of – & contribute to – this important work on nutrition & health.	Chapter 1: Introduction; Chapter 2: Consumer choice? Chapter 3: Improving Food Production & Manufacture; Chapter 4: Improving Food Supplied by Retailers, Caterers & the Workplace; Chapter 5: Improving Nutrition in Pregnancy & the Early Years; Chapter 6: Improving Nutrition in Schools; Chapter 7: Improving Nutrition in the National Health Service; Chapter 8: Improving Nutrition in Local Communities; Chapter 9: The process of consultation & how to contribute; Annexes	http://www.dh.gov.uk/assetRoot/04/08/14/56/04081456.pdf
Transformation & Sustainability – Future Support, Management & Monitoring of the New Deal for Communities Programme (May 2004)	The NDC Programme is innovative, & puts communities in the forefront. It is unusual in that Government has committed resources for 10 years to tackle the problems of the 39 communities. A range of work has been undertaken to evaluate the programme. Whilst not half way into the 10 year programme, the results gave grounds for cautious optimism. But much remains to be done, particularly in the poor performing NDCs. This paper aimed to address areas that need to be strengthened.	Summary; New Deal for Communities Programme; Background; Taking Stock; What has been achieved so far; Tackling Barriers to Effective Delivery; Responsibilities; Communities & Stakeholder Relationships; NDC Boards – What Works; Governance; Mainstreaming & Sustainability; LSPs (Local Strategic Partnerships); Linkages to other key programmes; Financial Reporting Framework for the NDC programme; Accountable Body Role; Skills & Knowledge; Race & Diversity; Community Engagement; Government Office Support; Employment; Performance Management Framework; Rationalising Outcomes; Publicising Success; Conclusion	http://www.neighbourhood.gov.uk/formatteddoc.asp?id=682
Crime in England & Wales 2003/2004 (July 2004)	This bulletin brings together statistics from the British Crime Survey (BCS) and the numbers of crimes recorded by the police to provide a comprehensive account of the latest patterns & trends in the main high volume crimes.	Chapter 1 – Summary; Chapter 2 – Levels & trends; Chapter 3 – Reporting & recording crime; Chapter 4 – Property crime; Chapter 5 – Violent crime; Chapter 6 – Patterns of crime; Chapter 7 – Detection of crime; Appendices	http://www.homeoffice.gov.uk/rds/pdfs04/hosb1004.pdf

Continued

Table 9.5.3 UK Level Social Issues Policy. *Continued*

Policy Type	Document Title	Objective	Contents	Website
Strategies & Reports	The English Indices of Deprivation 2004 (revised) (July 2004)	This report rehearses the conceptualisation underpinning the model of multiple deprivation used & outlines the indicators and domains that go to make up the ID 2004. For the first time domains to cover the deprivation suffered as a result of crime & a poor living environment are included. The Index has been produced at Super Output Area Lower Layer, which allows better identification & targeting of areas where small pockets of deprivation exist. These have been major achievements, which were identified by consultation respondents as being important in any measure of deprivation.	Introduction; Chapter 1 – Measuring Multiple Deprivation at the small area level: a conceptual framework; Chapter 2 – Domains & Indicators; Chapter 3 – Combining the Domains into an Index of Multiple Deprivation; Chapter 4 – Presentation of results & interpretation; Chapter 5 – The geography of deprivation; Chapter 6 – Comparing the Indices of Deprivation 2004 with the Indices of Deprivation 2000; Annexes	http://www.odpm.gov.uk/stellent/groups/odpm_urbanpolicy/documents/page/odpm_urbpol_029534.pdf
	Get a Public Life: Evaluation of the 2002 National Outreach Campaign (Sept 2004)	The research discussed in this report was concerned with evaluating the factors which encourage women in Britain to apply for & successfully take up a wide range of public appointments. The research included an evaluation of the public appointments workshops, & their impact on encouraging women to apply for & obtain public appointments; focus groups with women most likely to be eligible for & interested in public appointments e.g. women professionals; & interviews with women currently holding public appointments.	Executive summary; Introduction; The public appointment seminars & their impact; Awareness & perceptions of public appointments; The application process; Experience of Public Appointments; Barriers to application; Factors that would encourage women to take up public appointments; Annex A – Case studies; Annex B – Bibliography; Further contacts & information	http://www.womenandequalityunit.gov.uk/research/getapubliclife_2002noc_evaluation.pdf
	Sustainable Solutions to Anti-social Behaviour: Local Government's Joined-up Approaches to Tackling Anti-social Behaviour (Sept 2004)	In this briefing the LGA (Local Government Association) expands its earlier response to the government's approach to anti-social behaviour, & provides examples of ways in which local government is tackling anti-social behaviour of all types. The case studies in this briefing, & countless other examples across the UK, help to demonstrate the vital role that these longer & medium-term solutions can play in tackling anti-social behaviour locally & promote good behaviour.	Foreword; Introduction; Enforcement – important but only part of the picture; Sustainable solutions; Serving children well; Education; Youth crime services; New Deal programmes; Community enterprises; Supporting ex-offenders; Leisure activities; Drugs programmes; Helping families; Physical environment; Conclusions; Further Information	http://www.lga.gov.uk/Documents/Publication/ASB%20FINAL%20WEB.pdf

Releasing the Potential for the Public's Health (Sept 2004)	The purpose in collaborating to develop this paper was to identify the key principles & the common ground that might underpin a joint vision between organisations for a new approach to improving & sustaining the public's health. It was not a detailed implementation programme but a high-level consideration of the key drivers for such an initiative, which could be undertaken with confidence by the constituencies represented.	Foreword; Introduction; Summary & recommendations; Our rationale; Our shared principles; Our vision; A more comprehensive way of thinking about health; Public health & sustainable development; Achieving the vision: our proposals; Government leadership; Cross-government co-ordination; Releasing the potential of communities; Releasing the potential of local government; Releasing the potential of the NHS; Partnership; Beyond the rhetoric; Making local partnerships work; Conclusions; References	http://www.lga.gov.uk/Documents/Publication/releasingthepotential.pdf
Breaking the Cycle: Taking Stock of Progress & Priorities for the Future – A report by the Social Exclusion Unit (Sept 2004)	This report takes stock after seven years of the Social Exclusion Unit. Based on a new programme of research & analysis, it outlines: ▪ the progress made to date in tackling the causes & consequences of social exclusion; ▪ the challenges that remain & those that might arise in the future; ▪ the lessons learned about delivering services to disadvantaged people; ▪ the priorities for future action. The full report offers a detailed exploration of what drives social exclusion, including those factors which transmit poverty & disadvantage from one generation to the next. It identifies the main groups at risk and assesses the scale of the problem & outlines the Government's strategy since 1997.	Summary; Introduction; Chapter 1: What is social exclusion? Chapter 2: What the Government has done; Chapter 3: Overall progress in tackling social exclusion; Chapter 4: Sustaining progress; Chapter 5: Equalising opportunity; Chapter 6: Reaching the most disadvantaged; Chapter 7: Facing the future; Chapter 8: Conclusions; Annex A: Key new policies introduced since 1997 & policies in the pipeline to tackle social exclusion; Annex B: Floor & PSA (Public Service Agreement) targets	http://www.socialexclusion.gov.uk/downloaddoc.asp?id=262
What Works in Community Involvement in Area-based Initiatives? A Systematic Review of the Literature (Oct 2004)	This report was commissioned by the Home Office to build the evidence base for Civil Renewal. Increasing community involvement is seen as key to this agenda which seeks to put active citizens at the heart of tackling social problems. This report is the product of a systematic review of research evidence on the effectiveness of community involvement in Area Based Initiatives (ABIs).	Executive summary; Introduction; Methodology; An overview of the literature; What the literature says; What Works? Lessons & good practice advice; Conclusions & future research; Annex & appendices	http://www.homeoffice.gov.uk/rds/pdfs04/rdsolr5304.pdf

Continued

Table 9.5.3 UK Level Social Issues Policy. *Continued*

Policy Type	Document Title	Objective	Contents	Website
Strategies & Reports	Choosing Health: Making Healthier Choices Easier – White Paper (Nov 2004)	This White Paper sets out the key principles for supporting the public to make healthier and more informed choices in regards to their health. The Government will provide information and practical support to get people motivated and improve emotional wellbeing and access to services so that healthy choices are easier to make.	Foreword & contents; Chapter 1: The time for action on health & health inequalities; Chapter 2: Health in the consumer society; Chapter 3: Children & young people – starting on the right path; Chapter 4: Local communities leading for health; Chapter 5: Health as a way of life; Chapter 6: A health-promoting NHS; Chapter 7: Work & Health; Chapter 8: Making it happen – national & local delivery; Annex A: Choosing Health? Consultation; Annex B: Making it happen;	http://www.dh.gov.uk/PublicationsAndStatistics/Publications/PublicationsPolicyAndGuidance/PublicationsPolicyAndGuidanceArticle/fs/en?CONTENT_ID=4094550&chk=aN5Cor
	Towards Self-Governing Communities: The Role of Local Government in Civil Renewal (Dec 2004)	In this position paper, the case is made for putting local government at the heart of civil renewal. It is shown what councils are doing both to build stronger communities, better able to look after themselves, & to strengthen their capacity to participate in local governance. It also identifies where councils can do more.	Foreword; Introduction; What is civil renewal & why is it important? The role of local government; Building stronger communities; Widening the involvement in local governance; An agenda for civil renewal: future steps	http://www.lga.gov.uk/Documents/Publication/towardsselfgoverningcommunities.pdf
Consultation Papers	Analysis of Consultation Responses on the Review of Community Participation (Sept 2003)	This report summarises & considers the 109 responses received by the Neighbourhood Renewal Unit (NRU) on the consultation paper proposing changes to the funding arrangements for neighbourhood renewal community participation programmes published in May 2003.	Executive summary; Introduction; Who responded? Summary characteristics & key messages; Conclusions & recommendations; Annex A: Summary of responses.	http://www.neighbourhood.gov.uk/formatteddoc.asp?id=570

Guidance Notes	Factsheet No.11: Employment & Neighbourhood Renewal (Oct 2002)	The strength of the economy depends on the number of people in work & how productive they are. The problem of 'worklessness' is a constraint on Britain's economic growth potential – & also one of the most important causes of poverty & deprivation. This factsheet charts progress on employment & neighbourhood renewal within the UK.	Policy highlights; Key issues; What are the solutions? Floor targets; New Deal; Employment Zones; Over-50s; Helping the inactive; Help for lone parents; Help with childcare; Help for people with Disabilities; Support for people at a disadvantage through the misuse of drugs; Help for people with low skills; Action Teams for Jobs; Step Up; Ethnic minorities; Asylum seekers & refugees; Transport; Housing Benefit; National Minimum Wage; New Deal for Communities; Encouraging enterprise; Business Brokers; Regional Development Agencies; Further information	http://www.renewal.net/Documents/RNET/Frequently%20Asked%20Questions/Employment neighbourhoodrenewal.pdf
	Guidance on Community Cohesion (12/12/02)	This guidance is designed to assist all Local Authorities & their partners in strengthening & building community cohesion. It provides advice on ways to review existing policies & practices so that they help to build more cohesive communities.	Foreword; Introduction; What is community cohesion?; What is the role for central government?; Ownership of the community cohesion agenda; community & political leadership; Vision & values; Local context, baseline assessment & monitoring progress; The value of local strategic partnerships; Community & voluntary organisations; Faith communities; Young people; Asylum seekers, refugee & travelling communities; Regeneration; Sports & cultural services; Education; Housing & planning; Employment & economy; Community safety & policing; Press & media; References	http://www.homeoffice.gov.uk/docs2/cc_guidance.html
	Community Cohesion Our Responsibility (Sept 2002)	This briefing paper summarises some of the recent initiatives implemented to help build cohesive communities. In particular, it looks at what various stakeholders, including the National Association of Citizen's Advice Bureaux (NACAB) & the Commission for Racial Equality (CRE), are doing to foster a multicultural society in which each person enjoys equal rights while also sharing the responsibility for strengthening communities.	Introduction; What is community cohesion? Government initiatives; NACAB Initiatives; CRE initiatives; Joint Initiatives by NACAB & the CRE; Useful websites	http://www.cre.gov.uk/pdfs/cohesion.pdf

Continued

Table 9.5.3 UK Level Social Issues Policy. *Continued*

Policy Type	Document Title	Objective	Contents	Website
Guidance Notes	Health & Neighbourhood Renewal Guidance from the Department of Health & the Neighbourhood Renewal Unit (2002)	This guide was designed for all those partnerships who are currently or would like to tackle poor health in their area. It should be of particular use for Local Strategic Partnerships (LSPs) wanting to build on their health programmes, & for other initiatives such as Healthy Living Centres (HLCs). It was particularly designed for those whose area of expertise is not health, but as it attempts to draw out the connections between health & the wider factors influencing health, it should be useful for those in the health field as well.	Introduction; Health & its Determinants; Baselines & Setting Targets; Health Impact Assessment; Improving Health; Annexes	http://www.neighbourhood.gov.uk/formatteddoc.asp?id=781
	How to Make Your Neighbourhood A Better Place to Live – Sustainable Communities Programme Handbook (Sept 2004)	This is a practical book written to help everyone work towards making his or her own neighbourhood a better place to live. A word often used to describe this process is sustainability. Sustainability is not only about your community becoming a better place to live, it is also about looking to the future & finding ways of living that will last for future generations. Some people will be happiest working on very practical tasks while others will want to concentrate on issues of policy. Others will want a combination of both. This handbook is written to help those who want to work in these ways but within their community. It is possible to carry out very practical local tasks, but there are many factors affecting the sustainability of a community which come from outside. Many of these can only be addressed by operating at a policy level.	Making your neighbourhood a better place to live & how to achieve your aims; Practical ways in which a community can play a part in achieving greater sustainability; Techniques, processes & skills; Appendices	http://www.encams.org/information/publications/litteranddogfouling/scp_handbook.pdf

Building Community Cohesion into Area Based Initiatives: A guide for residents and practitioners (Oct 2004)	The guide suggests concrete ways in which diverse groups of people – for example of different classes, ages, faiths, races & ethnicities – can be encouraged to live & work together cohesively within ABIs (Area Based Initiatives), & looks at the factors that either contribute to or undermine this process. The guide draws extensively on good practice found during research into community cohesion initiatives. The ideas in this publication will help encourage ownership by communities of the changes occurring in their neighbourhoods.	Executive Summary; Understanding the local context; Defining needs; Setting geographical boundaries; Flexibility – working beyond the boundaries; Relationships between communities; Supporting new residents; Consultation & community capacity building; Communication; Leadership & accountability; Monitoring & evaluation; Mainstreaming community cohesion; Annex A: Case Studies; Annex B: Useful websites & publications	http://www.neighbourhood.gov.uk/formatteddoc.asp?id=793
Community Cohesion – An Action Guide: Guidance for Local Authorities (Nov 2004)	This practitioner guidance includes descriptions of the application of community cohesion in key policy areas, pointers for successful implementation and case studies drawn from different local authority & other contexts. This is essential reading for all heads of service delivery & programme managers – community cohesion principles must be embedded in mainstream services, not just in special projects. Obviously, it is also aimed at those practitioners specifically concerned with building community cohesion.	Foreword; Introduction; Defining community cohesion; Vision, values & strategy; Measuring community cohesion; Working with the local strategic partnership; Conflict resolution; Working with black & minority ethnic (BME) communities; Working with faith communities; Working with the voluntary & community sector; Working with the media; Embedding community cohesion in delivering services; Working with the police & the fire service; Housing; Regeneration; Arts, sport & other cultural services; Education; Children & young people; Older people & intergenerational cohesion; Asylum seekers & refugees; Gypsies & Travellers; Case studies on other aspects of community cohesion	http://www.lga.gov.uk/Documents/Publication/communitycohesionactionguide.pdf

Table 9.5.4 Northern Ireland Level Social Issues Policy.

Policy Type	Document Title	Objective	Contents	Website
Strategies & Reports	Community Safety Auditing Sound Options – Hard facts Guidance Paper III: A Model for Reviewing Neighbourhood Levels of Crime, Anti-social Behaviour & the Concerns of Local People (2000)	The purpose of this document is to recommend a model for community safety audits by all community safety partnerships across Northern Ireland. Co-ordinating local community safety strategies is a new approach in Northern Ireland, providing opportunities for people living in neighbourhoods to become directly involved in finding solutions to community safety problems. Their concerns can be addressed effectively through a process of review or audit in a consultation & analysis process.	Introduction; Stage 1 – Scoping information & desktop search; Stage 2 – Community consultation; Stage 3 – Audit conclusions – bringing your information together; Conclusion; Appendices	http://www.communitysafetyni.gov.uk/documents/Publications/csm_evaluation_report.pdf
	Unlocking Creativity – A Strategy for Development (March 2001)	This paper outlines a co-ordinated strategy for the development to the full of the creative & cultural resources of the people of NI. In a world of rapid economic & social change, the future prosperity & well being of NI will depend increasingly on the creativity & adaptability of its people. Developing these abilities in the best way calls for new approaches in education, at all levels, & new approaches to economic & cultural development.	Foreword; Introduction; A new context; What is creative & cultural education? A strategy for development; An agenda for action; Consultation; Appendix 1: Membership of future search creativity in education group; Appendix 1a: Membership of future search creative industries group; Appendix 2: Bibliography; Glossary	http://www.dcalni.gov.uk/ContMan/uploads/unlocking%20creativity,%20a%20strategy%20for%20development.pdf
	Face-to-Face: A Vision for Arts & Culture in Northern Ireland (June 2001)	This document presents a vision for arts & culture in Northern Ireland. It is a result of a consultative process that gives voice to, & is owned by, the arts & culture community.	Mind map; Background & information; Creativity in education; Creative industries; Cultural diversity; International; Infrastructure; Universal accessibility; Support for the individual artist; Review & the way forward	http://www.dcalni.gov.uk/ContMan/uploads/face_to_face.pdf

Unlocking Creativity – Making it Happen (June 2001)	This new document takes into account the responses to the original *Unlocking Creativity* document & the consultation process; it presents the policy & development context & Action Agenda. The revised plan provides an initial framework that is achievable, sustainable & measurable, thus enabling discernible progress without lessening the commitment to the creative imperative, & the radical approaches that are required to support it.	Foreword; Consultation; Introduction; Making the connections; Making it happen; Definition; Vision; Mission & aims; Perspectives & needs; A creativity framework; Action agendas; Acknowledgements; Bibliography; Glossary	http://www.dcalni.gov.uk/ContMan/uploads/unlocking_creativity2.pdf
New TSN Action Plan – Department of Culture, Arts & Leisure (April 2002)	To achieve the New TSN (Targeting Social Needs) objectives in enhancing the quality of life for those in society who are socially disadvantaged DCAL (Department of Culture Arts and Leisure) is committed to effectively targeting its resources & in particular to those in most need. The Department made excellent progress in taking forward the actions within this New TSN Action Plan. As a new Department DCAL had little baseline information and relevant indicators to facilitate the targeting of resources. The DCAL's New TSN Revised Action Plan contains a number of actions relating to the identification of indicators to be used for targeting and for monitoring & evaluation.	Section 1: Foreword; Preface; New TSN policy; DCAL's main responsibilities & key programmes & services Commitment to & the drivers in place to implement new TSN; Approaches being taken to targeting new TSN, relevant to specific programmes & services; Description & intended use of Noble indicator; Section 2: Explanation of Annex A; Set of individual tables that constitute the key objectives of the new TSN revised action plan 2001–03; Section 3: Annexes	http://www.dcalni.gov.uk/ContMan/uploads/NTSN%20Action%20Plan%202003-04%20Final%20Draft%20-%20pdf.pdf
Weak Community Infrastructure Report (Nov 2002)	This is the report of the Weak Community Infrastructure Conference organised by the Voluntary Activity Unit 1 in the Department for Social Development. The conference took place on 10 December 2001. It was facilitated by the Sustainable Northern Ireland Programme (SNIP) and held in the Northern Ireland Council for Voluntary Action (NICVA) conference facility. There were almost 100 participants at the conference and 23 workshop discussions. These produced a wide range of views and a wealth of ideas. This report endeavours to present the essence and flavour of these wide-ranging discussions for the benefit of both the participants and other readers.	Introduction; Setting the agenda; Prioritising the issues; Identifying action points; Some common themes; What has happened since the conference; Appendix 1: Conference programme; Appendix 2: Conference attendees; Appendix 3: Opening address	http://www.dsdni.gov.uk/publications/documents/Community%20Report.pdf

Continued

Table 9.5.4 Northern Ireland Level Social Issues Policy. *Continued*

Policy Type	Document Title	Objective	Contents	Website
Strategies & Reports	Report on Research into Evaluating Community-Based & Voluntary Activity in NI (March 2003)	The current project undertaken by Community Evaluation Northern Ireland (CENI) was designed to re-look at the issue of measuring the outcomes of voluntary & community activity in the light of key developments of the late 1990s. Of some importance was the renewed emphasis on partnerships amongst government, the private sector & civil society advocated by the New Labour government. Equally important were the establishment of the Assembly & Executive in NI & the consequent production of the Programme for Government. This also highlighted the partnership mechanism as a means of delivering modern, integrated services in NI. Moreover, given the specific context of NI, slowly recovering from 30 years of political conflict, there is a vital need to maximise the potential of organisations working in civil society to contribute to the process of reconciliation. The task set for CENI was to build on the work already done, but to develop an approach to measuring outcomes that reflected this new context.	Introduction; A new approach to funding: investing in community capability; Using the concept of social capital in an evaluation framework; Constructing an evaluation template; Building evaluation indicators using social capital; Conclusions.	http://www.dsdni.gov.uk/publications/documents/Indicators_final_Summary_Report_Dec_02.pdf
	Creating a Safer Northern Ireland through Partnership: A Strategy Document (March 2003)	This strategy document sets out a range of proposed structures & actions, to be delivered during the period April 2002–March 2007, designed to create the conditions for improving community safety throughout NI.	Foreword; Executive summary; Chapter 1: An introduction to community safety; Chapter 2: Crime in NI – the context; Chapter 3: Issues & crimes of most concern; Chapter 4: Delivering the strategy – structures & accountabilities; Chapter 5: Delivering the strategy – resource implications; Annex A: Organisations consulted; Annex B: Equality aspects; Annex C: Extract from Action Plan	http://www.communitysafetyni.gov.uk/documents/Publications/CSU%20Strategy%20Doc.pdf

Young People, Unemployment Duration & the New Deal in Northern Ireland (April 2003)	In September 2001 the Department for Employment & Learning (DEL) commissioned a research team from the Northern Ireland Economic Research Centre (NIERC) & the University of Southampton to examine the effects of the introduction of the New Deal for 18–24 Year Olds on the duration of unemployment spells for the 18–24 year old age group in Northern Ireland. This revised report sets out the findings of the study.	Executive summary; Introduction; Background; The model; The data; Unemployment duration pre & post new deal; Conclusions; References; Appendices.	http://www.delni.gov.uk/docs/pdf/newdealRevisedReport.pdf
Entrepreneurship & Education Action Plan (April 2003)	This Action Plan has been developed by our three Departments, Enterprise, Trade and Investment (DETI), Education (DE) & Employment & Learning (DEL) & is targeted primarily at key stakeholders in the business & education sectors. It establishes an initial framework of agreed & co-ordinated actions to ensure greater synergy, co-operation & understanding between the worlds of business & education. This Action Plan aims to ensure commonality of understanding & approach & to describe the steps to be taken over the short to medium term to bridge any gaps & identify & address weaknesses in existing systems. It represents a first step towards a longer-term goal of ensuring that all young people in Northern Ireland, regardless of their academic ability, have equal access to a wide range of high quality enterprise learning experiences which help them to develop the skills & attributes needed to reach their full potential in life, & particularly in the world of work.	Foreword; Purpose; Background; Approach; Objectives; Monitoring & reporting arrangements; Annexes.	http://www.delni.gov.uk/docs/pdf/DETI%20Action%20Plan.pdf
People & Place – A Strategy for Neighbourhood Renewal (June 2003)	This document is the first of several that will collectively map out the approach to urban regeneration & Neighbourhood Renewal. Collectively, these strategies will be known as 'People & Place' to emphasise their inter-related nature & common themes.	Executive summary; The context; Learning the lessons; Goals & objectives; Delivering neighbourhood renewal; Measuring success; Conclusions & next steps; Annexes.	http://www.dsdni.gov.uk/publications/displayitemSections.asp?ID=343

Continued

Table 9.5.4 Northern Ireland Level Social Issues Policy. *Continued*

Policy Type	Document Title	Objective	Contents	Website
Strategies & Reports	Think, Create, Innovate – The Regional Innovation Strategy for Northern Ireland (July 2003)	The Strategy sets out the research and development & innovation context & challenges for Northern Ireland & proposes a vision & a framework for action comprising four key Priorities & eight Action Areas with a series of recommendations aimed at addressing research, development & innovation policy & activity in Northern Ireland; & considers how best to implement & monitor the RIS.	Foreword; Executive summary; Background; Challenge; Vision; Actions; Anexes.	http://www.detini.gov.uk/cgi-bin/downutildoc?id=9
	Position Paper Policy on Homelessness in the UK & Republic of Ireland (2003)	The purpose of this document is to provide a context & starting point for the work of the Promoting Social Inclusion (PSI) Working Group on Homelessness (the Working Group). It sets out: ▪ Current policy on homelessness in Northern Ireland, England, Wales, Scotland & the Republic of Ireland; ▪ Recent statistics on homelessness; ▪ A list of recent publications which address preventative & cross-sectoral strategies for dealing with homelessness within the UK & the Republic of Ireland; ▪ Issues that the Working Group might wish to address.	Introduction; Policy provision for homelessness; Housing roles & responsibilities across the UK; Housing roles & responsibilities in the Republic of Ireland; Current levels of homelessness; Recommendations for future work; Areas to be addressed by the Working Group; Summary; Sources; Annex A: Definitions of Priority Need across the UK; Annex B: Roles and Responsibilities of the NIHE (Northern Ireland Housing Executive)/Local Authorities; Annex C: NIHE Recommendations with a cross-agency dimension; Annex D: Decisions taken on homelessness applications (UK)	http://www.dsdni.gov.uk/publications/documents/PSI%20HMLS%2003%20003(1).pdf
	Unlocking Creativity – A Creative Region (Jan 2004)	This action plan builds on the learning of the past three years & continues to encompass education, lifelong learning, the arts, the creative industries, business, & the individual excellence witnessed, especially through the work of the Creativity Seed Fund which has supported around 90 projects across sectors, interest groups & regions.	Introduction; Mission & aims; Definition; Refreshing the strategy; Themes: learning, innovation, enterprise & connecting; Equality impact assessment; Consultation questions; Unlocking creativity action plan; Learning; Innovation; Enterprise; Connecting; Further reading & selected websites; Glossary; Appendix – Equality impact assessment	http://www.investni.com/print/unlocking_creativity_2004.pdf

People & Place – Neighbourhood Renewal in Belfast Draft Implementation Plan (Jan 2004)	The Implementation Plan will set out: ▪ agreed target neighbourhoods; ▪ broad objectives & priority actions; ▪ agreed implementation structures; ▪ a timetable for implementation, including the transition from existing expenditure programmes; ▪ resourcing implications; ▪ equality screening arrangements. The Implementation Plan will also explain links of the Neighbourhood Renewal approach with other urban regeneration activities. It will build upon the policies that have been effective in the past & ensure that all urban regeneration policy instruments work in ways that reinforce each other.	Introduction; People & place: A strategy for neighbourhood renewal; Implementation proposals for Belfast; Identifying neighbourhood renewal areas; Proposed delivery structures; Making neighbourhood renewal work; Equality impact assessment; Letting us know your views; Annexes	http://www.dsdni.gov.uk/consultation-zone/documents/Belfast%20Plan%20a4.pdf
Equality Impact Assessment (EQIA) Urban Regeneration – Neighbourhood Renewal Implementation Plan For Belfast (Jan 2004)	This document is an Equality Impact Assessment (EQIA) of the Department's Draft Belfast Regeneration Office, *Neighbourhood Renewal Implementation Plan* & conforms to the requirement to have due regard to the need to promote equality of opportunity & the need to provide equality of opportunity in developing new policy.	EQIA; Context; Reasons for Changing Existing Policy; Main Elements of New Policy on Neighbourhood Renewal; How will target Neighbourhoods be identified? Strategy Implementation; Resourcing the Strategy; Neighbourhood Action Plans (NAPS); Consultation on the Policy; Consideration of Available Data & Research; Assessment of Impact; Impact on Section 75 Groups; Monitoring Arrangements; Your comments on EQIA; Annexes	http://www.dsdni.gov.uk/consultation-zone/documents/DraftEqualityScheme1.pdf
People & Place – Neighbourhood Renewal in Regional Towns & Cities Draft Implementation Plan (Jan 2004)	This Draft Implementation Plan is designed to obtain views on the Department's proposals for how Neighbourhood Renewal should be delivered in the regional towns & cities. Since the major theme of Neighbourhood Renewal is the need to work closely with others in the public sector, local government, the private sector & in local communities, the Department wants to see an open, meaningful, informed & realistic discussion of the priorities & proposals outlined in this Draft Implementation Plan & therefore urges all interested parties to fully contribute to the discussion.	Introduction/foreword; People & place – A strategy for neighbourhood renewal; The Proposed Neighbourhood Renewal Areas in Northern Ireland's Towns & Cities; Implementation Process & Timetable; Proposed Roles & Responsibilities; Funding & Potential Activities; What Do You Think? Appendices	http://www.dsdni.gov.uk/consultation-zone/documents/NR.pdf

Continued

Table 9.5.4 Northern Ireland Level Social Issues Policy. *Continued*

Policy Type	Document Title	Objective	Contents	Website
Strategies & Reports	Evaluation of Partners for Change – Final Report (June 2004)	This report details the independent evaluation of *Partners for Change*. It comments on the actions taken by government departments & agencies during 2001–04 to further their relationship with the voluntary & community sector. It highlights resultant impacts & outcomes, examples of good practice & identifies areas for improvement.	Executive summary; Introduction; Research approach; Developing the relationship between government & the sector; Facilitating improvement & good practice; Progress towards core aims; Measuring progress & performance; Promoting equality of opportunity & the shared values & principles of the Social Compact; Added value of partners for change & future directions; Appendices	http://www.dsdni.gov.uk/evaluation_pfc_final_report.pdf
	People & Place: Neighbourhood Renewal in Derry/Londonderry Implementation Plan – Discussion Paper (Aug 2004)	In line with the overarching strategy for Northern Ireland, the Department intends to engage with the public, private & community/voluntary sectors in Londonderry on how it proposes to implement Neighbourhood Renewal in the city. This Paper has been prepared by the North West Development Office to facilitate discussions on the Strategy.	Introduction; People and place – A strategy for neighbourhood renewal; Implementation proposals for Derry/Londonderry; Identifying neighbourhood renewal areas; Proposed delivery structures; Making neighbourhood renewal work; Equality impact assessment; Letting us know your views	http://www.dsdni.gov.uk/consultation-zone/Document.asp?ID=31
	Ending Fuel Poverty: A Strategy for Northern Ireland (Nov 2004)	This strategy provides the overall policy context for government to eradicate fuel poverty & the direction for all our partners to help achieve that aim. Ending fuel poverty is therefore not simply a strategy for housing or raising income, or for addressing energy prices. It is a document that addresses fuel poverty in a holistic way.	Foreword; Executive Summary; Introduction; The causes & definition of fuel poverty; Extent & analysis of fuel poverty in Northern Ireland; Vision statement; Action required & delivery mechanisms; Assessment of resources; Targets & milestones; Monitoring, reporting & review; Research & evaluation; Equality & rural proofing; Appendices	http://www.dsdni.gov.uk/publications/documents/FUEL_POVERTY.pdf

	First Review & Progress Report of the Taskforce on Employability & Long Term Unemployed (Dec 2004)	For the first time in Northern Ireland, the Taskforce carried out an analytical study of the concept of employability. Key themes were identified & the consultation process reinforced the view that there are many complex factors that impinge on an individual's ability to enter and remain in employment ranging from transport, childcare & health to the prevailing labour market. The report sets out what has been achieved against the action plan published in the report of the Taskforce on employability & long-term unemployment in December 2002.	Foreword; Introduction; Update on the Economic and Labour Market Context; Progress Report – Improving Employability in Practice; Looking Ahead – Key Areas; Performance against target; Appendices	http://www.delni.gov.uk/docs/pdf/tasforcedec04revision.pdf
Guidance Notes	Guidance to District Councils on the Development of Local Cultural Strategies (July 2001)	This guidance is intended to embed the excellent work undertaken at local level by District Councils, enabling them to deliver on their objectives for the creation of Integrated Area Plans. It can serve as a planning tool to help Councils to plan strategically for culture, arts & leisure in such a way as to improve services to local communities & to maximise funding opportunities. The guidance is intended to allow for local circumstance & in particular achieve recognition for the essential services provided through culture, arts & leisure at local level & the positive impact it has on the lives of communities.	Introduction; Background; Scope; Value of culture, arts & leisure; Local cultural strategies; Developing local cultural strategies; The need for partnership & community engagement; PEACE II: locally-based regeneration & development strategies; The link to best value & Local Agenda 21; Guidance; Specific & thematic approaches; Developing a local cultural strategy; Monitoring; Review; Conclusion; Appendix 1: DCALs responsibilities; Appendix 2: Common elements of Local Area Strategies (Priority 3, PEACE II); Appendix 3: Membership of the cultural forum	http://www.dcalni.gov.uk/ContMan/uploads/guidance_to_district_councils_on_the_development_of_local_cultural_strategies.pdf

10 Sustainable Development Policy and Practice

10.1 Sustainable Development Policy and Practice and the Built Environment

Several principles for sustainable development (SD) have been developed since the 1980s with these based on growing global awareness of ecological degradation issues and the need for environmental protection during the latter part of the twentieth century. The headline principles were established and consolidated at the UN Conference on Environment and Development (Earth Summit I) in 1992 and at the World Summit on Sustainable Development (Earth Summit II) in 2002. Action was to be delivered through environment and development principles, an action plan to achieve a sustainable pattern of development, forest conservation and management principles, a framework for reducing the risks of global warming, and an agreement on protecting the diversity of species and habitats.

The link between SD principles and strategies has been demonstrated at UN and national levels, while translating these into policies and subsequent practice with appropriate implementation action plans remains less evident. Hence, there has been high rhetoric about sustainability principles and aspirations tainted by considerable criticism about the lack of practical progress. Such policies and action plans need to be set alongside progress measurements and appropriate indicators. The richness in developing principles and expressing these in strategies has been essential and visionary, but a lot of hard work remains to be done.

The European Union Strategy for Sustainable Development (2001) addressed SD principles but also recognised the need for policy coherence across disciplines with sustainable development becoming the central objective of all its sectors and policies. This required policy-makers to identify likely cross-cutting issues with other policy areas and take these into account, so that each policy proposal included estimates of its economic, environmental and social impacts inside and outside the EU. This is to include, where relevant, the effects on gender equality and equal opportunities and identify the groups who bear the burden of change so that policy-makers can judge the need for measures to help these groups to adapt.

Sustainable development policies should retain the high-level objectives of sustainability while the detailed policies, connected with themes or disciplines, have the opportunity and challenge to embed the SD principles while dealing with theme-specific aspects. This connectivity can be seen, for example, in waste management strategies which require an emphasis on recycling waste and reducing the need to deposit inert and putrescible waste into landfill sites. The achievement of these goals has considerable impact on environmental protection, employment generation within the recycling industry, community integration with waste minimisation and reduced dependency on raw materials. However while the waste industry addresses specific targets it is largely unaware that its success will make a considerable and positive response to SD principles.

There are several opportunities for SD principles to be translated into sustainability related policies within the built environment disciplines and themes. However, the understanding and response to key legislative drivers is equally important, in that high-level policies are enshrined in law at European level through Directives, and normally at national level in local guidance notes, consultation papers, strategy reports and legal instruments. These high-level policy issues are normally integrative and can be embedded into all sectoral policies, covering areas such as equality, environmental justice, labour and community human rights, health and safety, company law and government accountability.

Professional bodies associated with the built environment also play key roles in designing and delivering sustainable development policies. It is normal that these professional body members will have agreed to a code of ethics or charter which requires a sustainable development response. This could be achieved through understanding the role of other professionals, development and delivery of relevant net-working activities, formal continuing professional development courses, formulation and dissemination views on SD in appropriate fora, and addressing SD in all core business activities.

Corporate Social Responsibility (CSR) is a concept whereby companies integrate social and environmental concerns into their business operations and their interaction with stakeholders on a voluntary basis as they become increasingly aware that responsible behaviour leads to sustainable business success. CSR is also about managing change at company level in a socially responsible manner. This approach ensures that a sustainable development policy is encapsulated into business, as it embraces a range of policy issues such as transport, the environment, enterprise, employment, social affairs, project procurement and external interactions.

The concept of 'greening government' impacts on aspects of the built environment. This initiative is aimed at establishing guidelines for all government departments on how to integrate environmental considerations into the management of their operations, and embraces functions and activities at several levels, such as operations, funding, housekeeping, procurement, transport and waste. The UK government has developed this concept by embedding sustainable development across all government departments, with each having a Green Minister, who acts as a champion for sustainable development in their own department. Green Ministers are members of an environment committee, which considers the impact of government policies on sustainable development and aims to improve all departments' sustainable development performance, reporting back at Cabinet level.

10.2 Key Sustainable Development Policy and Practice Elements

The elements of sustainable development policies are founded on outputs from conferences, conventions, agreements and protocols; directives, decisions, communications and recommendations; strategies, reports, consultation papers and guidance notes, and as a result of discussions, consultations, negotiations, political advocacy and leadership. Each of these elements are applied at the appropriate level of decision-making processes and in accordance with the jurisdiction involved. These should contribute to the continuum of procedural development involving conceptual thinking, high-level strategy development, policy drafting, legal and equality appraisal, implementation plan production, and the introduction of measurement, monitoring and reporting mechanisms.

There are overarching influences on policy development such as regional cultural practice, the availability of funds, government ethos, a nation's state of readiness, government prioritisation issues and the impact of unexpected and major incidents. At a lower but equally important level, there are softer issues which influence the progress of the above elements of sustainable development policy. These issues influence human behaviour and attitude, respond to the guiding principles which may not require legal compliance, and secure a voluntary and supportive approach to the long-term implications of project development. They pertain primarily to sustainable development practice, the need for behavioural change and for enhanced sustainability consciences.

The need for balanced sustainable development policy can be appreciated by examining a theme such a climate change. High-level energy strategies establish the need to reduce production of harmful gases and set suitable targets. In turn, policies dealing with waste management may identify the need for a thermal treatment plant as part of the solution to meeting waste targets, with such a plant increasing carbon dioxide emissions, contrary to the energy targets. Simultaneously, an urban development policy may permit substantial development in a low-lying region resulting in population movements and increases in several items such as waste, carbon dioxide, transport pollution, child illness and rising water levels. Hence it is essential to establish a balance across a range of policy and practice elements to ensure that the causes and effects of any action are fully understood and that policies are in support of the wide-reaching principles of sustainable development.

The success and viability of policy and practice will depend on the application of suitable performance measurements. Sustainable development indicators have been developed to assess if practice has attained sustainability standards and supported the credibility and suitability of the policy strategy. A range of indicators are used and these have been largely developed to meet specific regional or themed issues. At a higher level, sustainable development indicators should address a balanced range of parameters and measures. A comprehensive European indicator set covers social issues by measuring aspects of equity, health, education, housing, security and population; environmental issues by measuring aspects of atmosphere, land, oceans/seas/coasts, fresh water and biodiversity; economic issues by measuring aspects of economic structure and consumption and production patterns; and institutional issues by measuring aspects of the institutional framework and its capacity.

10.3 The Sustainability Challenge

Sustainable development policies should be congruent and co-exist with the key objectives of sustainable development. They should fall from the key guiding principles and provide the basis for the development, design, production and long-term appraisal of all built environment projects, utilising a balanced and conciliatory approach in reaching compromise and balanced solutions. However, the full delivery of sustainable development practice is even more challenging, demanding further balance and negotiation. Hence it is important to secure full engagement by all parties and for each to subscribe to the aspirations of 'living within environmental limits', 'achieving a sustainable society', 'promoting good governance', 'using sound science responsibly', and 'ensuring a strong, healthy and just society', as these underwrite and encapsulate the triple bottom line of environmental protection, economic stability and social inclusion.

It is essential to address the quest for high-level successful delivery of sustainable development policy and practice, so that this can be cascaded down to each theme or discipline. There are several challenges which reside primarily at this high-level and are critical to global sustainability:

- *Develop a clearer understanding of the factors which motivate, shape and constrain pro-sustainability attitudes* by assessing human behaviour, examining key triggers and securing robust data to advise on the best ways to solicit voluntary responses to environmental best practice and deliver sound sustainable development practices.
- *Reconcile and balance the demands of economic stability, environmental protection and social progress* by developing models to calculate the trade-off between competing objectives and by the use of integrating decision-making tools which consider the sustainability consequences of several actions.
- *Maintain international momentum for sustainability aspirations* by consolidating international protocols and agreements, reinforcing robust assessment systems, ensuring that sustainable development is grounded in all national policies and political processes, and by encouraging politicians to champion the sustainable development agenda.
- *Integrate sustainable development concepts into developing nations* by responding to their poor environmental problems, recognising that many problems such as pollution, drought, famine and floods are rarely self-inflicted, developing action programmes to reduce low life expectancy and eradicate high instances of diseases such as malaria, HIV/Aids, starvation and cholera, and by securing funding and support systems to deliver basic food, water, sanitation, and appropriate shelter to achieve poverty elimination.
- *Establish wholesale greening of government* across all departments by setting an example of how changes can be made to support sustainability, 'greening' all operations, setting performance targets and choosing environmentally friendly solutions, and developing staff awareness on issues such as energy use, water conservation, green procurement practices and economic travel plans.
- *Engage with business and industry* by developing corporate social responsibility, encouraging company sustainability (reporting on measurables such as waste, water, gas emissions and transport usage),

setting targets for financial savings, enforcing 'polluter pays' initiatives, reviewing and reducing consumption and production, identifying resource productivity which contributes to business success and company reputation, and by researching how innovation can create a competitive advantage over rivals.

- *Evaluate and develop holistic sustainable measurement systems* by establishing robust and communicable headline indicators and relevant national indicators, by linking target-setting to performance monitoring, and by encouraging good practice benchmarking.
- *Develop coherent and internationally acceptable sustainable development policies and practice* for a large range of themes, by addressing key challenging issues including:
  - *Urban development*, especially brown-field development and demolition waste recycling;
  - *Sustainable communities*, especially local community involvement in combating urban problems and changing attitudes and lifestyles;
  - *Sustainable agriculture*, especially crop over-production, the effects of pesticides and water shortage for irrigation;
  - *Sustainable fisheries*, especially conservation of the marine environment and sustaining the fishing industry as a means of employment;
  - *Sustainable forestry*, especially forest management, control of tree felling and tree replacement programmes;
  - *Sustainable transportation*, especially public transport improvements, alternatives to private transport, impacts of air travel, development of green travel plans, school transport development and attention to travel safety;
  - *Sustainable development awareness programmes*, especially information and communication strategies and the use of environmentally technological action plans.

10.4 Sustainable Development Policy and Practice Theme Index

Jurisdiction	Tier	Title	Date	Page No.
EUROPEAN	Council Communications	Strategy for Integrating Environmental Considerations into the Common Fisheries Policies *(COM (99) 363)*	1999	400
		Integrating the Environment into Air Transport *(COM (1999) 640)*	1999	400
		Sustainable Urban Development in the European Union: A Framework for Action	1999	400
		Integrating Sustainable Development into Community Co-operation Policy *(COM (2000) 264)*	2000	401
		Promoting Sustainable Development in the Non-Energy Extractive Industry *(COM (2000) 265)*	2000	401
		Integration of the Environment into Economic Policy *(COM (2000) 576)*	2000	401
		Sixth Environment Action Programme *(COM (2001) 31)*	2001	401
		Ten years after Rio: Preparing for the World Summit on Sustainable Development in 2002 *(COM (2001) 53)*	2001	402
		A sustainable Europe for a Better World: a European Union Strategy for Sustainable Development *(COM (2001) 264)*	2001	402
		The Lisbon Strategy – Making Change Happen *(COM (2002) 14)*	2002	402
		Towards a Global Partnership for Sustainable Development *(COM (2002) 82)*	2002	403
		Impact Assessment *(COM (2002) 276)*	2002	403
		Corporate Social Responsibility: A business contribution to sustainable development *(COM (2002) 347)*	2002	404
		Towards a Thematic Strategy on the Sustainable Use of Natural Resources *(COM (2003) 572)*	2003	404
		The World Summit on Sustainable Development One Year on: Implementing our Commitments *(COM (2003) 829)*	2003	404
		Stimulating Technologies for Sustainable Development: An Environmental Technologies Action Plan for the European Union *(COM (2004) 38)*	2004	405
		Towards a Thematic Strategy on the Urban Environment *(COM (2004) 60)*	2004	406
	Miscellaneous	Treaty of Amsterdam *(Treaty Agreement)*	1997	406
		Conclusions of the Council on integrating environment & sustainable development into the industry policy of the EU	2001	406
		Integration of the Environmental Dimension in Developing Countries *(Reg No. 2493/2000)*	2000	407
		Integrating Environment & SD into Energy & Transport Policies *(SEC (2001) 502)*	2001	407
		Consultation Paper for the preparation of a European Union Strategy for Sustainable Development *(SEC (2001) 517)*	2001	407
		Environmental Technology for Sustainable Development *(COM (2002) 122)*	2002	407

Jurisdiction	Tier	Title	Date	Page No.
EUROPEAN	Miscellaneous	Impact of Information & Communication Technologies (ICT) on Sustainable Development	2004	408
		National Sustainable Development Strategies in the European Union	2004	408
		Assessing the EU Sustainable Development Strategy – Exploratory Opinion	2004	408
		Approaches to Integrating Sustainability into Community Policies	2004	409
		Impact Assessment: Next Steps – In Support of Competitiveness & Sustainable Development *(SEC (2004) 1377)*	2004	409
UK	Strategies & Reports	Analysis of the Responses to the UK Government's Consultation Paper on Sustainable Development: 'Opportunities for Change' – Final Report	1998	410
		A Better Quality of Life: A Strategy for Sustainable Development for the UK	1999	410
		First Annual Report of the Green Ministers Committee	1999	410
		Quality of Life Counts	1999	411
		Increasing Sustainable Development Awareness Across Government: Preparing Strategies for Action	2000	411
		Achieving Sustainability: Poverty Elimination & the Environment – Strategies for Achieving the International Development Targets	2000	412
		DTI Sustainable Development Strategy – A Strategy for Simultaneously Securing Economic Growth & Environmental Protection	2000	412
		Second Annual Report of the Green Ministers Committee	2000	412
		Achieving a Better Quality of Life – Review of Progress Towards Sustainable Development (Government annual report 2000)	2001	413
		Greening Government Third Annual Report	2001	413
		Implementing Agenda 21 – Report of the Secretary General of the United Nations	2002	414
		The London Principles: Finance & Insurance for Sustainable Development – Interim Report	2002	414
		Achieving a Better Quality of Life – Review of Progress Towards Sustainable Development (Government annual report 2001)	2002	414
		Foundations for our Future – DEFRA's Sustainable Development Strategy	2002	415
		Framework for Sustainable Development on the Government Estate	2002	415
		The Government's Response to the Environmental Audit Committee Third Report (2001–02) – UK Preparations for the WSSD	2002	415
		Sustainable Development & the 2002 World Summit	2002	416
		UK Indicators of Sustainable Forestry	2002	416

Jurisdiction	Tier	Title	Date	Page No.
UK	Guidance Notes & Circulars	Guidance on Preparing Regional Sustainable Development Frameworks	2000	426
		Guidelines for Company Reporting on Waste	2000	426
		Value, Growth, Success – How Sustainable is Your Business (A briefing note for directors)	2000	426
		Environmental Reporting Guidelines for Company Reporting on Water	2001	427
		Guidelines for Company Reporting on Greenhouse Gas Emissions	2001	427
		Environmental Reporting – General Guidelines	2001	427
NORTHERN IRELAND	Strategies & Reports	EHS Corporate & Business Plan 2002/2005	2002	428
		Putting Sustainable Development at the Centre in Northern Ireland – A Response from the SDC to the Discussion Paper on Proposals for a Sustainable Development Strategy for Northern Ireland	2002	428
		RSPB's Manifesto for a Sustainable Northern Ireland	2003	429
		EHS Corporate & Business Plan 2003/2006	2003	429
	Consultation Papers	Promoting Sustainable Living – A Discussion Paper on Proposals for a Sustainable Development Strategy for NI	2002	429

10.5 Sustainable Development Policy and Practice Tables

Table 10.5.1 International Level Sustainable Development Policy and Practice.

Title	Venue	Purpose	Key Outcome(s)	Website
UN Conference on Environment & Development (UNCED) – 'Earth Summit'	Rio de Janeiro, Brazil (1992)	To organise a conference to take stock of the global environment twenty years after Stockholm. First official world summit on Sustainable Development, with the goal of establishing a new and equitable global partnership through the creation of new levels of co-operation among states, key sectors of society and people.	▪ The Rio declaration – statement of SD principles. ▪ Agenda 21 – SD programme of action (40 chapters). ▪ Statement of Forest Principles – non-binding agreement on development, preservation & management of the earth's remaining forests. ▪ Framework Convention on Climate Change – binding international agreement that seeks to limit/reduce emissions of gases associated with global warming. ▪ Biodiversity Treaty – binding international agreement aimed at strengthening national control & preservation of biological resources. ▪ The creation of a new Commission for Sustainable Development.	http://www.johannesburgsummit.org/html/documents/unceddocs.html

Global Conference on the Sustainable Development of Small Island Developing States	Bridgetown Barbados (1994)	The Global Conference on the Sustainable Development of Small Island Developing States translates Agenda 21 into specific policies, actions & measures to be taken at the national, regional & international levels to enable small island developing states to achieve sustainable development. This is the first global conference on sustainable development & the implementation of Agenda 21.	Key Outcomes included: ▪ The Declaration of Barbados ▪ The Programme of Action for the Sustainable Development of Small Island Developing States.	http://www.un.org/documents/ga/conf167/aconf167-9.htm
World Summit on Sustainable Development (Earth Summit II)	Johannesburg South Africa (2002)	The Johannesburg Summit 2002 was an opportunity to rejuvenate the quest to build a more sustainable future, with the intention of bringing the world together, and forging more cohesive global partnerships for the implementation of Agenda 21.	Governments agreed to and reaffirmed a wide range of concrete commitments and targets for action to achieve more effective implementation of sustainable development objectives. The concept of partnerships between governments, business and civil society was given a large boost by the Summit and the Johannesburg Plan of Implementation. Areas covered by conference key commitments were poverty eradication; water & sanitation; sustainable production & consumption; energy; chemicals; management of the natural resource base; corporate responsibility; health; sustainable development of small island developing states; and sustainable development for Africa.	http://www.johannesburgsummit.org/index.html

Table 10.5.2 European Level Sustainable Development Policy and Practice.

Policy Type	Programme Title	Policy Objective	Community Measure	Document Reference	Website
Council Decisions	Fifth European Community Environment Programme: Towards Sustainability	Presentation of the new Community Strategy on the Environment & the measures to be taken towards SD for the period 1992–2000.	Resolution of the Council & the representatives of the Governments of the Member States, meeting with the Council on a community programme of policy & action in relation to the environment & SD (01/02/93).	COM (95) 624 Off Jnl C, 17/05/93	http://europa.eu.int/scadplus/leg/en/lvb/l28062.htm
			Decision No. 2179/98/EC of the European Parliament & the Council on the review of the European Community Programme of policy & action in relation to the environment & SD 'towards sustainability' (24/09/98).	COM (99) 543 Off Jnl L, 10/10/98	
	Thematic Programme: Energy, Environment & Sustainable Development	Maintaining sustainable development to create new opportunities for citizens & companies, making the latter more efficient and less polluting. In other words, the objective is to reconcile economic growth & protection of the environment.	Council Decision 1999/170/EC adopting a specific programme for research, technological development and demonstration on energy, environment and sustainable development (1998 to 2002) (25/01/99).	C Dec 1999/175/ Euratom Off Jnl L64, 12/03/99	http://europa.eu.int/scadplus/leg/en/lvb/i23005.htm
			Council Decision 1999/175/ Euratom adopting a research and training programme (Euratom) in the field of nuclear energy (1998 to 2002).		
	Integrating the Environmental Dimension into the Sustainable Development of the Urban Environment	To encourage local authorities to rise to the challenge of environmental issues & to provide them with financial & technical assistance in working towards sustainability.	Decision No. 1411/2001/EC of the European Parliament & the Council on a Community Framework for co-operation to promote sustainable urban development (2001).	Off Jnl L191, 13/07/01	http://europa.eu.int/scadplus/leg/en/lvb/l28106.htm

	Laying Down the Sixth Community Environment Action Programme	The Programme aims at a high level of protection of the environment & human health & at a general improvement in the environment & quality of life. It indicates priorities for the environmental dimension of the Sustainable Development Strategy & should be taken into account when bringing forward actions under the Strategy.	Dec. No. 1600/2002/EC of the European Parliament & of the Council laying down the sixth community environment programme (22/07/02).	Off Jnl L242/1	http://europa.eu.int/eur-lex/pri/en/oj/dat/2002/l_242/l_24220020910en00010015.pdf
Council Communications	Environment & Employment	To devise a strategy to generate synergies between environmental & employment policy.	Communication from the Commission on the environment & employment (Building a Sustainable Europe) (18/11/97).	COM (97) 592	http://europa.eu.int/scadplus/leg/en/cha/c11709.htm
	A Strategy for Integrating the Environment into EU Policies	To introduce a horizontal approach to environmental policy by integrating it into all Community policies.	Communication from the Commission to the European Council – A partnership for integration: a strategy for integrating the environment into EU policies (27/05/98).	COM (98) 333	http://europa.eu.int/scadplus/leg/en/lvb/l28075.htm
	Integrating the Environment into Community Energy Policy	To propose measures for further integrating of environmental considerations in Community energy policy & to review progress made so far.	Communication from the Commission – Strengthening environmental integration within Community energy policy (14/10/98).	COM (1998) 571	http://europa.eu.int/scadplus/leg/en/lvb/l28071.htm
	Approaches to Sustainable Agriculture	To pursue the approach adopted by the Commission to the integration of the environmental aspects of agriculture into the reform of the Common Agricultural Policy (CAP) under Agenda 2000.	Communication from the Commission to the Council, the European Parliament, the Economic & Social Committee & the Committee of the Regions – Directions towards sustainable agriculture (27/01/99).	COM (99) 22	http://europa.eu.int/scadplus/leg/en/lvb/l28101.htm
	Strategy for Integrating the Environment into the Single Market	To develop the synergies between the single market & Community environment policy following the strategy formulated by the European Council in Vienna.	Communication from the Commission to the European Parliament & Council on the single market & environment (08/06/99).	COM (99) 263	http://europa.eu.int/scadplus/leg/en/lvb/l28090.htm

Continued

Table 10.5.2 European Level Sustainable Development Policy and Practice. *Continued*

Policy Type	Programme Title	Policy Objective	Community Measure	Document Reference	Website
Council Communications	Strategy for Integrating Environmental Considerations into the Common Fisheries Policies	To define priority objectives & implementing measures for improving the co-ordination & consistency of the environment policy & the common fisheries policy.	Commission Communication to the Council & the European Parliament on fisheries management & nature conservation in the marine environment (14/07/99).	COM (99) 363	http://europa.eu.int/scadplus/leg/en/lvb/l28094.htm
	Integrating the Environment into Air Transport	To improve the environmental performance of air transport activities so as to offset the environmental impact of growth in this sector.	Commission Communication to the Council, the European Parliament, the Economic & Social Committee & the Committee of the Regions on air transport & the environment (01/12/99).	COM (1999) 640	http://europa.eu.int/scadplus/leg/en/lvb/l28116.htm
	Sustainable Urban Development in the European Union: A Framework for Action	In this Framework for Action the Commission is taking a step towards increasing the effectiveness of EU policies provided for by the Treaty by making them more 'urban sensitive' & ensuring that they facilitate integrated urban development. The EU Framework for action for sustainable urban development aims at better co-ordinated & targeted community action for urban problems.	Communication from the Commission to the Council, the European Parliament, the Economic & Social Committee & the Committee of the Regions (28/10/99).	Commission Communication	http://europa.eu.int/comm/regional_policy/sources/docoffic/official/communic/pdf/caud/caud_en.pdf

Integrating Sustainable Development into Community Co-operation Policy	To define a strategy to ensure that the environment plays a key role in the EU's economic & development aid to partner countries, to enable those countries to assume their environmental responsibilities in the long term.	Communication from the Commission to the Council, the European Parliament & the Economic & Social Committee on integrating the environment & sustainable development into economic & development co-operation policy, elements of a comprehensive strategy (2000).	COM (2000) 264	http://europa.eu.int/scadplus/leg/en/lvb/l28114.htm
Promoting Sustainable Development in the Non-Energy Extractive Industry	To set broad policy lines for promoting sustainable development in the EU non-energy extractive industry, while reconciling the competitiveness of the industry with environmental protection.	Communication from the Commission promoting sustainable development in the EU non-energy extractive industry (2000).	COM (2000) 265	http://europa.eu.int/scadplus/leg/en/lvb/l28113.htm
Integration of the Environment into Economic Policy	To define a Community strategy to integrate environmental issues into economic policy by using existing instruments such as the Member States annual reports on structural reform & the broad economic policy guidelines.	Communication from the Commission to the Council & the European Parliament on bringing needs & responsibilities together – integrating environmental issues & economic policy (2000).	COM (2000) 576	http://europa.eu.int/scadplus/leg/en/lvb/l23018.htm
Sixth Environment Action Programme	To define the priorities & objectives of Community environmental policy up to 2010 & beyond, & to describe the measures to be taken to help implement the European Union's SD strategy.	Communication from the Commission to the Council, European Parliament, the Economic & Social Committee & the Committee of the Regions on the Sixth Environmental Action Programme of the EC – 'Environment 2010: Our future, our choice' (24/01/01).	COM (2001) 31	http://europa.eu.int/scadplus/leg/en/lvb/l28027.htm

Continued

Table 10.5.2 European Level Sustainable Development Policy and Practice. *Continued*

Policy Type	Programme Title	Policy Objective	Community Measure	Document Reference	Website
Council Communications	Ten years after Rio: Preparing for the World Summit on Sustainable Development in 2002	With a view to ensure an effective EU contribution to the 2002 Summit, this Communication aims to set in motion a co-ordinated EU preparatory process, by: ■ Suggesting the EU's strategic objectives & key issues for the Summit; ■ Highlighting synergies with related processes, (notably the EU sustainable development strategy, the forthcoming Environment Action Programme, the Cardiff Integration process and other related sectoral strategies (e.g. energy & transport)) as a basis for a credible & decisive EU contribution to the Summit; ■ Indicating preparatory measures to take in the lead-up to the Summit.	Communication from the Commission to the Council & European Parliament – Ten years after Rio: Preparing for the World Summit on Sustainable Development in 2002 (06/02/01).	■ COM (2001) 53	http://europa.eu.int/eur-lex/en/com/cnc/2001/com2001_0053en01.pdf
	Strategy for Sustainable Development	To establish a long-term strategy to dovetail policies for economically, socially & environmentally sustainable development.	Communication from the Commission – A sustainable Europe for a Better World: a European Union Strategy for Sustainable Development (15/05/01).	■ COM (2001) 264	http://europa.eu.int/eur-lex/en/com/cnc/2001/com2001_0264en01.pdf
	The Lisbon Strategy – Making Change Happen	In Barcelona, the European Council will assess the progress that has been made. It should consolidate successes, highlight weaknesses & set priorities for the next twelve to eighteen months. It will do this at	Communication from the Commission to the Spring European Council in Barcelona – The Lisbon Strategy: Making Change Happen (15/01/02).	■ COM (2002) 14	http://europa.eu.int/comm/barcelona_council/14_en.pdf

	an important moment for the Lisbon strategy: ▪ The Commission has by now tabled most of the main policy proposals. ▪ The second phase of agreeing & adopting those policies is well underway. ▪ The final phase, where agreed policies are implemented & start to have an impact on the ground, has barely begun.			
Global Partnership for Sustainable Development	To highlight how the European Union contributes to global sustainable development and the action aimed at establishing a global deal for sustainable development.	Communication from the Commission to the European Parliament, the Council, the Economic and Social Committee and the Committee of the Regions – Towards a Global Partnership for Sustainable Development (2002).	▪ COM (2002) 82	http://europa.eu.int/scadplus/leg/en/lvb/l28015.htm
Impact Assessment	The Commission intends to launch Impact Assessment as a tool to improve the quality & coherence of the policy development process. It will contribute to an effective & efficient regulatory environment & to a more coherent implementation of the European strategy for Sustainable Development. Impact Assessment identifies the likely positive & negative impacts of proposed policy actions, enabling informed political judgements to be made and identifying trade-offs in achieving competing objectives. It also permits application of the subsidiarity & proportionality protocol annexed to the Amsterdam Treaty.	Communication from the Commission (05/06/02).	▪ COM (2002) 276	http://europa.eu.int/comm/sustainable/docs/2002_08492_impact_assessment_en.pdf

Continued

Table 10.5.2 European Level Sustainable Development Policy and Practice. *Continued*

Policy Type	Programme Title	Policy Objective	Community Measure	Document Reference	Website
Council Communications	Corporate Social Responsibility: A business contribution to sustainable development	In the present Communication, which constitutes a follow-up to previous year's Green paper, the Commission presents a EU strategy to promote CSR, its place in the sustainable development context & its impact on business & society.	Communication from the Commission concerning corporate social responsibility: a business contribution to sustainable development (02/07/02).	COM (2002) 347	http://europa.eu.int/comm/employment_social/soc-dial/csr/csr2002_en.pdf
	Towards a Thematic Strategy on the Sustainable Use of Natural Resources	This Communication is a first step towards the Thematic Strategy on the Sustainable Use of Natural Resources (Resources Strategy), called for in the EU's Sixth Environment Action Programme. After analysing the environmental issues associated with the use of natural resources, it outlines the main features that a future strategy should comprise, building on existing policies. Although it sets out basic ideas on how the EU should target its efforts to reduce the environmental impacts of resource use, it does not actually propose specific measures to this end. This will be done in the final strategy to be presented in 2004.	Communication from the Commission to the Council & the European Parliament – Towards a Thematic Strategy on the Sustainable use of Natural Resources (01/10/03).	COM (2003) 572	http://europa.eu.int/eur-lex/en/com/cnc/2003/com2003_0572en01.pdf
	The World Summit on Sustainable Development One Year on: Implementing our Commitments	Without prejudging the review of the Sustainable Development Strategy & the cross-sectoral headline objectives of this strategy, this communication takes stock of progress achieved so far in implementing the WSSD	Communication from the Commission to the Council & the European Parliament on the World Summit on Sustainable Development One Year on: Implementing Our Commitments (23/12/03).	COM (2003) 829	http://europa.eu.int/eur-lex/en/com/cnc/2003/com2003_0829en01.pdf

	commitments & outlines the actions undertaken by the EU, both internally and externally in this regard.			
Stimulating Technologies for Sustainable Development: An Environmental Technologies Action Plan for the European Union	The Environmental Technologies Action Plan (ETAP) aims to reduce pressures on our natural resources, improve the quality of life of European citizens & stimulate economic growth. It is an important means to implement the EU Sustainable Development Strategy & to pursue the Lisbon Strategy, while also helping developing countries. It is based on the significant untapped technological potential for improving the environment while contributing to competitiveness & growth. The Action Plan's objectives are: ▪ to remove the obstacles so as to tap the full potential of environmental technologies for protecting the environment while contributing to competitiveness & economic growth; ▪ to ensure that over the coming years the EU takes a leading role in developing & applying environmental technologies; ▪ to mobilise all stakeholders in support of these objectives.	Communication from the Commission to the Council & the European Parliament – Stimulating technologies for sustainable development: An Environmental Technologies Action Plan for the European Union (28/01/04).	▪ COM (2004) 38	http://europa.eu.int/eur-lex/en/com/cnc/2004/com2004_0038en01.pdf

Continued

Table 10.5.2 European Level Sustainable Development Policy and Practice. *Continued*

Policy Type	Programme Title	Policy Objective	Community Measure	Document Reference	Website
Council Communications	Towards a Thematic Strategy on the Urban Environment	A high level of environmental protection is one of the key elements to achieving the sustainable development of cities and to providing a high quality of life for Europe's urban citizens. This Strategy builds on a series of initiatives that have contributed to the development of Europe's policy on the urban environment (Annex 1). It is part of the European Community's Sixth Environment Action Programme 'Environment 2010: Our Future, Our Choice', and is one of seven thematic strategies in the Programme introduced to provide a holistic approach to key environmental issues that are characterised by their complexity, the diversity of actors concerned & the need for innovative & multiple solutions.	Communication from the Commission to the Council, the European Parliament, the Economic & Social Committee & the Committee of the Regions – Towards a Thematic Strategy on the Urban Environment (11/02/04).	COM (2004) 60	http://europa.eu.int/eur-lex/en/com/cnc/2004/com2004_0060en01.pdf
Miscellaneous	Treaty of Amsterdam	Enshrines the principle of Sustainable Development as one of the European Community's aims & makes a high degree of environmental protection one of its absolute priorities.	Treaty agreement (17/06/97).	Treaty Agreement	http://europa.eu.int/scadplus/leg/en/lvb/a15000.htm
	Strategy for Integrating the Environment into Industry	To extend the strategy for integrating environmental policy into European Union policies in the industrial sector.	Conclusions of the Council on integrating environment & sustainable development into the industry policy of the EU (29/04/99).	Conclusions of the Council (29/04/99)	http://europa.eu.int/scadplus/leg/en/lvb/l28093.htm

Integration of the Environmental Dimension in Developing Countries	To contribute to the effective integration of the environmental dimension in measures of co-operation between the Community & Developing Countries.	Regulation (EC) No. 2493/2000 of the European Parliament & of the Council on measures to promote the full integration of the environmental dimension in the development process of developing countries (07/11/00).	Regulation (EC) No. 2493/2000 Off Jnl L288, 15/11/00	http://europa.eu.int/scadplus/leg/en/lvb/l28086.htm
Integrating Environment & SD into Energy & Transport Policies	To evaluate recent energy & transport market trends, describing policy initiatives & outlining future actions that contribute to SD.	Commission Working Paper (21/03/01).	SEC (2001) 502	http://europa.eu.int/comm/energy_transport/library/integr_report_en.pdf
Consultation Paper for the preparation of a European Union Strategy for Sustainable Development	This consultation paper is designed to provide the analytical underpinnings for this strategy. It sets out initial views on the challenges & opportunities of sustainable development. It identifies some important trends that pose a threat to sustainable development in the EU, & presents a policy toolkit for tackling these problems. This consultation paper does not include specific objectives & measures. These will be contained in the Commission's proposal for a sustainable development strategy to the Gothenburg European Council.	SEC (2001) 517 – Commission staff working paper (27/03/01).	SEC (2001) 517	http://europa.eu.int/comm/environment/eussd/consultation_paper_en.pdf
Environmental Technology for Sustainable Development	This Communication, a review of the contribution that the environment technology sector can make to promoting growth & employment, is a response to commitments made in 2001. It sets the scene for the development of a future Action Plan. It has to be seen in the context of the strategic goals set for the EU at the recent Lisbon & Gothenburg European Councils.	Report from the Commission (13/03/02).	COM (2002) 122	http://europa.eu.int/eur-lex/en/com/rpt/2002/com2002_0122en01.pdf

Continued

Table 10.5.2 European Level Sustainable Development Policy and Practice. *Continued*

Policy Type	Programme Title	Policy Objective	Community Measure	Document Reference	Website
Miscellaneous	Impact of Information & Communication Technologies (ICT) on Sustainable Development	The purpose of this report is to find out in what areas ICT could best contribute to sustainable development in Europe & what actions should be carried out in order to promote a positive impact, given that recent research shows that use of ICT has many negative effects that have to be balanced with the advantages it provides.	EC Report (Feb 2004).	EC Report	http://europa.eu.int/information_society/topics/ework/information/our_publications/documents/publications/ImpactOfIct.pdf
	National Sustainable Development Strategies in the European Union	This review of the 15 Member States & the 10 acceding countries aims to contribute to the fulfilment of the recommendations of Earth Summits I & II that governments draw up national sustainable development strategies (NSDS) & begin their implementation by 2005. It is also an input to the forthcoming 2004 review of the European Union's own Sustainable Development Strategy (EU SDS).	Commission Staff Working Document – National Sustainable Development Strategies in the European Union: A first analysis by the European Commission (April 2004).	Commission Working Document	http://europa.eu.int/comm/sustainable/docs/sustainable_development_strategies.pdf
	Assessing the EU Sustainable Development Strategy – Exploratory Opinion	Efforts have been under way for many years to promote sustainable development in the EU & curb non-sustainable trends. In a bid to consolidate & step up these efforts, a sustainable development strategy was adopted for the Union at the EU summit in Gothenburg. However, the most recent studies by European Commission indicate that efforts so far remain inadequate & that, on this front, Europe continues to face	Opinion of the European Economic & Social Committee on Assessing the EU sustainable development strategy – exploratory opinion (28/04/04).	NAT/229	http://europa.eu.int/comm/sustainable/docs/ces661-2004_ac_en.pdf

	major challenges. This exploratory opinion, drawn up by the Committee at the Commission's request, examines the range of problems facing the EU on the road towards sustainable development & considers how the EU should strengthen its sustainable development strategy.			
Approaches to Integrating Sustainability into Community Policies	This report summarises the results of the study to evaluate approaches to integrating sustainability into Community policies. The study aimed to identify problems most frequently encountered with a view to identifying relevant recommendations for future action. The study was launched shortly after the adoption of the EU strategy for sustainable development in Gothenburg in 2001. Thus, this timing did not provide for an assessment that was based on formally established mechanisms to realise the Gothenburg strategy.	European Commission Secretariat General – Evaluation of Approaches to Integrating Sustainability into Community Policies: Final summary report (Sept 2004).	European Commission Secretariat General Report	http://europa.eu.int/comm/sustainable/docs/Report_Evaluation_of_Approaches.pdf
Impact Assessment: Next Steps – In Support of Competitiveness & Sustainable Development	The Commission's new Impact Assessment procedure cuts across all sectors & provides policy-makers with a better & more coherent analysis of all relevant impacts across the various policy dimensions. This report summarises the Commission's assessment of how the Impact Assessment process is working & outlines the steps that the Commission intends to take in order to improve current practice.	Commission staff working paper – Impact Assessment: Next Steps – In support of competitiveness and sustainable development (21/10/04).	SEC (2004) 1377	http://europa.eu.int/comm/secretariat_general/impact/docs/progress_report_council_sec_1377.pdf

Table 10.5.3 UK Level Sustainable Development Policy and Practice.

Policy Type	Document Title	Objective	Contents	Website
Strategies & Reports	Analysis of the Responses to the UK Government's Consultation Paper on Sustainable Development: 'Opportunities for Change' – Final Report (July 1998)	This report covers the main issues and ideas raised as a result of a rapid review of some 950 of the written responses to 'Opportunities for Change', the consultation paper on a revised UK strategy for sustainable development.	Preface; Summary; Introduction; Broad issues; Government's approach to sustainable development; Sustainable goals & services; Building sustainable communities; Managing the environment & resources; Sending the right signals; International co-operation & development; Annex A: Abbreviations.	http://www.sustainable-development.gov.uk/uk_strategy/quality/analysis/index.htm
	A Better Quality of Life: A Strategy for Sustainable Development for the UK (May 1999)	This Strategy is a catalyst for change. It identifies priority areas for action, and indicators and targets to measure progress. It sets out action that the Government has already taken and further initiatives that are planned, and highlights what others can do. The Government will use the Strategy as a framework to guide its policies. It will encourage others to do the same.	Foreword; Summary; Chapter 1: The need for change; Chapter 2: Producing a strategy; Chapter 3: Progress & priorities; Chapter 4: Guiding principles & approaches; Chapter 5: Sending the right signals; Chapter 6: A sustainable economy; Chapter 7: Building sustainable communities; Chapter 8: Managing the environment & resources; Chapter 9: International co-operation & development; Chapter 10: Action & future reporting.	http://www.sustainable-development.gov.uk/uk_strategy/content.htm
	First Annual Report of the Green Ministers Committee (Sept 1999)	This report clarified the role of Green Ministers, explained the actions being taken & aimed to make the whole process more transparent. The establishment of a Parliamentary Environmental Audit Committee was one of the early steps taken to make sure that the promise of putting the environment at the heart of decision-making was kept. It exists to scrutinise the environmental performance of all departments and their associate bodies, together with their contribution to the wider goal of sustainable development. This provides the baseline for government-wide action, although individual departments will want (& will be encouraged) to go further wherever possible.	Foreword; Executive summary; Setting the scene – the role & activities of green ministers; The systems; Better policy appraisals; Environmental management systems (EMSs); Greening operations; The wider agenda; Annexes.	http://www.sustainable-development.gov.uk/sdig/reports/ar9899/index.htm

Quality of Life Counts (Dec 1999)	The sustainable development strategy included 14 'headline' indicators, to give a broad overview of trends; and outlined a national set of about 150 indicators, to focus on specific issues and identify areas for action. This report provided a baseline assessment, looking in more detail at those indicators and providing a benchmark against which future progress could be measured. Most of the indicators include data up to 1997 or 1998, prior to the publication of A Better Quality of Life. Annex D considers the DPSIR framework for assessing vulnerability to water shortage – the causal chain of 'driving forces', 'pressures', 'states', 'impacts' and 'responses'.	Foreword; Summary; Chapter 1: Indicators & the sustainable development strategy; Chapter 2: Criteria for sustainable development – the framework & models; Chapter 3: The headline indicators – assessing the implications for sustainable development; Chapter 4: The core set of indicators; Chapter 5: Further analysis of the indicators: integration & uncoupling; Chapter 6: International, national, regional & local indicators; Chapter 7: Consultation to develop the indicators; Chapter 8: Future work, in the UK, International activities; Packages – indicators by sector: Agriculture, business, individuals, international, resource, transport & water; Annex A: Complete list of objectives & indicators within the framework of the sustainable development strategy; Annex B; Index of all the indicators by keyword; Annex C: Glossary of abbreviations, acronyms & symbols used; Annex D: DPSIR, EEA (European Environment Agency) typology & comparison with the 1996 core set of indicators; Annex E: Bibliography & websites; Annex F: Fast track to individual indicators & to indicator families.	http://www.sustainable-development.gov.uk/sustainable/quality99/content.htm
Increasing Sustainable Development Awareness Across Government: Preparing Strategies for Action (Mar 2000)	Green Ministers made a commitment in their first annual report to have strategies to raise awareness of sustainable development in place by 31 March 2000. This leaflet, which was prepared with the help of the Sustainable Development Education Panel, aims to help them achieve that.	What is sustainable development? How can my department contribute? Planning: the aim; Delivery: How can we achieve this? Keep staff informed; Make it relevant; Get everyone involved; Evaluation: monitor your progress; Publications; Further advice; Environmental management systems (EMS); Investors in People (IIP); Wider campaigns, initiatives & resources.	http://www.sustainable-development.gov.uk/sdig/promoting/increase/pdf/increase.pdf

Continued

Table 10.5.3 UK Level Sustainable Development Policy and Practice. *Continued*

Policy Type	Document Title	Objective	Contents	Website
Strategies & Reports	Achieving Sustainability: Poverty Elimination & the Environment – Strategies for Achieving the International Development Targets (01/10/00).	This paper is about the environment & its contribution to sustainable development. It looks at the critical need to integrate the environment into all aspects of development. It concludes that protection of the environment is not a technical issue, it is about establishing truly participatory systems of governance, particularly giving an effective voice to the poor & marginalised.	Foreword; Executive summary; Target statement; Present position [illegible] the challenge; Experience to date; Meeting the challenge; Priorities for DfID; Monitoring progress; Annex 1: Country examples of strategic action for sustainable development; Annex 2: International environmental agreements; Annex 3: Global & regional indicators of development progress for the international development targets.	http://www.dfid.gov.uk/Pubs/files/tspenvironment.pdf
	DTI Sustainable Development Strategy – A Strategy for Simultaneously Securing Economic Growth & Environmental Protection (Oct 2000)	This paper sets out top level priorities, key objectives and outcomes for: ▪ Improving resource productivity (one of the key sustainability challenges facing business) ▪ Reducing greenhouse gas emissions ▪ Reducing waste generation ▪ Encouraging greater corporate social responsibility (one of the key business means of moving towards sustainability).	Foreword; Introduction; Sustainable development; The strategy. Focus; Priorities; Improving resource productivity – Reducing greenhouse gas emissions; Reducing waste generation; Encouraging corporate social responsibility; Modernising DTI; Reporting & monitoring progress.	http://www.dti.gov.uk/sustainability/strategy/docs/strategy2.pdf
	Second Annual Report of the Green Ministers Committee (Nov 2000)	Increased information on the progress by government – collectively and at individual department level – against its commitment to integrate sustainable development and environmental concerns into all that it does. The work programme at the end of the report sets out the scope of activities across policy making, awareness raising and management of the government estate. It shows progress against existing targets and there is the addition of new commitments to underline greening government as an on-going process. The work programme also indicates a continued focus on increasingly broader	Introduction; Background; The challenge of sustainable development; Training & raising awareness; Better policy-making; Environmental management systems; Greening operations; The wider public sector; Annexes.	http://www.sustainable-development.gov.uk/sdig/reports/ar2000/index.htm

	sustainable development issues while not neglecting the objectives for greener government operations.		
Achieving a Better Quality of Life – Review of Progress Towards Sustainable Development (Government Annual Report 2000) (Jan 2001)	This report was the first in a series intended to meet the commitment made in *A Better Quality of Life – a Strategy for Sustainable Development for the UK* to review progress by the country as a whole towards sustainable development & report annually from 2000. The report reflected & developed thinking on sustainable development, by Government & others, & discussed the objectives, principles & approaches which define and underpin the concept. It identified priorities & highlighted practical action which contributes to sustainable development at home & abroad. The report also described progress against each of the 15 headline indicators.	Foreword; Introduction; Strategic direction; Principles, approaches & signals; Indicators; Priorities; Monitoring & reviewing.	http://www.sustainable-development.gov.uk/ann_rep/susdevel.pdf
Greening Government Third Annual Report (Nov 2001)	This report builds upon the previous two reports by the Green Ministers Committee, offering more detail than in previous years. It looks back at performance over the last Parliament and the substantial progress made in a number of areas: ■ Greater understanding of sustainable development across Government was an early priority for Green Ministers and 18 departments now have strategies to raise awareness of sustainable development amongst their staff. ■ Travel plans are another area where Government has made considerable progress with plans now in place for virtually all major sites across the Estate. ■ Biodiversity issues were an early priority for Green Ministers and there are now many good examples of biodiversity considerations being built into projects on the Government Estate.	Introduction; Background; Leadership & communication; Better policy making; Environmental management systems; Greening operations; Wider public sector; Conclusions & future priorities; Annexes.	http://www.sustainable-development.gov.uk/gm2001/part1/index.htm

Continued

Table 10.5.3 UK Level Sustainable Development Policy and Practice. *Continued*

Policy Type	Document Title	Objective	Contents	Website
Strategies & Reports	Implementing Agenda 21 – Report of the Secretary General of the United Nations (Feb 2002)	The purpose of this report is to review progress in implementing Agenda 21 and other UNCED outcomes, to provide an overview of key developments that affect implementation, and to suggest policies and programmes for strengthening the implementation of the outcomes of the Rio Summit and related processes. However the full scope of the work done can only be appreciated by reference to all reports submitted during the course of the preparatory process, particularly the Country Profiles which contain more detailed information on country initiatives.	Introduction; Major trends & developments since UNCED; Combating poverty & promoting sustainable livelihoods; Sustainable consumption & production; Protecting the integrity of life-supporting ecosystems; Institutional framework for sustainable development; The need for policy integration, knowledge-based decision-making & participation; Means of implementation; Strengthening implementation – global partnerships.	http://www.sustainable-development.gov.uk/wssd2/a21rsg/index.htm
	The London Principles: Finance & Insurance for Sustainable Development – Interim Report (22/02/02)	This interim report sets out the current position of a project to develop a set of broad principles and recommendations for better practice in 'sustainable finance'. The project, commissioned and chaired by the Corporation of London in partnership with DEFRA Environmental Protection International, is based on detailed examples of best practice and the recent experiences of UK financial institutions.	Introduction; The London principles; Frequently asked questions: What is sustainable finance? Which principles apply to which business area? Why another set of principles? The approach taken by this project; Good practice sustainable finance in the UK; Pricing assets & exercising responsibility; Supplying new finance; Providing risk management; Products; Regulatory framework; Workshop comments & responses; Workshop acceptance list; Institutions & individuals consulted; A view of financial services & sustainable development.	http://www.sustainable-development.gov.uk/wssd2/lp/
	Achieving a Better Quality of Life – Review of Progress Towards Sustainable Development (Government	This is the second annual report in the series. As with the previous report, it reflects & develops thinking on sustainable development, by Government & others, & discusses the objectives, principles & approaches which define & underpin the concept. It identifies	Foreword; Introduction; Strategic direction; Principles, approaches & signals; Indicators; Priorities; World Summit on Sustainable Development; Monitoring & reviewing.	http://www.sustainable-development.gov.uk/ar2001/pdf/ar2001.pdf

annual report 2001) (Mar 2002)	priorities & highlights practical action contributing to sustainable development at home & abroad, & describes progress against each of the 15 headline indicators.		
Foundations for our Future – DEFRA's Sustainable Development Strategy (18/06/02)	This document sets out the intention to make sustainable development fundamental to DEFRA & embedded in everything it does. It identifies a set of indicators by which to measure progress, together with a mechanism for regular review & reporting.	Foreword; Chapter 1: New department, new opportunity; Chapter 2: New thinking, better solutions; Chapter 3: Indicators of progress; Chapter 4: DEFRA's own impacts; Chapter 5: Reviewing progress.	http://www.defra.gov.uk/corporate/sdstrategy/sdstrategy.pdf
Framework for Sustainable Development on the Government Estate (23/07/02)	This is the main vehicle for improving the performance of the Government Estate and will eventually cover all key environmental and social impacts of the running of departments. The Framework also contains guidance to support Departments in achieving targets. It sets out a new approach to managing and reporting the key sustainable development impacts of the operation of the Government Estate, and a common agenda and priorities, agreed by all Departments.	Part A: Overarching statement; Part B: Travel; Part C: Water; Part D: Waste; Part E: Energy; Part F: Procurement; Part G: Estates management; Part H: Biodiversity; Part I: Social impacts.	http://www.sustainable-development.gov.uk/sdig/improving/index.htm
The Government's Response to the Environmental Audit Committee Third Report (2001–02) – UK Preparations for the WSSD (24/07/02)	An inquiry into UK preparations for the World Summit on Sustainable Development (WSSD), which helped to raise the profile of the Summit & provided an opportunity to highlight the work that has been done, both domestically & internationally, as well as identifying what further action needed to be taken before the Summit. It also provided an opportunity to critically assess what activity has taken place, across all departments, to meet the commitments agreed at the UN Conference on Environment & Development in Rio in 1992.	Introduction; Progress since Rio; The WSSD; A true SD summit; Developing a new global framework; Delivery; Financing the Rio commitments; Leadership; Facilitating the full participation of other nations; Developing the summit's agenda; The emerging global wish list; UK priorities for the summit; Influencing the EU agenda; UK progress; Reporting to the UN; Structures & processes; Generating enthusiasm & action at home; Presenting clear messages; Communications strategy; Conclusions.	http://www.sustainable-development.gov.uk/wssd2/cm5558/pdf/cm5558.pdf

Continued

Table 10.5.3 UK Level Sustainable Development Policy and Practice. *Continued*

Policy Type	Document Title	Objective	Contents	Website
Strategies & Reports	Sustainable Development & the 2002 World Summit (10/10/02)	This paper looks at the concept of sustainability & the action being taken by the UK to address some of the commitments made in 1992 at the UN conference on Environment & Development (Earth Summit).	Part 1: The concept of sustainability; Spaceship Earth – temperature control; Provision of food; Provision of water & air; Sustaining planet Earth – atmosphere; Soil; Water; Biodiversity; Part 2: Development; The need to develop; The nature of development; Part of a larger, mainstream process; Part 3: Sustainable development; Defining SD; Implementation of SD; Implementing SD in the UK; SD indicators; Part 4: The world summit on SD; Comment prior to the summit; The summit; Commentary on summit; Outcomes; A sustainable summit?	http://www.parliament.uk/commons/lib/research/rp2002/rp02-055.pdf
	UK Indicators of Sustainable Forestry (Oct 2002)	Indicators of sustainable forestry help measure the contribution of woodland towards quality of life, & the ability to sustain their contribution into the future. This document contains a set of UK indicators of sustainable forestry. The indicators mostly provide information about the present state, & trends over time, of woodlands & their management, rather than pressures or responses.	Foreword; Introduction; Indicators; A: Woodland; B: Biodiversity; C: Condition of forest & environment; D: Timber & other forest products; E: People & forests; F: Economic aspects; Acronyms; Bibliography; Websites.	http://www.forestry.gov.uk/website/pdf.nsf/pdf/fullsfi.pdf/$FILE/fullsfi.pdf
	International Financial Institutions: Enhancing their Role in Promoting Sustainable Development – Report by the Royal Institute for International Affairs & Forum for the Future (Oct 2002)	This is a report of a workshop commissioned by the UK's Department of Environment, Food and Rural Affairs (DEFRA) & Foreign & Commonwealth Office (FCO) in order to generate insights about the role of International Financial Institutions (IFIs) in promoting sustainable development. The workshop was intended to be primarily focused on the environmental dimension of sustainable development, with sessions on how the sustainable use of environmental resources could contribute to poverty reduction, & on	Executive summary; Chapter 1: Introduction; Chapter 2: An overview of the role of IFIs; Chapter 3: IFIs' role in promoting sustainability in selected areas – Poverty reduction; Energy; Water; Cleaner production; Chapter 4: Conclusions; Chapter 5: Participants; Glossary.	http://www.sustainable-development.gov.uk/wssd2/ifi/index.htm

	how IFIs could promote the sustainable use of energy & water resources & cleaner production processes. However, what emerged was a rich consideration of sustainable development in all its complexity, & how IFIs could act to promote it.		
Sustainable Development in Government – First Annual Report 2002 (Nov 2002)	This report provides information on progress over the last year across central government departments & their agencies. Progress on cross-government initiatives is reported as well as performance by individual departments. The overarching aim of the new Sustainable Development in Government (SDiG) Initiative is to broaden consideration of the economic, environmental & social impacts of Government departments through their policies & operations.	Foreword; Introduction; Integrating sustainable development into decision-making; Improving the performance of the Government Estate; Summary of environmental performance of the Government Estate; Social impacts – towards sustainability reporting; Promoting understanding of sustainable development; Conclusions & future priorities; Annexes.	http://www.sustainable-development.gov.uk/sdig/reports/ar2002/part1/pdf/ar2002.pdf
Reaching the Summit – Johannesburg, the UK & Sustainable Development (2002)	This booklet provides an overview of UK & international efforts to promote sustainable development, & outlines the UK's priorities for the World Summit.	The challenge ahead; In wildness is the preservation of the world; The G-word; Making a difference; So what's the UK doing about it? Johannesburg – what's it all about; UK priorities.	http://www.sustainable-development.gov.uk/wssd2/reaching/pdf/reaching.pdf
Realising the Value, Enhancing Business Success (Jan 2003)	This document takes those issues identified as key & prioritises them. Aimed at the senior management team, it describes a straightforward process of risk management. It is not a complete methodology for risk evaluation (many such tools already exist), rather it is a prompt for factoring emerging issues into existing risk management procedures. Such issues have historically fallen outside the remit of traditional risk management, but are having an increasing bottom line impact. A number of key current & emerging issues are listed in this document as a starting point for risk analysis.	Sustainability issues impact business; Guidance; Strategic review; What to do next? Evaluating risk/opportunity; References; Charting the impact of climate change through the value chain.	http://www.defra.gov.uk/environment/acbe/pubs/pdf/acbe-realisevalue.pdf

Continued

Table 10.5.3 UK Level Sustainable Development Policy and Practice. *Continued*

Policy Type	Document Title	Objective	Contents	Website
Strategies & Reports	Achieving a Better Quality of Life – Review of Progress Towards Sustainable Development (Government annual report 2002) (Feb 2003)	This report, the third in the series, sets out progress made in 2002. In addition to bringing together the latest data available up to the end of 2002 on the 15 headline indicators & highlighting the main strategic developments in 2002, it also contains a wide range of case-studies showing some of the cross-cutting action being carried out across the UK to further sustainable development.	Foreword; Introduction; Strategic direction; Guiding principles & approaches; Headline indicators; International co-operation & development.	http://www.sustainable-development.gov.uk/ar2002/pdf/ar2002.pdf
	Where next for the UK on Sustainable Development? Final Report (April 2003)	To assist the UK government in identifying priority areas for action, & initiating the review of the sustainable development strategy it was seen as beneficial to liaise with stakeholders at the earliest possible opportunity. The aims & objectives of the exercise were to secure initial input from a broad range of UK stakeholders & gain, where possible, consensus views on UK priorities/key challenges for sustainable development (taking into account the domestic follow-up action to WSSD) & the effectiveness & challenges of the current UK sustainable development strategy 'A Better Quality of Life'.	Introduction; The consultation process; Priority issues & key challenges of the UK government & stakeholders; 'A Better Quality of Life'; Where next?	http://www.sustainable-development.gov.uk/sd_strategy/final/pdf/2003-04-16_public_final_report.pdf
	Redefining Prosperity: Resource Productivity, Economic Growth & Sustainable Development (June 2003)	As far as the Sustainable Development Commission is concerned, sustainable development provides the best possible framework for redefining progress & redirecting economies to enable all people to meet their basic needs & improve their quality of life. This has to be done whilst	Introduction; The heart of the problem; Resource productivity; Sustainable consumption; Redefining prosperity; Recommendations; Annex 1: Redefining the debate; References.	http://www.sd-commission.org.uk/news/resource_download_search.php?attach_id=BWZ5VHT-3O2M0Y9-550DVUH-TL5P6U4

	simultaneously ensuring that the natural systems, resources & diversity upon which all depend are maintained & enhanced for everyone's benefit & for that of future generations. The main point of this publication is to stimulate debate, to get other branches of government to think about the issues, & to begin to put in place processes which should change the way economic growth & wellbeing is considered.		
Policies for Sustainable Consumption (Sept 2003)	In pursuit of its advocacy remit, the Sustainable Development Commission is currently seeking to influence government policy on sustainable consumption, specifically with regard to DEFRA's forthcoming Sustainable Consumption and Production strategy and other UK policy developments. The purpose of this document is to enhance that process. It aims to provide a guide to the complexity of the sustainable consumption debate, an overview of the extensive literatures on consumer behaviour and lifestyle change, and an analysis of the policy opportunities suggested by these literatures.	Executive summary; Introduction; The sustainable consumption debate; Understanding consumption; Changing consumption patterns; Policy opportunities; References.	http://www.sd-commission.org.uk/news/resource_download_search.php?attach_id=OA492PF-KAOQ5IH-EAICDK6-H10C8GA
Sustainable Development in Government – Second Annual Report 2003 (Nov 2003)	This SDiG Report is only available electronically; it provides an online report & searchable database of information relating to the management of the Government's land, buildings & staff. It covers data from the twenty central Government Departments & their executive agencies for the year April 2002–March 2003. More detailed information should now be covered by individual departmental reporting.	Foreword; Executive summary; Introduction; Part A: Overarching commitments – environmental management systems & public reporting; Part B: Travel; Part C: Water services; Part D: Waste; Part E: Energy; Part F: Procurement; Part G: Estates management; Part H: Biodiversity; Part I: Social impacts; Conclusions & future priorities; Glossary.	http://www.sustainable-development.gov.uk/sdig/reports/ar2003/index.htm

Continued

Table 10.5.3 UK Level Sustainable Development Policy and Practice. *Continued*

Policy Type	Document Title	Objective	Contents	Website
Strategies & Reports	Sustainability & Business Competitiveness – Measuring the Benefit for Business Competitive Advantage from Social Responsibility & Sustainability (Dec 2003)	The question of whether or not there is a strong business case for corporate sustainability was the central issue for the workshop commissioned by the DTI, which brought together experts from academia, business schools, think-tanks & business. The workshop drew on a wider group than is normal in discussion on sustainability and Corporate Social Responsibility (CSR). Instead researchers in business intangible value & experts in business strategy were invited to discuss the insights their work could bring to research into the contribution sustainability performance makes to business success.	Executive summary; Introduction; The importance of CSR for business competitiveness; Sustainability & the foundations of corporate success; Sustainability & intangible business value; Sustainability & competitive advantage; Corporate responsibility clusters & innovation; Working with companies on sustainable advantage; Measuring business intangible value; Sustainability & the capacity to innovate; Sustainability & reputation; Sustainability & key internal & external relationships; Sustainability & strategic assets; Conclusions; List of workshop participants; Glossary.	http://www.csr.gov.uk/pdf/e2businesscompetitiveness.pdf
	Assessment of Progress Against the Headline Indicators (Feb 2004)	The Sustainable Development Commission commissioned an independent analysis of the UK's progress against the 15 headline indicators of sustainable development adopted in 1999. This is the report. There was access to unpublished drafts of the 2004 annual update of these indicators, & also of the update of the 150-odd core indicators which had mostly not been updated since 1999. Extensive internet searches were also carried out for examples of different achievements & good practice from different parts of the UK and other countries.	Introduction; Summary – Progress on the individual headline indicators; Overall picture; Some broader issues; Future use of the indicators; Traffic and mobility patterns; Gender inequalities in employment in Europe; Eco-efficiency; Indicator assessments; Indicator H1: Economic Output; Indicator H2: Investment; Indicator H3: Employment; Indicator H4: Poverty and Social Exclusion; Indicator H5: Education; Indicator H6: Health; Indicator H7: Housing conditions; Indicator H8: Crime; Indicator H9: Climate Change; Indicator H10: Air Quality; Indicator H11: Road Traffic; Indicator H12: River Water Quality; Indicator H13: Wildlife; Indicator H14: Land Use; Indicator H15: Waste; References.	http://www.sd-commission.org.uk/news/resource_download.php?attach_id=6P5C3U5-K8XRKLT-JV2TBRC-LUQNXNV

Achieving a Better Quality of Life – Review of Progress Towards Sustainable Development: Government Annual Report 2003 (March 2004)	This is the UK Government's fourth & last annual report on progress towards sustainable development against the 1999 Sustainable Development Strategy. As well as covering some of the key developments during 2003, it also provides a stock-take & review of Government action & progress since publication of the Strategy; including checks against the 15 headline indicators of sustainable development. Alongside this report the full set of 147 indicators of sustainable development which provided a baseline measurement in 1999 are being updated.	Foreword; Introduction; Chapter 1 – Strategic direction; Chapter 2 – UK government action; Chapter 3 – The Devolved Administrations, & English Regional & Local Government; Chapter 4 – Indicators; Chapter 5 – Progress on achieving a sustainable economy; Chapter 6 – Progress on sustainable communities; Chapter 7 – Progress on managing the environment & resources; Chapter 8 – Progress on International co-operation & development.	http://www.sustainable-development.gov.uk/ar2003/pdf/ar2003.pdf
Shows Promise. But Must Try Harder (April 2004)	The Sustainable Development Commission has taken this opportunity to conduct its own review of the last five years, & now publishes this report as a commentary on the Government's own assessment, & as a contribution to defining priorities & objectives for the forthcoming review of the UK strategy & of the Scottish, Welsh & Northern Irish reviews.	Introduction; Chapter 1 – The commission's overall assessment; Chapter 2 – Progress against headline indicators; Chapter 3 – UK government action; Chapter 4 – The devolved administrations & English regional & local government; Chapter 5 – A sustainable economy; Chapter 6 – sustainable communities; Chapter 7 – Managing the environment & resources; Chapter 8 – Progress on international co-operation & development; Chapter 9 – Conclusions; Annex – SDC's principles for sustainable development.	http://www.sd-commission.org.uk/news/resource_download.php?attach_id=SMDOB79-7JQMZFE-1OHI1PA-UFRDL3I
A Selection of the UK Government's Indicators of Sustainable Development: Sustainable Development Indicators in Your Pocket 2004 (April 2004)	The aim of this booklet is to provide an easily accessible set of indicators covering a number of key sustainable development themes. It includes a variety of economic, social & environmental issues of everyday concern including health, housing, jobs, crime, education & our environment. It is hoped this handy-sized publication will be a useful reference to experts but also to others less familiar with the concept of sustainable development.	Foreword; Introduction; Indicators setting the scene; International context; A sustainable economy; Economic stability & competitiveness; Resource use & waste; Employment & education; Building sustainable communities; Poverty, health & housing; Crime & social cohesion; Travel & access; Managing the environment; Climate change & energy; Air, water & radioactive waste; Landscape & wildlife; Taking it on: the consultation to develop new UK sustainable development strategy; Enquiries about indicators or this publication.	http://www.sustainable-development.gov.uk/indicators/sdiyp/sdiyp04a4.pdf

Continued

Table 10.5.3 UK Level Sustainable Development Policy and Practice. *Continued*

Policy Type	Document Title	Objective	Contents	Website
Strategies & Reports	Corporate Social Responsibility: A Government Update (May 2004)	There is a need to encourage business practices that help to ensure successful & profitable companies & that also contribute fully to achieving sustainable development goals. In this report the aim is to provide a summary of progress against the set priorities as well as look ahead to future challenges. It is not intended to be an exhaustive review of all the activity that Government has undertaken or supported. Rather it aims to illustrate the breadth of the agenda & range of action that can be & is being taken.	Chapter 1 – Priorities, recent trends & future directions; Chapter 2 – Progress check: CSR delivering objectives; Chapter 3 – Government demonstrating CSR values: the framework for sustainable development on the government estate; Chapter 4 – Looking ahead: future priorities.	http://www.csr.gov.uk/pdf/dti_csr_final.pdf
	The Impact of Sustainable Development on Public Behaviour Report 1 of Desk Research (May 2004)	This first report represents a study of the research evidence available on the concept of Sustainable Development (SD) in relation to the general public & asked what impact communications on the concept of SD would have on public behaviour. The study was designed to provide communicators in government and civil society with the evidence base on which to develop effective Sustainable Development communications.	Background and Methodology; Key Findings; Public Awareness and Understanding of Sustainable Development; Driving Public Behaviour Change for Sustainability; Sustainability and Public Behaviours; Effective Applications of Sustainable Development; Recommendations; Appendix: List of Selected Sources.	http://www.sustainable-development.gov.uk/taking-it-on/pdf/desk-research1.pdf
	Driving Public Behaviours for Sustainable Lifestyles Report 2 of Desk Research (May 2004)	Where the first report asked what impact communications on the concept of SD would have on public behaviour, this second report approaches that question from the other end. Thus this report asks 'What behaviours could the public undertake which would deliver the objectives of Sustainable Development?' In attempting to answer this question, this report builds on an understanding of the concept of SD developed in the first report, & interprets that concept into a set of everyday behaviours which the public could adopt to achieve positive impacts in terms of sustainability.	Background & Methodology; Key Findings; The Behaviours in a Sustainable Lifestyle; Barriers to Behaviour Change for Sustainability; Drivers of Behaviour Change for Sustainability; Public Lifestyles, by Audience; Barriers and Drivers to the Behaviours in a Sustainable Lifestyle; Recommendations; Appendix: List of Selected Sources.	http://www.sustainable-development.gov.uk/taking-it-on/pdf/desk-research2.pdf

Environmental Audit – Eighth Report (July 2004)	This report focuses on the second SDiG (Sustainable Development in Government) annual report, published in November 2003, & covering the period from April 2002 to March 2003. The previous work on 'Greening Government' has been mainly based on an analysis of departmental data carried out internally. This year the National Audit Office (NAO) has carried out the analysis, & its memorandum is printed as an appendix. It is anticipated that NAO analysis of future SDiG reports & associated departmental data will become a regular feature & some specific areas have been identified.	Terms of reference; Conclusions & recommendations; The Sustainable Development in Government 2003 annual report; Greening operations; Conclusion; Introduction; SDiG Second annual report; A narrower scope; Reporting elsewhere; The role of ENV(G) (the Cabinet sub-committee of UK Green Ministers); The need for a structured approach; The framework; Departmental performance; Reporting on the Government Estate; Other NAO observations; Energy; Conclusions; Appendices.	http://www.publications.parliament.uk/pa/cm200304/cmselect/cmenvaud/881/88102.htm
Taking it on – Developing UK Sustainable Development Strategy Together Summary of Responses from Online General Access Consultation (Aug 2004)	This document summarises the responses received through the publicly accessible consultation. The purpose of this summary is to illustrate the wealth of knowledge & opinion contained in the 8149 responses received from 444 participants to the questions in 'Taking it on', & to encourage participants to read more of the responses.	Background; Introduction; Purpose of this document; Consultation questions & summary of responses; Chapter 1 – Setting the scene; Chapter 2 – The approach to a new strategy; Chapter 3 – What is sustainable development & how do we do it? Chapter 4 – Setting priorities; Chapter 5 – Climate change & energy; Chapter 6 – Sustainable consumption, production & use of natural resources; Chapter 7 – Environment & social justice; Chapter 8 – Helping communities to help themselves; Chapter 9 – Changing behaviour; Chapter 10 – Beyond the UK – sustainable development in Europe & internationally; Chapter 11 – Getting the structures right for leadership; Chapter 12 – The business contribution to sustainable development; Chapter 13 – Measuring our progress.	http://www.sustainable-development.gov.uk/taking-it-on/pdf/general.pdf

Continued

Table 10.5.3 UK Level Sustainable Development Policy and Practice. *Continued*

Policy Type	Document Title	Objective	Contents	Website
Strategies & Reports	Taking it On: Developing UK Sustainable Development Strategy Together – Overall Summary of Responses Report (Oct 2004)	This document sets out the results and analysis of the 'Taking it on' consultation to develop new UK sustainable development strategy. The consultation was launched by the UK Government, Scottish Executive, Welsh Assembly Government & the Northern Ireland Administration in April 2004, the deadline for comments was reached on the 31 July 2004.	Introduction; Executive summary; Guide to the document; The approach to a new strategy; What is sustainable development & how do we do it? Setting priorities; Climate change and energy; SCP (Sustainable Consumption and Production) & use of natural resources; Environment and social justice; Helping communities to help themselves; Changing behaviour; Beyond the UK – sustainable development in Europe & internationally; Getting the structures right for leadership; Business contribution to sustainable development; Measuring our progress; Devolved Administration responses; Sustainable Development Commission (SDC) initial submission; Summary of other events; Session two responses; Further information.	http://www.sustainable-development.gov.uk/taking-it-on/pdf/finalsummary.pdf
	Sustainable Development in Government: Third Annual Report 2004 (Nov 2004)	This Report presents information about the way in which Government Departments & their executive agencies manage their land, buildings & staff. It does not set out to provide comprehensive data on each Department's operational performance since such information is more appropriately provided at Departmental level. What it provides is a snapshot of the way in which Departments have performed in areas covered by the targets & commitments in the Framework for Sustainable Development on the Government Estate, for the period April 2003–March 2004.	Foreword; Introduction & scope; Part A: Overarching commitments; Part B: Travel; Part C: Water services; Part D: Waste; Part E: Energy; Part F: Procurement; Part G: Estates management & construction; Part H: Biodiversity; Part I: Social impacts; Glossary & abbreviations.	http://www.sustainable-development.gov.uk/sdig/reports/ar2004/index.htm
Consultation Papers	Sustainable Development: Opportunities for Change	This document sets out the Government's vision of sustainable development, and what is being done to put it into practice; explores what further action might be taken, by	Foreword; Introduction; What is sustainable development; Setting a framework; Achieving sustainable development; Keeping track of progress; Sustainable goods & services;	http://www.sustainable-development.gov.uk/sustainable/consult/pdf/oppsforchange199802.pdf

Consultation Paper on a Revised UK Strategy (Feb 1998)	everyone, to achieve sustainable development; and poses some questions on which comments and suggestions are invited.	Building sustainable communities; Managing the environment & resources; Sending the right signals; International co-operation & development; Seeking your views: the consultation process.	
Sustainability Counts (Nov 1998)	The consultation document promised that a vital part of the strategy would be devising indicators to keep track of progress, and reporting in a way that is clear, comprehensive and useful to a wide audience. A first set of sustainable development indicators was published in 1996. It comprised some 120 measures, and was widely welcomed. However, there was pressure for a much more limited set of indicators which would allow an overall picture to be seen of whether the development is sustainable. This document contains specific proposals on what the headline indicators might comprise.	Foreword; Introduction; Why have indicators? What is sustainable development? Links with other indicators; Selection of indicators; Aggregated indicators; Will the indicators be fixed? How often will the indicators be updated? Geographical coverage; Seeking views on the proposals; The proposed headline indicators; Annex 1: Expected release dates of the headline indicators of sustainable development; Annex 2: Background & technical information; References.	http://www.sustainable-development.gov.uk/sustainable/consult/pdf/SusCounts199802.pdf
Taking it on – Developing UK sustainable development strategy together: Consultation Document (April 2004)	This joint consultation paper by the United Kingdom Government, Scottish Executive, Welsh Assembly Government & the Northern Ireland Administration sought views on what should be in a new strategy for sustainable development for the UK, & what needed to be done and by whom to make it happen. For the first time the intention was to build a UK strategic framework for sustainable development to 2020. This would provide the background for separate strategies & plans in each administration to take forward action, & for action by others, an opportunity to kick-start the debate on the challenges faced in achieving sustainable development, & on what more needs to be done to meet them – as government in the UK, & more widely across society.	Foreword; Introduction; Setting the scene; The approach to a new strategy; What is sustainable development, and how do we do it? Setting priorities; Climate change and energy; Sustainable consumption, production, and use of natural resources; Environment and social justice; Helping communities to help themselves; Changing behaviour; Beyond the UK – sustainable development in Europe and internationally; Getting the structures right for leadership; The business contribution to sustainable development; Measuring our progress; Getting your views: the consultation process.	http://www.sustainable-development.gov.uk/taking-it-on/consult/document/pdf/consult.pdf

Continued

Table 10.5.3 UK Level Sustainable Development Policy and Practice. *Continued*

Policy Type	Document Title	Objective	Contents	Website
Guidance Notes	Guidance on Preparing Regional Sustainable Development Frameworks (10/05/00)	This guidance sets out the policy context and Government priorities for the preparation of regional sustainable development frameworks. In its strategy for sustainable development, *A Better Quality of Life*, the Government said that it wished to see frameworks in place in all English regions by the end of 2000. The Government envisages that regional sustainable development frameworks will be high level documents that set out a vision for sustainable development in each region, and the region's contribution to sustainable development at national level. In doing so frameworks should take a wide overview of regional activity and the regional impact of Government policy.	Summary; The Government's policy on regional sustainable development frameworks; What will be the outcome of producing frameworks?; Ownership; Consideration of other initiatives; Indicators; London issues; Annex 1: Menu of other government initiatives (by indicator); Annex 2: Indicator sources & contacts for statistical information; Annex 3: Contacts; Annex 4: Websites & further reading; Annex 5: Table of activity on frameworks.	http://www.defra.gov.uk/environment/sustainable/rsdf/guidance2000/index.htm
	Guidelines for Company Reporting on Waste (06/06/00)	The purpose of these Guidelines is to help businesses to: ▪ Measure all the waste they produce. ▪ Identify how their waste management could be improved. ▪ Set targets to reduce their waste or where this is not practicable to re-use, recycle, or recover their waste to the benefit of the company and the environment. ▪ Achieve savings and improve their performance. ▪ Report on their performance to meet current and future legislative requirements.	Why business waste is important; The waste reporting guidelines; How the waste reporting guidelines work; How to start reporting on waste; Useful information; Annex 1: Example of proforma for describing waste streams; Annex 2: Most significant waste streams by industrial sector; Annex 3: Typical container weights & waste conversion factors; Annex 4: Top 10 hit list of waste streams for review; Annex 5: Hazardous or special wastes; Response form.	http://www.defra.gov.uk/environment/envrp/waste/pdf/cowaste.pdf
	Value, Growth, Success – How Sustainable is Your Business (A briefing note for directors) (Dec 2000)	This note poses some key questions to a company board, the responses to which can help provide a valuable insight into a company's attitude to sustainable development.	Seeing the big picture; Sustainable development; Getting our priorities right.	http://www.defra.gov.uk/environment/acbe/pubs/directors/directors.pdf

Environmental Reporting Guidelines for Company Reporting on Water (23/01/01)	These guidelines are designed to help businesses: ▪ Measure their water usage. ▪ Review water costs, use and discharges and detect possible leaks. ▪ Set targets to reduce water & wastewater usage. ▪ Achieve savings & improve performance & competitiveness. ▪ Report on their performance to demonstrate to shareholders, consumers & other stakeholders that theirs is a well-run, environmentally responsible business & to build commitment within the business. ▪ Meet current and future legislative requirements.	Why reporting on water is important; The purpose & scope of these guidelines; How these guidelines work; Annex 1: Converting between systems & units; Annex 2: Constructing a water use inventory; Annex 3: The Mogden formula for trade effluent charges; Useful information; Response form.	http://www.defra.gov.uk/environment/envrp/water/pdf/water.pdf
Guidelines for Company Reporting on Greenhouse Gas Emissions (22/03/01)	The purpose of the Guidelines is to help businesses to: ▪ Measure and aggregate the emissions produced from energy use in commercial & industrial buildings; industrial processes, including refrigeration & air conditioning; & transport. ▪ Set targets to reduce emissions. ▪ Achieve savings & report on performance.	Why greenhouse gases are important to us all; Guidelines for company reporting on greenhouse gas emissions; Comparisons with other reporting requirements; How to start reporting on greenhouse gases; Annex 1: Fuel conversion factors; Annex 2: CHP – imports & exports; Annex 3: Electricity conversion factors 1990–97; Annex 4: Typical process emissions; Annex 5: Conversion factors for greenhouse gas process (including emissions from refrigerators & air conditioning systems); Annex 6: Transport conversion tables; Useful information.	http://www.defra.gov.uk/environment/envrp/gas/index.htm
Environmental Reporting – General Guidelines (22/11/01)	These guidelines aim to help by cutting through some of the confusion by providing an introduction to environmental reporting. They explain how to produce an environmental report, outline its main contents & suggest key indicators to report against. They also include a possible incremental approach to help organisations plan staged improvements in their reporting.	Foreword; Introduction; The reporting process; The contents of an environmental report; Environmental performance indicators; Other issues.	http://www.defra.gov.uk/environment/envrp/general/index.htm

Table 10.5.4 Northern Ireland Level Sustainable Development Policy and Practice.

Policy Type	Document Title	Objective	Contents	Website
Strategies & Reports	EHS Corporate & Business Plan 2002/2005 (Jan 2002)	The Environment and Heritage Service's Corporate & Business Plans outline the agenda for the next three years. They support the DoE's plans for the same period & the NI Executive's Programme for Government. The main themes include protecting & improving the quality of air, water & land; promoting sustainable waste management; conserving biodiversity; protecting historic monuments & buildings of special architectural or historic interest; & raising awareness & appreciation of the environment & heritage.	Introduction; Strategic direction, aims & values; Key performance measures; Policy environment – strategic analysis; Corporate strategy; EHS service structure; Business plan for year ending 31st March 2003; 2002/03 key performance targets & key management tasks; Resources; Efficiency; Key performance targets.	http://www.ehsni.gov.uk/pubs/publications/CPBP0205.pdf
	Putting Sustainable Development at the Centre in Northern Ireland – A Response from the SDC to the Discussion Paper on Proposals for a Sustainable Development Strategy for Northern Ireland (Oct 2002)	The Sustainable Development Commission welcomes the decision of the Northern Ireland Executive to develop a sustainable development strategy for the Province, & to issue a consultation document on it. The following response does not seek to address many of the specific questions raised in the document, which are for the people & authorities of Northern Ireland themselves to determine. Instead, it offers observations on what appear to the SDC to be the key issues of a sustainable development strategy for Northern Ireland.	Defining sustainable development; Principles of sustainable development; Putting sustainable development at the centre; Valuing nature; Fair shares; Polluter pays; Good governance; Adopting a precautionary approach.	http://www.sd-commission.org.uk/news/resource_download_search.php?attach_id=JV4MDRE-LTN7F5H-7J57TUF-ZYEZMUI

	RSPB's Manifesto for a Sustainable Northern Ireland (May 2003)	A recent survey of public opinion and attitudes conducted for the RSPB by Ulster Marketing Surveys found that 60 per cent of those interviewed named the countryside as the most important aspect of quality of life in Northern Ireland. The RSPB has set out Northern Ireland's key environmental concerns in this document.	Sustainable development; Education for sustainable development; Sustaining & enhancing biodiversity; Sustainable management of wetlands; A sustainable planning system; A sustainable vision for agriculture.	http://www.rspb.org.uk/Images/NIagenda_tcm5-44694.pdf
	EHS Corporate & Business Plan 2003/2006 (Nov 2003)	The Corporate Plan covers the period up to 31 March 2006 & outlines the main themes of EHS work. These include protecting & improving the quality of air, water & land; promoting sustainable waste management; conserving biodiversity; protecting historic monuments & buildings of special architectural or historic interest; & raising awareness & appreciation of the environment & heritage.	Foreword; Introduction; Aims & objectives; key performance measures; The policy context; Corporate strategy; Business plan; Key performance targets & key management tasks; Resources; Key performance targets; Annexes.	http://www.ehsni.gov.uk/pubs/publications/CorporateBusinessPlan0306.pdf
Consultation Papers	Promoting Sustainable Living – A Discussion Paper on Proposals for a Sustainable Development Strategy for NI (May 2002)	The aim of this document is to seek views on how best to achieve sustainable development in NI. This document sets out a definition of sustainable development; the main objectives & principles which follow from that definition; & a possible framework for achieving it in NI. This publication is the start of a process of discussion & debate which will be encouraged over the coming months across all sectors & at all levels in society.	Foreword; Introduction; What is sustainable development? How are we planning to do this? A framework for SD; The executive's commitment; A new way of taking decisions; Business & resource planning; Strategic policies & programmes; Indicators; Open reporting on progress; High level co-ordination; Local co-ordination; Sharing information & best practice; Conclusion; Annex A.	http://www.ehsni.gov.uk/pubs/publications/Promoting_Sustainable_Living.pdf

11 Transport

11.1 Transport and the Built Environment

Transport is part of a wider communications concept in which messages, people and goods are transported by a range of devices and methods. Within the context of the built environment, transport issues, at a basic level, are about safety of people and properties and, at a strategic level, are addressed against efficiency and effectiveness measures. At both levels, transport provision facilitates the dynamics of urban and rural development and is a key element in the quest for sustainable communities. Although the quality and quantity of transport systems are viewed as an indicator of the growth and development of nations and their wealth, there are underlying aspects of transport systems which go against a balanced approach to sustainable development.

Transport systems should meet the needs of business and communities as they provide an appropriate service across a full range of modes of travel including movements by sea, air, road, rail and by foot within publicly or privately funded organisations. The meteoric growth in travel resulting from increased wealth within developed nations and innovations in engineering and infrastructure developments has brought transport into sharp focus. Transport is now perceived as an essential for life and hence a burden on both the natural environment and on finances. A variety of approaches may be used to resolve the conflict between the growing freedom of some members of society with their associated transport independence, and the battle to protect the environment from transport emissions and excessive fossil fuel use. The production and application of balanced transport strategies is essential, while the use of fiscal restricting measures may help stimulate behavioural change. However of more importance is the need for a change in the psychological attitudes to the use of transport systems, ensuring that transport impacts at the heart of personal and corporate decision-making. There is strong resonance between this transport debate and humanity's need to balance conscience and choice in other areas such as water conservation, reduced energy consumption and waste minimisation.

The built environment has evolved because of the need for communities to live collectively in the economically beneficial units of cities, towns and villages, with the converse needs of rural communities driven by industries such as agriculture, fishing and tourism. Transport is central to the planning and development of both rural and urban zones, but the mode of delivery is usually dictated by historical provision, availability of finance and local travel and occupancy trends. The scale and scope of any transport system is dictated by the critical mass of people and industry in each conurbation. Economic factors strongly influence the provision of transport systems, with large towns and cities normally well served with a balance between public and private systems. Whereas rural communities, even in the developed world, suffer badly, frequently resulting in private transport dominating the scant provision of public transport modes in rural and even some urban areas. Social equity is not always fully considered, with the consequent need to have a coherent integrated transport vision and strategy.

The EC White Paper (2001) *European Transport Policy for 2010* sets a framework for national transportation policies by proposing to strike a balance between economic development and the quality and safety demands made by society resulting in a modern, sustainable transport system for 2010. A number of common themes have emerged in subsequent national strategies including:

- moves towards greater integration between transport and other government policies;
- greater integration between transport modes;
- a shift in emphasis away from the car towards more sustainable modes such as walking, cycling and public transport; and
- public and private sectors working together to boost investment.

Current transportation networks represent substantial assets to individuals, businesses and community interests, but these networks, and the services operating on them, need continuous management and financial support to ensure that they operate efficiently and are not allowed to deteriorate. Transport system integration has been shown to have benefits exceeding the basic requirements of efficient freight and people movement, as anecdotal evidence shows that environmental conditions, health, quality of life and physical fitness are all influenced by the need to interact with non-motorised transport. The use of air travel is important for both industry and tourism, but there are severe negative environmental impacts, while sea travel is less time efficient but much less environmentally damaging. This debate between use of power energy for travel with its polluting impacts and a need to sustain economic growth and business competitiveness resides at the heart of the quest for sustainable development and the right to maintain freedom of choice. Some heat can be released from this debate by the ingenuity of engineers and scientists who are engaged in the development of low-energy road, rail and air vehicles, with safety rather than cost, remaining paramount to the success of any new innovation.

11.2 Key Transport Elements

The elements of transportation networks, driven by legislation and strategies, comprise essential hardware, management systems, and long-term planning and maintenance facilitation. There are peripheral and equally relevant effects and considerations such as safety, environmental impact, social and community cohesion, transport economics, future planning, a vision for future developments and an appreciation of how transport networks impinge on issues such as health, energy consumption, tourism and general quality of business and community life. This complex mix is judged by a perceptive public who have high expectations for a high quality service and do not necessarily want to engage in peripheral matters.

Air travel has the most far-reaching implications as well as the greatest distances travelled. These include: the highest initial investment in both aircraft and airports; significant environmental and energy consumption impacts; and the highest international relevance. However it is rarely cited in transportation strategies as it is not viewed as an alternative to land-based modes. Similarly sea or ocean travel, by ferries, liners and hovercraft, offers considerable options for short-distance travel while supporting the tourism and freight industries. Sea travel has been developed into viable and economic transport business, weathering the storm of air travel competition.

Road networks continue to be the flagship of transportation systems by serving communities and business in both urban and rural settings. The road infrastructure has been designed and constructed by civil and highway engineers, who have sought efficiency and effectiveness through innovation, research and development. Despite financial pressures, road maintenance and construction has been sustained allowing roads of various types to service cars, freight vehicles, taxis and bus services, thereby contributing to transport integration. Rail networks have only been properly financed and developed in some inner cities through underground or light rail systems, or in the large trans-national high speed networks. Several nations attempt to support local rail networks, but these are unlikely to meet the demands of communities and industry against stiff competition from road freight travel or bus networks. Hence, there remains a need for a consolidated vision to help integrate all systems to increase service efficiency and to meet the long-term aspirations of sustainable development.

Cycle pathways and designated pedestrian zones can facilitate short journeys, contribute to the health and fitness agenda, reduce road traffic congestion, improve air quality, contribute to social cohesion, and minimise damage to the landscape and streetscape. However, these transport modes are not common, with the exception of some European cities which have a culture of cycling, and there needs to be a greater understanding of the psychological barriers to cycling and walking. There remains a compelling argument for children to walk to school to enhance health and fitness levels, reduce pollution levels and minimise potential road congestion.

The safety of people is paramount to any transport mode, and this has exercised the minds, finance and publicity campaigns of all professional transportation providers. These concerted efforts have involved road material improvement, better design, superior construction and maintenance, advances in vehicle development and in-depth research into human activity and the effects of drugs, sleep deprivation and excessive speeding. The EC Communication *Saving 20 000 Lives on our Roads – A Shared Responsibility (2003)* highlights the need for better understanding of road safety and the need for more research and development through case studies. There is evidence that road casualties are steadily reducing despite increased road transport volumes, but this statistic masks the fact that there are a high proportion of deaths in the under 15 year old bracket and a higher proportion of pedestrian deaths. Also the increasing popularity of motorcycling has resulted in a percentage increase in casualties from that sector.

Environmental damage and increases in energy consumption from powered-transport systems cannot be ignored, and in Europe this is being tackled at a fiscal level by increased levies on petrol and diesel fuels, while rewarding manufacturers who have developed efficient low-energy engines. Speed restrictions on motorway systems and in residential areas provide additional support, while integration of transport modes can further reduce pollution and excessive energy usage. These factors are closely related to health and climate issues with improved air quality, due to fewer harmful emissions, reducing child illness and reducing global warming effects. Similarly, individual health and fitness can be improved by greater access to walking and cycling.

There are correlations between access to public transport systems and social exclusion. Traditionally transport policy-makers have had concerns with car dependence and the journey to work, thereby ignoring poverty and the consequences of non-car ownership in terms of access to new employment centres, retail and educational facilities. The link between transport and social exclusion is further demonstrated by other key factors such as differences in education and training opportunities, socio-economic circumstances, a restrictive local environment and poor access to essential services like shopping, recreation, culture, employment and personal development.

11.3 Sustainability Challenges for Transport

Transport networks and operations are perceived as a service to communities and industry rather than a fundamental service provision such as water or energy. However, the efficient provision of integrated transport systems will greatly enhance quality of life, and this requires full understanding of the cause and effect of all elements of transport networks. These can be better understood when set against the objectives of sustainable development and this is a high level deliverable of most transportation strategies. The transport links with the sustainable development objectives can be described as follows:

- *Effective protection of the environment* arises from a combination of reducing emissions from fossil fuel vehicles and introducing incentives to encourage the use of non-motorised transport such as walking or cycling. Development of road and rail systems should avoid designated preservation zones and habitats of protected species, pollution of surface and groundwater systems during both construction and operation and minimise adverse acoustic and air quality impacts. Increased energy efficiency, integrated transport planning and improved building management systems at transport nodes will all contribute to environmental protection.
- *Social progress which recognises the needs of everyone* can be enhanced by providing transport services which fully account for the social needs of all communities as well as industry, thereby ensuring that transport policies contribute to social inclusion. Road and rail network planning must meet the needs of major transport movements but avoid severance of existing communities. Sensitive use of town or city bypasses can provide improved quality of life and safety, but could also impact

on business viability due to loss of passing trade. Increased safety measures will make a valuable contribution to society by reducing deaths and injuries from transport related sources.

- *Prudent use of natural resources* can be achieved through construction, operation and maintenance systems which encourage use of recycled materials, adoption of life cycle analysis and the use of low-energy vehicles. These principles should also be applied to vehicle fleets, transport portals, harbours and airports, and in road construction.
- *Maintenance of high and stable levels of economic growth and employment* requires detailed attention to the development of integrated transportation strategies and the allocation of appropriate funding to support this integration and its added benefits to social cohesion, human health and business viability. The application of pricing equality across all transport sectors will allow fair competition, with government subsidies used to drive forward the sustainable transport modes. The use of a suitable range of procurement methods is encouraged with due attention to the design build finance and operate system. Green procurement will enable whole life costing to be applied, while toll or congestion charges may be used to facilitate and ensure financial viability of an integrated transport strategy. Design quality must not be compromised in new procurement methods, with designer and contractor partnerships being promoted.

The design, development, operation and maintenance of a range of transport modes within an agreed strategy should be set alongside the aspirations of sustainable development. The following list describes the critical challenges which contribute to this broader sustainability agenda:

- *Increased accessibility to transport systems*, by reducing barriers which prevent the elderly and disabled from using current transport systems because of cost, service frequency, service coverage or physical access. Access for school children should be seen in light of the health benefits of walking to school, while reducing congestion and air pollution and improving safety issues. Siting of public transport services should relate to other public services and help prevent congestion.
- *Greater integration of sustainable transport strategies and systems* can reduce the need for and demand to travel, possibly through mixed use development, reducing the need to use transportation, while encouraging walking and cycling. An increase in public transport use will reduce cost and improve reliability, frequency and coverage, and make better use of existing networks.
- *Promotion of bio-fuels and cleaner technologies* will reduce the need for fossil fuels such as petrol, diesel and oil, which are heavy polluters.
- *Encouragement for walking and cycling* provides for the most sustainable forms of transport with no pollution, less congestion and positive health benefits.
- *Increased use of green travel plans* can offer incentives to those using sustainable transportation, while allocation of home and neighbourhood zones will give preference to pedestrians and cyclists on shared surfaces. This requires traffic management which safely integrates walking and cycling with other transport infrastructure.
- *Improvement in use of and understanding of transport modes* can increase efficiency and improve connections between different modes, and avoid unnecessary competitiveness. Air travel needs to be fully assessed by considering the benefits from low air fares, increasing air travel, and the associated undesirable pollution from aviation fuels. Water and maritime travel needs environmental and financial evaluation as it has caused water pollution problems in the past necessitating the introduction of double hull oil tankers. The context and long-term shape of road travel needs clarification as it continues to face increased demand, resulting in more road networks being constructed with possible loss of habitat and other environmental damage, while increasing the volume of traffic. To reduce private transport and its adverse environmental impacts requires initiatives to encourage car-sharing and promote greater public transport usage.
- *Public transport* needs financial and conceptual support as it has suffered from poor coverage, due to excessive cost, questionable safety, poor frequency of service and poor reliability. Voluntary transport should be developed to address equality issues by providing access and assistance for disabled and elderly people, especially when seeking health and social care services.

- *Specific attention to road safety* should address the causes of accidents and the influence of alcohol, drugs and excessive speeding. Measures to reduce accidents may include speed cameras, penalty points, traffic calming measures, stringent testing procedures, breathalyser and drug testing.
- *Measurement of the environmental impact of transport* is crucial as data for benchmarks and should cover pollution from exhaust emissions, oil and petrol on road surfaces and their infiltration into water supplies through surface run-off. Local environmental quality assessments will address landscaping from main highways and loss of habitats due to increasing road capacities.
- *The social impact of transport must be included* in all transportation strategies by fully considering the role of public transport in reducing social exclusion and the sense of isolation felt by low income households. Relationships between transport and health should be fully investigated to cover areas such as bronchial problems caused by exposure to car fume emissions and the positive impact of walking and cycling on healthier lifestyles. Similarly, suburbanised settlement patterns, often resulting from gentrification, make increasing demands on private transport and leave the low-income households to face the high pollution consequences. Hence, the transport system should not disadvantage any groups in society or adversely impact on the environment. Pedestrian needs are to be covered by safety, lighting, signage, infrastructure and security.
- *Travel awareness campaigns* must challenge the sustainability of travel journeys such as home to office whilst promoting public transport information, the use of car clubs and car sharing, and the development of tele-working and tele-conferencing.
- *An assessment of the economic impact of transport* is needed to consider the use of fiscal measures such as congestion charges, vehicle taxes and road tolls. Infrastructure development is essential and could attract inward investment and improve accessibility for commercial and industrial business interests. Road construction and surface quality can improve driveability, prevent accidents and use recycled materials, while being designed and constructed for longer life-spans as assessed by whole life costing.
- *Promotion of high-powered automobile industry research and development* to create the next generation vehicles which minimise the impact on the environment, integrate durable and recyclable materials, are accessible to low income families, and are used within an integrated transportation strategy to unleash the entrepreneurial skills of engineers through advanced technologies. This 'Hypercar' approach can synthesise many of the emerging automobile technologies, deliver the 'Factor Four' idiom of gaining four times more return for effort and can call on the hydrogen fuel cell revolution to produce lightweight sustainable transport.

11.4 Transport Theme Index

Jurisdiction	Tier	Title	Date	Page No.
EUROPEAN	Council Directives	Development of the Community's Railways *(CD 91/440/EEC)*	1991	442
		Maritime Safety: Requirements for Vessels Carrying Dangerous or Polluting Goods *(CD 93/75/EEC Amended by Comm Dir 96/39/EC; Comm Dir 97/34/EC;CD 98/55/EC; CD 98/74/EC)*	1993 1996 1997 1998	442
		Maritime Safety: Enhancement of International Standards for Pollution Prevention & Shipboard Living & Working Conditions (Port State Control) *(CD 95/21/EEC Amended by CD 98/25/EC; Com Dir 98/42/EC)*	1995 1998	443
		Taxation on the Carriage of Goods by Road: II *(Dir 1999/62/EC)*	1999	443
		The Allocation of Railway Infrastructure Capacity & the Levying of Charges for the use of Railway Infrastructure & Safety Certification *(Dir 2001/14/EC)*	2001	443
		Reduction of the Level of Pollutant Emissions from 2 & 3-Wheeled Motor Vehicles *(Dir 2002/51/EC)*	2002	444
		Use of Bio-fuels *(Dir 2003/30/EC)*	2003	444
	Council Decisions	Community Database on Road Traffic Accidents *(C Dec 93/704/EEC)*	1993	444
		Community Guidelines for the Development of the Trans-European Transport Network *(Dec No. 1692/96/EC)*	1996	444
	Council Communications	Action Programme on Road Safety I *(COM (93) 246)*	1993	444
		Action Programme on Road Safety II *(COM (97) 131)*	1997	445
		Connecting the Infrastructure Network *(COM (97) 172)*	1997	445
		Developing the Citizens Network *(COM (1998) 431)*	1998	445
		Sustainable Mobility: 2000–2004 Action Programme *(COM (1998) 716)*	1998	445
		Cohesion & Transport *(COM (1998) 806)*	1998	446
		Integrating the Environment into Air Transport *(COM (1999) 640)*	1999	446
		Safety of the Sea-borne Oil Trade (Erika I) *(COM (2000) 142)*	2000	446
		Towards a Safer & More Competitive High Quality Road Transport System *(COM (2000) 364)*	2000	446
		Promoting Sustainable Transport in Development Co-operation *(COM (2000) 422)*	2000	447
		Safety of the Sea-borne Oil Trade (Erika II) *(COM (2000) 802)*	2000	447
		Saving 20 000 Lives on our Roads – A Shared Responsibility *(COM (2003) 311)*	2003	447
	Miscellaneous	Community Railway Policy	1981	448
		The Development of a European High-speed Rail Network	1990	448
		Towards Fair & Efficient Pricing in Transport Policy *(COM (95) 691)*	1995	449
		White Paper: A Strategy for Revitalising the Community Railways *(COM (96) 421)*	1996	449

Jurisdiction	**Tier**	Title	**Date**	**Page No.**
EUROPEAN	Miscellaneous	Towards Sustainable Transport Infrastructure – A Sectoral Approach in Practice	1996	449
		Green Paper: Seaports & Maritime Infrastructure *(COM (97) 678)*	1997	449
		White Paper: Common Charging Framework *(COM (1998) 466)*	1998	450
		Accelerated Phasing in of Double Hulled Oil Tankers *(COM (2000) 142-4* Updated by *Reg No. 417/2002)*	2000 2002	450
		Council resolution on the Improvement of Road Safety	2000	450
		Maritime Safety: Committee on Safe Seas *(COM (2000) 489)*	2000	451
		Maritime Safety: Compensation Fund for Oil Pollution Damage in European Waters *(COM (2000) 802)*	2000	451
		European Maritime Safety Agency *(COM (2000) 802)*	2000	451
		Integrating Environment & SD into Energy & Transport Policies *(SEC (2001) 502)*	2001	451
		White Paper: Transport Policy for 2010 *(COM (2001) 370)*	2001	451
		Use of Bio-fuels *(COM (2001) 547)*	2001	452
		Railway Safety *(COM (2002) 21)*	2002	452
		European Railway Agency *(COM (2002) 23)*	2002	452
		Organisation for International Carriage by Rail (OTIF) *(COM (2002) 24)*	2002	453
		European Energy & Transport Trends to 2030	2003	453
		Europe at a Crossroads – The Need for Sustainable Transport	2003	453
		Financial Assistance to Improve the Environmental Performance of the Freight Transport System (Marco Polo Programme) *(Reg No.1382/2003)*	2003	454
		Charging Heavy Goods Vehicles for the use of Certain Infrastructure *(COM (2003) 448)*	2003	454
		Integration of Environment into Transport Policy – From Strategies to Good Practice	2003	455
		Market Development of Alternative Fuels	2003	455
		European Union: Energy & Transport in Figures	2003	456
		Enforcement in the Field of Road Safety	2004	456
		Establishing the 2nd Marco Polo Programme for the Granting of Community Financial Assistance to Improve the Environmental Performance of the Freight Transport System (Marco Polo II) *(COM (2004) 478)*	2004	456
		The Single European Sky – Implementing Political Commitments	2004	457
UK	Strategies & Reports	National Cycling Strategy	1996	458
		A New Deal for Transport, Better for Everyone	1998	458
		Sustainable Distribution: A Strategy	1999	458
		Review of Voluntary Transport: Main Report	1999	459

Jurisdiction	Tier	Title	Date	Page No.
UK	Guidance Notes	Inclusive Mobility – A Guide to Best Practice on Access to Pedestrian & Transport Infrastructure	2002	471
		Making Travel Plans Work – Lessons from UK Case Studies	2002	471
		Travelling to School: A Good Practice Guide	2003	471
		In Town Without my Car – Good Practice Guide	2004	472
		The Benefits of Green Transport Plans: The Guide	2004	472
		Guidance on Accessibility Planning in Local Transport Plans	2004	472
		Full Guidance on Local Transport Plans Second Edition	2004	472
NORTHERN IRELAND	Strategies & Reports	Transportation in NI – The Way Forward	1995	473
		Environmental Handbook	1998	473
		Moving Forward – NI Transport Policy Statement	1998	473
		Northern Ireland Cycling Strategy	2000	474
		Road Safety Engineering Report – Annual Report	2000	474
		Summary of Transport Problems, Issues & Opportunities	2002	474
		Regional Transportation Strategy for NI 2002–2012	2002	475
		Northern Ireland Road Safety Strategy 2002–2012	2002	475
		Regional Transportation Strategy – Annual Monitoring Report 2002–2003	2003	475
		Regional Strategic Transport Network Transport Plan: Emerging Plan	2003	476
		Regional Strategic Transport Network Transport Plan: Report of Working Conference	2003	476
		Travel Survey for Northern Ireland 2000–2002	2003	477
		Recommendations of Northern Ireland Inquiry into 'Home to School Bus Transport' – The Development of Terms of Reference for a Cost Benefit Analysis for use by DOE (NI)	2003	477
		Walking in Northern Ireland: An Action Plan	2003	477
		Northern Ireland Transport Statistics 2003–2004	2004	478
		Belfast Metropolitan Transport Plan 2015	2004	478
		Equality Impact Assessment: Belfast Metropolitan Transport Plan 2015	2004	478
	Consultation Papers	Developing a Regional Transportation Strategy – A Consultation Paper	2001	479
		Proposed Regional Transportation Strategy – A Consultation Paper	2002	479
		A New Start for Public Transport in NI – A Consultation Paper	2002	480
		Belfast Metropolitan Transport Plan – Working Conference Papers	2003	480
		Proposed Regional Strategic Transport Network Transport Plan 2015: A Consultation Paper for Equality Impact Assessment	2004	480
		An Accessible Transport Strategy for Northern Ireland: Draft for Public Consultation	2004	481
	Guidance Notes	Transport Assessment – Traffic Impact Assessment Guidelines	2001	481

11.5 Transport Policy Tables

Table 11.5.1 International Level Transport Policy.

Title	Venue	Purpose	Key Outcome(s)	Website
OECD Conference – Towards Sustainable Transportation	Vancouver, Canada (1996)	The Conference, held in Vancouver, Canada, from 24–27 March 1996 brought together over 400 stakeholders in the transport sector (automobile & alternative vehicle manufacturers, fuel producers, government officials, regional & local planners, etc.) from 25 countries, in order to develop a vision of & chart a course towards sustainable transport. This document is the Conference Report.	Vancouver principles for sustainable transport – nine principles being: ▪ Access; ▪ Equity; ▪ Individual & community responsibility; ▪ Health & safety; ▪ Education & public participation; ▪ Integrated planning; ▪ Land & resource use; ▪ Pollution prevention; & ▪ Economic well-being.	http://www.oecd.org/dataoecd/28/54/2396815.pdf
First Ministerial Conference on Infrastructure	New Delhi, India (1996)	This conference recognised the vital role being played by infrastructure in economic & social development & the urgent need to solve the infrastructure deficiencies & bottlenecks hindering further economic & social development.	Key outcomes included: ▪ New Delhi Declaration on infrastructure development in Asia & the Pacific; ▪ The New Delhi Action Plan on Infrastructure Development in Asia & the Pacific; ▪ Approval of the operational set of activities for Phase I (1997–2001) of the Regional Action Programme proposed under the Plan; & ▪ The launching of the Asia Infrastructure Development Alliance (AIDA). AIDA is a concrete initiative aimed at developing a closer partnership between governments, the private sector & multi-lateral & intergovernmental agencies active in infrastructure development.	http://www.unescap.org/tctd/pubs/files/ndap.pdf

Continued

Table 11.5.1 International Level Transport Policy. *Continued*

Title	Venue	Purpose	Key Outcome(s)	Website
UN Regional Conference on Transport & Environment	Vienna, Austria (1997)	This conference highlighted various ways which the assembled parties felt that society and governments could help minimise the impact of transportation on the environment.	Vienna Declaration included recommendations on: ▪ Towards sustainable transport; ▪ Promoting energy efficient & less polluting vehicles & fuels; ▪ Promoting efficient & sustainable transport systems; ▪ Protection of sensitive areas; ▪ Promoting sustainable urban transport; ▪ Promoting safe transport of dangerous goods; ▪ Promoting the prevention of water pollution; ▪ A programme of joint action.	http://www.unece.org/doc/ece/rcte/ece.rcte.conf..2.final.e.pdf http://www.unece.org/doc/ece/rcte/ece.rcte.conf..8.final.e.pdf
International Euro-Asian Conference on Transport	St Petersburg, Russia (1998)	This conference brought together representatives of the governments & the parliaments of European & Asian countries, international organisations having a bearing on transport & international financial institutions. The purpose of the conference was to put forward a number of recommendations for developing an integrated Euro-Asian transport system.	Declaration adopted by the International Euro-Asian Conference on Transport.	http://www1.oecd.org/cem/online/region/EurAsia98.pdf
OECD Conference Environmentally Sustainable Transport (EST): Futures, Strategies & Best Practice	Vienna, Austria (2000)	In response to the problems posed by transport for sustainability, the OECD (Organisation for Economic Co-operation and Development) initiated the Environmentally Sustainable Transport (EST) project in 1994. Since then, this major project has involved participants in 25 countries directly & indirectly. The conference held in Vienna in October, was a culmination of work on the EST project. This document is the report on that conference.	The EST project sought to determine how transport in OECD countries could become more sustainable. An important outcome of the project is a set of *EST Guidelines*, endorsed at the conference, that can help decision-makers formulate policies for moving towards sustainable transport.	http://www.olis.oecd.org/olis/2001doc.nsf/43bb6130e5e86e5fc12569fa005d004c/5ebb4a5e203badc5c1256b5a003f3dd7/$FILE/JT00120803.PDF

Second Ministerial Conference on Infrastructure	Seoul, South Korea (2001)	The Ministerial Conference on Infrastructure organised by the United Nations Economic & Social Commission for Asia & the Pacific (UN-ESCAP) was structured into two segments, namely, the Senior Officials Meeting, and the Meeting of Ministers.	The key outcomes included: ▪ the Seoul Declaration on Infrastructure Development in Asia & the Pacific; & ▪ the Regional Action Programme (2002–2006) of the New Delhi Action Plan on Infrastructure Development in Asia & the Pacific.	http://www.unescap.org/tctd/seoul2001/files/MCI2_Rep.pdf
International Conference on Environmentally Sustainable Transport (EST) – Is Rail on Track?	Eskilstuna, Sweden (2001)	This conference, organised by the OECD Environment Directorate & the International Union of Railways (UIC), sought to address barriers to the implementation of EST Guidelines & explore policy options to overcome them, focusing on rail's contribution to EST. The objectives of the conference were: ▪ to address present & future demands for rail transport in the context of progress towards EST; ▪ to highlight innovative solutions, promising options & approaches in the fields of technology, mobility services & freight logistics; & ▪ to provide new insights & to highlight opportunities for business & industry in the development & deployment of sustainable transport systems by examining practical examples & innovations in passenger & freight movement from different OECD countries.	The Eskilstuna Declaration on Rail Transport was discussed, revised & adopted at the conference. Use of the Declaration will help keep rail on track towards achieving Environmentally Sustainable Transport. The Declaration includes the advantages, challenges, areas to be improved, possible solutions & recommendations concerning rail transport in the context of EST.	http://www.olis.oecd.org/olis/2001doc.nsf/43bb6130e5e86e5fc12569fa005d004c/50ffd6af42251557c1256c550049e37b/$FILE/JT00133526.PDF
London Charter on Transport, Environment & Health	London, UK (1999)	This Charter sets out the principles, strategies & a plan of action to guide our policies towards achieving transport sustainable for health & the environment.	The key issues that the Charter addresses are: ▪ Why health is an issue in transport & environmental policies: the concerns; ▪ What has happened so far: the state of play; ▪ Where do we want to go: principles, approaches & guiding strategies for transport sustainable for health & the environment; ▪ How will the goods be achieved: a plan of action for moving towards transport sustainable for health & the environment; ▪ Implementation & follow-up process.	http://www.who.dk/document/peh-ehp/charter_transporte.pdf

Table 11.5.2 European Level Transport Policy.

Policy Type	Programme Title	Policy Objective	Community Measure	Document Reference	Website
Council Directives	Development of the Community's Railways	The Directive's aim is to facilitate the adoption of the Community railways to the needs of the Single Market and to increase their efficiency: ▪ By ensuring the management independence of railway undertakings; ▪ By separating the management of railway operation & infrastructure from the provision of railway transport services, separation of accounts being compulsory & organisational or institutional separation being optional; ▪ By improving the financial structure of undertakings; ▪ By ensuring access to the networks of Member States for international groupings of railway undertakings & for railway undertakings engaged in the international combined transport of goods.	Council Directive 91/440/EEC on the development of the Community's railways (29/07/91).	▪ Off Jnl L237, 24/08/91	http://europa.eu.int/smartapi/cgi/sga_doc?smartapi!celexapi!prod!CELEXnumdoc&lg=EN&numdoc=31991L0440&model=guichett
	Maritime Safety: Requirements for Vessels Carrying Dangerous or Polluting Goods	To avoid conditions likely to cause accidents during the transport of dangerous or polluting goods and to reduce the resulting damage when such accidents occur.	Council Directive 93/75/EEC concerning minimum requirements for vessels bound for or leaving Community ports and carrying dangerous or polluting goods (13/09/93). Amended by the following measures: ▪ Commission Dir 96/39/EC (19/06/96) ▪ Commission Dir 97/34/EC (06/06/97)	▪ Off Jnl L247, 05/10/93 ▪ Off Jnl L196, 07/08/96 ▪ Off Jnl L158, 17/06/97	http://europa.eu.int/scadplus/leg/en/lvb/l24068.htm

		■ CD 98/55/EC (17/07/98) ■ CD 98/74/EC (01/10/98).	■ Off Jnl L215, 01/08/98 ■ Off Jnl L276, 13/10/98	
Maritime Safety: Enhancement of International Standards for Pollution Prevention & Shipboard Living & Working Conditions (Port State Control)	To establish the legal framework needed for introducing a harmonised port state control system to reduce the number of substandard vessels operating in Community waters &, with the aid of these preventive measures, enhancing safety at sea & protection of the marine environment.	Council Directive 95/21/EC concerning the enforcement, in respect of shipping using Community ports and sailing in the waters under the jurisdiction of the Member States, of international standards for ship safety, pollution prevention and shipboard living and working conditions (port State control) (19/06/95). Amended by: ■ CD 98/25/EC (27/04/98) ■ Comm Dir 98/42/EC (19/06/98).	■ Off Jnl L157, 07/07/95 ■ Off Jnl L133, 07/05/98 ■ Off Jnl L184, 27/06/98	http://europa.eu.int/scadplus/leg/en/lvb/l24072.htm
Taxation on the Carriage of Goods by Road: II	To harmonise levy systems (vehicle taxes, tolls and charges relating to the use of road infrastructure) and introduce fair mechanisms for charging infrastructure costs to hauliers.	Directive 1999/62/EC of the European Parliament and of the Council on the charging of heavy goods vehicles for the use of certain infrastructures (17/06/99).	■ Off Jnl L187, 20/07/99	http://europa.eu.int/scadplus/leg/en/lvb/l24045b.htm
Allocation of Railway Infrastructure Capacity & the Levying of Charges	This Directive concerns the principles & procedures to be applied with regard to the setting & charging of railway infrastructure charges & the allocation of railway infrastructure capacity. Member States shall ensure that charging & capacity allocation schemes for railway infrastructure follow principles set down in this Directive & this allows the infrastructure manager to market & make optimum effective use of the available infrastructure capacity.	Directive 2001/14/EC of the European Parliament & the Council on the allocation of railway infrastructure capacity and the levying of charges for the use of railway infrastructure and safety certification (26/02/01).	■ Off Jnl L75, 15/03/01	http://europa.eu.int/eur-lex/pri/en/oj/dat/2001/l_075/l_07520010315en00290046.pdf

Continued

Table 11.5.2 European Level Transport Policy. *Continued*

Policy Type	Programme Title	Policy Objective	Community Measure	Document Reference	Website
Council Directives	Reduction of the Level of Pollutant Emissions from 2 & 3-Wheeled Motor Vehicles	This Directive aims at reducing the level of pollutant emissions from two- and three-wheeled motor vehicles by tightening the limit values for such emissions.	Directive 2002/51/EC of the European Parliament & of the Council on the reduction of the level of pollutant emissions from two- and three-wheeled motor vehicles – amending Directive 97/24/EC (19/07/02).	▪ Off Jnl L252, 20/09/02	http://europa.eu.int/eur-lex/pri/en/oj/dat/2002/l_252/l_25220020920en00200030.pdf
	Use of Biofuels	This Directive aims at promoting the use of biofuels or other renewable fuels to replace diesel or petrol for transport purposes in each Member State, with a view to contributing to objectives such as meeting climate change commitments, environmentally friendly security of supply & promoting renewable energy sources.	Directive 2003/30/EC of the European Parliament & of the Council on the promotion of the use of biofuels or other renewable fuels for transport (08/05/03).	▪ Off Jnl L123/43, 17/05/03	http://europa.eu.int/eur-lex/pri/en/oj/dat/2003/l_123/l_12320030517en00420046.pdf
Council Decisions	Community Database on Road Traffic Accidents	To counter road traffic accidents and the consequences suffered by accident victims.	Council Decision 93/704/EEC on the setting-up of a Community database on road traffic accidents (30/11/93).	▪ Off Jnl L329, 30/12/93	http://europa.eu.int/scadplus/leg/en/lvb/l24054.htm
	Community Guidelines for the Development of the Trans-European Transport Network	To establish the broad lines of action envisaged for the establishment of the trans-European transport network, and identify projects of common interest, the implementation of which should contribute towards the development of the network.	Decision No. 1692/96/EC of the European Parliament and the Council on Community guidelines for the development of the trans-European transport network (23/07/96).	▪ Off Jnl L228, 09/09/96	http://europa.eu.int/scadplus/leg/en/lvb/l24094.htm
Council Communications	Action Programme on Road Safety I	To improve road safety in order to reduce the number of accidents in the Community.	Commission communication to the Council for an action programme on road safety (1993).	▪ COM (93) 246	http://europa.eu.int/scadplus/leg/en/lvb/l24055a.htm

Action Programme on Road Safety II	To improve road safety with a view to reducing the number of accidents in the European Union.	Communication from the Commission to the Council, the European Parliament, the Economic and Social Committee and the Committee of the Regions: Promoting road safety in the European Union – the programme for 1997–2001 (1997).	COM (97) 131	http://europa.eu.int/scadplus/leg/en/lvb/l24055b.htm
Connecting the Infrastructure Network	To promote a pan-European transport networks partnership for co-ordinated transport planning, studying regulatory measures & facilitating financial arrangements.	Communication from the Commission on connecting the Union's transport infrastructure network to its neighbours: Towards a co-operative pan-European transport network policy (1997).	COM (97) 172	http://europa.eu.int/scadplus/leg/en/lvb/l24185.htm
Developing the Citizens Network	To develop a system of local and regional passenger transport by providing the public authorities, operators and user groups with appropriate tools and establishing a policy framework which promotes sustainable mobility.	Communication from the Commission to the Council, the European Parliament, the Economic and Social Committee and the Committee of the Regions: 'Developing the citizens' network – Why local and regional passenger transport is important and how the European Commission is helping to bring it about' (10/07/98).	COM (1998) 431	http://europa.eu.int/scadplus/leg/en/lvb/l24215.htm
Sustainable Mobility: 2000–2004 Action Programme	To implement a Common transport policy which is safe, efficient, competitive and socially and environmentally friendly.	Communication from the Commission to the Council, the European Parliament, the Economic and Social Committee and the Committee of the Regions on the common transport policy – 'Sustainable mobility: perspectives for the future' (1998).	COM (1998) 716	http://europa.eu.int/scadplus/leg/en/lvb/l24208.htm

Continued

Table 11.5.2 European Level Transport Policy. *Continued*

Policy Type	Programme Title	Policy Objective	Community Measure	Document Reference	Website
Council Communications	Cohesion & Transport	To develop an efficient and sustainable European transport system taking account of economic development prospects at regional level, the enlargement of the Union to include new States and the importance of public transport.	Communication from the Commission to the Council, the European Parliament, the Economic and Social Committee and the Committee of the Regions: 'Cohesion and Transport' (14/01/99).	■ COM (1998) 806	http://europa.eu.int/scadplus/leg/en/lvb/l24207.htm
	Integrating the Environment into Air Transport	To improve the environmental performance of air transport activities so as to offset the environmental impact of growth in this sector.	Commission communication to the Council, the European Parliament, the Economic and Social Committee and the Committee of the Regions on air transport and the environment (01/12/99).	■ COM (1999) 640	http://europa.eu.int/scadplus/leg/en/lvb/l28116.htm
	Safety of the Seaborne Oil Trade (Erika I)	To improve safety in the seaborne petroleum products trade.	Commission communication to Parliament and the Council on the safety of the seaborne oil trade (21/03/00).	■ COM (2000) 142 ■ Off Jnl C212E, 25/07/00	http://europa.eu.int/scadplus/leg/en/lvb/l24230.htm
	Towards a Safer & More Competitive High Quality Road Transport System	To indicate the measures necessary to deploy a coherent, global policy so as to ensure the development of a safer and more competitive high-quality road transport system.	Communication from the Commission to the European Parliament and Council – Towards a Safer and More Competitive High Quality Road Transport System in the Community (21/06/00).	■ COM (2000) 364	http://europa.eu.int/scadplus/leg/en/lvb/l24233.htm
	Promoting Sustainable Transport in Development Co-operation	This Communication guides the Community & Member States support to transport in the developing countries for achieving the main development goals laid down in the Treaty establishing the European Community. The	Communication from the Commission to the Council & the European Parliament on promoting sustainable transport in development cooperation (06/07/00).	■ COM (2000) 422	http://europa.eu.int/eur-lex/en/com/cnc/2000/com2000_0422en01.pdf#zoom=100

	Communication therefore outlines a comprehensive sectoral approach that is valid for all transport modes – roads, railways, air, maritime & waterway transport – as well as the services to facilitate movement of goods & people. This approach builds on a framework of principles to achieve the wider development goals, & provides a strategy that will deliver sustainable transport which is safe, economically, financially & institutionally sustainable as well as environmentally sound & socially aware.			
Safety of the Seaborne Oil Trade (Erika II)	To improve the safety of the seaborne oil trade.	Communication from the Commission to the Council and the European Parliament on a second set of Community measures on maritime safety following the sinking of the oil tanker Erika (06/12/00).	COM (2000) 802	http://europa.eu.int/scadplus/leg/en/lvb/l24242.htm
Saving 20 000 Lives on our Roads – A Shared Responsibility	This communication describes a number of direct & accompanying measures which the Commission plans to implement to enhance the benefit of the activities undertaken by the European Union, in particular the development of new safety technologies under the research framework programmes to add value to the efforts made by the Member States. It describes in particular specific measures for establishing a methodological framework to identify & disseminate best practices, through the drafting of technical guides, improving the collection & analysis of data on accidents & physical injuries, & pursuing research & development to find solutions for the future.	Communication from the Commission on saving 20 000 lives on our roads (2003).	COM (2003) 311	http://europa.eu.int/comm/transport/road/library/rsap/rsap_en.pdf

Continued

Table 11.5.2 European Level Transport Policy. *Continued*

Policy Type	Programme Title	Policy Objective	Community Measure	Document Reference	Website
Miscellaneous	Community Railway Policy	Recognises that Member States, have a direct interest in improving the situation of the railways within the framework of the common transport policy, with particular attention paid to: ▪ Optimum use of available railway capacity; ▪ Closer international co-operation, particularly on commercial matters; ▪ Emphasis of the common interest in international transport; ▪ Organisation of relations between railway undertakings & governments so as to give such undertakings sufficient independence, to enable them to adapt to market requirements & technical developments; ▪ Elimination of distortions of competition affecting railways; ▪ Development of combined transport; ▪ Facilitation & speeding up of frontier crossings; ▪ Adaptation of the organisation, infrastructure, & equipment of the railways to enable them to play their part in meeting the present-day needs of the economy & the population; ▪ Financial situation of the railways.	Council Resolution on Community railway policy (15/12/81).	▪ Off Jnl C157, 22/06/82	http://europa.eu.int/smartapi/cgi/sga_doc?smartapi!celexapi!prod!CELEXnumdoc&lg=EN&numdoc=31982Y0622(01)&model=guichett
	The Development of a European High-speed Rail Network	Requests the Commission to look in greater detail at: ▪ The socio-economic impact of the network on the integrated transport market & the development of the Community;	Council Resolution on the development of a European high-speed rail network (17/12/90).	▪ Off Jnl C33, 08/02/91	http://europa.eu.int/smartapi/cgi/sga_doc?smartapi!celexapi!prod!CELEXnumdoc&lg=EN&numdoc=31991Y0208(01)&model=guichett

	▪ The impact of the network on the environment in & how it compares with other modes of transport in this regard; ▪ Economic studies, including, the commercial aspects of the key links & other crucial points in the network & the problem of financing them.			
Green Paper: Fair & Efficient Pricing	To launch a debate about how pricing instruments can contribute to solving the principal transport problems (congestion, accidents and pollution).	Commission Green Paper entitled 'Towards fair and efficient pricing in transport – policy options for internalising the external costs of transport in the European Union' (20/12/95).	▪ COM (95) 691	http://europa.eu.int/scadplus/leg/en/lvb/l24177.htm
White Paper: A Strategy for Revitalising the Community Railways	To lay down a strategy to revitalise the Community's railways by creating a sound financial basis, ensuring freedom of access to all traffic and public services and by promoting the integration of national systems and social aspects.	Commission White Paper: 'A strategy for revitalising the Community's railways' (30/07/96).	▪ COM (96) 421	http://europa.eu.int/scadplus/leg/en/lvb/l24014.htm
Towards Sustainable Transport Infrastructure – A Sectoral Approach in Practice	Over-sized transport networks are a major financial burden for many developing countries. As a result of deferred maintenance, roads, railways, ports & airports are deteriorating, in some cases beyond repair. Recipient countries & donors are thus now seeking ways of sustaining these vital national assets. The European Commission is adopting a sectoral approach to sustainability. These guidelines are an important step towards improving our effectiveness in the transport infrastructure sector.	DG Environment Report (July 1996).	▪ DG XI Report	http://europa.eu.int/comm/development/body/theme/transport/guidelines/en/entoc.htm
Green Paper: Seaports & Maritime Infrastructure	To launch a debate on the efficiency of ports and maritime infrastructure, their integration into the multimodal trans-European network and the application of competition rules to this sector.	Green Paper on seaports and maritime infrastructure (10/12/97).	▪ COM (97) 678	http://europa.eu.int/scadplus/leg/en/lvb/l24163.htm

Continued

Table 11.5.2 European Level Transport Policy. *Continued*

Policy Type	Programme Title	Policy Objective	Community Measure	Document Reference	Website
Miscellaneous	White Paper: Common Charging Framework	To show the need for gradual harmonisation at Community level of the charging principles applied in the various commercial modes of transport.	White Paper, presented by the Commission: Fair payment for infrastructure use: A phased approach to a common transport infrastructure charging framework in the EU (22/07/98).	▪ COM (1998) 466	http://europa.eu.int/scadplus/leg/en/lvb/l24138.htm
	Accelerated Phasing in of Double Hulled Oil Tankers	To speed up the replacement of single hull oil tankers by double hull or design equivalent oil tankers.	Proposal for a Regulation of the European Parliament and the Council on the accelerated phasing-in of double hull or equivalent design requirements for single hull oil tankers Updated by Regulation (EC) No. 417/2002 of the European Parliament & Council repealing CR (EC) No. 2978/94 (21/04/00).	▪ COM (2000) 142–4 ▪ Off Jnl C212E, 25/07/00 ▪ Off Jnl L64, 07/03/02	http://europa.eu.int/scadplus/leg/en/lvb/l24231.htm http://europa.eu.int/eur-lex/pri/en/oj/dat/2002/l_064/l_06420020307en00010005.pdf
	Improvement of Road Safety	Aim of reducing accidents as much as possible, by encouraging Member States national, regional & local authorities to calculate expenditure on road safety measures & monitor their effects & to increase their investment in road safety projects.	Council resolution on the improvement of road safety (26/06/00).	▪ Off Jnl C218, 31/07/00	http://europa.eu.int/eur-lex/pri/en/oj/dat/2000/c_218/c_21820000731en00010004.pdf

Maritime Safety: Committee on Safe Seas.	To establish a committee on safe seas centralising the duties of the existing committees. To facilitate the amendment of Community legislation in the light of developments in the main conventions, resolutions or other agreements that have entered into force at international level.	Proposal for a Regulation of the European Parliament and of the Council concerning the Committee on Safe Seas (15/09/00).	▪ COM (2000) 489 Final ▪ Off Jnl C365E, 19/12/00	http://europa.eu.int/scadplus/leg/en/lvb/l24241.htm
Maritime Safety: Compensation Fund for Oil Pollution Damage in European Waters.	To ensure fair compensation for oil pollution damage occurring in European waters by complementing the existing international liability and compensation regime. To provide for financial penalties to be imposed on any person liable for causing oil pollution.	Proposal for a Regulation of the European Parliament and of the Council on the establishment of a fund for the compensation for oil pollution damage in European waters and related measures (06/12/00).	▪ COM (2000) 802	http://europa.eu.int/scadplus/leg/en/lvb/l24238.htm
European Maritime Safety Agency	To set up a European Maritime Safety Agency in order to ensure a high, uniform and effective level of maritime safety & pollution prevention within the Community.	Proposal for a Regulation of the European Parliament and of the Council establishing a European Maritime Safety Agency (06/12/00).	▪ COM (2000) 802	http://europa.eu.int/scadplus/leg/en/lvb/l24245.htm
Integrating Environment & SD into Energy & Transport Policies	To evaluate recent energy & transport market trends, describing policy initiatives & outlining future actions that contribute to SD.	Commission Working Paper (21/03/01).	▪ SEC(2001) 502	http://europa.eu.int/comm/energy_transport/library/integr_report_en.pdf
White Paper: Transport Policy for 2010	To strike a balance between economic development and the quality and safety demands made by society in order to develop a modern, sustainable transport system for 2010.	White Paper submitted by the Commission: 'European transport policy for 2010: time to decide' (12/09/01).	▪ COM (2001) 370	http://europa.eu.int/scadplus/leg/en/lvb/l24007.htm

Continued

Table 11.5.2 European Level Transport Policy. *Continued*

Policy Type	Programme Title	Policy Objective	Community Measure	Document Reference	Website
Miscellaneous	Use of Bio-fuels	To create a Community framework that will promote the use of bio-fuels in order to reduce greenhouse gas emissions and the environmental impact of transport, and to increase security of supply.	Proposal for a Directive of the European Parliament and of the Council on the promotion of the use of bio-fuels for transport (2001).	■ COM (2001) 547	http://europa.eu.int/scadplus/leg/en/lvb/l21061.htm
	Railway Safety	To create a common regulatory framework for railway safety in the EU in order to facilitate the exchange of information and of experiences and co-ordination at Community level.	Proposal for a Directive of the European Parliament and of the Council on safety on the Community's railways and amending Council Directive 95/18/EC on the licensing of railway undertakings and Directive 2001/14/EC on the allocation of railway infrastructure capacity and the levying of charges for the use of railway infrastructure and safety certification (2002).	■ COM (2002) 21	http://europa.eu.int/scadplus/leg/en/lvb/l24201a.htm
	European Railway Agency	To establish a European Railway Agency responsible for formulating common solutions on matters concerning safety and inter-operability, in order to help create a safe, integrated and competitive railway area without frontiers.	Proposal for a Regulation of the European Parliament and of the Council establishing a European Railway Agency (2002).	■ COM (2002) 23	http://europa.eu.int/scadplus/leg/en/lvb/l24013.htm

Organisation for International Carriage by Rail (OTIF)	Council recommendation authorising the Commission to negotiate an agreement setting out the conditions for Community accession to the Intergovernmental Organisation for International Carriage by Rail.	Recommendation for a Council Decision authorising the Commission to negotiate the conditions for Community accession to the Convention concerning International Carriage by Rail (COTIF) of 9 May 1980, as amended by the Vilnius Protocol of 3 June 1999 (2002).	COM (2002) 24	http://europa.eu.int/scadplus/leg/en/lvb/l24011.htm
European Energy & Transport Trends to 2030	This publication on energy & transport trends to 2030 extends the analysis presented in previous editions of the annual energy review. In addition to a statistical analysis of the last 10 years, this publication gives detailed projections on energy & transport for the current EU & the enlarged Union of 25. It also addresses the wider European & world developments.	Report for the DG Transport & Energy (Jan 2003).	DG TREN Report	http://europa.eu.int/comm/dgs/energy_transport/figures/trends_2030/1_pref_en.pdf
Europe at a Crossroads – The Need for Sustainable Transport	This report highlights the future of EU transport policy focusing closely on the need for the Union to push the move towards sustainable transportation. The report is set within the context of a global approach involving all players – local, national & European. It states that the main challenge is to encourage a better balance between the different forms of transport & how to make a better use of existing networks.	DG Transport & Energy Report (June 2003).	DG TREN Report	http://europa.eu.int/comm/publications/booklets/move/39/en.pdf

Continued

Table 11.5.2 European Level Transport Policy. *Continued*

Policy Type	Programme Title	Policy Objective	Community Measure	Document Reference	Website
Miscellaneous	Financial Assistance to Improve the Environmental Performance of the Freight Transport System (Marco Polo Programme)	This Regulation establishes a financing instrument the objective being to reduce congestion, to improve the environmental performance of the transport system and to enhance inter-modal transport. This is to contribute to an efficient and sustainable transport system, for the period from 1 January 2003 to 31 December 2010, in order to achieve a traffic shift of the expected yearly aggregate increase of international road freight traffic, measured in tonne-kilometres, to short sea shipping, rail and inland waterways or to a combination of modes of transport in which road journeys are as short as possible.	Reg (EC) No. 1382/2003 of the European Parliament & of the Council on granting community financial assistance to improve the environmental performance of the freight transport system (Marco Polo Programme) (22/07/03).	▪ Off Jnl L196/1, 02/08/03	http://europa.eu.int/eur-lex/pri/en/oj/dat/2003/l_196/l_19620030802en00010006.pdf
	Charging Heavy Goods Vehicles for the use of Certain Infrastructure	This proposal amends Directive 1999/62/EC, the 'Eurovignette Directive'. Amendment is all the more urgent as most Member States are examining the reforms needed to include the cost to society in the prices paid by infrastructure users. Some countries are planning or have already undertaken to introduce new road charging systems linked to the distance travelled.	Proposal for a Directive of the European Parliament & of the Council amending Directive 1999/62/EC on the charging of heavy goods vehicles for the use of certain infrastructure (23/07/03).	▪ COM (2003) 448	http://europa.eu.int/comm/transport/infr-charging/library/dir_com_2003_448_en.pdf

Integration of Environment into Transport Policy – From Strategies to Good Practice	The Report aims at providing inspiration & insights about how to promote integration of environmental considerations into transport policy. It focuses on messages & issues discussed during conference held in October 2002 on 'Good Practice in Integration of Environment into Transport Policy'. The publication: ▪ Briefly summarises the background & main developments in the EU policy framework for integration of environment into transport policy; ▪ Highlights selected 'lessons learned' & cases presented at the conference; & ▪ Highlights selected conclusions from the various thematic workshops.	Directorate-General Environment Report (Sept 2003).	▪ DG Environment Report	http://www.europa.eu.int/comm/environment/gpc/pdf/transp_policy_en.pdf
Market Development of Alternative Fuels	The European Commission, following the Communication on alternative fuels in 2001, set up a stakeholder Contact Group in 2002 to advise on technical & economic status & developments of alternative fuels for road transport, with priority on natural gas & hydrogen, and on measures by which the Community could promote their use. Topic Groups on natural gas & hydrogen, respectively, & a Working Group on Biomass-to-Liquid fuels deepened the assessment of these alternative fuels. The Contact Group published an Interim Report in March 2003 & now presents this comprehensive report, offering the basis for a more detailed strategy on these fuels.	Report of the Alternative Fuels Contact Group (Dec 2003).	▪ Alternative Fuels Contact Group Report	http://europa.eu.int/comm/energy_transport/envir/2003_report_en.pdf

Continued

Table 11.5.2 European Level Transport Policy. *Continued*

Policy Type	Programme Title	Policy Objective	Community Measure	Document Reference	Website
Miscellaneous	European Union: Energy & Transport in Figures	The main aim of this pocket-book is to provide energy and transport statistics for the EU Member States in a user-friendly way. Most of the data cover the period 1970–2001.	Directorate-General for Energy & Transport Statistical Report (Dec 2003).	■ DG Energy & Transport Report	http://europa.eu.int/comm/dgs/energy_transport/figures/pocketbook/doc/etif_2003.pdf
	Enforcement in the Field of Road Safety	This Recommendation aims at improving road safety & reducing the number of deaths & injuries on the roads in the European Union, through the implementation of national enforcement plans of the Member States which contain measures of best practice in enforcement & information of the public in the field of speeding, drink–driving & non-use of seat belts.	Commission Recommendation on enforcement in the field of road safety (06/04/04).	■ Off Jnl L111/75, 17/04/04	http://europa.eu.int/eur-lex/pri/en/oj/dat/2004/l_111/l_11120040417en00750082.pdf
	Establishing the 2nd Marco Polo Programme for the Granting of Community Financial Assistance to Improve the Environmental Performance of the Freight Transport System (Marco Polo II)	The forecast increase of road freight by 2013 is more than 60% in the European Union, & a doubling is foreseen for the 10 new member states by 2020. The results are congestion, environmental degradation, accidents, & a danger of loss of competitiveness of European industry, which needs to rely, for the management of its supply chains, on cost-efficient & reliable transportation systems. Road freight transport is also fully dependent on fossil fuel, which makes the transport system vulnerable to	Proposal for a Regulation of the European Parliament & of the Council establishing the second Marco Polo programme for the granting of Community financial assistance to improve the environmental performance of the freight transport system (Marco Polo II) (14/07/04).	■ COM (2004) 478	http://europa.eu.int/comm/transport/marcopolo/legal/doc/com_2004_0478_en.pdf

	the changing global supply patterns. Fossil fuel is a major CO_2 contributor. A more energy efficient transport system is therefore in the interest of the Union, both in terms of better environmental performance, & in terms of a more robust transport economy. In this situation, a stronger reliance on intermodality is necessary. Intermodality makes better use of existing infrastructure & service resources through integrating short sea shipping, rail & inland waterways into the logistics chain. One thus does not have to wait until the necessary large-scale network infrastructures are in place for a better transport system.			
The Single European Sky – Implementing Political Commitments	The European Union's single European sky initiative seeks to promote a more rational organisation of European airspace, increasing capacity while ensuring uniformly high safety standards throughout Europe. It aims to put in place a framework for decision-making & operational improvement that will enhance the efficiency, safety & cost-effectiveness of the system.	DG Transport & Energy Report (Dec 2004).	▪ DG TREN Report	http://europa.eu.int/comm/transport/air/single_sky/materials/doc/publications/2004_brochure_ses_en.pdf

Table 11.5.3 UK Level Transport Policy.

Policy Type	Document Title	Objective	Contents	Website
Strategies & Reports	National Cycling Strategy (23/09/96)	This National Cycling Strategy represents a major breakthrough in transport thinking in the UK. It will promote cycling priority on the highway in the centre of towns, at the workplace and in new developments. By so doing, the Strategy will generate a culture change for cycling.	Foreword; Introduction; Key strategic outputs; Context; Principal outputs; Cycling & sustainable transport; Integrating cycling with traffic management; Cycle security; Changing attitudes – a communication programme; The local framework; The strategy at a glance; Annex 1: Acknowledgements; Annex 2: A model 'local cycling strategy'; Annex 3: Innovative cycling schemes – the cycle challenge project.	http://www.dft.gov.uk/stellent/groups/dft_susttravel/documents/pdf/dft_susttravel_pdf_503877.pdf
	A New Deal for Transport, Better for Everyone (20/07/98)	This is a United Kingdom White Paper. It sets out a new approach to transport policy which has relevance throughout the United Kingdom, and it embodies new, modern thinking on integrating transport with other aspects of government policy. The paper sets the framework within which detailed policies will be taken forward.	Foreword; Acknowledgements; Scope of the White Paper; Part I – Chapter 1: A new deal for transport; Part II – Chapter 2: Sustainable transport; Chapter 3: Integrated transport; Part III – Chapter 4: Making it happen; Chapter 5: Sharing responsibility; Annex A: Future publications; Annex B: Consultation on integrated transport policy; Annex C: Royal Commission on Environmental Pollution; Annex D: 'Transport: the way forward'; Annex E: Trunk road network; Annex F: Rail network pinch-points.	http://www.dft.gov.uk/stellent/groups/dft_about/documents/pdf/dft_about_pdf_503890.pdf
	Sustainable Distribution: A Strategy (02/03/99)	In the White Paper (*A New Deal for Transport, Better for Everyone*) the Government promised to set out a range of policies for moving goods sustainably. This document sets out how the Government, working in partnership with industry, will deliver a strategy for the efficient movement of goods, supporting a strong economy with minimum harm to the environment & people's health.	Foreword; Summary; Chapter 1: Introduction; Chapter 2: The role of distribution & logistics; Chapter 3: Towards 'sustainable distribution'; Chapter 4: A sustainable market; Chapter 5: Strategic planning; Chapter 6: Raising standards; Chapter 7: Indicators for sustainable distribution; Chapter 8: The Government's aims: in conclusion; Appendix 1: Responses to the White Paper consultation process.	http://www.dft.gov.uk/stellent/groups/dft_freight/documents/pdf/dft_freight_pdf_503891.pdf

Review of Voluntary Transport: Main Report (July 1999)	A study into the nature & scale of voluntary transport in England, Scotland & Wales was commissioned by the Department of the Environment, Transport & the Regions. In particular it considered: ▪ the role voluntary transport plays in enabling disabled & elderly people to sustain independent living & to access health & social care services; & ▪ the involvement of voluntary transport in meeting the needs of others who do not have use of a car.	Introduction; Context; Project findings; Key issues; Proposals for change; Appendices; Bibliography.	http://www.dft.gov.uk/stellent/groups/dft_mobility/documents/pdf/dft_mobility_pdf_503229.pdf
A Travel Plan Resource Pack for Employers (Feb 2000)	The development of travel plans by organisations will help to encourage staff, & others visiting site(s) to use more environmentally-friendly alternatives than driving alone, at least for some of their journeys. They should contain a mix of incentives & disincentives to be successful e.g. car-sharing, promoting more use of public transport, encouraging walking & cycling, restricting on-site car parking spaces & supporting alternative work practices which reduce the need to travel. This document highlights some of the key elements employers should incorporate into their travel plans.	Section 1: Travel plans – the background; Section 2: Roles & responsibilities; Section 3: Understanding current travel patterns; Section 4: Objectives, targets & indicators; Section 5: Identifying measures; Section 6: Awareness raising & marketing; Section 7: Maintaining the momentum; Section 8: Financial issues; Section 9: Useful information sources & contacts.	http://www.dft.gov.uk/stellent/groups/dft_susttravel/documents/pdf/dft_susttravel_pdf_504132.pdf
Transport 2010: The 10 Year Plan (20/07/00)	The strategy for transport is to tackle congestion and pollution by improving all types of transport – rail and road, public and private – in ways that increase choice. It is a strategy for investment in the future to create prosperity and a better environment. This requires a new approach, based on: ▪ Integrated transport: looking at transport as a whole, matching solutions to specific problems by assessing all the options. ▪ Public and private partnership: government and the private sector working more closely together to boost investment. ▪ New projects: modernising the transport network in ways that make it bigger, better, safer, cleaner and quicker.	Foreword; Introduction; Summary; Chapter 1: Vision; Chapter 2: Progress; Chapter 3: Challenge; Chapter 4: Integration; Chapter 5: Partnerships & resources; Chapter 6: Investment; Chapter 7: Safety; Chapter 8: Modern, clean, efficient; Chapter 9: Future choices; Chapter 10: Monitoring; Conclusion: Serving people better; Annex 1: Investment figures; Annex 2: Targets & indicators; Annex 3: Investment & charging assumptions; Annex 4: Facts & figures.	http://www.dft.gov.uk/stellent/groups/dft_about/documents/pdf/dft_about_pdf_503944.pdf

Continued

Table 11.5.3 UK Level Transport Policy. *Continued*

Policy Type	Document Title	Objective	Contents	Website
Strategies & Reports	Transport 2010: The Background Analysis (20/07/00)	This paper provides an overview of the modelling and analytical work that has informed the Government's 10 year Plan for transport.	Introduction; Chapter 1: Key influences & core assumptions; Chapter 2: The models used & analysis of the challenges; Chapter 3: Analysis of policy options; Chapter 4: Assessment of the Plan's contribution to sustainable development; Annex A: Econometric modelling of non-commuting rail demand; Annex B: Modelling using the national road traffic forecasts framework; Annex C: Rail emissions modelling; Annex D: Commentary on appraisal summary table; Annex E: Evidence from project appraisals.	http://www.dft.gov.uk/stellent/groups/dft_about/documents/pdf/dft_about_pdf_503943.pdf
	Social Exclusion & the Provision & Availability of Public Transport (27/10/00)	In December 1998, Transport Research and Consultancy, and the University of North London, working with Social and Transport Research Services, were commissioned by the DETR (through the Department's Mobility and Inclusion Unit) to examine the relationship between social exclusion and transport and to identify the contribution that public and community transport can make to reducing levels of exclusion.	Foreword; Acknowledgements; Executive Summary; Chapter 1: Introduction; Chapter 2: Current knowledge on transport & social exclusion; Chapter 3: Carrying out the fieldwork; Chapter 4: The study areas – urban; Chapter 5: The study areas – rural; Chapter 6: The people (effects of transport on different social groups); Chapter 7: The adequacy of public transport; Chapter 8: Transport provision; Chapter 9: Overview; Chapter 10: Conclusions, policy implications & recommendations; Bibliography; Appendix 1: Definitions of adequacy; Appendix 2: Area research summary; Appendix 3: A comparative analysis of journey characteristics; Appendix 4: Profile of respondents.	http://www.dft.gov.uk/stellent/groups/dft_mobility/documents/pdf/dft_mobility_pdf_506795.pdf

Older People: Their Transport Needs & Requirements – Main Report (Feb 2001)	The report is an important contribution in understanding the transport needs & requirements of older people. These needs are not just about the ability to travel from A to B – travel serves a number of important functions for older people. The report proves a useful guide for local authorities & transport operators in developing affordable, safe, accessible & available transport systems which meet these needs.	Foreword; Executive summary; Introduction; Demographic changes & the policy context; Previous research; Qualitative findings; The household survey; conclusions – key issues; Study recommendations; Appendices.	http://www.dft.gov.uk/stellent/groups/dft_mobility/documents/pdf/dft_mobility_pdf_506792.pdf
Transport 2010: Meeting the Local Transport Challenge (12/04/01)	This document describes how the settlement will help to implement the Government's vision of a modern local transport system. Chapter 1 explains the LTP Settlement (Local Transport Plan) for England in more detail, and provides examples of the schemes that will be progressed and the achievements that are expected over the next five years. Chapter 2 explains how the settlement, & local transport plans in general, will help to deliver the outcomes and targets set out in the 10 Year Plan.	Foreword; Introduction; Chapter 1: The local transport settlement for England; Chapter 2: Meeting the local transport challenges; Better public transport; Better integration; Better accessibility; Reducing the environmental impact of traffic; Easing urban congestion; Safer roads; Properly maintained roads; Encouraging cycling, walking & Home Zones; Centres of excellence; Conclusion; Further information.	http://www.dft.gov.uk/stellent/groups/dft_localtrans/documents/pdf/dft_localtrans_pdf_503949.pdf
Transport Trends: 2001 Edition (29/05/01)	The intention of this section is to provide transport planners, operators & policy makers with an insight into the major trends in domestic transport. The Government's 1998 White Paper, *A New Deal for Transport*, undertook to continue to publish *Transport Trends* so that progress & the impacts of polices can be monitored. There is as yet no single, definitive, set of indicators being used nationally or internationally to monitor transport trends & policies. However, in 2000 the Government published its 10 Year Plan for Transport, which includes a number of quantified indicators & targets. The information in *Transport Trends* reflects these indicators and initiatives, & represents a starting point for the development of a national set of key indicators for use by government, regional & local authorities, businesses & others involved in the transport sector.	List of indicators/Data tables; Commentary & indicators; Notes & definitions; Article 1: Monitoring the 10 year plan for transport; Article 2: Understanding attitudes to transport policy; Article 3: The effect of the 'fuel protest' on road traffic; Article 4: Travel by taxi & minicab; Article 5: Motorcycling in Great Britain; Article 6: Partially responding households in the National Travel surveys; Article 7: The use & characteristics of vehicle stock data; Article 8: How do passengers travel from UK airports.	http://www.dft.gov.uk/stellent/groups/dft_control/documents/contentservertemplate/dft_index.hcst?n=7683&l=3

Continued

Table 11.5.3 UK Level Transport Policy. *Continued*

Policy Type	Document Title	Objective	Contents	Website
Strategies & Reports	Crime & Disorder on Public Transport (Nov 2001)	This briefing paper is for Crime and Disorder Partnerships. Its aim is to encourage them to address the issue of public transport safety. It provides information about: ▪ public transport use ▪ crime-related problems ▪ what can be done to encourage security, and ▪ how to access useful data.	Introduction; Context; Who uses public transport? Crime, disorder & fear on public transport; What can be done to prevent crime & disorder on public transport? Who's who on public transport? Sources of information; Contacts.	http://www.dft.gov.uk/stellent/groups/dft_mobility/documents/pdf/dft_mobility_pdf_503806.pdf
	Perceptions of Congestion: Report on Qualitative Research Findings (10/12/01)	The purpose of the research was to explore drivers' experiences and perceptions of traffic congestion, and to help evaluate possible approaches to congestion measurement.	Perceptions of congestion; Report on qualitative research findings; Executive summary; The study; Problems for drivers; Congestion; Perceived trends; Importance of congestion; Coping with congestion; Tackling congestion; Other perspectives on congestion; Measuring congestion; Public transport; Conclusion; Appendices.	http://www.dft.gov.uk/stellent/groups/dft_roads/documents/pdf/dft_roads_pdf_503854.pdf
	Tomorrow's Roads: Safer for Everyone – The Government's Road Safety Strategy & Casualty Reduction Targets for 2010 (26/03/02)	This document sets out the Government's framework for improving road safety. The strategy contains many specific recommendations, but these should not be seen as strict guidelines for the future. Inevitably in the course of 10 years, new ideas and technology will emerge which will help to achieve the target.	Summary; Foreword; Introduction; Safer for children; Safer drivers – training & testing; Safer drivers – drink, drugs & drowsiness; Safer infrastructure; Safer speeds; Safer vehicles; Safer motorcycling; Safety for pedestrians, cyclist & horseriders; Better enforcement; Promoting safer road use.	http://www.dft.gov.uk/stellent/groups/dft_rdsafety/documents/pdf/dft_rdsafety_pdf_504644.pdf
	Flexible Transport Services (24/07/02)	This booklet describes the flexible services currently available in more detail, with examples where available, including car sharing schemes. Because all may not be aware of these flexible options, this booklet is aimed principally at existing or would-be bus, taxi and private hire operators; at local authorities; at transport partnerships; and at community transport operators and voluntary groups who would like to develop their own schemes.	Foreword; Part 1 – Buses & taxis in the countryside; Part 2: Flexible options; Special bus & taxi licences; Running public service vehicles (buses); Community & group bus services; Part 3: Sharing schemes; Taxis & private hire; Community & other car schemes; Annexes.	http://www.dft.gov.uk/stellent/groups/dft_localtrans/documents/pdf/dft_localtrans_pdf_504004.pdf

Child Road Safety: Achieving the 2010 Target (Sept 2002)	Section one introduces the issue of child road safety; Section 2 focuses on what the casualty data tells us; Section 3 reviews strategy progress so far; Section 4 outlines the Department's research programme; Section 5 refers to our demonstration projects; Section 6 describes publicity initiatives; and Section 7 sets out how all road safety partners can contribute in taking the strategy forward.	Section 1: Introduction; Section 2: Casualties – What the Data tells us; Section 3: Strategy Progress; Section 4: Research; Section 5: Demonstration Projects; Section 6: Publicity; Section 7: Taking the Strategy Forward – an Action Plan for Delivery; Annexes.	http://www.dft.gov.uk/stellent/groups/dft_rdsafety/documents/page/dft_rdsafety_507867.pdf
Department of Transport – Greening Operations Policy Statement (Oct 2002)	This policy statement sets out the Department's aims and objectives for greening its operations in line with the Framework for Sustainable Development on the Government Estate which was launched in July 2002.	Introduction; Responsibility – Corporate level; Local management level; Individual level; Associate bodies; Aims; Objectives; Energy; Waste; Water; Wood; Paper; Horticulture; Climate change; Ozone-depleting substances; Traffic congestion & vehicle emissions; Pesticides & artificial fertilisers; Asbestos; Hazardous substances; Batteries, solvents & paints; Bio-degradable substances; Litter; Noise; What this means for the procurement of goods & services; What this means for estate management; Monitoring & auditing; Guidance.	http://www.environmental-management.gov.uk/envoy/System/Uploads/20021019019_DfT%20Greening%20Operations%20Policy%20Statement%20-%201002.pdf
Sustainable Development in the Department for Transport (Nov 2002)	This report describes the contribution of the Department for Transport to Sustainable Development in the Government Estate.	Introduction; Estate impacts; Policy impacts; social impacts; Useful links; Contacts.	http://www.dft.gov.uk/stellent/groups/dft_about/documents/pdf/dft_about_pdf_507663.pdf
Walking – the Way Ahead: Report from the National Seminar Series (July 2003)	The main purpose of this publication is to present the good practice case studies in a way which is informative, helpful &, it is hoped, interesting to policy makers, practitioners & campaigners. There are many different aspects to the promotion of walking & the improvement of the pedestrian environment. The case studies are arranged in sections under fifteen different aspects of walking policy. Under each aspect are one or more case studies. In this way the reader should be able quickly to find the case studies which are of most interest to her/him.	Preface; Introduction; The good practice casebook; Evaluation & audit of pedestrian needs; Community involvement; Road crossings; Signs & lighting; Home zones; Road charging & the needs of the pedestrian; Walking & regeneration; Walking & health promotion; Walking in rural areas; Walking to school; Walking to work; Walking to leisure facilities; Walking – links with public transport; Joined up management of pedestrian facilities; Environmental enforcement; Seminar series case study list; Bibliography.	http://www.dft.gov.uk/stellent/groups/dft_localtrans/documents/pdf/dft_localtrans_pdf_022858.pdf

Continued

Table 11.5.3 UK Level Transport Policy. *Continued*

Policy Type	Document Title	Objective	Contents	Website
Strategies & Reports	Travelling to School: An Action Plan (Sept 2003)	Government policy wants to bring about a step change in home to school travel patterns to cut congestion & pollution, but also to allow many more pupils to take regular exercise. This document sets out ideas and incentives to help & reward schools that commit themselves to increasing the proportion of pupils walking & cycling or – for those living some distance from school – catching the bus. There are already 2000 schools that have adopted this agenda & many have found that it is surprisingly easy to engage their local transport authority in setting up walking buses & changing road layouts.	Foreword; Getting to school; Safely & healthily; Takes vision & commitment; Each School; Every local education authority & transport authority; With our support; And piloting changes to the rules on statutory school transport; Can help pupils travel independently; Useful contacts; Guidance on school travel plans & related issues.	http://www.dft.gov.uk/stellent/groups/dft_susttravel/documents/page/dft_susttravel_024011.pdf
	The Announce System Monitoring Report (Dec 2003)	The Announce project is a partnership formed between the Department for Transport (DfT), the Disabled Passenger Transport Advisory Committee (DPTAC), and others. The aim of the project was to introduce & evaluate the benefits of audible & visual next stop information on buses. The monitoring programme carried out consisted of focus groups, passenger interviews, analysis of passenger complaints, mystery traveller surveys, feedback from driving & supervisory staff, & analysis of patronage. Appropriate techniques were used to analyse results.	Introduction; Pre Scheme Monitoring – Focus Group; Pre Scheme Monitoring – Passenger Survey; Pre Scheme Conclusions; Announce Post Scheme Monitoring; Post Scheme Passenger Survey; Post Scheme Survey Conclusions; Complaints Received by ARRIVA and Metro; Disability Discrimination Act Focus Groups; Patronage Analysis; Reliability of Audible and Visual Next Stop Announcements; Driver Focus Groups; A bus company perspective; Costs of wide scale deployment of the Announce system; Cost–Benefit Analysis; Conclusions.	http://www.dft.gov.uk/stellent/groups/dft_mobility/documents/page/dft_mobility_507581.pdf
	Tomorrow's Roads – Safer for Everyone: The First Three Year Review (April 2004)	The primary objective of the review was to provide a realistic evaluation of the effectiveness of the strategy (& its implementation) & of the likelihood of delivering the 2010 targets. This report presents the conclusions of the review.	Foreword; Chapter 1: Introduction; Chapter 2: Progress towards the targets; Chapter 3: The policy review; Theme 1: Safer for children; Theme 2: Safer drivers – training & testing; Theme 3: Safer drivers – drink, drugs & drowsiness; Theme 4: Safer infrastructure; Theme 5: Safer speeds; Theme 6: Safer vehicles; Theme 7: Safer	http://www.dft.gov.uk/stellent/groups/dft_rdsafety/documents/pdf/dft_rdsafety_pdf_028165.pdf

		motorcycling; Theme 8: Safer for pedestrians, cyclists & horseriders; Theme 9: Better enforcement; Theme 10: Promoting safer road use; Chapter 4: Conclusions & next steps; Chapter 5: Beyond 2010.	
Smarter Choices – Changing the way we Travel (July 2004)	This report draws on earlier studies of the impact of soft measures, new evidence from the UK & abroad, case study interviews relating to 24 specific initiatives, & the experience of commercial, public & voluntary stakeholders involved in organising such schemes. Each of the soft factors is analysed separately, followed by an assessment of their combined potential impact.	Introduction; Overview of other soft factor studies; Workplace travel plans; School travel plans; Personalised travel planning; Public transport information & marketing; Travel awareness campaigns; Car clubs; Car sharing schemes; Teleworking; Teleconferencing; Home shopping; Projections of the potential traffic impacts of soft factors & associated costs; Conclusions.	http://www.dft.gov.uk/stellent/groups/dft_control/documents/contentservertemplate/dft_index.hcst?n=10689&l=1
The Future of Rail – White Paper (July 2004)	This White Paper sets out the railway review's conclusions. They have been based on analysis of the past & current performance of the railways, & on detailed submissions from & discussions with key players from the industry & other relevant organisations. As agreed with the Chairman of the Transport Committee, this White Paper also serves as the formal response to the Committee's recent report on the railways.	Foreword; Executive summary; Chapter 1 – The case for reform; Chapter 2 – Where we are now; Chapter 3 – The role of Government; Chapter 4 – A better industry structure for delivery; Chapter 5 – Better decision-making; Chapter 6 – Making change happen.	http://www.dft.gov.uk/stellent/groups/dft_railways/documents/pdf/dft_railways_pdf_031105.pdf
The Future of Transport – White Paper (July 2004)	This White Paper looks at the factors that will shape travel & transport over the next thirty years & sets out how the Government will respond to the increasing demand for travel, maximising the benefits of transport while minimising the negative impact on people & the environment. It represents a long-term strategy for a modern, efficient & sustainable transport system backed up by sustained high levels of investment over the next 15 years.	Executive summary; The challenge: demand for travel & under-investment; What we have achieved; Roads: smarter travel; Transforming our railways; Buses: better services, easier access; Walking & cycling: a positive choice; Aviation & shipping: delivering sustainable growth; Freight: moving goods more efficiently; Making choices locally & regionally; Protecting the environment; Safety & security: protecting people & networks; Annexes.	http://www.dft.gov.uk/stellent/groups/dft_about/documents/divisionhomepage/031259.hcsp

Continued

Table 11.5.3 UK Level Transport Policy. *Continued*

Policy Type	Document Title	Objective	Contents	Website
Strategies & Reports	In-car Safety & the Personal Security Needs of Female Drivers & Passengers (Aug 2004)	The project had four aims: ▪ Review the accident statistics injury patterns through STATS 19 & the Co-operative Crash Injury Study (CCIS). ▪ Recognise the social & behavioural reasons underlying them. ▪ Identify any specific concerns of female car users including safety & security. ▪ Develop an optimal car design to address their concerns & reduce injuries. This paper concentrates on accident data, & women's stated requirements reflected in the optimal car design.	Background; Introduction; Male & female car access & travel patterns; Female car user's concerns & priorities; Optimal car design; Conclusions; References; Acknowledgements.	http://www.dft.gov.uk/stellent/groups/dft_mobility/documents/pdf/dft_mobility_pdf_030414.pdf
	Encouraging Walking & Cycling: Success Stories (Oct 2004)	*Walking and Cycling: an action plan*, which was published on 16 June 2004, sets out measures from across government to increase levels of active travel. This companion guide contains 50 examples of successful schemes from across England. They have all achieved significant results with often modest, but always well-chosen, initiatives which have improved local conditions for walking & cycling & encouraged people to get around on foot and by bike. These schemes illustrate clear benefits in terms of reduced congestion, improved public health & enhanced quality of local streets & spaces.	Introduction; Chapter 1: Improving the walking and cycling environment; Chapter 2: Better facilities for walking and cycling; Chapter 3: Influencing travel behaviour.	http://www.dft.gov.uk/stellent/groups/dft_susttravel/documents/divisionhomepage/031517.hcsp

	Transport Trends: 2004 Edition (Dec 2004)	This publication presents an overview & analysis of trends in transport & travel in GB over the past twenty years, & highlights some of the key issues. It is intended as a companion volume to *Transport Statistics Great Britain*, which contains reference tables containing more detailed figures &, in some cases, longer time trends. This is the seventh annual edition of *Transport Trends*, published to fulfil the commitment in the Government's 1998 Transport White Paper (Cm 3950).	Summary; Section 1: Roads, vehicles & congestion; Section 2: Personal travel by mode; Section 3: Public transport; Section 4: Variation in personal travel & access to services; Section 5: Freight & logistics; Section 6: Ports & airports; Section 7: Safety; Section 8: Health & the environment; Annex 1: Transport & sustainable development; Annex 2: Abbreviations.	http://www.dft.gov.uk/stellent/groups/dft_transstats/documents/pdf/dft_transstats_pdf_508294.pdf
Consultation Papers	Transport Safety (April 1999)	This consultation forms the first part of the review process. The review is not concerned with specific safety strategies or measures, but with the principles which should govern how transport safety is organised, with any institutional obstacles to better safety regulation, & with potential conflicts of interest between the different elements of the safety regime. Views are invited on these issues.	Summary; Part 1: Introduction; Part 2: Transport safety responsibilities – issues for consideration; Part 3: Accident inquiries & legal proceedings; Annexes.	http://www.dft.gov.uk/stellent/groups/dft_localtrans/documents/pdf/dft_localtrans_pdf_504095.pdf
	The Flexible Future – A Consultation Paper on Changes to Local Bus Service Registration & Bus Service Operators Grant Requirements to Allow more Flexibly Routed Services (Aug 2002)	Following the consultation, it is intended to make the necessary regulatory changes quickly so that the public can benefit from more flexible buses as soon as possible. With this in mind, the ideas put forward in this document are based on adapting two existing regimes: registration of local bus services with the Traffic Commissioners & financial support for bus services through the Bus Service Operators' Grant (BSOG), formerly called Fuel Duty Rebate (FDR). In the longer term, there may be a case for a legal base other than registration for these services. And the BSOG is currently under review.	Introduction & summary of issues; The case for change; The present statutory position; 'Prescribed particulars' & the scope for flexibility; Constraints on flexibility; The relationship between buses & taxis; The guiding principles; Specific amendments to allow the registration of flexible services; Particulars of registration; Alternative registration particulars for service type (c); Operation of flexible, registered local bus services by PSV vehicles with fewer than 9 passenger seats (small PSVs); Modifying the BSOG rules to extend it to flexible, registered services; Other changes to the registration rules; Responses to this document.	http://www.dft.gov.uk/stellent/groups/dft_localtrans/documents/page/dft_localtrans_503970.pdf

Continued

Table 11.5.3 UK Level Transport Policy. *Continued*

Policy Type	Document Title	Objective	Contents	Website
Consultation Papers	On the Move: By Foot – A Discussion Paper (July 2003)	This discussion paper was aimed primarily at practitioners in local authorities & the health sector. However, views were also wanted from the much wider range of people & organisations which influence the street & surrounding environment (such as telecommunication & utility companies), also voluntary groups & the communities themselves. This document suggests how to both improve conditions for walking & increase the number of journeys made on foot. Some ideas had already been suggested at a series of seminars run for local authority experts throughout England during 2002. It was the aim to build on these views & the responses to this discussion document to produce the new strategy document to be a part of the Department for Transport's (DfT) contribution to the broader 'liveability' agenda being developed across government.	Introduction; The case for walking; The key challenges; The right planning policies & practice; A better physical environment for walkers; Supporting the choice to walk; Improving the institutional framework; Indicators; Bringing it all together; Your views; Annex: the evolving context; Bibliography.	http://www.dft.gov.uk/stellent/groups/dft_localtrans/documents/downloadable/dft_localtrans_022250.pdf
	Working in Partnership with the Voluntary Sector (Oct 2003)	The Department for Transport is strongly committed to creating a framework in which the voluntary & community sector can flourish, be strong & independent. The Compact will be used as the basis for effective partnership working with the sector & to underline respect for its independence. The DfT recognises the contribution voluntary & community organisations make to developing & delivering policies, to service delivery, to reducing social exclusion & to achieving sustainable development. In 1998 the Government agreed a Compact which recognised the important role of the Voluntary & Community Sector in the engagement of volunteers, the services it provides, the support it gives to individuals & groups & its immense contribution to community & civil life. A summary of the main provisions of the Compact & its codes are at Annex 1.	Consultation Information; Foreword; Part 1: Introduction; Personal Mobility; Volunteering within the Department for Transport; Part 2: Action Plan; Annex 1: About the Compact; Annex 2: About the Department for Transport; Annex 3: About the Agencies; Annex 4: Department for Transport: Non Departmental Public Bodies (NDPBs).	http://www.dft.gov.uk/stellent/groups/dft_mobility/documents/pdf/dft_mobility_pdf_024869.pdf

Guidance Notes	Encouraging Walking: Advice to Local Authorities (28/03/00)	This document is a working guide for the people who will put policy into action. It is based on the work of an advisory group drawn together from a wide range of organisations with interests in the issues. This strategy therefore sets out ways in which walking can be made easier, more pleasant & safer.	Introduction; Chapter 1: Changing the priorities; Chapter 2: Planning & partnership; Chapter 3: Practical actions; Chapter 4: Keeping up the momentum; Appendix A: Checklist for developing a strategic approach to walking; Appendix B: Checklist for the local walking environment; Appendix C: Outline of recommendations of advisory group & membership of advisory & working groups; Appendix D: Publications.	http://www.dft.gov.uk/stellent/groups/dft_susttravel/documents/page/dft_susttravel_504172.pdf
	Guidance on Full Local Transport Plans (March 2000)	This document replaces Guidance on Provisional Local Transport Plans, issued in April last year. It advises local authorities on producing full local transport plans (LTPs) which are due to be submitted to Government Offices by 31 July 2000. Separate guidance will be issued for Scotland and Wales.	Introduction; Part I – The local transport plan process; Timetable; Statutory local transport plans; London; Best value & transport; Working in partnership; Joint LTPs; Public participation; What LTPs should contain; Monitoring; Assessment of LTPs; The allocation of resources; Level of available resources; Annual progress reports; Part II – The coverage of local transport plans; An integrated approach; Traffic management & demand restraint; Integrated transport; Planning & managing the highways network; Rural transport; Sustainable distribution; Integration with wider policies; Part III – Road Traffic Reduction Act 1997: guidance to local authorities; Purpose of this guidance; Assessing levels of traffic; Setting targets; Consultation; Production & publication of reports; Annex A: The Road Traffic Reduction Act; Annex B: National road traffic forecasts definitions; Annex C: References; Annex D: Assessment criteria for LTPs; Annex E: Appraisal; Annex F: Finance forms; Annex G: Assessing the applicability of PFI (Public Finance Initiative) options.	http://www.dft.gov.uk/stellent/groups/dft_localtrans/documents/page/dft_localtrans_033303.pdf http://www.dft.gov.uk/stellent/groups/dft_localtrans/documents/page/dft_localtrans_033304.pdf

Continued

Table 11.5.3 UK Level Transport Policy. *Continued*

Policy Type	Document Title	Objective	Contents	Website
Guidance Notes	A Good Practice Guide for the Development of Local Transport Plans (27/04/00)	This guide highlights examples of good practice from provisional LTPs so that authorities can learn from each other. It does not replace the Department's guidance for local authorities on preparing LTPs, but should be read in conjunction with it. This guide has been developed primarily to help local authority officers responsible for developing & producing LTPs. However, we also hope it will be of interest to the wider audience now involved in the LTP process, including transport operators & providers, local communities, businesses and transport interest groups.	Foreword; Chapter 1: Introduction; Chapter 2: The local transport planning process; Chapter 3: Establishing partnerships; Chapter 4: Vision; Chapter 5: Identifying & analysing problems; Chapter 6: Setting objectives; Chapter 7: Strategy; Chapter 8: Measuring success; Chapter 9: Local transport plan presentation; Appendix: Other sources of good practice.	http://www.dft.gov.uk/stellent/groups/dft_localtrans/documents/pdf/dft_localtrans_pdf_504010.pdf
	Home Zones & Quiet Lanes (Aug 2001)	This document sets out the Department's thinking on statutory guidance and regulations covering home zones and quiet lanes in England. It also invites your comments. The same primary legislation covers both home zones and quiet lanes – section 268 of the Transport Act 2000. But any guidance will distinguish between the two, and following this consultation the Department may wish to make separate regulations for each.	The consultation criteria; List of consultees; Consultation on statutory guidance & regulations; Introduction; The primary legislation; Statutory guidance on designation; Regulations governing designation procedures; Regulations authorising speed orders & use orders; Procedure regulations for making use orders & speed orders; Transport Act 2000.	http://www.dft.gov.uk/stellent/groups/dft_localtrans/documents/pdf/dft_localtrans_pdf_503971.pdf
	A Road Safety Good Practice Guide (08/10/01)	This is the second publication in the DTLR's good practice advice series on local transport. This guide shares good practice to help achieve the targets set out in DTLR's road safety strategy *Tomorrow's Roads – Safer for Everyone* (2000). This guide has been developed primarily as a reference for local authority staff with an interest in road safety engineering and associated issues.	Acknowledgements; Introduction; Planning, liaison & consultation; General principles; Treatment selection – engineering measures; Measuring effectiveness; Bibliography; Abbreviations; Feedback; Appendix A: Road accident countermeasures; Appendix B: Statistical tests; Appendix C: MOLASSES (Monitoring of Local Authority Safety Schemes).	http://www.dft.gov.uk/stellent/groups/dft_rdsafety/documents/pdf/dft_rdsafety_pdf_504564.pdf

Making Travel Plans Work – Lessons from UK Case Studies (Aug 2002)	This guide has been written for employers who want to reduce traffic congestion around their sites, improve the travel options available for their staff – and save money at the same time. This guide points to key success factors and features of good practice for setting up an effective travel plan. It is based on a recent review of the experience of 20 UK organisations that have successfully brought about a change in the travel patterns of their staff.	Foreword; Introduction; Part 1: Travel plan strategy & design; Part 2: Measures for change; Supporting bus & rail; Supporting walking; Supporting cycling; Supporting car sharing; Managing parking; Reducing the need to travel & other strategies; Part 3: Funding travel plans; Further information.	http://www.dft.gov.uk/stellent/groups/dft_susttravel/documents/page/dft_susttravel_504107.pdf
Inclusive Mobility – A Guide to Best Practice on Access to Pedestrian & Transport Infrastructure (1/11/02)	Although the main purpose of these guidelines is to provide good access for disabled people, designs that satisfy their requirements also meet the needs of many other people. Those who are travelling with small children or are carrying luggage or heavy shopping will all benefit from an accessible environment, as will people with temporary mobility problems (e.g. a leg in plaster) and many older people. Thus, the overall objective of this guide is to provide inclusive design and through that achieve social inclusion.	Introduction; Basic human factors information; Footways, footpaths & pedestrian areas; Tactile paving surfaces; Car-parking; Bus stops; Taxi ranks; Access to & within transport-related buildings; Transport buildings – facilities; Signage & information; Lighting; Access in the countryside; Consultation, training & management; Glossary; Bibliography; Useful addresses; Summary dimensions sheet.	http://www.dft.gov.uk/stellent/groups/dft_mobility/documents/pdf/dft_mobility_pdf_503282.pdf
Travelling to School: A Good Practice Guide (Sept 2003)	Each day during the school term millions of pupils and their parents travel from home to school in the morning, & make the return trip in the afternoon. Many pupils living close to school walk, with those living further away travelling mainly by bus or by car. The school journey affects public transport patterns, causes localised congestion around schools & contributes to the sharp road traffic peak around nine o'clock each morning. This document describes what schools, local authorities & bus operators around the country have been doing to promote walking, cycling & public transport & combat increasing car use.	Foreword; Introduction; More walking & cycling; Encouraging bus use; Car sharing; Transport provision for children with special educational needs; Sustainable travel & road safety – within & beyond the curriculum; School travel plans & accessibility planning; Annexes.	http://www.dft.gov.uk/stellent/groups/dft_susttravel/documents/page/dft_susttravel_023992.pdf

Continued

Table 11.5.3 UK Level Transport Policy. *Continued*

Policy Type	Document Title	Objective	Contents	Website
Guidance Notes	In Town Without my Car – Good Practice Guide (Mar 2004)	This guide promotes a vision of town centres less dominated by cars. It is about using the streets as social spaces and for public transport, walking and cycling, allowing us to see what our town centre streets were like without through motor traffic for one day. Town centre streets were closed to cars and lorries, and opened up for people to enjoy other activities like walking, cycling, street theatre, live music, dancing, street cafés, public art and children's play areas.	Foreword; Introduction; How to participate in their town, without my car day; In town without my car 2002 UK case studies; UK opinion poll results; Lessons learnt; Case notes from some major European cities; European mobility week; Annexes.	http://www.dft.gov.uk/stellent/groups/dft_susttravel/documents/pdf/dft_susttravel_pdf_507956.pdf
	The Benefits of Green Transport Plans: The Guide (Mar 2004)	This guide is for businesses or other organisations that want to save money by using transport more efficiently. It explains the benefits of setting up a Green Transport Plan. By following the advice in this guide, it is suggested that one should be able to save money, make life better for staff and help the environment.	Foreword; Section 1: Green transport plans; Section 2: What does a green transport plan consist of? Section 3: Typical transport costs.	http://www.dft.gov.uk/stellent/groups/dft_susttravel/documents/pdf/dft_susttravel_pdf_027688.pdf
	Guidance on Accessibility Planning in Local Transport Plans (July 2004)	This is a consultation on draft guidance on accessibility planning in Local Transport Plans. Authorities will be expected to pay greater attention to accessibility within their second Local Transport Plans following the Social Exclusion Unit report *Making the Connections*. This report lays out the relationship between transport, social exclusion & the location of jobs & key services & forms a cross-government strategy for reducing social exclusion by improving access to jobs & key services.	Chapter 1: Introduction – background & purpose of the Guidance; Chapter 2: Accessibility strategies & plans in local transport plans; Chapter 3: Policy context for accessibility; Chapter 4: Accessibility assessments – identifying & analysing accessibility problems; Chapter 5: Option appraisal & identifying resources; Chapter 6: Measures to tackle accessibility barriers; Chapter 7: Measuring success – performance monitoring framework; Appendices.	http://www.dft.gov.uk/stellent/groups/dft_localtrans/documents/page/dft_localtrans_030119.pdf
	Full Guidance on Local Transport Plans Second Edition (Aug 2004)	This document sets out in draft form the Government's advice to local authorities on the approaches & methods they should adopt when drawing up their next Local Transport Plans (LTPs). Local transport authorities in England, outside London, will be required by the Transport Act 2000 to have regard to the final version of this guidance in drawing up their next LTPs.	Introduction; Summary: Key Elements of A Good LTP; Part 1 – The Government's Approach to Delivering Better Local Transport; Part 2 – Developing A High Quality LTP; Part 3 – Priorities for Local Transport Planning; Part 4 – Value for Money; Part 5 – Local Transport Plan Practicalities; Annexes.	http://www.dft.gov.uk/stellent/groups/dft_localtrans/documents/pdf/dft_localtrans_pdf_030145.pdf

Table 11.5.4 Northern Ireland Level Transport Policy.

Policy Type	Document Title	Objective	Contents	Website
Strategies & Reports	Transportation in NI – The Way Forward (Oct 1995)	In announcing a rethink of transportation policy in Northern Ireland the Minister for the Environment outlined seven principles that would underpin future transportation planning.	Developments in transportation policy; NI response; NI framework for action; Influencing the future use of the car; Protecting the environment; Improving public transport; Reducing the mobility gap; Providing for pedestrians & cyclists; Making better use of the road network; Developing strategic links; Altering travel behaviour; Summary.	http://www.nics.gov.uk/transport/wayf/wayf-fra.htm
	Environmental Handbook (Oct 1998)	This handbook helps to identify the impact of roads on the environment & offers advice on how to control or minimise the effects. Everyone within Roads Service with a role in designing or managing roads & verges should read this handbook. Initial advice is given in this handbook on the planting of trees and the treatment of road boundaries as well as on matters relating to the protection of wildlife & the built heritage.	Foreword; Introduction; General principles; Roadside verges; Roadside boundaries; Weed control; Trees & shrubs; Flora & fauna; Pollution; Areas of conservation value; Bridges; Street lighting; Recycling; Appendices.	http://www.roadsni.gov.uk/Publications/specific/envhandb.pdf
	Moving Forward – NI Transport Policy Statement (Nov 1998)	This Policy Statement is an important step in developing a strategy to enable a move from a transport system which is dominated by, & to some extent geared towards, car use to a more balanced & integrated transport system, in which public transport & non-motorised transport are attractive options fcr many trips.	Foreword; Executive summary; Chapter 1: An integrated transport policy framework; Chapter 2: Regional transport plan; Chapter 3: Interim regional transport plan; Chapter 4: Conclusions.	http://www.nics.gov.uk/transport/itp/

Continued

Table 11.5.4 Northern Ireland Level Transport Policy. *Continued*

Policy Type	Document Title	Objective	Contents	Website
Strategies & Reports	Northern Ireland Cycling Strategy (June 2000)	The Northern Ireland Cycling Strategy identifies a range of measures that will seek to improve conditions for cyclists & establish a pro-cycling culture. If more people are to be encouraged to cycle, a fundamental shift in attitude will be needed in every part of our society. Raising public awareness of the benefits of cycling is, therefore, one of the Strategy's key objectives.	Introduction; Strategic context; Health; Increasing cycle use; Planning for sustainable access; Improving cyclists' safety; Creating a cycle friendly infrastructure; Integrating with other modes; Providing cycle parking; Reducing cycle theft; Travel plans; School travel plans; Tourism; Sport & leisure; Raising public awareness; Community participation & consultation; The voluntary sector; Research; Resources; Monitoring progress; Annexes.	http://www.roadsni.gov.uk/publications/Cycling/contents.pdf
	Road Safety Engineering Report – Annual Report (Sept 2000)	This report monitors the effectiveness of the accident remedial & traffic calming programmes implemented from 1995/96 to 1997/98 & provides details of those schemes undertaken in 1998/99. The success of the accident remedial programme is set against the background of an overall increase in accidents that has occurred on Northern Ireland's public roads during the monitoring period.	Introduction; Accident remedial measures; Traffic calming measures; Appendix I: Accident Remedial Schemes; Appendix II: Traffic Calming Schemes; Appendix III: 1998/99 Accident Remedial and Traffic Calming Scheme; Definitions.	http://www.roadsni.gov.uk/Publications/specific/Road%20Safety%20Engineering%20Report.pdf
	Summary of Transport Problems, Issues & Opportunities (Feb 2002)	The Belfast Metropolitan Transport Plan (BMTP) is being developed by the Department for Regional Development (DRD), NI, to deliver a modern integrated transport system to serve the needs of the Belfast Metropolitan Area (BMA) Plan Area. This report presents a summary of the transport problems, issues & opportunities that need to be considered in developing the BMTP. The identification of problems, issues & opportunities is an essential first stage in the systematic process of plan development that is being followed by DRD.	Introduction; The policy context; The area covered by the plan; Personal travel in the Belfast Metropolitan Area; Freight movements in the BMA; Key generators & attractors; The road network; The public transport network; Walking & cycling; Transport impacts; Perceptions of transport problems, constraints & opportunities; Next steps.	https://pronet.wsatkins.co.uk/Bmtp/docs/BMTP_Summary.pdf

Regional Transportation Strategy for NI 2002–2012 (July 2002)	The Regional Transportation Strategy (RTS) for NI 2002–2012 identifies strategic transportation investment priorities & considers potential funding sources & affordability of planned initiatives over the next 10 years. The RTS is a 'daughter document' of the Regional Development Strategy (RDS) which sets out the spatial development framework for NI up to 2025. The purpose of the RTS is to support the RDS and to make a significant contribution over the 10 years towards achieving the longer-term vision for transportation contained within the RDS.	Executive summary; Introduction; The need for the regional transportation strategy; Regional transportation strategy contribution to the transportation vision; Developing the regional transportation strategy. Content of the regional transportation strategy; Funding & affordability of the strategy; Expected outcomes of the regional transportation strategy; Implementation, targets, monitoring & review; Annexes.	http://www.drdni.gov.uk/rts/pdf_files/final_pdfs/RTS_main_doc.pdf
Northern Ireland Road Safety Strategy 2002–2012 (Nov 2002)	The Northern Ireland Road Safety Strategy seeks to promote an integrated approach to the planning, co-ordination and delivery of the Government's road safety activities over the next decade, aimed at achieving the casualty reduction targets set out in the next section. It represents the commitment of the departments and agencies, whose activities contribute to road casualty reduction, to addressing the strategic objectives set out in the consultation document.	Foreword; Vision, key principles & background to the strategy; Targets; Strategic objectives; Co-ordination; Monitoring & reporting; Annex A: Equality, human rights & new targeting social need; Annex B: Consultees responding to the consultation document.	http://www.doeni.gov.uk/roadsafety/pdfs/roadsafetystrategy_051102.pdf
Regional Transportation Strategy – Annual Monitoring Report 2002–2003 (Aug 2003)	This document is the first report on the work to date in implementing the RTS. The Department has initiated mechanisms to monitor & report on progress in implementing the RTS & will draw upon these sources to report more fully in future years. The Department is fully committed to demonstrate transparently how effectively the RTS is being progressed & the difference that the strategy is making, even at this early stage in its implementation. This report is in two sections, Section A provides an overview of the main highlights during 2002/03, and Section B represents a financial summary.	Introduction; Section A: Background; Funding; M1/Westlink Strategic investment programme; Private sector involvement; Transport plans; Accessible transport strategy; Delivery structures; Pilot rapid transit scheme; Integrating transport with land use & other policies; Monitoring tables; Section B: Funding.	http://www.drdni.gov.uk/shapingourfuture/pdf/rts_annual_%20monitoring_report_2002_2003%20.pdf

Continued

Table 11.5.4 Northern Ireland Level Transport Policy. *Continued*

Policy Type	Document Title	Objective	Contents	Website
Strategies & Reports	Regional Strategic Transport Network Transport Plan: Emerging Plan (Sept 2003)	The Regional Strategic Transport Network Transport Plan (RSTNTP) is currently being prepared by the Department for Regional Development based upon the guidance set out in the Regional Development & Regional Transportation Strategies. The Plan will set out how Northern Ireland's Regional Transportation Strategy (RTS), will be implemented on the Regional Strategic Transport Network (RSTN). The development of the Transport Plan is now at an important stage & this document is intended to: ▪ inform delegates of the process which was used to develop the Emerging Plan; ▪ provide detail of the Emerging Plan; & ▪ to seek comments and stimulate discussion on the Emerging Plan at the Conference. The conference will help us consult the key stakeholders & obtain informed comment and opinion on the proposals presented in this Emerging Plan.	Foreword; Executive Summary; Introduction; Regional Strategic Transport Network Transport Plan; The Working Conference; Proposed Initiatives by Mode; Implementation, Targets, Monitoring & Review; Annexes.	http://www.drdni.gov.uk/foi/search/details.asp?docid=842
	Regional Strategic Transport Network Transport Plan: Report of Working Conference (Oct 2003)	A Working Conference on the contents of the Emerging Regional Strategic Transport Network Transport Plan was held in September 2003. Over 170 delegates attended it from all 26 local councils, political parties, business, private developers, academic institutions, consumer groups, community & environmental groups, trade unions, government departments & statutory agencies.	Executive Summary; Introduction; Background; Conference Format; Workshop Findings; Open Forum Discussion; Appendix – Conference Facilitators.	http://www.drdni.gov.uk/foi/search/details.asp?docid=894

Travel Survey for Northern Ireland 2000–2002 (Nov 2003)	This is the second Travel Survey for Northern Ireland (TSNI) report. The first reported on the years 1999–2001, whilst this one covers the 2000–2002 time period. The Travel Survey for Northern Ireland is based on the National Travel Survey (NTS), as used in Great Britain. The TSNI is the only source of information on how, over the region as a whole, people as individuals or family groups use different forms of transport to meet their travel needs. It is & will continue to be an important source for many years to come & will provide information to influence government policy, set objectives & monitor performance in relation to transport & travel in NI.	Key points; Section 1: Introduction; Section 2: Trends in personal travel; Section 3: How people travel; Section 4 Why people travel; Section 5: Other factors affecting travel; Appendix A: Travel survey for NI definitions & survey roles.	http://www.roadsni.gov.uk/Publications/specific/Travel_Survey_for_NI_V2%20.pdf
Recommendations of Northern Ireland Inquiry into 'Home to School Bus Transport' – The Development of Terms of Reference for a Cost–Benefit Analysis for use by DOE (NI) (Nov 2003)	In September 2000, and prompted by general concerns about the safety & comfort of children travelling to & from school & specifically about overcrowding on buses, the Northern Ireland Assembly Environment Committee began a public inquiry into home to school transport. The objective of this, the Phase 1 Study, is: ▪ to carry out an initial, structured, appraisal of the 28 Inquiry recommendations & then to prioritise them, & ▪ to draft appropriate Terms of Reference for the cost-benefit analysis assignment. This document constitutes Volume I of the report on the Phase 1 Study and presents the prioritisation of the inquiry recommendations and the development of the cost-benefit analysis framework for the terms of reference. Volume II presents the draft Terms of Reference.	Executive summary; Introduction; Findings emerging from the inquiry recommendations; Assessment & prioritisation of inquiry recommendations; Conclusions & recommendations; Appendix A: Bibliography; Appendix B: Nata (New approach to appraisal framework; Appendix C: Inquiry recommendations; Appendix D: Development of a Cost–Benefit Analysis Framework.	http://www.doeni.gov.uk/uploads/NI_DOE_School_Bus_TOR_Vol1_FR211103.pdf
Walking in Northern Ireland: An Action Plan (Dec 2003)	The purpose of this Action Plan is to set out a comprehensive list of actions that will assist in the delivery of improved conditions for walking. These actions complement the substantial commitment to measures aimed at 'Making it easier to walk' identified in the Regional Transportation Strategy.	Introduction; Background; Objectives, outcomes & targets; Planning for pedestrians; Design for pedestrians; Improving existing walking routes; Improving pedestrian safety & security; Walking for health; Walking for leisure; Tourism; Raising public awareness; Monitoring progress Appendices.	http://www.roadsni.gov.uk/publications/specific/Walking_Action_Plan.pdf

Continued

Table 11.5.4 Northern Ireland Level Transport Policy. *Continued*

Policy Type	Document Title	Objective	Contents	Website
Strategies & Reports	Northern Ireland Transport Statistics 2003–2004 (Sept 2004)	This publication presents Northern Ireland statistics on various transportation issues including vehicle registrations, driver & vehicle testing, the road network, freight, road safety, public transport, air transport & other general transport statistics. The publication is updated on an annual basis to provide an accurate picture of transportation within Northern Ireland.	Commentary; Chapter 1 Vehicle Registrations; Chapter 2 Driver and Vehicle Testing; Chapter 3 Road Network; Chapter 4 Road Freight; Chapter 5 Road Safety; Chapter 6 Public Transport; Chapter 7 Air Transport; Chapter 8 General Transport Statistics; References.	http://www.drdni.gov.uk/foi/search/details.asp?docid=1505
	Belfast Metropolitan Transport Plan 2015 (Nov 2004)	This document presents the Belfast Metropolitan Transport Plan: a comprehensive & realistic local transport plan for the Belfast Metropolitan Area (BMA) up to 2015. The transport plan has been prepared by the Department for Regional Development. It takes forward the strategic initiatives of the Regional Transportation Strategy for Northern Ireland 2002–2012, setting out detailed transport proposals for the BMA in a form that will enable them to move towards implementation (subject to detailed economic appraisal, funding availability & statutory processes).	Executive summary; Introduction; Developing the plan; An overview of the 2015 plan; Walking & cycling; Public transport; The highway network; Management; Proposals by council area; Key topic areas; Implementation programme; Outcomes, targets, monitoring & review; Annexes.	https://pronet.wsatkins.co.uk/Bmtp/
	Equality Impact Assessment: Belfast Metropolitan Transport Plan 2015 (Nov 2004)	Section 75 of the 1998 Northern Ireland Act requires the Department for Regional Development, in carrying out its functions relating to Northern Ireland, to have due regard to the need to promote equality of opportunity between: ▪ Persons of different religious belief, political opinion, race, age, marital status & sexual orientation; ▪ Men & women generally; ▪ Persons with a disability & persons without; & ▪ Persons with dependants and persons without.	Executive Summary; Introduction; The Approach to Assessment; Consideration of Available Data and Research; Assessment of Impacts; Conclusions of Analysis; Mitigation/Alternatives; Consultation; Conclusions; Appendix A – List of Consultees.	https://pronet.wsatkins.co.uk/Bmtp/final/finaldoc/BMTPEqIA291004.pdf

		This document sets out the Final EQIA undertaken for the BMTP in accordance with the Department's Equality Scheme and framed by the EQIA undertaken for the RTS.		
Consultation Papers	Developing a Regional Transportation Strategy – A Consultation Paper (Jan 2001)	This consultation paper provides background information & seeks views on the key issues affecting the movement of people & goods in NI. It is part of the consultation process which the Department for Regional Development is conducting to inform the development of a 10 year Regional Transportation Strategy (RTS) for NI. The paper outlines the need for a RTS, sets it within the wider policy context, outlines the methodology being used in the development of the strategy, gives guidance on the consultation process & explains how views on the key issues can be submitted.	Introduction; Why we need a regional transportation strategy; Policy context for the regional transportation strategy; The transportation vision; Methodology for the development of the regional transportation strategy; Funding; Forwarding your views; Annex 1: Glossary; Annex 2: Potential solutions.	http://www.drdni.gov.uk/rts/pdf_files/rts.pdf
	Proposed Regional Transportation Strategy – A Consultation Paper (Feb 2002)	This consultation paper seeks views on the Department for Regional Development's proposals for a 10 year Regional Transportation Strategy (RTS) for NI. The Department would also welcome comments on the draft Equality Impact Assessment that has been prepared in respect of the Proposed Strategy. The Proposed Strategy is set out in Chapter 5, the summary findings of the draft Equality Impact Assessment are at Annex A & the draft Equality Impact Assessment is published, in full, as an accompanying report to this document. This paper also outlines the need for the RTS, sets it within the wider policy context, considers funding issues and highlights expected outcomes against the strategic objectives and for particular groups of people. Finally, proposals are suggested for implementation, monitoring & review of the Proposed Strategy.	Foreword; Introduction; The need for the regional transportation strategy; Proposed regional transportation strategy contribution to the transportation vision; Developing the regional transportation strategy; Content of the proposed regional transportation strategy; Expected outcomes of the proposed regional transportation strategy; Implementation, monitoring & review proposals; Annexes.	http://www.drdni.gov.uk/rts/pdf_files/RTS%20Draft%20report.pdf

Continued

Table 11.5.4 Northern Ireland Level Transport Policy. *Continued*

Policy Type	Document Title	Objective	Contents	Website
Consultation Papers	A New Start for Public Transport in NI – A Consultation Paper (Sept 2002)	The purpose of this paper is to provide a basis for consultation on the future planning & delivery of public transport in NI. Responses to the paper will guide DRD in preparing the necessary legislative & other measures to give effect to the proposed new institutional & regulatory framework.	Introduction; Policy context; Current institutional arrangements; Proposed institutional arrangements; Next steps.	http://www.drdni.gov.uk/transportpolicy/pdfs/new%20start.pdf
	Belfast Metropolitan Transport Plan – Working Conference Papers (Feb 2003)	Consultation during the development of the Plan is a key element of this process. A number of consultation meetings have already been held with Elected Representatives, a Key Stakeholders Group & with twenty Reference Groups. The Working Conference forms another & very important element of the consultation on the priorities & context of the emerging Plan. This document sets out the proposals in the form of the 2025 Draft Preferred Strategy, explaining its rationale by drawing upon past consultation & the work undertaken by the BMTP team in assessing alternative long-term transport strategies.	Introduction; Developing a transport plan – guiding principles; A draft preferred strategy for 2025; An emerging plan for 2015; Annex A: Definitions of BMTP objectives; Annex B: BMTP guidance from RDS & RTS; Annex C: Glossary.	https://pronet.wsatkins.co.uk/Bmtp/docs/WCP_Website.pdf
	Proposed Regional Strategic Transport Network Transport Plan 2015: A Consultation Paper for Equality Impact Assessment (July 2004)	The purpose of the RSTN TP is to plan the maintenance, management & development of Northern Ireland's Strategic Transport Network in accordance with the strategic direction & underlying principles of the RTS.	Executive Summary; Introduction; Regional strategic transport network transport plan; Walking & cycling; Public transport; Highways; Outcomes, targets, monitoring & review; Annexes.	http://www.drdni.gov.uk/foi/rs/search/details.asp?docid=1377

	An Accessible Transport Strategy for Northern Ireland: Draft for Public Consultation (Oct 2004)	Older people & people with disabilities make up a significant proportion of the Northern Ireland population & they face a number of barriers when trying to make use, or make full use of the transport system. Transport accessibility is about providing transport services, infrastructure & information that provides for equitable access. This Strategy aims to address the barriers that prevent older people & people with disabilities from using the transport system. The Accessible Transport Strategy stems from the Regional Transportation Strategy & has been developed in the context of the Priorities & Budget 2004–2006.	Executive Summary; Introduction; Policy context for the development of an Accessible Transport Strategy; Why an accessible transport system is needed; Components of an Accessible Transport System; The ATS Action Plan; Proposed funding & affordability of the Strategy; Progress review & monitoring; Appendices.	http://www.drdni.gov.uk/foi/search/details.asp?docid=1554
Guidance Notes	Transport Assessment – Traffic Impact Assessment Guidelines (Mar 2001)	This report provides brief notes relating to various aspects of a traffic impact assessment that have been prepared to assist developers & consultants in understanding the roads service's technical requirements.	General guidance note; Technical guidance note; Existing conditions; Proposed development; Modal choice/trip attraction; Trip distribution; Trip assignment; Assessment years; Highway impact; Environmental impact; Road safety; Internal layout; Parking layout; Public transport, cycling & walking; Access for the disabled; Travel impact of major retail developments; Safety audit; Other relevant documents; Appendices.	http://www.roadsni.gov.uk/Publications/specific/TransportAssess.pdf

12 Urban Development

12.1 Urban Development and the Built Environment

Sustainable development is the backdrop or framework by which urban development can be organised for those who live, work and play in a built-up zone or area. Urban living is increasing across the world due to rapid population growth and the movement of people from rural communities into the burgeoning cities. This is especially problematic in the developing nations where the concept of 'mega-cities' is emerging. These are sometimes referred to as 'mega-slums' due to the inadequacy of infrastructure, housing, city management, community structures and an urban culture, and the significant increase in environmental degradation, disease and violence. There are over 400 cities in the world with a population of over 1 million people, while Mexico City has in excess of 30 million inhabitants. None of these statistics are congruent with sustainable communities and the aspirations of sustainable development.

There has been an international vision for urban development for several years, succinctly expressed in the UN Conference on Human Settlements (1976), known as Habitat I. This conference produced the Vancouver Declaration which 'established the concept of human settlements to consist of several elements that had been previously considered separately from one another'. These elements included housing, building, planning and the relationship of these with other activities such as environmental change, national and international development. This vision was reinforced at Habitat II (1996) in the Istanbul Declaration on Human Settlements, in which nations subscribed their commitment to a political, economic, environmental, ethical and spiritual vision of human settlements based on the principles of equality, solidarity, partnership, human dignity, respect and co-operation. This was supplemented by an agreement for 'adequate shelter for all' and for 'sustainable human settlements' in an urbanising world.

The built environment, in the urban context, is essentially larger in scale and complexity than rural towns, villages and communities, requiring a more sophisticated approach to a range of urban development processes. The integration of concepts, services, management and operation are well established in the Habitat statements, and these can be translated into four key and co-existing tenets or objectives of an urban development:

- *Building Prosperity* to facilitate sustainable economic growth which provides increased opportunities for all residents. This can be achieved through building economic partnerships, providing quality environments which are attractive to investors, supporting a competitive business environment, facilitating access to employment opportunities, promoting a strong rural economy, facilitating the expansion of tourism, and ensuring a responsive education service for industry and commerce.
- *Caring for the Environment* by creating a clean, safe and secure environment which contributes to the improvement of health and well-being in communities. This can be achieved by maintaining and enhancing the natural and built environments, reducing global warming and energy use, protecting and enhancing biodiversity, and maintaining or increasing tree coverage.

- *Improving Communications* by providing reasonable accessibility for all residents to the full range of resources, services and opportunities and developing an efficient and sustainable transport system.
- *Valuing People* by harnessing the resources of all residents to achieve social progress for the benefit of all. This can be achieved by endorsing equality of opportunity and equity of treatment and by meeting future housing demand.

These objectives can be assessed by measurements, in light of the UN Conference on Environment and Development (UNCED) and the *Agenda 21* document agreed in Rio de Janeiro (1992), when it became imperative to assess the state of sustainable development using appropriate indicators. These covered the three pillars of sustainable development – economics, environment and society. Urban development should be created within a sustainable development framework and is governed by a range of economic, business and environmental decision-making processes. In an urban environment, the planning system plays a critical role and has a direct impact through the restrictions it places on the market value of land.

The Åalborg +10 European Conference (1994) confirmed a shared vision of a sustainable future for communities. This vision advocated cities and towns that are inclusive, prosperous, creative and sustainable, and that they provide a good quality of life for all citizens, enabling their participation in all aspects of urban life. Such an approach required appropriate and innovative design, a theme adopted at the European Urban Design for Sustainability Conference (2004), which identified good practice models and strategies of urban design to support sustainability in EU and EU-accession countries, presenting a review of best practice and making recommendations for action at all levels. It explored the theme of re-designing and retro-fitting existing urban areas, designing for greenfield sites, and knitting the urban fabric together to achieve an integrated citywide vision.

12.2 Key Urban Development Elements

Urban development comprises several elements relating to planning, design, operation, maintenance and management, and these provide the urban legacy and the culture of the conurbation for its residents. Its generic aspiration should be to secure, create and develop a compact, mixed use area which is a desirable location for people, business and industry to function in line with the co-existing objectives given above. Each development should demonstrate the detailed elements of design excellence, economic strength, environmental and social well-being and good governance. Furthermore each development should be positively influenced by the current driving factors of the technological revolution, an ecological imperative and the desire for social transformation.

The creation of a city or town, involves some or all of a number of indicative elements of process and activity. Each development should be allowed to self-create within legal and planning parameters, thereby utilising innovative processes as required. A successful urban development process strives for balanced integration of the following key elements:

- *Urban Governance* is the activity that makes the city engine work, despite political differences, bureaucracies and arbitrary limits. It requires leadership, imagination, flexibility, participation, and negotiation abilities. The objective of good governance is to optimise the performance of a city according to shared ambitions, excluding particular interests, and prioritising the residents' living conditions. This can be delivered by City Councils, Urban Alliances or government officials.
- *Strategic Urban Planning* deals with the future image and shape of a city and how residents are going to cope with a changing environment. It involves urban stakeholders in the design and implementation of the city strategy, as well as strong leadership, social and political support, and a detailed analysis of the spatial-economic conditions which are fundamental in strategic planning.
- *Economic Development* refers to the processes which occur at a city level to underpin the local economy and urban competitiveness, including attracting inward investment, encouraging public–private partnerships and facilitating job creation and local enterprises.
- *Infrastructure Provision* meets anticipated demand for transportation, utilities and telecommunication networks. Public transport, road, rail, airport, port, and the broader concept of mobility are essential elements of urban management, enabling people and goods to move at an inter/intra-metropolitan

level. Similarly its effectiveness becomes a fundamental condition for urban competitiveness and social cohesion. The efficient supply of utilities includes water distribution, sewerage systems and treatment, water amenities and features, energy supply through electricity and gas and telecommunications through telephones and network coverage.

- *Habitat Protection and Housing Provision* implies creating a balance between building more residential accommodation and maintaining or protecting the limited amount of green-field land available. It also refers to meeting the basic needs and rights of society by supplying adequate living conditions, guaranteeing secure tenure, preventing or avoiding housing slums and appropriate levels of local service provision, all at an affordable cost.
- *Urban Regeneration Delivery* requires a pro-active, integrated approach that incorporates environmental, economic and social dimensions. Free trade movement and the globalisation of the world economy have resulted in cities having to compete for inward investment and economic growth. This will result in rapid decline of urban areas unless they demonstrate a willingness to compete and an inbred survival instinct to sustain positive economic forces.
- *Consolidation of Sustainable Communities* requires a set of guiding principles and an action plan which will provide a zonal or community development which attempts to secure zero waste output, uses sustainable integrated transport, sources local construction material, sources local food, promotes water efficiency, conserves flora and fauna, provides adequate education, seeks high levels of employment, delivers social services, respects cultural heritage, considers equality and fairtrade and secures healthier lifestyles.
- *Partnership Development* involves teamwork and links between government, private sector industries and non-governmental organisations (NGOs) across all phases of strategic planning, project procurement, delivery and operation, regeneration and at high strategic levels within themes such as waste management, infrastructure provision and energy. Team building, transparent accountability and shared visions are essential for these partnerships to operate effectively.

The performance of the urban environment should be monitored against a set of sustainable development indicators, drawn up by stakeholders from the city or town, based on the *Local Agenda 21* process. The following list gives an indicative set of indicators under the three pillars of sustainable development:

- *Environmental*: air, water, natural environment, built environment, waste, energy and transport.
- *Social*: education, housing, physical health, mental health, drug use, crime rate, community involvement.
- *Economic*: business sustainability, economic structures, tourism, economic activity, wealth, poverty, employment.

12.3 Sustainability Challenges for Urban Development

Urban development is a high level complex activity which can be used to help facilitate and implement sustainable development. As it relates to and interacts with several processes it needs to respond to the economic, environmental and social aspects of sustainability in equal measure. Urban development has links to and can have a key impact on environmental protection, social issues and economic structures while also contributing to the resource management objective of sustainable development, as follows:

- *Effective protection of the environment* through the appropriate design, development, use and management of the natural and built heritage. Urban development can prevent environmental damage through reconciling green-field and brown-field development, preserving urban green space, stipulating the need for environmental management systems in new buildings, and ensuring that spatial planning encourages and champions the use of public transport through the promotion of a compact urban form.
- *Social progress which recognises the needs of everyone* is at the heart of quality urban development as urban structures and urban living can only be successful if people want to live there. This means

creating a vibrant, safe and exciting space which not only encourages social integration between culturally diverse communities but also facilitates equal opportunities across all religious, racial, and ethical boundaries. Providing the appropriate social infrastructure such as community focus groups and local-level partnerships can help improve quality of life, and greatly enhance social inclusion at all levels of governance and democracy.

- *Prudent use of natural resources* is an essential part of sustainable urban development and can be achieved through appropriate land uses and mixed development. Reconciling limited land availability and site productivity can be achieved through increased urban densities and compact urban form, while the reuse of brown-field sites and buildings, the use of recycled materials, green procurement and citywide support for waste minimisation will restrict the scavenging of virgin materials.
- *Maintenance of high and stable levels of economic growth and employment* can be delivered by the job creation associated with the construction industry, an area's competitiveness in attracting inward investment and the business acumen of the region. Economic growth is often created in an urban development scheme by using public funding to pump-prime infrastructure which will lever substantially more private investment. Furthermore economic growth can be delivered through a quality urban development and the opportunity for planning gain, to aid long-term employment and profitability.

The pursuit of sustainability has placed significant challenges on urban development, and a positive response is essential to secure the sustainable development objectives. The following list summarises the key sustainability challenges that are likely to be faced, now and in the future, by urban development:

- *Design and development of sustainable communities* which meet the aspirations of sustainable development and the issues raised at both Habitat I and II. This could be achieved by addressing the local and global issues of the physical and spatial organisation of human life, accommodating growing populations in urban and rural communities and avoiding urban sprawl, all within the framework of *Local Agenda 21*.
- *Driving urban design excellence* through building quality, creativity, and the incorporation of public spaces and transport systems which creates a sustainable urban paradigm where mixed-use development strengthens social integration and civic life while respecting private needs. Further issues such as suburbanisation, pedestrian movement and preserving the historic character of urban areas will challenge the architects, planners and engineers.
- *Deliver an environmentally friendly and appropriate infrastructure* which meets long-term needs, by providing the basic needs of clean water, sanitation, pollution reduction and affordable housing.
- *Secure economic stability, strength and identity* through committed businesses, targeted public sector investment and support for areas of decline, regions with social exclusion and for key activities such as education, arts and culture.
- *Invest in local urban governance* to establish a clear vision for democratic participation and delivery of best value by developing local partnerships and facilitating local empowerment and community participation.
- *Develop urban and rural links* by preventing rural isolation while encouraging countryside protection.
- *Encourage social inclusion* for all of society through economic development, social integration and sustainable communities, including consideration of social cohesion, securing inward investment and seeking poverty reduction with better health care and adequate education.
- *Encourage deeper studies and understanding for a range of professionals on the role of cities* to ensure competitiveness, productive cities, globalisation of local trade and development, reduced industrial, transportation and domestic use polluters, efficient and effective infrastructure, improved spatial development planning, and increased attention on healthy cities for all ages and interests.
- *Protect and improve the urban environment through environmental responsibility* by reducing reliance on natural resources, minimising air pollution, avoiding land contamination, seeking energy efficiency, enhancing biodiversity, and reusing or cleansing brown-field sites. This will improve quality of life, stimulate greater awareness and promotion of cultural, built and natural heritage, assist in

the development of a green space strategy, and encourage engagement with slum and wasteland clearances.

- *Govern and plan for sustainable regeneration* by using brown-field land in preference to green-field alternatives, building reuse and recycling materials or down-taking.
- *Set all proposals within Regional Development Strategies* to consolidate an urban-wide vision to meet future population and housing demands. This could be achieved by securing involvement and use of integrated transport, negotiating sustainable growth, employment and consolidation of spatial development and settlement trends through designing for and avoiding peripherality issues.

12.4 Urban Development Theme Index

Jurisdiction	Tier	Title	Date	Page No.
INTERNATIONAL	Conferences	UN Conference on Human Settlements (Habitat I)	1976	490
		International Conference on Population & Development (ICPD)	1994	490
		Ninth UN Conference on Trade & Development (UNCTAD IX)	1996	490
		Second UN Conference on Human Settlements (Habitat II)	1996	491
		International Meeting on Urban Poverty	1996	491
		International Conference on Urban Poverty	1997	492
		Third International Conference on Social Integration & Security for the Urban Poor – Towards Cities for All	1999	492
		Tenth session of the United Nations Conference on Trade & Development (UNCTAD X)	2000	492
		UN Special Session of the General Assembly for an Overall Review & Appraisal of the Implementation of the Habitat Agenda (Istanbul + 5)	2001	493
		Fourth International Conference on Productive & Inclusive Cities – Towards Cities for All	2001	493
		First Session of the World Urban Forum	2002	493
		2003 International Healthy Cities Conference	2003	494
		The International Conference on Sustainable Urbanisation Strategies	2003	494
		Second International Federation of Surveyors Conference – 'Urban–Rural Inter-relationships for Sustainable Environment'	2003	495
		Eleventh Session of the United Nations Conference on Trade & Development (UNCTAD XI)	2004	495
		Second World Urban Forum	2004	495
		Inter-Regional Conference on Urban–Rural Linkages	2004	496
	Conventions	UNESCO World Heritage Convention (WHC)	1972	496
EUROPEAN	Council Decisions	Integrating the Environmental Dimension into the Sustainable Development of the Urban Environment *(Dec No. 1411/2001/EC)*	2001	497
	Council Communications	Green Paper on the Urban Environment *(COM (90) 218)*	1990	497
		Towards an Urban Agenda in the EU *(COM (97) 197)*	1997	497

Jurisdiction	Tier	Title	Date	Page No.
UK	Strategies & Reports	Moving Towards Excellence in Urban Design & Conservation	2003	508
		Sustainable Communities: Building for the Future	2003	509
		Cities, Regions & Competitiveness	2003	509
		Making it Happen: Thames Gateway & Growth Areas	2003	510
		Searching for Solid Foundations: Community Involvement & Urban Policy	2003	510
		Living Places: Caring for Quality	2004	510
		Creating Sustainable Communities: Greening the Gateway – A Greenspace Strategy for Thames Gateway	2004	511
		Competitive European Cities: Where do the Core Cities Stand?	2004	511
		Making it Happen: The Northern Way	2004	511
		Evaluating the Impact of Design Awards for Housing	2004	512
		Our Cities Are Back: Competitive Cities make Prosperous Regions & Sustainable Communities – Third Report of the Core Cities Working Group	2004	512
	Consultation Papers	Urban Policy Evaluation Strategy – Consultation Paper	2002	513
	Guidance Notes	Sustainable Regeneration Good Practice Guide	1998	513
		Running & Sustaining Renewal Areas: Good Practice Guide	1999	514
		Regeneration that Lasts	2000	514
		Local Strategic Partnerships – Government Advice	2001	515
		Assessing the Impacts of Spatial Interventions	2003	515
NORTHERN IRELAND	Strategies & Reports	Shaping Our Future – Towards a Strategy for the Development of the Region	1997	516
		Shaping Our Future – Population & Housing Technical Report	1998	516
		Shaping Our Future – Family of Settlements Report	1998	516
		Shaping Our Future – Regional Baseline Report	1998	517
		Strategic Environmental Appraisal Report – A Sustainability Assessment of the Draft Regional Strategic Framework for NI	1999	517
		The CRISP Story	1999	517
		A Strategic Review of the Impact of District Partnership Expenditure in NI – Executive Summary	2000	518
		Reinvigoration of Town Centres Report	2000	518
		URBAN Community Initiative in NI 1994–1999	2001	518
		Programme for Government – Making a Difference (2001–2004)	2001	519
		Draft Regional Development Strategy for NI 2025	2001	519
		Final Regional Development Strategy for NI 2025	2001	519
		Belfast Metropolitan Area Plan 2015 (BMAP) Issues Paper	2001	520

Jurisdiction	Tier	Title	Date	Page No.
NORTHERN IRELAND	Strategies & Reports	Implementation of the Regional Development Strategy & Monitoring & Evaluation Arrangements	2002	520
		Programme for Government – Making a Difference (2002–2005)	2002	520
		Programme for Government Annual Report 2001/02	2002	521
		Building on Progress – Priorities & Plans 2003–06	2002	521
		The Executive's Position Report to the Assembly: Developing the Programme for Government & the Budget for 2003–04	2002	521
		Equality Impact Assessment (EQIA): Urban Regeneration – Neighbourhood Renewal	2003	522
		Final Closure Report by the Department for Social Development on the Urban Operational Programme in Northern Ireland (1994–1999)	2003	523
		Belfast Regeneration Office Draft Strategy	2003	523
		People & Place – A Strategy for Neighbourhood Renewal	2003	523
		Implementation and Monitoring of the Regional Development Strategy for Northern Ireland 2025	2003	524
		People & Place: A Strategy for Neighbourhood Renewal – Urban Regeneration in Northern Ireland: Policy Measurement	2003	524
		Environmental Assessment of the URBAN II Community Initiative Programme for Northern Ireland	2003	524
		Economic Development Forum – Working Together for a Stronger Economy – First Review	2004	525
		Monitoring Report – Third Annual Report Regional Development Strategy for Northern Ireland 2025: April 2003–March 2004	2004	525
		Third Annual Report Regional Development Strategy: April 2003–March 2004	2004	526
	Consultation Papers	Shaping Our Future – An Invitation to Consultation	1997	526
		Shaping Our Future – Public Consultation Report by the Consortium	1998	526
		Shaping Our Future – Draft Strategy	1998	527
		URBAN II Programme 2000–2006 for NI – Consultation Document	2000	527
		Urban Regeneration in NI – Neighbourhood Renewal a Consultation Document	2001	527

12.5 Urban Development Policy Tables

Table 12.5.1 International Level Urban Development Policy.

Title	Venue	Purpose	Key Outcome(s)	Website
UN Conference on Human Settlements (Habitat I)	Vancouver, Canada (1976)	In 1976 in Vancouver, the United Nations held its first conference on the issue of physical & spatial organisation of human life on this planet, & on the national & international actions needed to accommodate the growing population in urban & rural communities. This conference established the concept of human settlements to consist of several elements that had been previously considered separately from one another – housing, building, planning & the relationship of these & such other activities as environmental change & national & international development.	The Vancouver Declaration on human settlements which highlighted 19 General Principles & 10 guidelines for action.	http://www.unchs.org/declarations/The_Vancouver_Declaration.pdf
International Conference on Population & Development (ICPD)	Cairo, Egypt (1994)	To consider the broad issues of and interrelationships between population, sustained economic growth and sustainable development, and advances in the education, economic status and empowerment of women.	The present Programme of Action recommends to the international community a set of important population and development objectives, as well as qualitative and quantitative goals that are mutually supportive and of critical importance to these objectives. Among these objectives and goals are: sustained economic growth in the context of sustainable development; education, especially for girls; gender equity and equality; infant, child and maternal mortality reduction; and the provision of universal access to reproductive health services, including family planning and sexual health.	http://www.unhchr.ch/html/menu5/popdev94.htm
Ninth UN Conference on Trade & Development (UNCTAD IX)	Midrand, South Africa (1996)	To give recognition to idea that only through a global partnership for development can the international community eliminate disparities between and within nations.	The Midrand Declaration gives strong political boost to international co-operation for development and to UNCTAD as the focal point within the United Nations for the integrated treatment of development and inter-related	http://www.unctad.org/Templates/Meeting.asp?m=4290&intItemID=1942&lang=1

			issues in the areas of trade, finance, technology, investment & sustainable development. The Declaration calls for greater partnership between developed, developing countries and the least developed countries. It also emphasised the benefit of involving the civil society in the work of the organisation. A Partnership for Growth and Development: ▪ Part I – promoting growth and sustainable development in a globalising and liberalising world economy ▪ Part II – the contribution of UNCTAD to sustainable development ▪ Part III – the future work of UNCTAD: institutional implications.	
Second UN Conference on Human Settlements (Habitat II)	Istanbul, Turkey (1996)	The two major themes of the conference were adequate shelter for all & sustainable human settlements development in an urbanising world.	The Habitat Agenda: a commitment to a political, economic, environmental, ethical and spiritual vision of human settlements based on the principles of equality, solidarity, partnership, human dignity, respect and co-operation. The Istanbul Declaration on Human Settlements.	http://www.unchs.org/unchs/english/hagenda/index.htm
International Meeting on Urban Poverty	Recife, Brazil (1996)	Participants carried out an intensive programme of discussion in sectoral groups (focusing on employment & the urban informal sector, access to land & basic services, & social integration), in groups working on poverty reduction issues at different levels (community, municipal, & national), as well as in plenary sessions, addressing general cross-cutting problems & operational proposals. The Recife Meeting discussed strategies for urban poverty reduction to provide recommendations to the City Summit (Habitat II), to community organisations & to institutions at the local, national & global levels.	Recife Declaration – Six key areas are covered in the Declaration: ▪ Globalisation & the challenge of urban poverty; ▪ Defining poverty: Understanding unity & diversity; ▪ Transforming public & private action: forging new relationships with the poor; ▪ Towards enabling policies for cities: investing in the poor; ▪ Organising for action: articulating public & private actors; & ▪ The future of our cities: our common future.	http://www.unhabitat.org/programmes/ifup/rde.asp

Continued

Table 12.5.1 International Level Urban Development Policy. *Continued*

Title	Venue	Purpose	Key Outcome(s)	Website
International Conference on Urban Poverty	Florence, Italy (1997)	The objectives of the Florence Conference were two-fold: ▪ to formally launch the Forum & to put in place mechanisms for it to function; ▪ to bring together representatives from the North & South to deliberate on the theme: governance & participation – practical approaches to urban poverty reduction.	The key thematic areas covered at the conference included: ▪ Governance & participation; ▪ From urban violence & exclusion to safer cities; ▪ Shelter, employment & the informal city; ▪ Access to affordable transport for the urban poor; ▪ Urban children.	http://www.unhabitat.org/programmes/ifup/fc.asp
Third International Conference on Social Integration & Security for the Urban Poor Towards Cities for All	Nairobi, Kenya (1999)	The main objectives of the Conference were to examine the relationships between social integration, poverty alleviation & human security in cities, to identify the main constraints & obstacles for their achievement, to learn from current practices designed to achieve these objectives, to share the lessons from those practices, & to promote & facilitate action plans seeking to eradicate poverty & increase social integration & security in cities.	The conference output was based around four separate workshops: ▪ Informal cities & security of tenure; ▪ Urban safety for the poor & crime prevention; ▪ Safety in urban transport; ▪ Safety of children & youth in urban areas.	http://www.unhabitat.org/programmes/ifup/ifupen.asp http://www.unchs.org/programmes/ifup/DOWNLOAD/dpu.doc
Tenth session of the United Nations Conference on Trade & Development (UNCTAD X)	Bangkok, Thailand (2000)	The States members of the United Nations Conference on Trade & Development (UNCTAD) deliberated on developmental strategies in an increasingly interdependent world & on how to make globalisation an effective instrument for development. In the context of the opportunities created by the technological revolution, the opening of markets & globalisation, the paramount objectives of UNCTAD have been the promotion of growth & sustainable development, the quest for equity & participation by all, translating broadly agreed concepts into effective action, & effective co-operation & co-ordination in dealing with global interdependence & development.	Member States agreed on the Bangkok Declaration & Plan of Action	http://www.unctad.org/en/docs/ux_td390.en.pdf http://www.unctad.org/en/docs/ux_td387.en.pdf http://www.unctad.org/en/docs/ux_td386.en.pdf

UN Special Session of the General Assembly for an Overall Review & Appraisal of the Implementation of the Habitat Agenda (Istanbul – 5)	New York, USA (2001)	At the special session the Assembly adopted a political declaration reaffirming the Istanbul Declaration on Human Settlements & the Habitat Agenda; a review & assessment of the implementation of the Habitat Agenda; & proposals for further actions for achieving the goals of adequate shelter for all & sustainable development of human settlements.	Declaration on Cities & Other Human Settlements in the New Millennium.	http://www.unhabitat.org/istanbul+5/declaration.PDF http://www.unhabitat.org/istanbul+5/A_56_477.pdf
Fourth International Conference on Productive & Inclusive Cities – Towards Cities for All	Marrakech, Morocco (2001)	The objectives of the conference were as follows: ▪ To present & discuss the achievements since the Nairobi Conference; ▪ To draw the lessons from activities undertaken under the two thematic issues & in each region; ▪ To prepare the creation of the Urban Forum.	The output of the conference was organised around two key areas – Inclusive cities; & Productive cities. There were also a series of recommendations developed on: ▪ Competitiveness & equity; ▪ Sustainability & profit; ▪ Exclusion against peace & development.	http://www.unhabitat.org/programmes/ifup/conf/morocco1.htm
First Session of the World Urban Forum	Nairobi, Kenya (2002)	Building on the very valuable experiences of the Urban Environment Forum & the International Forum on Urban Poverty, the unified World Urban Forum has international co-operation in shelter & urban development as its substantive focus, & serves as an advisory body to the Executive Director of the United Nations Human Settlements Programme on these issues. Towards this end, the meetings of the World Urban Forum facilitate the exchange of experiences & the advancement of collective knowledge among cities & their development partners. The meetings of the World Urban Forum also place strong emphasis on the participation of Habitat Agenda Partners & relevant international programmes, funds & agencies, thus ensuring their inclusion in the identification of new issues, the sharing of lessons learnt, the exchange of best practices & good policies.	Some of the key issues considered at the first session of the WUF included: ▪ Preparations for the WSSD (World Summit on Sustainable Development) (Rio +10); ▪ The role of local authorities & other Habitat Agenda partners; ▪ Cities without slums; ▪ Global campaigns; ▪ Monitoring & assessment of urban conditions & trends.	http://www.unhabitat.org/uf/report.pdf

Continued

Table 12.5.1 International Level Urban Development Policy. *Continued*

Title	Venue	Purpose	Key Outcome(s)	Website
2003 International Healthy Cities Conference	Belfast, Northern Ireland (2003)	The conference aimed to demonstrate the successes of Healthy Cities across the world, & also to provide a meeting point & platform for cities & organisations. It also aimed to build strong partnerships at all levels, & draw up goals & strategies for the future. Parallel sessions covered a range of themes from healthy urban planning & city health profiles to promoting the health of young people, mental health & healthy ageing. Each session featured a range of experiences from different parts of the world, which acted as a basis for debate.	The main output from this conference was the Belfast Declaration, adopted by mayors & senior politicians from cities in Europe attending the International Healthy Cities Conference.	http://www.healthycitiesbelfast2003.com/Belfast/BelfastDeclarationFinalUK.pdf http://www.healthycitiesbelfast2003.com/pdf/Conferencereport.pdf
The International Conference on Sustainable Urbanisation Strategies	Weihai, China (2003)	The conference brought together policy makers, managers & practitioners from local & national governments, international organisations & agencies & representatives of civil societies, especially from the Asian & African regions. The conference was designed to facilitate the sharing of best practices, good policies & lessons learned in addressing the social, economic & environmental dimensions of sustainable urbanisation. The conference also focused on issues connected with the realisation of Millennium Development Goals to reduce the proportion of people without sustainable access to safe drinking water & sanitation, & to improve significantly the lives of slum dwellers.	The following themes represent the key outputs of this conference: ▪ Theme one: Sustainable Urbanisation & Economic Development. It will cover issues related to urbanisation and regional economic development, strategic development planning & urban–rural linkages. ▪ Theme two: Sustainable Urbanisation & Social Development. It will cover issues of slum upgrading, social inclusion & poverty reduction in rapidly growing cities, especially in metropolitan areas. ▪ Theme three: Sustainable Urbanisation & Environmental Management. It will cover issues of provision of safe drinking water & sanitation & pollution reduction.	http://www.unhabitat.org/conference/weihai.pdf

Second International Federation of Surveyors Conference – 'Urban-Rural Inter-relationships for Sustainable Environment'	Marrakech, Morocco (2003)	The aim of the conference was to shed some light on selected issues of urban–rural interrelations & to raise awareness of this complex topic. It also tried to explain the close linkage of land policy and land administration.	The main outcome of this conference was the Marrakech Declaration which reflected the findings & recommendations of the 2nd FIG (International Federation of Surveyors) Regional Conference. It is intended to support politicians, senior managers, professional organisations & decision makers in their efforts to enable a balanced co-existence of a sustainable nature & a liveable habitat.	http://www.fig.net/pub/figpub/pub33/figpub33.pdf
Eleventh Session of the United Nations Conference on Trade & Development (UNCTAD XI)	São Paulo, Brazil (2004)	UNCTAD XI constituted an opportunity to identify new developments & issues in the area of trade & development since Bangkok, & to generate greater understanding of the interface & coherence between international processes & negotiations on the one hand & the development strategies & policies that developing countries need to pursue on the other. UNCTAD can play an important role in helping to ensure that coherence for development. Advancing this objective is the overarching goal of the São Paulo Conference.	The Member States of UNCTAD gathered in São Paulo, agreed on a Declaration that they called 'The Spirit of São Paulo' as well as the São Paulo consensus.	http://www.unctad.org/en/docs/td412_en.pdf http://www.unctad.org/en/docs/tdl382_en.pdf
Second World Urban Forum	Barcelona, Spain (2004)	The World Urban Forum was organised & convened by the Executive Director of the United Nations Human Settlements Programme, UN-Habitat, pursuant to paragraph 10 of resolution 18/5 of the Commission on Human Settlements in which the Commission requested the Executive Director 'to promote a merger of the Urban Environment Forum & the International Forum on Urban Poverty into a new urban forum, with a view to strengthening the co-ordination of international support to the implementation of the Habitat Agenda'.	The Forum presented reports on both partner & thematic dialogues as follows: ■ Reports on partner dialogues – Urban Cultures – Urban Realities – Urban Governance – Urban Renaissance. ■ Thematic dialogues: – Urban poor – Urban resources – Urban sustainability – Urban services – Urban disasters and reconstruction.	http://www.unhabitat.org/wuf/2004/default.asp http://www.unhabitat.org/wuf/2004/documents/WUF-FINAL_Report.pdf

Continued

Table 12.5.1 International Level Urban Development Policy. *Continued*

Title	Venue	Purpose	Key Outcome(s)	Website
Inter-Regional Conference on Urban–Rural Linkages	Nairobi, Kenya (2004)	The objective of the conference is to create & promote awareness among national economic & social development decision-makers & planners, their policy advisers & urban-rural development actors, of the need for enhancing the balanced territorial development of both rural & urban areas. The Conference envisages to do this by pooling, collating, synthesising & sharing experiences & lessons learned, with a view to stimulating mutual learning, capacity-building & utilisation. The participants are expected to share their countries' & regions' experiences & review current practices in this area. The theme of the World Habitat Day this year – *Cities: engines of rural development* reflects the substance & significance of the conference. The WHD will provide an opportunity to send & publicise the outcome of the conference to all cities & countries participating in the WHD.	The conclusions of the conference include statements: ▪ underlining critical urban-rural development ▪ calling on governments to adopt a more holistic & integrated regional-planning approach to investment decision making ▪ calling for improved rural urban communication links & the stimulation of growth of small & medium-sized towns to help alleviate the intensification of slum development. A cross-sectoral institutional structure/framework was also created to support effective environmental planning & management together with economic & social programmes and development of physical infrastructure.	http://www.unhabitat.org/whd/2004/urbanrural_overview.asp
UNESCO World Heritage Convention (WHC)	Paris, France (1972)	This Convention recognised a duty of ensuring the identification, protection, conservation, presentation and transmission to future generations of the cultural and natural heritage.	To date, more than 170 countries have adhered to the Convention, making it one of the most universal international legal instruments for the protection of the cultural and natural heritage. Establishment of the World Heritage Committee to establish, keep up to date and publish, under the title of 'World Heritage List' a list of properties forming part of the cultural heritage and natural heritage, which it considers as having outstanding universal value in terms of such criteria as it shall have established. An updated list shall be distributed at least every two years. A trust fund for the protection of the World Cultural and Natural Heritage of Outstanding Universal Value, called 'the World Heritage Fund' is established.	http://whc.unesco.org/nwhc/pages/doc/main.htm

Table 12.5.2 European Level Urban Development Policy.

Policy Type	Programme Title	Policy Objective	Community Measure	Document Reference	Website
Council Decision	Integrating the Environmental Dimension into the Sustainable Development of the Urban Environment	To encourage local authorities to rise to the challenge of environmental issues & to provide them with financial & technical assistance in working towards sustainability.	Decision No. 1411/2001/EC of the European Parliament & the Council on a Community Framework for co-operation to promote sustainable urban development (2001).	Off Jnl L191, 13/07/01	http://europa.eu.int/scadplus/leg/en/lvb/l28106.htm
Council Communications	Green Paper on the Urban Environment	The problems of cities are an early warning signal of a more deep-seated crisis which will force rethinking of current models of organisation & urban development. Of course not all European cities suffer identical problems, given their different levels of economic & social development, different economic functions, & geography. This green paper is intended as a first step towards debate & reflection & attempts to identify possible lines of action.	Communication from the Commission to the Council & Parliament – green paper on the urban environment (27/06/90).	COM (90) 218	http://europa.eu.int/comm/environment/urban/pdf/com90218final_en.pdf
	Towards an Urban Agenda in the EU	This Communication examines possibilities for improving urban development & for increasing the effectiveness of existing Community intervention in urban areas. The intention is not to develop European wide urban policies for matters which are best dealt with at a local or regional scale. However, since it is clear that cities in the EU are facing a number of common problems, there are also opportunities at the European scale to share & facilitate potential solutions. This would not require additional powers at the European level. Rather, much can be achieved through a more focused approach using existing instruments at national & Community level & enhanced co-operation & co-ordination at all levels.	Communication from the Commission on an urban agenda for the European Union (06/05/97).	COM (97) 197	http://europa.eu.int/comm/regional_policy/sources/docoffic/official/communic/pdf/urban/urban_197_en.pdf

Continued

Table 12.5.2 European Level Urban Development Policy. *Continued*

Policy Type	Programme Title	Policy Objective	Community Measure	Document Reference	Website
Council Communications	Sustainable Development in the EU: A Framework for Action	The EU Framework for action for sustainable urban development aims at better co-ordinated & targeted community action for urban problems & is organised under 4 interdependent policy aims: ▪ Strengthening economic prosperity & employment in towns & cities ▪ Promoting equality, social inclusion & regeneration in urban areas ▪ Protecting & improving the urban environment – towards local & global sustainability ▪ Contributing to good urban governance & local empowerment.	Communication from the Commission to the Council, the European Parliament, the ESC & the Committee of the Regions on Sustainable Development in the EU: A Framework for Action (28/10/99).	▪ COM (1999) 605	http://europa.eu.int/comm/regional_policy/sources/docoffic/official/communic/pdf/caud/caud_en.pdf
	Towards a Thematic Strategy for Soil Protection	The purpose of this Communication is to build on the political commitment to soil protection in order that it be achieved more fully & systematically in coming years. As it is the first communication addressing soil protection, it is both descriptive & action orientated in order that a full picture of the complexity of the issue can be fully understood & it can serve as a basis for future work. A distinction is made between soil, which is the subject of this communication & land use, which will be the subject of a separate communication addressing the territorial dimension to be published in 2003.	Communication from the commission to the council, the European Parliament, the Economic & Social Committee & the Committee of the Regions on a thematic strategy for soil protection (16/4/02).	▪ COM (2002) 179	http://europa.eu.int/eur-lex/en/com/pdf/2002/com2002_0179en01.pdf
	Towards a Thematic Strategy on the Urban Environment	Planning for a high level of environmental protection is one of the key elements to achieving the sustainable development of cities and to providing a high quality of life for Europe's urban citizens. The Thematic Strategy on the Urban Environment constitutes an important step towards realising this aim, building on a series of initiatives that have contributed to the	Communication from the Commission to the Council, the European Parliament, the Economic & Social Committee & the Committee of the Regions – Towards a	▪ COM (2004) 60	http://europa.eu.int/eur-lex/en/com/cnc/2004/com2004_0060en01.pdf

		development of Europe's policy on the urban environment (see Annex 1). It is part of the European Community's Sixth Environment Action Programme 'Environment 2010: Our Future, Our Choice', and is one of seven thematic strategies in the Programme introduced to provide a holistic approach to key environmental issues that are characterised by their complexity, the diversity of actors concerned & the need for innovative & multiple solutions.	thematic strategy on the urban environment (11/02/04).		
Miscellaneous	Åalborg Charter of European Cities & Towns Towards Sustainability	The Conference adopted the Åalborg Charter, which provides a framework for the delivery of local sustainable development & calls on local authorities to engage in Local Agenda 21 processes. Nearly 2000 local authorities in Europe participate in the Campaign. A group of 10 networks of cities & towns active in different aspects of sustainable development have joined together to form the Campaign & support local authorities in their work on local sustainable development.	The Åalborg Charter (May 1994).	■ Charter Agreement	http://europa.eu.int/comm/environment/urban/pdf/aalborg_charter.pdf
	European Sustainable Cities Report	In the final years of the 20th Century, the increasing urbanisation of the world coupled with global issues of climate change, water shortage, environmental degradation, economic restructuring & social exclusion, demand a deeper look at the future of cities in Europe. The European Commission Green Paper on the Urban Environment, the Treaty on EU, the Fifth Environmental Action Programme 'Towards Sustainability', the UN World Earth Summit at Rio, the series of UN conferences concluding with Habitat II, all have common themes & recommendations that invite urgent action about sustainability, the future of cities & how these will contribute both locally & globally. The European Sustainable Cities Report expresses how these ideas have been developed & how they should be pursued further in European urban settings.	Report by the Expert Group on the Urban Environment for the DG XI (Mar 1996).	■ Expert Group on the Urban Env Report	http://europa.eu.int/comm/environment/urban/pdf/rport-en.pdf

Continued

Table 12.5.2 European Level Urban Development Policy. *Continued*

Policy Type	Programme Title	Policy Objective	Community Measure	Document Reference	Website
Miscellaneous	Response of the EC Expert Group on the Urban Environment on the Communication 'Towards an Urban Agenda in the European Union'	The formal response of the EC Expert Group on the Urban Environment is intended to, from the point of view of sustainable development, influence & provide direction to the European Commission's efforts to increase the 'effectiveness of Community intervention in urban areas' & maximise the 'opportunities at European scale to share & facilitate potential solutions'.	Response of the EC Expert Group on the Urban Environment on the Communication 'Towards an Urban Agenda in the European Union' (07/05/98).	▪ Response of the EC Expert Group on the Urban Environment	http://europa.eu.int/comm/environment/urban/pdf/respons_en.pdf
	Towards a European Peripherality Index – Final Report	The purpose of this study is to undertake, for the fifteen member states of the European Union & the twelve candidate countries, the calculation of an index of peripherality of the 'potential' type (sometimes also called 'gravity-model' type). In effect, it is assumed that the potential for economic activity at any location is a function both of its proximity to other economic centres & of its economic size or 'mass'. The economic potential of a given location is found by summing the influence on it of all other centres in the system.	Report for General Directorate XVI Regional Policy of the European Commission (Nov 2000).	▪ GD XVI Regional Policy Report	http://europa.eu.int/comm/regional_policy/sources/docgener/studies/pdf/periph.pdf

Towards more Sustainable Urban Land Use: Advice to the European Commission for Policy & Action	The Working Group on Sustainable Land Use, steered by the Expert Group on the Urban Environment, was given the task of assessing whether the European Commission, through a range of activities, is helping or hindering the achievement of sustainable land use in urban areas, which especially requires reducing urban sprawl, the environmentally-sound re-use of previously-developed land & buildings & making more effective use of infrastructure.	Report by the Expert Group on the Urban Environment (Jan 2001).	▪ Expert Group on the Urban Environment Report	http://europa.eu.int/comm/environment/urban/pdf/landuse_wg.pdf
Multi-Level Governance: Linking & Networking the Various Regional & Local Levels	Sustainable & balanced development of the European Union's territory is the subject area that was selected in order to examine new ways of linking various decision-making levels. The objective is to examine & propose co-operation mechanisms that would improve the ways in which the different levels of decision-making are interlinked. The underlying rationale is that powers are shared & not hierarchically organised. The challenge is to find ways of co-operating & organising new complementarities between the various levels, to provide them with better access to the decision-making process & also to help each level to strengthen the effectiveness, coherence & transparency of the whole system.	Report by Working Group 4C on Multi-Level Governance: Linking & Networking the Various Regional & Local Levels (May 2001).	▪ Working Group 4C Report	http://europa.eu.int/comm/governance/areas/group10/report_en.pdf
Towards more Integrated Implementation of Environmental Legislation in Urban Areas – Final Report	The Working Group on integrated implementation of environmental legislation (WG – IIEL) has been set up by the Expert Group and by the European Commission, with the specific task to define & propose the further steps which the Commission could take to encourage more integrated approaches to the implementation of environmental legislation in urban areas. The Working Group was established to give some recommendations on how the integration of the implementation of environmental laws could be made easier, better & more common.	Working Group on Integrated Implementation of Environmental Legislation – WG/IIEL (March 2002).	▪ WG/IIEL Report	http://europa.eu.int/comm/environment/urban/pdf/0302finalreport.pdf

Continued

Table 12.5.2 European Level Urban Development Policy. *Continued*

Policy Type	Programme Title	Policy Objective	Community Measure	Document Reference	Website
Miscellaneous	European Common Indicators – Towards a Local Sustainability Profile	The ECI initiative was started off in May 1999 with the setting up of a Working Group on Sustainable Indicators (for initiative of and under the supervision of the Expert Group on the Urban Environment & led by the French Environmental Ministry) with the task to develop common (harmonised) indicators for local sustainability, in close collaboration with a wider Group of Local Authorities. Since the beginning, the aim of the initiative has been to develop & test indicators reflecting local actions towards sustainability in as integrated way as possible. The outcome of the initial phase was a proposal, suggesting a set of indicators on a limited number of themes, in order to allow the strengthening of some core methodologies through effective implementation. The set is however intended to remain flexible & open to include other relevant topics.	Directorate General Environment sponsored Report (May 2003).	■ DG Environment Report	http://europa.eu.int/comm/environment/urban/pdf/eci_final_report.pdf
	Stakeholder Consultation Event on the Development of the Thematic Strategy on the Urban Environment Overview Report	The aim of the event was to collect the views of stakeholders, through a series of workshops, in order to inform the scope & content of the Strategy, particularly looking at actions at the EU level which the Strategy could include. The event was attended by over 100 delegates from across Europe, representing a wide range of interests. The EC is undertaking extensive consultation to inform the Strategy, one element of which was the Stakeholder Consultation Event. This report presents the outcomes of this part of the consultation process.	Report prepared for the European Commission Directorate-General Environment by Land Use Consultants (July 2003).	■ DG Environment Report	http://europa.eu.int/comm/environment/urban/stakeholder_consultation.htm http://europa.eu.int/comm/environment/urban/pdf/stakeholder_consultation/overview_report.pdf
	Working Group on Sustainable Urban Management Final Report	The Working Group (WG) on Sustainable Urban Management was set up by the EU Expert Group on the Urban Environment to contribute to the preparation & definition of the Thematic Strategy on the Urban Environment foreseen by the Sixth	Sustainable Urban Management Working Group Report (Jan 2004).	■ Working Group Report	http://europa.eu.int/comm/environment/urban/pdf/0401_finalreport.pdf

	Environment Action Programme of the European Union. The targets of the WG can be summarised as follows: ▪ To define major gaps & weaknesses in the management procedures of urban systems, with focus on environmental & health protection; ▪ To propose tools & strategies for an improved sustainable management of European urban areas.			
Urban Design for Sustainability	The report identifies models & strategies of good practice in urban design to support sustainability in EU & EU-accession countries, & presents a review of best practice & recommendations for action at all levels. It explores the themes of re-designing & retro-fitting existing urban areas, designing for greenfield sites, & knitting the urban fabric together to achieve an integrated city-wide vision. These themes are explored within the broader context of achieving sustainable urban development in Europe. The report sets out the main issues to be faced on a Europe-wide scale in response to a common set of 'mega trends'.	Final Report of the Working Group on Urban Design for Sustainability to the European Union Expert Group on the Urban Environment (Jan 200[illegible]).	▪ Working Group Report	http://europa.eu.int/comm/environment/urban/pdf/0404final_report.pdf
The Place of Peri-urban Natural Spaces for a Sustainable City	This report has been drawn up at the request of the European Commission by FEDENATUR, Fédération Européean des Espaces Naturels et Ruraux Métropolitains et Périurbains. It is about natural spaces in Urban areas. This report has two main objectives: ▪ To make a table of values for peri-urban natural spaces & classify their typology. ▪ To formulate recommendations for an improved evaluation of these spaces. The aim of this work remains a modest one. It is a first contribution to the subject, based on a study with numerous parks & exchanges between the members of this network. It is not its aim to deal with all the aspects of this particularly complex subject.	FEDENATUR Report to the European Commission, DG Environment (Jan 2004).	▪ FEDENATUR Report	http://europa.eu.int/comm/environment/urban/pdf/report_en.pdf

Continued

Table 12.5.2 European Level Urban Development Policy. *Continued*

Policy Type	Programme Title	Policy Objective	Community Measure	Document Reference	Website
Miscellaneous	Reclaiming City Streets for People: Chaos or Quality of Life?	The challenge facing urban planners & politicians in many European towns & cities is that of balancing the demand for increasing personal mobility & economic growth, with the need to respect the environment & provide an acceptable quality of life for all citizens. This handbook illustrates the concept of traffic evaporation using case studies from a selection of European cities. The case studies demonstrate the importance of well planned integrated strategies, combined with effective public consultation & communication.	Report for the DG Environment (Sept 2004).	▪ DG Environment Report	http://europa.eu.int/comm/environment/pubs/pdf/streets_people.pdf
	Åalborg +10 – Inspiring Futures	European local governments united in the European Sustainable Cities & Towns Campaign, assembled at the Åalborg+10 Conference, confirm a shared vision of a sustainable future for communities. The vision is of cities & towns that are inclusive, prosperous, creative & sustainable, & that provide a good quality of life for all citizens & enable their participation in all aspects of urban life. Since the Rio Summit in 1992 & the adoption in 1994 of the sustainability principles embodied in the Charter of European Cities & Towns Towards Sustainability – the Åalborg Charter – the vision has grown. The 2004 'Inspiring Futures – Åalborg+10' Conference is a milestone in this ongoing process.	Conference Proceedings (2004).	▪ Conference Proceedings	http://www.aalborgplus10.dk/media/finaldraftaalborgcommitments.pdf

Table 12.5.3 UK Level Urban Development Policy.

Policy Type	Document Title	Objective	Contents	Website
Strategies & Reports	Towards an Urban Renaissance (Urban Task Force Report) (29/06/99)	Achieving an urban renaissance is not only about numbers & percentages. It is about creating the quality of life & vitality that makes urban living desirable. A change in urban attitudes is needed so that towns & cities once again become attractive places to live, work & socialise. The Task Force's report contains over 100 recommendations for change. They cover design, transport, management, regeneration, skills, planning and investment. This Executive Summary presents the main findings & recommendations.	Mission statement; Introduction; Urban fact-file; The key themes & measures; Designing the urban environment; Making the connections; Managing the urban environment; Delivering urban regeneration; Investing in skills & innovation; The 60% target – our assessment; Planning for change; Managing the land supply; Cleaning up the land; Recycling the buildings; Making the investment; Sustaining the renaissance; Supporting reports.	http://www.odpm.gov.uk/stellent/groups/odpm_urbanpolicy/documents/pdf/odpm_urbpol_pdf_608462.pdf
	Urban Exchange Initiative (12/10/99)	The basic aim of this initiative is to draw together some key principles based on good practice experience in dealing with various issues in urban areas, which are home to most of Europe's population. Member States have much to learn from each other & it is important that approaches which have been tried & found to work well are disseminated for the wider benefit. This should help Member States to develop more effective policies within their own particular legislative & administrative frameworks.	Chapter 1: Introduction; Chapter 2: Multiple deprivation – selected approach: comprehensive (integrated approach); Chapter 3: Environment & urban quality – selected approach: town centre management; Chapter 4: Local democracy – selected approach: good governance & community involvement; Case study annex.	http://www.odpm.gov.uk/stellent/groups/odpm_urbanpolicy/documents/pdf/odpm_urbpol_pdf_608469.pdf
	Millennium Villages & Sustainable Communities (18/08/00)	The purpose of this study was to facilitate the subsequent evaluation of the Millennium Villages initiative & to draw out the transferable lessons which could be fed back into their development. The research would also aim to support the achievement of sustainable communities more generally in a wider range of contexts, including local housing & regeneration programmes in urban areas, small towns & rural areas.	Executive summary; Chapter 1: Introduction; Chapter 2: Creating sustainable settlements & sustainable communities; Chapter 3: Focusing on outcomes; Chapter 4: The evaluation framework; Chapter 5: Applying the framework; Chapter 6: Lessons for promoting sustainability in settlements; Chapter 7: Delivering sustainability – processes & procedures; Chapter 8: Conclusions & recommendations; Bibliography; Appendix.	http://www.odpm.gov.uk/stellent/groups/odpm_urbanpolicy/documents/pdf/odpm_urbpol_pdf_608096.pdf

Continued

Table 12.5.3 UK Level Urban Development Policy. *Continued*

Policy Type	Document Title	Objective	Contents	Website
Strategies & Reports	Urban Regeneration Companies: A Process Evaluation (Oct 2000)	Three pilot Urban Regeneration Companies (URCs) were established in Liverpool, East Manchester & Sheffield during 1999/2000. Each was formed as a partnership between three bodies: the respective City Councils, the relevant Regional Development Agencies (RDAs), & English Partnerships (EP). This evaluation has had two principal aims: first to develop a dialogue with the Companies & those involved with their work in order to provide an independent sounding board; & second to produce an initial commentary on the issues facing the Companies and the potential that they offer as a way of tackling integrated regeneration in large & complex areas of big cities.	Executive summary & policy recommendations; Introduction; Part 1: Policy lessons; Part 2: URCs in Liverpool, East Manchester & Sheffield; Sheffeld One; Conclusions; Annexes.	http://www.odpm.gov.uk/stellent/groups/odpm_urbanpolicy/documents/pdf/odpm_urbpol_pdf_608086.pdf
	Our Towns & Cities: The Future Delivering an Urban Renaissance (Urban White Paper) (16/11/00)	This White Paper explains how towns & cities can function as economic powerhouses, helping to achieve the Government's core objective of increasing sustainable growth & employment for all & bringing benefits not just to their own population but to the surrounding region. It also targets those areas that are striving to renew their economic identity and which are seeking to take full advantage of the rapid growth in the new industries. The guiding principle is that people must come first. Policies, programmes & structures of governance are based on engaging local people in partnerships for change with strong local leadership. This inclusive approach is at the heart of work on tackling social exclusion, & is central to achieving sustainable economic growth. And it is the approach which underlies both this White Paper & the parallel White Paper on rural communities.	Foreword; Summary; Chapter 1: Towns & cities today; Chapter 2: Towns & cities – issues & vision; Chapter 3: Delivering better towns & cities; Chapter 4: Places for people; Chapter 5: Creating & sharing prosperity; Chapter 6: Quality services & opportunities for all; Chapter 7: Making it happen – action for all; Annex: Recommendations of the Urban Task Force Report; Bibliography.	http://www.odpm.gov.uk/stellent/groups/odpm_urbanpolicy/documents/pdf/odpm_urbpol_pdf_608358.pdf

Indices of Deprivation 2000 (15/12/00)	This report outlines the indicators and domains which were used to construct the IMD 2000. It also summarises the procedures taken to create the overall Index and provides summaries of the Index at the local authority level.	Introduction; Chapter 1: Background; Chapter 2: Domains & indicators; Chapter 3: Population estimates; Chapter 4: Methodology; Chapter 5: Presentation of results & interpretation; Chapter 6: Guidance; Chapter 7: Comparing the indices of deprivation 2000 with the 1998 ILD; Appendix A: Consultation; Appendix B: Data sources; Appendix C: Table of district level summaries; Appendix D: Flow chart of index of multiple deprivation methodology.	http://www.odpm.gov.uk/stellent/groups/odpm_control/documents/contentservertemplate/odpm_index.hcst?n=3103&l=2
Urban Regeneration Companies (URCs) – Learning the Lessons (05/09/01)	URCs represent a major new development in current regeneration policy. They are a recognition of market failures in many areas of our towns and cities and of the inability of past public sector interventions to correct these failures and create lasting improvements. This research provides policy and practice lessons for the future development of the URC and other regeneration programmes. However, although it is too early to be able to identify the impact of the pilot URCs on their areas, it is possible to build upon the earlier process evaluation to critically assess progress to date.	Introduction; The pilot URCs – a progress review; Good Practice lessons; Policy lessons; Implications for future URCs; An overall monitoring & evaluation framework; Conclusions; Appendix A: Method of approach; Appendix B: A review of the pilot URCs; Appendix C: Baseline indicators; Appendix D: Monitoring & evaluation – proposed methodological framework & implementation arrangements.	http://www.odpm.gov.uk/stellent/groups/odpm_urbanpolicy/documents/pdf/odpm_urbpol_pdf_608152.pdf
A Modern Regional Policy for the UK (Mar 2003)	This paper outlines the Government's approach to supporting the economic development of our nations & regions & in this context launches a consultation on an EU Framework for Devolved Regional Policy. This would allow UK nations & regions & all Member States to pursue effective policies for regional growth within a framework which is built on the key principles of effective regional policy. It would devolve policy responsibility & accountability to the most effective level & establish a clear framework of long term goals. Finally, it would be flexible enough to meet the differing needs of the nations & regions of the EU.	Foreword; Executive summary; The Government's approach to developing the UK's nations & regions; Delivery of the vision; Extending the vision; The UK government's proposals for EU regional policy; Annexes.	http://www.odpm.gov.uk/stellent/groups/odpm_urbanpolicy/documents/page/odpm_urbpol_609285.pdf

Continued

Table 12.5.3 UK Level Urban Development Policy. *Continued*

Policy Type	Document Title	Objective	Contents	Website
Strategies & Reports	Building Sustainable Communities: Actions for Housing Market Renewal (June 2003)	Five national organisations have come together to produce this paper, charged by Government to: ▪ promote sustainable development in practice; ▪ champion good design in the built environment; ▪ make the most of our built heritage; ▪ protect our natural environment; & ▪ advise on integrated transport. This paper defines what is considered to be the key success factors for turning around these areas, highlighting the most important issues or ambitions that should be attempted through housing market renewal strategies. It raises the debate about collective responsibility to help the Pathfinders succeed & to inform best practice elsewhere.	Introduction; The 7 key actions for successful housing market renewal; Raise the scale of the opportunity; Positively address heritage as an asset; Create places of distinction; Recognise the value of design & its role in renewal; Adopt policies & tools to deliver high quality urban neighbourhoods; Place sustainable development at the heart of thinking & action; Get ready for the challenge; The signatories; The pathfinders.	http://www.cabe.org.uk/pdf/Housing%20Market%20Renewal.pdf
	Building Sustainable Communities: Developing the Skills we Need (June 2003)	Our urban environment can only ever be as good as the teams that build it. This paper promotes a fundamental shift in gear needed within these professions, the lack of relevant skills being the single biggest barrier to the development of sustainable communities.	Foreword; The need for change; The skills deficit; The journey to date; Our programme for action; Creating a trained network of project leaders & deliverers; Improving our knowledge & evidence base; Young people; Responsive organisational frameworks; Overcoming institutional inertia; Conclusion.	http://www.cabe.org.uk/pdf/Skills%20Manifesto.pdf
	Moving Towards Excellence in Urban Design & Conservation (June 2003)	This paper aims to help local authorities put in place the practices & processes that can deliver excellent conservation & urban design through the planning process. However it is not a comprehensive specification for either a conservation or urban design service. There is much to do in both fields across the spectrum of local government, for example in outreach & in	Preface; Scope & purpose; Key issues; Towards an excellent service; Excellence matrix; Using the matrix; Matrix; Appendix.	http://www.cabe.org.uk/pdf/Matrix-web%20version.pdf

	the work of highway authorities, & effective linkages need to be sought. Nonetheless this paper does set out clearly the urban design & conservation components of an excellent planning service.		
Sustainable Communities: Building for the Future (June 2003)	This is a programme of action to tackle pressing problems in communities in England: homes are unaffordable in some areas, but are being abandoned in others. This document marks a step change in approach; a strengthened determination by Government to reverse, over the next 15–20 years, some damaging, deep-seated trends. It is part of the Government's wider drive to raise the quality of life in all communities through increasing prosperity, reducing inequalities, more employment, better public services, better health & education, tackling crime & anti-social behaviour, etc. It reflects key principles for public service reform: raising standards, devolving & delegating decision-making, providing greater flexibility over use of resources & choice for customers.	An Overview; The background in charts and figures; Part 1: Decent homes, decent places; Part 2: Low demand and abandonment; Part 3: A step change in housing supply; Part 4: Land, countryside and rural communities; Part 5: Sustainable growth; Part 6: Reforming the delivery; Annexes.	http://www.odpm.gov.uk/stellent/groups/odpm_communities/documents/pdf/odpm_comm_pdf_022184.pdf
Cities, Regions & Competitiveness (July 2003)	This report is about strengthening the UK's growth capacity in future global upturns. It uses evidence from competitor countries to demonstrate how stronger regional cities could 'add more cylinders to the UK's economic engine' & provide more space for London to further develop its unique global city role. The report contains an action plan for strengthening productivity & the urban renaissance in the major regional cities, as the essential foundation for progressive improvements in the performance of all regions.	Executive summary; A pivotal report for economic growth & prosperity; The working group & its remit; The Government's commitment to securing improvements in regional economic performance; Cities are the 'powerhouses' of modern economies; Competitive cities create prosperous regions; The key characteristics of competitive core cities; Innovation & knowledge transfer; Education & skills; Investment in high productivity functions; Spatial planning; Strategic transport connections; City renaissance – the roles of culture & sport; Measuring competitiveness; Fulfilling the potential of cities – the next steps.	http://www.odpm.gov.uk/stellent/groups/odpm_urbanpolicy/documents/pdf/odpm_urbpol_pdf_022847.pdf

Continued

Table 12.5.3 UK Level Urban Development Policy. *Continued*

Policy Type	Document Title	Objective	Contents	Website
Strategies & Reports	Making it Happen: Thames Gateway & Growth Areas (July 2003)	This is one of a series of progress reports on delivery of the *Sustainable Communities: Building for the Future* programme of action. This report dealt with proposals for sustainable growth & housing supply in the wider South East over the following 15 years & provided an overview of government action with partners to help deliver growth. It also contained announcements on funding & delivery to help partners finalise their own programmes. Future progress reports are planned on other key issues in *Sustainable Communities: Building for the Future.*	A progress report by the Deputy Prime Minister; What this report is about; Building sustainable communities; Developing the sustainable communities approach for the wider South East; Thames Gateway; Other growth areas; Milton Keynes – South Midlands; London-Stansted-Cambridge; Cambridge Area Development Frameworks; Ashford; London; Capturing the value of growth; Helping to accelerate development & increase housing supply; Boosting housing delivery – English Partnerships & the Housing Corporation; Quality & Design.	http://www.odpm.gov.uk/stellent/groups/odpm_communities/documents/pdf/odpm_comm_pdf_023301.pdf
	Searching for Solid Foundations: Community Involvement & Urban Policy (Sept 2003)	This study was commissioned by the Office of the Deputy Prime Minister (ODPM) to review guidance & other literature on community involvement with particular reference to urban renaissance, & to draw out implications for the implementation of urban policy. This report therefore focuses particularly on programmes within the remit of ODPM, but takes some account of the overall government context.	Chapter 1 – Introduction; Chapter 2 – Policy context and guidance landmarks; Chapter 3 – What enables people to get involved or stops them from doing so? Chapter 4 – Towards a unified approach to community involvement; Chapter 5 – Conclusions; Recommendations; Sources & references.	http://www.odpm.gov.uk/stellent/groups/odpm_urbanpolicy/documents/page/odpm_urbpol_023886.pdf
	Living Places: Caring for Quality (Jan 2004)	This paper reports on research that examined how local authorities & other local stakeholders, through the better management of public space, are rising to the challenge of reversing the dissatisfaction & pessimism identified by the Urban Task Force. This pursuit of more 'liveable' public space represents a key aspiration of government as set out in *Sustainable Communities: Building for the Future*. On the basis of extensive research evidence, a wide range of ideas is outlined in the following chapters to help move practice towards new more holistic approaches to public space management & therefore towards the delivery of more sustainable communities.	Foreword; Introduction; Part I: The challenge; The character of public space; Current practice; Taking responsibility for action; Part II: The delivery; Caring through better co-ordination; Caring through better investment; Caring through better regulation; Caring through better maintenance; Part III: Moving on; Questioning practice; Annexes.	http://www.odpm.gov.uk/stellent/groups/odpm_urbanpolicy/documents/downloadable/odpm_urbpol_027185.pdf

Creating Sustainable Communities: Greening the Gateway – A Greenspace Strategy for Thames Gateway (Jan 2004)	This document sets out the Government's vision for the landscape of Thames Gateway & the positive contribution which the network of green open spaces should make to the quality of life for all those who live & work there. It is a statement of intent which outlines our approach to the outdoor environment. The more detailed strategies & action plans of other organisations need to fit within it. It is not a spatial strategy & neither is it a prescription for land use on specific sites, but rather it is intended to be a clear indication of the way in which *Greening the Gateway* can help in delivering sustainable communities.	Summary; Introduction; Policy context; Operational context; Functional green infrastructure; Core principles; Delivery.	http://www.odpm.gov.uk/stellent/groups/odpm_communities/documents/downloadable/odpm_comm_026750.pdf
Competitive European Cities: Where do the Core Cities Stand? (Jan 2004)	Cities are high on the British policy agenda as the 'dynamos of the UK national economy'. This idea has been an increasingly significant dimension of national policy & in continental Europe there is equally growing interest in the contribution that cities can make to the national welfare – & to economic competitiveness in particular. This study of competitive European cities is firmly located in that wider context. It was commissioned & supported by the Core Cities Working Group, a partnership consisting of the eight UK Core Cities (Birmingham, Bristol, Leeds, Liverpool, Manchester, Newcastle, Nottingham & Sheffield) the nine Regional Development Agencies & the departments of central government primarily concerned with national economic performance (including the ODPM, the Treasury & the Departments of Trade & Industry, Transport & Culture, Media & Sport).	Executive summary; Chapter 1 – Comparing Competitive Cities – Why & how did we do it? Chapter 2 – Cities in Europe – what's going on? Chapter 3 – What is the Core Cities' current position? Chapter 4 – Urban Competitiveness: What do we mean, how do we measure it? Chapter 5 – How do core cities match up to their European competitors? Chapter 6 – What makes the difference? Chapter 7 – What's next and for whom? Appendices.	http://www.odpm.gov.uk/stellent/groups/odpm_urbanpolicy/documents/downloadable/odpm_urbpol_027437.pdf
Making it Happen: The Northern Way (Feb 2004)	This statement reports on the action being taken to regenerate our established communities in the North & the Midlands, & to ensure that they are sustainable. It looks at the successes since the Communities Plan was published in February 2003, & at the challenges to come. The need to learn from experiences, both good & bad, is at the heart of the Communities Plan.	Foreword; Setting the scene; Part One: Local solutions; Revitalised housing market; Thriving towns & cities; Coalfield communities; New Deal for communities; Other neighbourhood initiatives; Sustainable rural communities; Part Two: Decent homes, decent places; Decent homes; Action on liveability & quality of life; Conclusion.	http://www.odpm.gov.uk/stellent/groups/odpm_communities/documents/pdf/odpm_comm_pdf_027362.pdf

Continued

Table 12.5.3 UK Level Urban Development Policy. *Continued*

Policy Type	Document Title	Objective	Contents	Website
Strategies & Reports	Evaluating the Impact of Design Awards for Housing (April 2004)	The objective of this research was to consider how the *Housing Design Awards* & other design awards for housing contribute to encouraging better design & in so doing complement Government policy objectives. To do this it was necessary to find out how particular groups respond to the range of award programmes that run in England, & to look at the range of factors that encourage or discourage interest & participation. In particular, it was necessary to find where tensions occur between what people want from housing development & award programmes, & to consider a range of options which highlight how an award programme might be designed to be attractive to one group or another.	Chapter 1 Introduction; Chapter 2 Judging Housing Quality; Chapter 3 Government Policy for Housing Design & Layout; Chapter 4 Awards for Housing Design; Chapter 5 Do Awards for Housing Design Influence Industry Practice? Chapter 6 Choices in the Design of a Housing Design Award; Chapter 7 Conclusions & Recommendations; Appendices.	http://www.odpm.gov.uk/stellent/groups/odpm_urbanpolicy/documents/pdf/odpm_urbpol_pdf_029125.pdf
	Our Cities Are Back: Competitive Cities make Prosperous Regions & Sustainable Communities – Third Report of the Core Cities Working Group (Nov 2004)	This report is about the UK's future economic growth and prosperity. It will be of particular interest to all organisations & individuals concerned with urban & regional economic growth & regeneration. It makes a direct & unique contribution to the Government's Public Service Agreement Target aimed at improving the performance of all the English Regions & reducing the persistent gap in growth rates between regions. The report provides a fresh perspective on the economic roles of modern cities – evidencing their pivotal role in creating prosperous regions & setting out a comprehensive action plan for achieving a step change in the competitiveness of the largest English regional cities, the Core Cities.	Executive Summary; Core Cities – Powerhouses of the Knowledge Economy; What Makes Competitive Core Cities? 'Our Cities Are Back'; City-Regions: The Key Economic Building Blocks; Action Plan for Competitive Core Cities; Creating Sustainable Communities	http://www.odpm.gov.uk/stellent/groups/odpm_urbanpolicy/documents/downloadable/odpm_urbpol_033302.pdf

Consultation Papers	Urban Policy Evaluation Strategy – Consultation Paper (Feb 2002)	This paper seeks views on proposals for an Evaluation Strategy for the Government's urban policy in England. The agenda for an urban renaissance was set out in the Urban White Paper, *Our Towns and Cities: the Future*, published in November 2000, oversight of the implementation of which is the responsibility of the Urban Policy Unit (UPU) of the Department for Transport, Local Government and the Regions (DTLR). This consultation paper has been produced in discussion with other government departments whose programmes and initiatives are relevant to this agenda.	Introduction; Why do we need an evaluation strategy for urban policy? What are we trying to evaluate? What are towns & cities? What do we mean by getting better? How can we measure this? What is driving the changes? How do we evaluate the impact of policies? To what timescales should we be working? The consultation process.	http://www.odpm.gov.uk/stellent/groups/odpm_urbanpolicy/documents/pdf/odpm_urbpol_pdf_608044.pdf
Guidance Notes	Sustainable Regeneration Good Practice Guide (05/10/98)	It is in the nature of an integrated regeneration policy to establish the widest cross-programme linkage between differing aspects of any particular problem facing urban & rural communities. Achieving sustainable development represents perhaps the most challenging of such cross-sectoral problems & it is the aim of this guide to assist in highlighting both the strategic issues facing regeneration partnerships & some of the practical ways in which these issues may be addressed. In some localities good practice now reflects a very positive interpretation of the relationship between local economic development policies & environmental action. Sustainability – economic, social, environmental & institutional – is a key value that informs all local action.	Preface; Introduction; The policy context; Principles & processes; The role of SRB (Single Regeneration Budget) partnerships; Sustainable regeneration in practice; Appendix 1: Sustainability checklist; Appendix 2: Useful contacts; References.	http://www.odpm.gov.uk/stellent/groups/odpm_urbanpolicy/documents/pdf/odpm_urbpol_pdf_608058.pdf

Continued

Table 12.5.3 UK Level Urban Development Policy. *Continued*

Policy Type	Document Title	Objective	Contents	Website
Guidance Notes	Running & Sustaining Renewal Areas: Good Practice Guide (Aug 1999)	There has been a recommendation for new guidance on how to implement a Renewal Area programme successfully. This manual has been produced to meet this request. This guidance is divided into a two broad parts. Chapters 2 to 5 cover the partnerships & structures needed for working together to successfully implement the Renewal Area including: ▪ general comments on partnerships – partnerships in urban renewal; ▪ partnerships within the local authority – corporate working; ▪ partnerships with the Renewal Area team – Renewal Area team; ▪ partnerships with the residents, community & other stakeholders – stakeholder involvement. Chapter 6 on environmental sustainability provides extra guidance for consideration during the early stages of the Renewal Area. Chapters 7 to 14 cover the management & organisation of the Renewal Area.	Chapter 1: Introduction; Chapter 2: Partnerships in Urban Renewal; Chapter 3: The Corporate Approach; Chapter 4: Renewal Area Team; Chapter 5: Stakeholder Involvement; Chapter 6: Sustainable Development; Chapter 7: Managing a Renewal Area; Chapter 8: Communication and Promotion; Chapter 9: Programme and Projects; Chapter 10: Projects in Practice; Chapter 11: Private and Public Sector Investment; Chapter 12: Monitoring Renewal Areas; Chapter 13: Strategies for Sustainable Maintenance; Chapter 14: Forward Strategies; References; Appendices.	http://www.odpm.gov.uk/stellent/groups/odpm_housing/documents/pdf/odpm_house_pdf_603504.pdf
	Regeneration that Lasts (04/12/00)	This guide to good practice is the product of a research assignment designed to identify those factors in the design and implementation of estate regeneration programmes which appear to lead to 'sustainable' results.	Executive summary; Chapter 1: Introduction; Chapter 2: The estates; Chapter 3: Getting started; Chapter 4: Implementing programmes; Chapter 5: Building partnerships; Chapter 6: Achieving sustainability; Chapter 7: Conclusions; Annex 1: Estate summaries; Glossary; References.	http://www.odpm.gov.uk/stellent/groups/odpm_housing/documents/pdf/odpm_house_pdf_603915.pdf

Local Strategic Partnerships – Government Advice (March 2001)	This guidance takes into account the responses to the comments received on the draft guidance published last year (2000). In particular, the final guidance emphasises: ▪ The value of successful partnership working in tackling community problems & priorities; ▪ The importance of involving local people, including faith, black & minority ethnic communities; ▪ Central government's commitment to supporting LSPs & helping them to succeed.	Summary; Chapter1: LSPs: What are they & how should they work; Chapter 2: Central government's commitment to making LSP a success; Chapter 3: Cross-cutting responsibilities of LSPs; Chapter 4: LSP & neighbourhood renewal; Annex A: Rationalising plans & partnerships; Annex B: Neighbourhood renewal fund – supplementary guidance; Annex C: Membership of LSPs – involving the public, private, community & voluntary sectors; Annex D: Community empowerment fund; Annex E: Good practice in partnership working; Annex F: Examples of partnership structures; Annex G: Abbreviations; Annex H: Bibliography.	http://www.odpm.gov.uk/stellent/groups/odpm_localgov/documents/page/odpm_locgov_605602.pdf
Assessing the Impacts of Spatial Interventions (Sept 2003)	The aim of the guidance is to set out the broad framework within which the assessment of spatially targeted interventions should take place. The drafting recognises the need for flexibility & hence focuses on the broad principles that should be followed rather than defining rigid procedures. The guidance aims to provide advice on assessment techniques that can be applied to policies, programmes & projects in the 3R (Regeneration Renewal and Regional Development) area & for non-3R interventions that may nevertheless have such impacts. The guidance is not primarily about the *design* of interventions – but about the *assessment* of interventions in a way that rigorously identifies value for money & supports the development of interventions that provide value for money. This guidance, it is hoped, will ensure better appraisal & evaluation of projects thereby saving time & resources & increasing the benefits secured by each intervention & each £ of public money spent in pursuit of such policy aims.	Foreword; Executive Summary; Introduction: Scope, Audience & Underlying Principles; Aims, Actions, Areas & Actors: an Overview of 3R Interventions; Stage 1: Defining the Problem; Stage 2: Identifying alternatives & comparators; Stage 3: Identifying & measuring costs & other inputs; Stage 4: Identifying & measuring outputs & linking to outcomes; Stage 5: Presentation of results; Annexes.	http://www.odpm.gov.uk/stellent/groups/odpm_urbanpolicy/documents/pdf/odpm_urbpol_pdf_023746.pdf

Table 12.5.4 Northern Ireland Level Urban Development Policy.

Policy Type	Document Title	Objective	Contents	Website
Strategies & Reports	Shaping Our Future – Towards a Strategy for the Development of the Region (Nov 1997)	The purpose of the paper is to stimulate public debate & facilitate the widest possible community input on the key issues which should guide the development of the Region over the next 25 years. It is a consultative document that highlights a number of the key challenges confronting NI, & suggests a possible shape for a planning framework for the future development of the Region.	Part 1: Introduction; Part 2: Taking stock – NI in a changing world; Part 3: Building the strategy; Part 4: Towards a preliminary strategy; Part 5: The way forward; Annex 1: Regional profile key facts; Annex 2: Evaluation & Monitoring.	http://www.drdni.gov.uk/shapingourfuture/pdf/reports/nov97_intro.pdf . . . _part1.pdf . . . _part2.pdf . . . _part3.pdf . . . _part4.pdf . . . _part5.pdf
	Shaping Our Future – Population & Housing Technical Report (Dec 1998)	The purpose of this report is to set out the population & housing technical data which informed judgements on household growth & housing need for the Regional Strategic Framework.	Introduction; Regional population; Components of regional population change; Age structure of regional population; Distribution of regional population; Population projections; Regional households; Regional housing stock; The dynamics of housing stock change; Projected regional housing need 1996–2025; Housing growth in the settlement groups; Regional housing targets for the main urban centres; Housing growth in the rural community; Appendices.	http://www.drdni.gov.uk/shapingourfuture/pdf/reports/pop_housing_tech_report.pdf
	Shaping Our Future – Family of Settlements Report (Dec 1998)	The purpose of this report is to set out how the dwellings required to meet housing need in the Region up to 2010 will be distributed across the Family of Settlements comprising urban & rural communities, in accordance with the Spatial Development Strategy for NI which reflects the Guiding Principles & Shared Vision which emerged from the extensive public consultation process.	Part 1: Future housing growth in the family of settlements; Chapter 1: Introduction; Chapter 2: Regional housing projections; Chapter 3: Analysis of development trends in the settlement groups; Chapter 4 : Regional housing targets 2010; Part 2: Profiles of the main urban centres.	http://www.drdni.gov.uk/shapingourfuture/pdf/reports/doe2_family_settlements_rep_nopics.pdf

Shaping Our Future – Regional Baseline Report (Dec 1998)	The Regional Baseline is one of three background reports which accompany the publication of the Draft Regional Strategic Framework for NI. The Regional Strategic Framework (RSF) seeks to set directions for the future development of the Region to 2025 through a shared vision which emerged from an extensive public consultation exercise. While the Strategy takes account of a number of driving forces for change, it is not based on past statistical trends but seeks to influence future patterns of development in accordance with a number of guiding principles. The Regional Baseline report is not therefore intended as part of the justification of the draft RSF. Neither is it an exhaustive fact file of the current state of the Region. It is, rather, a compendium of key information collected during the RSF process which will provide a series of statistical snapshots of the major topics covered by the Strategy.	Introduction; Chapter 1: An outward looking region; Chapter 2: Population; Chapter 3: Housing; Chapter 4: The economy; Chapter 5: Tourism; Chapter 6: Transport; Chapter 7: Education & skills; Chapter 8: Environment; Chapter 9: Social trends; Chapter 10: Infrastructure; Information sources.	http://www.drdni.gov.uk/shapingourfuture/pdf/reports/regional_baseline.pdf
Strategic Environmental Appraisal Report – A Sustainability Assessment of the Draft Regional Strategic Framework for NI (Feb 1999)	A sustainability assessment has been carried out on the Draft Regional Strategic Framework to ensure that its effects are as beneficial as possible & that any negative effects are minimised. This document explains: ▪ What sustainability assessment is; ▪ How the sustainability assessment for NI's Regional Strategic Framework was carried out & how it influenced the Regional Strategic Framework; ▪ What the proposed Regional Strategic Framework's impacts are; & ▪ Future steps.	Introduction; Background: the process of sustainability assessment; Application of sustainability assessment; Appraisal of the draft regional strategic framework; Future steps; Appendix: Sustainability analysis matrices.	http://www.drdni.gov.uk/shapingourfuture/pdf/reports/strat_env_assessment.pdf
The CRISP (Community Regeneration and Improvement Special Programme) Story (Feb 1999)	This booklet tells the story of CRISP. It is a success story with 58 towns having benefited from CRISP funding. This has provided new jobs, improved facilities & the physical improvement of buildings & public spaces.	Foreword; Introduction; Strategy & key characteristics of CRISP; CRISP since 1990; The benefits of CRISP; A personal perspective from a senior planner; A community perspective from Councillor Jim Canning; District Council perspectives; The future of CRISP; Appendix – CRISP projects assisted.	http://www.dsdni.gov.uk/publications/documents/crispstory.pdf

Continued

Table 12.5.4 Northern Ireland Level Urban Development Policy. *Continued*

Policy Type	Document Title	Objective	Contents	Website
Strategies & Reports	A Strategic Review of the Impact of District Partnership Expenditure in NI – Executive Summary (May 2000)	This report seeks to identify the impact of the District Partnerships on the aims and objectives of the Special Support Programme for Peace and Reconciliation (SSPPR). In addition it will try to assess the additional process benefits which were expected to ensue from the use of this form of delivery mechanism.	Introduction; Our approach; Conclusions; Recommendations.	http://www.dsdni.gov.uk/publications/documents/exec.pdf
	Reinvigoration of Town Centres Report (Oct 2000)	This Study analyses the current town centres position in NI & sets out recommendations to ensure their reinvigoration as part of the ongoing process of economic & social renaissance. Town centres in NI have an individuality that many centres in the rest of the UK have lost – this is an asset that should be exploited for the future. This Study sets out practical proposals to exploit the strengths of town centres in NI as much as identifying the extent to which town centres need to be 'protected' from new development.	Executive summary; Introduction; The drivers of change in NI; Policy review; Strategic overview: planning & development; Strategic overview: accessibility; Case study overview; Best practice notes; Conclusions; Recommendations.	http://www.dsdni.gov.uk/publications/documents/ConDoc(1).pdf
	URBAN Community Initiative in NI 1994–1999 (March 2001)	The URBAN Community Initiative was officially launched in June 1996 in a total of 118 urban areas throughout the EU Member States. The Initiative targeted neighbourhoods in extreme deprivation, seeking to address problems of isolation, poverty & exclusion of their inhabitants through interventions that improved their physical & social surroundings. The output from this evaluation was an overall evaluation report, supplemented by three working papers providing a detailed analysis of the two Sub-Programmes funded under the Programme.	Background; Impacts of the URBAN Programme; Results of the meta evaluation; Implications for URBAN II; Conclusions.	http://www.dsdni.gov.uk/publications/documents/PwC.URBANreport.301.pdf

Programme for Government – Making a Difference (2001–2004) (March 2001)	This programme for government sets out the Executive's proposed strategic aims & priorities which will be pursued working with & for all the people. It will provide an important focus of co-operation, to enable the creation of a better future. The programme focuses on the three years from 1st April 2001 & will be reviewed on an annual basis, at which time there will be a wide consultation.	Foreword; Making a difference; Growing as a community; Working for a healthier people; Investing in education & skills; Securing a competitive economy; Developing North/South, East/West & international relations; Working together; Annex A: Financing our work; Annex B: Public service agreements/Departmental public service agreements; Annex C: Equality aspects; Annex D: Glossary.	http://www.ofmdfmni.gov.uk/publications/pfga/contents.htm
Draft Regional Development Strategy for NI 2025 (Aug 2001)	*Shaping Our Future* is a Strategy to guide the future development of NI to 2025 & help meet the needs of a fast growing Region with a population approaching two million. The vision is to create an outward-looking, dynamic & liveable Region & to sustain a high quality of life for all. A long-term perspective is taken keeping the needs of future generations in mind. Thus, the recurring theme of Sustainability runs through the Regional Development Strategy (RDS), with a strong emphasis on social cohesion & economic progress.	Chapter 1: Purpose & status; Chapter 2: Forces driving changes; Chapter 3: Vision & guiding principles; Chapter 4: Strengthening regional cohesion; Chapter 5: The spatial development strategy for NI; Chapter 6: The Belfast Metropolitan Area; Chapter 7: Londonderry – regional city for the North West; Chapter 8: Rural NI; Chapter 9: Meeting housing needs; Chapter 10: Supporting economic development; Chapter 11: Developing a regional transportation system; Chapter 12: Caring for the environment; Chapter 13: Implementation; Annexes; Appendices; Key statistics.	http://www.drdni.gov.uk/shapingourfuture/pdf/reports/shaping%20our%20future%20(Take%20Note)%2002jul01.pdf
Final Regional Development Strategy for NI 2025 (Sept 2001)	The Strategy paints an ambitious picture of what the Region might look like in 25 years' time. It sets out a demanding, though achievable, agenda. The next challenge is to make the Strategy a reality. If the quality of life is to be improved with economic opportunity for all citizens, there is a need to harness collective resources & effort right across the public, private, voluntary & community sectors.	Chapter 1: Purpose & status; Chapter 2: Forces driving changes; Chapter 3: Vision & guiding principles; Chapter 4: Strengthening regional cohesion; Chapter 5: The spatial development strategy for NI; Chapter 6: The Belfast Metropolitan Area; Chapter 7: Londonderry – regional city for the North West; Chapter 8: Rural NI; Chapter 9: Meeting housing needs; Chapter 10: Supporting economic development; Chapter 11: Developing a regional transportation system; Chapter 12: Caring for the environment; Chapter 13: Implementation; Annexes; Appendices; Key statistics.	http://www.drdni.gov.uk/shapingourfuture/pdf/reports/shaping%20our%20future(final).pdf

Continued

Table 12.5.4 Northern Ireland Level Urban Development Policy. *Continued*

Policy Type	Document Title	Objective	Contents	Website
Strategies & Reports	Belfast Metropolitan Area Plan 2015 (BMAP) Issues Paper (07/12/01)	A metropolitan plan which encompasses the entire built-up area & its immediate hinterland permits a strategic examination of the many inter-related issues – social, economic & environmental & transport which need to be addressed on a metropolitan wide basis. It provides a major opportunity to influence the quality of the environment, both urban & rural, not only for the citizens of today but generations to come.	Part 1: Introduction; Part 2: Recent trends; Part 3: Guiding principles; Part 4: Plan-wide issues; Part 5: Local issues; Part 6: Villages & small settlements; Part 7: The countryside; Part 8: Next steps; Appendix 1: Settlements in BMAP rural hinterland; Appendix 2: Existing conservation areas & areas of townscape character.	http://www.doeni.gov.uk/planning/Area_Plans/Belfast_Metropolitan_Area_Plan/BMAP2015.pdf
	Implementation of the Regional Development Strategy & Monitoring & Evaluation Arrangements (April 2002)	The purpose of this paper is to provide an overview of the arrangements for the implementation of the Regional Development Strategy (RDS) & to set out areas where indicators and targets should be developed for the purpose of monitoring & evaluation.	Purpose; The guiding approach to implementation; Background & overview; Partnership & participation; Learning lessons from elsewhere; Monitoring & review; Conclusion & recommendations.	http://www.drdni.gov.uk/shapingourfuture/pdf/final_implementation_paper240402.pdf
	Programme for Government – Making a Difference (2002–2005) (2002)	This update to the Programme for Government underlines the commitment to the principles & practice of open/accountable government. It sets out the actions the Executive is committed to in support of priorities & sub-priorities. The Executive has also set out in the public service agreements the key objectives, outcomes & outputs that each Department is working to deliver & the resources allocated to support these objectives. They will soon be complemented by new service delivery agreements for 2002/03 that will provide more detail on how Departments are delivering their targets & plans for improving customer service & raising service standards.	Foreword; Making a difference; Growing as a community; Working for a healthier people; Investing in education & skills; Securing a competitive economy; Developing North/South, East/West & international relations; Working together; Annex A: Actions completed; Annex B: Financing our work; Annex C: Public service agreements; Annex D: Equality aspects; Annex D(i): List of EQIAs under way/planned; Annex E: Glossary.	http://www.pfgni.gov.uk/report0205main.pdf

Programme for Government Annual Report 2001/02 (July 2002)	The purpose of this Report is to set out the progress made across all priorities. It is to highlight achievements, identify areas where progress has been slower & to provide explanations where this has happened. The outbreak of Foot & Mouth Disease & the global economic downturn have had a particular impact on the plans, but there have been other constraining factors too. This Report reflects these & identifies the steps being taken to minimise delays & overcome obstacles to deliver on all commitments.	Foreword; Introduction; Progress report on programme for government actions; Progress report on public service agreements; Looking forward: key challenges; Conclusions; Annexes.	http://www.pfgni.gov.uk/annreport0102.htm
Building on Progress – Priorities & Plans 2003–06 (24/09/02)	The purpose of this document is to set out the priorities & plans for government in Northern Ireland for the 2003/04 financial year & beyond. The document explains the policy priorities, the budgetary decisions & the key commitments being made by Government, setting out how these develop the proposals presented by the Executive in its draft Programme for Government. The document also reflects the results of the consultation that has taken place on the draft Programme for Government & the draft Budget & provides an overview of the equality impacts of the policy direction being set.	Foreword; Introduction; Results of consultation; The policy framework; Budget 2003/04 to 2005/06; The government's commitments; Appendix A: The commitments in full; Appendix B: Equality impact; Appendix C: Public service agreements; Appendix C(i): Departments.	http://www.pfgni.gov.uk/plans.pdf
The Executive's Position Report to the Assembly: Developing the Programme for Government & the Budget for 2003–04 (2002)	A key first phase in this year's Programme for Government & budget cycle is this Position Report in which the Executive sets out, for consideration by the Assembly & others, the key issues affecting public services which should influence the preparation of the Programme for Government & Budget for 2003–04. This allows the Executive, before it agrees its draft proposals for the Programme & Budget in September, to take account of others' views.	Section 1: Introduction; Section 2: Programme for Government; Section 3: Resources; Section 4: Departmental issues for 2003–04 & beyond; Annex: Summary of Departmental positions.	http://www.nics.gov.uk/execpositionreport/pdfs/epr.pdf

Continued

Table 12.5.4 Northern Ireland Level Urban Development Policy. *Continued*

Policy Type	Document Title	Objective	Contents	Website
Strategies & Reports	Equality Impact Assessment (EQIA): Urban Regeneration – Neighbourhood Renewal (Feb 2003)	The Department for Social Development (the Department) has decided to introduce a new urban regeneration policy the core aim of which is the regeneration of local neighbourhoods, targeting those communities experiencing the most serious economic & social deprivation. This lies at the heart of Government's policy for Targeting Social Need & addresses the commitment made in the draft Programme for Government 2002, & a related Priorities & Plans 2003/06 document, that renewal of the most disadvantaged urban neighbourhoods will be a priority issue. It will be a significant focus for the Department's urban regeneration activities & resources over the next 7–10 years. This document is an Equality Impact Assessment (EQIA) of the Department's Urban Regeneration, *Neighbourhood Renewal Strategy* & conforms to the requirement to have due regard to the need to promote equality of opportunity & the need to provide equality of opportunity in developing new policy.	Executive summary; EQIA; Aim of the proposed policy; Equality considerations; Existing policy/policy context; Reasons for changing existing policy; Main elements of new policy on neighbourhood renewal; Consultation on the policy; Equality implications for public authorities; Consideration of available data & research; Assessment of impact; Impact on section 75 groups; Formal consultation; Monitoring arrangements; Comments on EQIA.	http://www.dsdni.gov.uk/publications/documents/eqia_nrs.pdf

Final Closure Report by the Department for Social Development on the Urban Operational Programme in Northern Ireland (1994–1999) (Mar 2003)	An evaluation of the operational running of the Urban Programme in Northern Ireland, focusing on the delivery successes & the expenditure of the programme and its outcomes.	Introduction; Programme administration; Implementation; Programme delivery; Publicity; Evaluation; Final expenditure & final claim; Annex A – Publicity events.	http://www.dsdni.gov.uk/publications/documents/Urban1_closure_report.doc
Belfast Regeneration Office Draft Strategy (Mar 2003)	The thrust of the Belfast Regeneration Office (BRO) strategy is to create an urban environment where all & not just the few can aspire to enjoy a high standard of living & good quality of life. This will mean the BRO encouraging & supporting a strong partnership approach to tackling the causes of disadvantage & building communities in the most socially deprived areas of greater Belfast. The BRO will focus on four priority areas where it will concentrate its resources: ▪ Encouraging investment & physical regeneration; ▪ Raising educational achievement; ▪ Improving access to employment; ▪ Creating safe, healthy communities.	Foreword; Executive summary; Strategic policy context; Regeneration in Belfast – What's happened so far; Key strategic issues; BRO strategic aims & objectives; Performance targets; Making it happen; Annexes.	http://www.dsdni.gov.uk/publications/documents/BRO-Strategy-final.pdf
People & Place – A Strategy for Neighbourhood Renewal (June 2003)	The publication of the Neighbourhood Renewal Strategy marks both the end of a long process of extensive research, review & consultation & the beginning of the process of implementation. This document is the first of several that will collectively map out the approach to urban regeneration & Neighbourhood Renewal. Collectively, these strategies will be known as 'People & Place' to emphasise their inter-related nature & common themes.	Foreword; Executive summary; The context; Learning the lessons; Goals & objectives; Delivering neighbourhood renewal; Measuring success; Conclusions & next steps; Annexes.	http://www.dsdni.gov.uk/publications/displayitemSections.asp?ID=343

Continued

Table 12.5.4 Northern Ireland Level Urban Development Policy. *Continued*

Policy Type	Document Title	Objective	Contents	Website
Strategies & Reports	Implementation and Monitoring of the Regional Development Strategy for Northern Ireland 2025 (June 2003)	This annual report outlines progress under these headings & scopes the way forward. It also sets out progress on implementing the RTS, which has been agreed by the Assembly. In summary it: ▪ reports on the achievements since the RDS was formulated in September 2001; ▪ establishes a baseline against which future performance will be assessed; & ▪ sets out areas for action over the year to March 2004.	Summary of achievements; Introduction; Partnership & participation; Monitoring & review; Learning the lessons from elsewhere; The Regional Transportation Strategy; The way ahead; Glossary; Bibliography; Annex 1: Monitoring report on the strategic planning guidelines; Annex 2: Calculation of regional housing need 1998–2015.	http://www.drdni.gov.uk/shapingourfuture/pdf/reports/annreprds.pdf
	People & Place: A Strategy for Neighbourhood Renewal – Urban Regeneration in Northern Ireland: Policy Measurement (June 2003)	The Urban Regeneration and Community Development Group of the Department for Social Development commissioned a report to examine existing approaches to the measurement of urban regeneration policy in Northern Ireland & make recommendations for an enhanced performance measurement framework.	Preface; Executive summary; Introduction; The changing policy environment for urban regeneration; Requirements for monitoring & evaluation; Monitoring & evaluation for urban policy – the current position; Monitoring & evaluation for urban policy – the new position; Implementation issues & recommendations.	http://www.dsdni.gov.uk/publications/documents/WorkingPaper2(2).pdf
	Environmental Assessment of the URBAN II Community Initiative Programme for Northern Ireland (Sept 2003)	An ex-ante environmental appraisal has been commissioned of the 'URBAN II Community Initiative Programme for Northern Ireland 2000–2006'. The appraisal includes a Development Path Analysis an Environmental Criteria Analysis, guidance on project selection criteria and guidance on the monitoring and evaluation of the programme's environmental impacts over the life of the intervention.	Introduction; Executive summary; Methodology; Development path analysis; Environmental criteria analysis; Annex 1: Detailed development path analysis; Annex 2: Environmental criteria analysis; Annex 3: Environmental monitoring.	http://www.dsdni.gov.uk/publications/documents/URBAN%20Final%20EIA.doc

Economic Development Forum – Working Together for a Stronger Economy – First Review (Jan 2004)	In June 2002, *Working Together for a Stronger Economy* presented an agreed statement of the priorities to be addressed by all sectors if the Northern Ireland economy is to be significantly strengthened during the period to the year 2010. The purpose of this current statement is to present the findings of the Economic Development Forum's first review of Northern Ireland's economic performance & progress towards the agreed targets. This statement highlights some of the key findings of the review process, & identifies a number of key weaknesses to be resolved if the Forum's vision of significantly strengthening the Northern Ireland economy by the year 2010 is to become a reality.	Introduction; Purpose; Approach; Global context; Key findings; Key weaknesses; Specific comments on key target areas; EDF action plan; Way forward; Annex A: Review of progress against targets 2003; Annex B: Review of economic context.	http://www.drdni.gov.uk/shapingourfuture/pdf/Draft7-FINAL-21Jan04-Internet%20Ready.pdf
Monitoring Report – Third Annual Report Regional Development Strategy for Northern Ireland 2025: April 2003–March 2004 (Nov 2004)	This monitoring report provides statistical information in relation to the achievement of the supporting actions to the Strategic Planning Guidelines of the Regional Development Strategy within its 25 year timeframe. In consultation with the Interdepartmental Steering Group members & its supporting working group of officials from each department, a representative number of supporting actions have been selected, against which overall indicators can readily be set & whose outcome can be supported by an appropriate dataset. As stipulated in the 1st Annual Report, the monitoring report commenced tracking trends using the period 2001/2002 as a baseline. Due to the short timescale, analysis of data for the intervening years is not be robust enough to determine emerging trends. Some overall indicators have been included against which sponsoring departments have only initiated data collection, or where guarantees have been made that future information will be made available as part of the monitoring process.	Listing of policies; Supporting actions; Overall indicators & measured outcomes; Supporting statistics & graphs.	http://www.drdni.gov.uk/shapingourfuture/pdf/Monitoring%20Report%202004%20Nov.pdf

Continued

Table 12.5.4 Northern Ireland Level Urban Development Policy. *Continued*

Policy Type	Document Title	Objective	Contents	Website
Strategies & Reports	Third Annual Report Regional Development Strategy: April 2003–March 2004 (Nov 2004)	The purpose of this Third Annual Report is to provide information for the period April 2003 to March 2004 on actions taken towards achieving the objectives of the Regional Development Strategy for Northern Ireland 2025 (*Shaping our Future*), & in addition to report on the measurement of progress towards achieving selected key objectives of the RDS.	Introduction; Current activity; A participative sub-regional approach; Statementing; Update on policy work; Think Regionally – Act Locally; Planning Policy statements; Key Strategic Issues Emerging; Regional Housing Need; Significant Local Change in Housing Demand; Strategic Sites; Second Homes; Urban – Rural Balance; Focused Assessment; Monitoring process; The Evidence; Future Reporting; Appendices	http://www.drdni.gov.uk/shapingourfuture/pdf/3rd%20annual%20report.pdf
Consultation Papers	Shaping Our Future – An Invitation to Consultation (June 1997)	The purpose of this paper is to start the process of preparing a Regional Strategic Framework for NI. This document identifies the main themes which might underpin this framework & explains how everyone who wants to become involved in this process can play a positive role in shaping the development of the Region over the next quarter of a century.	Foreword; Towards a strategy for the development of the region; Why a regional strategic framework? What are the strategic issues? How will the regional strategic framework be prepared? Linkage to other strategies; What happens next?	http://www.drdni.gov.uk/shapingourfuture/speeches/jun97_intro.htm
	Shaping Our Future – Public Consultation Report by the Consortium (Nov 1998)	The report is in three sections. The first is a summary analysis of the main arguments emerging from the consultation. The second is a review of the opinions expressed in the consultation itself, distinguishing those points which carried wide consensus from those which were predominantly held by particular sectors or areas. The third offers further analysis & proposals, & draws upon comparative references to regional planning elsewhere.	The consortium; Introduction; Section 1: summary analysis; Section 2: review of consultation; Section 3: further commentary & suggestions; App 1: methodology; App 2: list of groups which participated.	http://www.drdni.gov.uk/shapingourfuture/pdf/reports/urban_public_consultation_reg_strat_frame_wpics.pdf

Shaping Our Future – Draft Strategy (Dec 1998)	The purpose of this report is to set out for public comment the Draft Regional Strategic Framework for the future development of NI to 2025 & the potential benefit of around nearly two million people who will be living in the Region. NI, in common with other Regions throughout Europe, is faced with the key challenge of providing a high quality of life for all its citizens. The Region needs to build on its strengths & look outwards to prosper in the face of the competitive challenges in the global market of the 21st Century. The Draft Regional Strategic Framework therefore addresses a range of economic, social, environmental & community issues which are relevant to delivering, through the New Assembly, the objectives of achieving sustainable development & social cohesion in NI.	Foreword; Part 1: Introduction; Part 2: Background; Part 3: The strategic planning guidelines; Part 4: The way forward; Appendices.	http://www.drdni.gov.uk/shapingourfuture/pdf/reports/doe_sof_drft_reg_strat_frm_pics.pdf
URBAN II Programme 2000–2006 for NI – Consultation Document (Aug 2000)	The first URBAN Initiative is now complete & the European Commission has issued Guidelines for the URBAN II Initiative that will apply across the EU for the period 2000–2006. Under these Guidelines, NI is required to submit a proposed strategy for URBAN II to the European Commission by Oct 2000. The DSD wishes to initiate a consultation process to identify the priorities for URBAN II in NI. This consultation will lead to decisions on the issues & areas to be addressed in the NI URBAN Programme 2000–2006.	Introduction; The scope of the document; URBAN II guidelines; NI urban policy context; How did URBAN I perform in NI; Ideas for discussion; Programme content; How to implement URBAN II; Responses & timescale.	http://www.dsdni.gov.uk/publications/documents/drafturbaniia.pdf
Urban Regeneration in NI – Neighbourhood Renewal a Consultation Document (July 2001)	This paper sets out for consultation a new strategic approach to urban regeneration across NI. In doing so, it reflects the mission statement of the DSD (Department for Social Development) which is: 'Together, tackling disadvantage, building communities'. It addresses the commitment made in the Programme for Government that renewal of the most disadvantaged urban neighbourhoods will be a priority for the NI Executive.	Foreword; Section 1: Introduction; Section 2: Background; Section 3: Key urban regeneration issues; Section 4: The proposed strategy; Section 5: Delivering the strategy; Section 6: Let us know your views on the strategy.	http://www.dsdni.gov.uk/publications/documents/URGStrategyConsult.pdf

13 Waste Management

13.1 Waste Management and the Built Environment

Waste is the by-product of many processes which exist to maintain a modern society involving activities across domestic or municipal, industrial, commercial and public sectors. Whilst much attention has been focused on the waste arising from human activity in the built environment, a larger proportion of waste is generated from the other sectors. Waste has been taken for granted in western societies but is frequently the source and pathway for pollution and damage to health in developing world nations. The use of everyday goods such as food, soap, clothing, cars and cleaning materials are considered to be major sources of waste, but these are dwarfed by industrial wastes such as ash, slurry, flue gases, construction debris, methane and wastes from manufacturing and extractive processes.

Human settlements ingest all manner of raw or processed materials such as food, domestic and household products, including luxury goods to meet modern human needs, with the requirements for shelter and communications necessitating infrastructure, buildings and public facilities. After processing, manufacturing and consumption, many by-products are discarded such as glass, paper, plastic, aluminium, wood and garden compost as well as complex products such as electrical goods, tyres, concrete, ash, metals and packaging. In turn, industry and commerce ingest energy, metals and minerals, water, forests, fisheries and agricultural products and eventually excrete liquids and solid waste.

Wastes have various degrees of biodegradability and polluting potential, with a possible range of gases in the form of molecular wastes. The solid wastes may be re-processed or be discarded to landfill sites, while the molecular wastes enter into the atmosphere, seas, rivers, groundwater, soil, plants and the flesh of wildlife and people. In either case, waste materials will follow the scientific laws of 'matter does not disappear' and 'matter tends to go back to its original state'.

Waste Management has been recognised internationally as being one of the key aspects of the built environment which needs to be controlled and regulated in order to minimise the consumption and discarding of unnecessary goods, and to divert waste away from landfill sites. The efficiency of waste systems has impacts on energy, construction and pollution systems and is driven by both social dynamics and industrial management and performance. The international drive to control waste was initially encapsulated in the European Directive 75/442/EEC on Waste Management and Disposal, and subsequent EC Directives have added further detail in a requirement to divert waste away from landfill.

13.2 Key Waste Management Elements

A hierarchical approach to the management of waste is internationally accepted. Its classical structure indicates that the *Reduction* of waste has the highest importance in encouraging the use of less materials, packaging and luxury goods, thereby preventing such waste streams from entering the waste systems. Awareness campaigns and various forms of incentive and reward are being used to influence human behaviour into waste reduction, also referred to as prevention, as the primary contributor in avoiding the steady growth of waste.

The second element of the hierarchy, *Reuse* of products in their original and clean state, can help deliver waste minimisation, aided by Municipal Collection Facilities, often referred to as Civic Amenity Sites, and innovative 'bring back' arrangements with manufacturers. Products which are not re-usable can be recycled by re-processing into useful secondary goods. This *Recycling* part of the hierarchy depends heavily on the financial and technical capability of a range of recycling markets, and by applying whole life costing to the recycled product.

The *Recovery* of energy from waste is already feasible for the composting of green waste and waste-waters, while stringent financial appraisals are needed for more complex energy recovery processes, across a range of options such as thermal treatment, anaerobic digestion and mechanical biological treatment. This sector of the hierarchy needs to be supported by technical research, government support and appropriate planning systems.

The *Disposal* of waste to landfill is the least preferred element of the hierarchy, but has historically been the preferred and sole option, thereby requiring a psychological switch from treating waste as rubbish to waste being seen as a resource. However, properly engineered landfill sites remain as essential recipients of biodegradable waste and residues from materials recovery, anaerobic digestion, thermal treatment and composting facilities. Similar sites are required to receive hazardous and clinical wastes, with specially designed units for agriculture and other potentially polluting matter.

Waste management strategies have been influenced by a raft of EC Directives for waste electrical and electronic equipment (WEEE), packaging, waste disposal and hazardous waste, end-of-life vehicles (ELV), and supported by the use of landfill taxes to reduce the reliance on landfill sites. Locally based initiatives are commonplace in raising awareness of the social, economic and environmental benefits of driving municipal waste habits up the waste hierarchy. In turn, the environmental and economic benefits to industry and commerce of more waste minimisation and recycling are driven through the use of accredited Eco-Management and Audit Schemes (EMASs). These systems are ultimately informed and directed by setting time-related waste reduction goals, typically in the categories of 'diversion away from landfill', recycling and reduction targets and requirements to provide facilities to recycle a range of products and to receive difficult wastes.

A stable and sustainable waste management industry requires strategic leadership, sound recycling market development, appropriate planning systems and infrastructure, regulations and guidance, well-grounded educational and awareness raising programmes and a robust marketing strategy, all built on the waste hierarchy. These will be guided by principles of resource management, whole life costing, 'self sufficiency', 'proximity' and 'polluter pays', best value and the use of appropriate decision making systems tools such as the 'Best Practicable Environmental Option' (BPEO).

13.3 Sustainability Challenges for Waste Management

Waste management challenges fall initially into both the 'social progress' and 'economic growth' sectors as it is fully recognised that the attitude of and response by society to the waste targets must be delivered by changes in human behaviour, while industrial responses can be enhanced by better understanding of the economic business case. In general, waste management contributes to all four objectives of sustainable development through its role in natural, manufactured and economic capitals. Waste management links to sustainable development objectives include:

- *Effective protection of the environment* through reducing pollution to surface water, groundwater and the landscape from waste contamination and fly tipping. Vulnerable habitats are also preserved by removal of illegal flows or deposits of processed wastes and from hazardous and clinical sources. Agricultural by-products can pollute without proper management and monitoring. Open cast mining and disposal of residues must be controlled to protect the environment.
- *Social progress which recognises the needs of everyone* relates directly to the attitude and behaviour of society to the waste management hierarchy, as it takes deliberate acts of intent for people to actively engage in – purchasing goods which generate a minimum of further waste and have been recycled, reusing products, regularly collecting and delivering recyclable goods to home-based or centrally located civic amenity sites or recycling facilities, supporting educational programmes in waste minimisation and encouraging supportive community initiatives.

- *Prudent use of natural resources* relates to sustainable waste management practices through the minimum extraction of raw materials, aggressive engagement with market development for recyclables, re-use of construction and demolition waste, ownership by manufacturers of all products and understanding the linkages from waste management to areas such as energy reduction, pollution incidents, economic impact and environmental protection. These matters are driven and changed through application of quality and environmental management accreditations by industry.
- *Maintenance of high and stable levels of economic growth and employment* is supported by efficient and effective management of several industries which have traditionally had high waste arisings. These include activities in the construction and demolition industry, goods packaging, extractive industries, agriculture, fisheries and tourism activities, as well as municipal and regional governance. Informed business judgements can contribute to the waste hierarchy and add to employment numbers and prospects, to improve business efficiency and public profile.

Several issues and challenges exist for environmentalists, economists, decision-makers and strategists in delivering sustainable waste management and related areas. This sector is being seen as central to the broader sustainable development agenda and an area which must be arrested within a timeframe of up to 2020. Similarly, sustainable development is viewed as the high level driver for the delivery of the principles of waste management. The critical challenges which arise for waste management to contribute to this broader sustainability agenda include:

- *Increasing Reduction, Reuse and Recycling of waste materials* by both public and private sectors in order to achieve the stringent recycling and 'diversion of waste from landfill sites' targets by the key timelines of 2010, 2013 and 2020. These can be achieved by typically delivering at least 45% recycling and composting and diverting 65% of waste away from landfill by 2020, based on the 1998 levels.
- *Consolidation of Market Developments* for primary dry recyclables to ensure waste supply, economic stability and the use of appropriate and reliable technology. The mix of solutions to achieve the targets can be best resolved by using a 'Best Practicable Environmental Option' approach, and the use of tools such as environmental impact assessment and life cycle costing.
- *Research and development* is needed to consolidate the technology, economics and environmental impact of processes connected with the recycling of industrial products such as tyres, electrical goods and end-of-life vehicles. There is also a need for improved understanding of mechanical and biological treatments, anaerobic digestion processes and thermal treatment.
- *Enforcement of legislation and guidance* to control trans-frontier shipments of waste, apply the 'polluter pays' principle and integrate the principles of waste minimisation into business plans, *Local Agenda 21* programmes and environmental management systems. The use of waste taxes should be fully investigated.
- *Waste management plans and strategic planning* need to be in place to support national waste strategies. The application of waste audits and monitoring, and the securing of reliable waste data will allow efficient assessment against targets.
- *Greater public awareness and participation* in waste reduction initiatives is essential requiring a deeper psychological analysis of human behaviour and attitudes to waste issues. Education must take place both in the classrooms for all school pupils and college students and as an integral part of lifelong learning programmes.
- *Use of waste minimisation examples* in government, business, community and domestic environments can act as catalysts for improved participation.

13.4 Waste Management Theme Index

Jurisdiction	Tier	Title	Date	Page No.
INTERNATIONAL	Conventions	Convention on the Prevention of Marine Pollution by Dumping Waste & Other Matter (London Dumping Convention)	1972	537
		Convention on the Control of Transboundary Movements of Hazardous Wastes & Their Disposal (Basel)	1989	537
		Joint Convention on the Safety of Spent Fuel Management & on the Safety of Radioactive Waste Management	1997	538
EUROPEAN	Council Directives	Disposal of Waste Oil *(CD 75/439/EEC Amended by CD 87/101/EEC; CD 91/692/EEC)*	1975 1987 1991	539
		Waste Management & Disposal *(CD 75/442/EEC Amended by CD 91/156/EEC; CD 91/692/EEC; Comm Dir 96/350/EC; CD 96/59/EC)*	1975 1991 1996	539
		Disposal of Titanium Dioxide Industrial Waste *(CD 78/176/EEC Amended by CD 82/883/EEC; CD 83/29/EEC CD 91/692/EEC)*	1978 1982 1983 1991	539
		Surveillance & Monitoring of Titanium Dioxide Waste *(CD 82/883/EEC)*	1982	540
		Use of Sewage Sludge in Agriculture *(CD 86/278/EEC)*	1986	540
		New Waste Incineration Plants *(CD 89/369/EEC)*	1989	540
		Existing Waste Incineration Plants *(CD 89/429/EEC)*	1989	540
		Disposal of Spent Batteries & Accumulators *(CD 91/157/EEC Amended by Comm Dir 93/86/EEC; Comm Dir 98/101/EC)*	1991 1993 1998	541
		Urban Waste Water Treatment *(CD 91/271/EEC)*	1991	541
		Controlled Management of Hazardous Waste *(CD 91/689/EEC Amended by Comm Dir 94/31/EC)*	1991 1994	541
		Transfer of Radioactive Waste: Supervision & Control *(CD 92/3/Euratom)*	1992	541
		Reduction of Pollution Caused by Waste from the Titanium Dioxide Industry *(CD 92/112/EEC)*	1992	541
		Basel Convention on the Control of Transboundary Movements of Hazardous Waste *(CD 93/98/EEC)*	1993	542
		Packaging & Packaging Waste *(CD 94/62/EC)*	1994	542
		Incineration of Hazardous Waste *(CD 94/67/EC)*	1994	542
		Disposal of Polychlorinated Biphenyls (PCBs) & Polychlorinated Terphenyls (PCTs) *(CD 96/59/EC)*	1996	542
		The Landfill of Waste *(CD 99/31/EC)*	1999	543
		End-of-Life Vehicles *(Dir 2000/53/EC)*	2000	543
		Waste Incineration *(Dir 2000/76/EC)*	2000	543
		The Restriction of the Use of Certain Hazardous Substances in Electrical & Electronic Equipment *(Dir 2002/95/EC)*	2002	543
		Waste Electrical & Electronic Equipment (WEEE Directive) *(Dir 2002/96/EC)*	2002	543

Jurisdiction	Tier	Title	Date	Page No.
EUROPEAN	Council Directives	Amendment to the WEEE Directive *(Dir 2003/108/EC)*	2003	543
		Amendment to Packaging & Packaging Waste Directive *(Dir 2004/12/EC)*	2004	544
	Council Decisions	The Convention on the Control of Transboundary Movements of Hazardous Wastes & their Disposal – Decision III/I of the conference of parties *(C Dec 97/640/EC)*	1993 1997	544
		Establishing the Identification System for Packaging Materials Pursuant to European Parliament & Council Directive 94/62/EC on Packaging & Packaging Waste *(C Dec 97/129/EC)*	1997	544
		Establishing a List of Wastes *(C Dec 2000/532/EC)*	2000	544
		Establishing Criteria & Procedures for the Acceptance of Waste at Landfills *(C Dec 2003/33/EC)*	2003	545
		Establishing Component & Material Coding Standards for Vehicles Pursuant to Directive 2000/53/EC of the European Parliament & of the Council on End-of-life Vehicles	2003	545
	Council Communications	Removal & Disposal of Disused Offshore Oil & Gas Installations *(COM (98) 49)*	1998	545
		Competitiveness of the Recycling Industries *(COM (98) 463)*	1998	545
		Present Situation & Prospects for Radioactive Waste Management *(COM (98) 799)*	1998	546
		Community Strategy for Dioxins, Furans & Polychlorinated Biphenyls *(COM (2001) 593)*	2001	546
		Towards a Thematic Strategy on the Prevention & Recycling of Waste *(COM (2003) 301)*	2003	546
	Miscellaneous	Community Plan of Action in the Field of Radioactive Waste (1980–99)	1980	547
		Supervision & Control of Transfrontier Shipments of Waste *(C Reg 259/93/EEC Amended by Dec 94/721/EC; Dec 96/660/EC; Reg 120/97/EC; Reg 2408/98/EC; Reg 1420/1999/EC; Reg 1547/1999/EC; Reg 2557/2001/EC)*	1993 1994 1996 1997 1998 1999 2001	547
		Shipments of Radioactive Substances *(C Reg No. 1493/93)*	1993	547
		End-of-Life Vehicles: Current Basic Data Reflecting the Overall Ecological & Economic Context of the ELV Issue	1996	548
		Waste Management Statistics *(COM (99) 31)*	1999	548
		Construction & Demolition Waste Management Practices, & Their Economic Impacts	1999	548
		European Conference on Waste Management Planning	1999	549
		EU Focus on Waste Management	1999	549
		Chemical Recycling of Plastics Waste (PVC & other Resins)	1999	549
		Implementation of Community Waste Legislation (1995–1997) *(COM (1999) 752)*	1999	550

Jurisdiction	Tier	Title	Date	Page No.
EUROPEAN	Miscellaneous	Waste Electrical & Electronic Equipment *(COM (2000) 347)*	2000	550
		The Behaviour of PVC in Landfill	2000	551
		Heavy Metals in Vehicles	2000	551
		Economic Evaluation of PVC Waste Management	2000	552
		Green Paper on Environmental Issues of PVC *(COM (2000) 469)*	2000	552
		Economic Evaluation of PVC Waste Management	2000	553
		A Study on the Economic Valuation of Environmental Externalities from Landfill Disposal & Incineration of Waste	2000	553
		Successful Stories on Composting & Separate Collection	2000	554
		European Packaging Waste Management Systems – Main Report	2001	554
		The Impact of Community Environmental Waste Policies on Economic & Social Cohesion	2001	554
		Waste Management Options & Climate Change	2001	555
		Heavy Metals in Vehicles II – Final Report	2001	555
		Applying Compost – Benefits & Needs	2001	556
		Critical Review of Existing Studies & Life Cycle Analysis on the Regeneration & Incineration of Waste Oils	2001	556
		Reuse of Primary Packaging – Final Report	2001	557
		Heavy Metals in Waste	2002	557
		Costs for Municipal Waste Management in the EU	2002	557
		Financing & Incentive Schemes for Municipal Waste Management – Case Studies	2002	558
		Study on Hazardous Household Waste (HHW) with a main Emphasis on Hazardous Household Chemicals (HHC)	2002	558
		Economic Analysis of Options for Managing Biodegradable Municipal Waste	2002	558
		Case Studies on Waste Minimisation Practices in Europe	2003	559
		Evaluation of Costs & Benefits for the Achievement of Reuse & Recycling Targets for the Different Packaging Materials	2003	559
		Preparing a Waste Management Plan – A Methodological Guidance note	2003	559
		Management of Waste from the Extractive Industries *(COM (2003) 319)*	2003	560
		Shipments of Waste *(COM (2003) 379)*	2003	560
		The Implementation of Community Waste Legislation for the Period 1998–2000 *(COM (2003) 250)*	2003	561
		Impact Assessment on Selected Policy Options for Revision of the Battery Directive Final Report	2003	561

Jurisdiction	Tier	Title	Date	Page No.
EUROPEAN	Miscellaneous	Proposal for a Directive on Batteries & Accumulators & Spent Batteries & Accumulators *(COM (2003) 723)*	2003	562
		Proposal for a Directive on Waste *(COM (2003) 731)*	2003	563
		Proposal for a Council Directive amending Directive 94/62/EC on Packaging & Packaging Waste *(COM (2004) 127)*	2004	563
		Amended Proposal for a Regulation on Shipments of Waste *(COM (2004) 172)*	2004	563
		Heavy Metals & Organic Compounds from Wastes used as Organic Fertilisers	2004	564
		Study on the Implementation of Directive 94/62/EC on Packaging & Packaging Waste & Options to Strengthen Prevention & Reuse of Packaging	2004	565
		A Cleaner, Greener Europe: LIFE & the European Union Waste Policy	2004	565
UK	Strategies & Reports	Response to the Environment, Transport & Regional Affairs Select Committee Report on Sustainable Waste Management	1998	566
		A Way with Waste: A draft waste strategy for England & Wales (Part I & II)	1999	566
		Waste Strategy for England & Wales (Part I & II)	2000	566
		The Implementation of Council Directive 1999/31/EC on the Landfill of Waste	2000	567
		Research Study on International Recycling Experience	2001	567
		Distributions of the Aggregates Levy Sustainability Fund – Summary of Responses	2002	568
		Model Forecasts of Recycling Participation Rates & Material Capture Rates for Possible Future Recycling Scenarios	2002	568
		Digest of Environmental Statistics & Municipal Waste Management Statistics	2002	568
		The Context for Emissions & Health Impacts Associated with Waste Management	2002	569
		Public Attitudes Towards Recycling & Waste Management – Quantitative & Qualitative Review	2002	569
		Waste Not, Want Not – A Strategy for Tackling the Waste Problem in England	2002	570
		Delivering the Landfill Directive: The Role of New & Emerging Technologies	2002	570
		Analysis of Household Waste Composition & Factors Driving Waste Increases	2002	570
		Collection Cost Projections	2002	571
		A Study to Estimate the Disamenity Costs of Landfill in Great Britain – Final Report	2003	571
		Municipal Waste Management Survey Results 2001/02	2003	571
		Mid-Term Evaluation of the Aggregates Levy sustainability Fund	2003	572
		'Towards a Thematic Strategy on the Prevention & Recycling of Waste' United Kingdom Government Response	2003	572

Jurisdiction	Tier	Title	Date	Page No.
UK	Strategies & Reports	Waste Implementation Programme – Factsheets	2003	573
		WEEE & Hazardous Waste	2004	573
		Review of Environmental & Health Effects of Waste Management: Municipal Solid Waste & Similar Wastes	2004	573
		Government Interpretation of the Landfill (England & Wales) Regulations 2002 (As Amended)	2004	574
		International Waste Prevention & Reduction Practice: Final Report	2004	574
		Clean Neighbourhoods & Environment Bill	2004	574
	Consultation Papers	Less Waste. More Value. Consultation Paper on the Waste Strategy for England & Wales	1998	575
		Limiting Landfill: A Consultation Paper on Limiting Landfill to Meet the EC Directive's Targets for the Landfill of Biodegradable Municipal Waste	1999	575
		Landfill Allowance Trading Scheme Consultation	2003	575
		Fly-Tipping Strategy – Consultation Document	2004	576
		Review of the Special Waste Regulations 1996 in England – Proposals for replacement Hazardous Waste Regulations & List of Wastes Regulations: A Consultation Paper	2004	576
		The Landfill (England & Wales) Regulations 2005 – A Consultation Paper	2004	576
		Consultation on Sustainable Waste Management	2004	577
	Guidance Notes & Circulars	Waste Strategy Guidance – Best Value & Waste Management	2000	577
		Strategic Planning for Sustainable Waste Management	2002	578
		Guidance on Directive 2000/76/EC on the Incineration of Waste	2002	578
		The Producer Responsibility Obligations (Packaging Waste) Regulations 1997 (as amended) – User's Guide Second edition	2003	579
		Draft Guidance for Waste Collection Authorities on the Household Waste Recycling Act 2003	2004	579
NORTHERN IRELAND	Strategies & Reports	A Guide to the Special Waste Regulations (Northern Ireland) 1998	1998	580
		Waste Management Strategy for NI	2000	580
		Guidelines for Company Reporting on Waste	2001	580
		Waste Arisings Survey for NI – 1999–2000	2001	581
		Evaluation of the First Phase of the Waste Management Public Awareness & Information Programme	2002	581
		Waste & Contaminated Land (Northern Ireland) Order 1997 – Article 5: Waste Management The Duty of Care – A Code of Practice	2002	581
		The NW Region Waste Management Group – Waste Management Plan	2002	582
		Southern Waste Management Partnership (SWaMP) – Waste Management Plan	2002	582

Jurisdiction	Tier	Title	Date	Page No.
NORTHERN IRELAND	Strategies & Reports	Interim Needs for Landfill Capacity 2003–2007	2003	583
		Draft Biodegradable Waste Strategy for NI	2003	583
		Compliance Monitoring Strategy for 2003	2003	584
		Wake up to Waste II Evaluation Report	2003	584
		Factors Affecting the Future of Hazardous Waste in Northern Ireland	2003	585
		Northern Ireland Waste Management Strategy Review Report	2004	585
		Hazardous Waste in Northern Ireland: An Action Plan for its Environmentally Sound Management	2004	585
		BPEO for the Management of Waste Asbestos	2004	586
		Baseline Waste Audit	2004	586
	Consultation Papers	ARC21: Eastern Region Waste Management Group – Consultation Waste Management Plan	2002	587
		Consultation Paper on Recovery & Recycling Targets in NI for Packaging Waste in 2002	2002	587
		Consultation Paper on a Proposal of the European Parliament & Council to Amend the Directive on Packaging & Packaging Waste	2002	587
		A Consultation Paper on Limiting Landfill to meet the EC Landfill Directive's Targets for Reducing the Landfill of Biodegradable Municipal Waste	2002	588
		Consultation Paper on the Implementation of Council Directive 2000/76/EC on the Incineration of Waste	2003	588
		Consultation Paper on Implementation of EC Directive 1999/31/EC on the Landfill of Waste	2003	588
		Consultation on the Transposition of Article 6 & Annex I of the End-of-Life Vehicles Directive (2000/53/EC) – Proposals for Regulations	2003	589
		Waste Management & Contaminated Land – Proposed Fees & Charging Scheme (Consultation Paper)	2003	589
	Guidance Notes	Tackling Fly Tipping – A Guidance Note & Response Protocol	1999	589
		Best Practicable Environmental Option (BPEO) – Decision-Makers Guide	2000	589
		Municipal Waste Data Monitoring & Reporting – Interim Guidelines	2003	590
		Hazardous Waste – Interpretation of the Definition & Classification of Hazardous Waste	2003	590
		De-polluting End-of-Life Vehicles Guidance for Authorised Treatment Facilities	2004	590

13.5 Waste Management Policy Tables

Table 13.5.1 International Level Waste Management Policy.

Title	Venue	Purpose	Key Outcome(s)	Website
Convention on the Prevention of Marine Pollution by Dumping Waste & Other Matter (London Dumping Convention)	London, UK (1972)	To improve protection of the marine environment by encouraging States with a common interest in particular geographical areas to enter into appropriate agreements supplementary to this Convention. (Superseded by the 1996 Protocol)	An agreement that Contracting Parties shall individually and collectively promote the effective control of all sources of pollution of the marine environment, and pledge themselves especially to take all practicable steps to prevent the pollution of the sea by the dumping of waste and other matter that is liable to create hazards to human health, to harm living resources and marine life, to damage amenities or to interfere with other legitimate uses of the sea. A comprehensive listing of substances which are prohibited from ocean disposal. Regulations for the control of incineration of wastes and other matter at sea.	http://www.londonconvention.org/documents/lc72/LC1972.pdf
Convention on the Control of Transboundary Movements of Hazardous Wastes & Their Disposal (Basel)	Basel, Switzerland (1989)	To ensure that the threat to human health & the environment caused by hazardous wastes & their transboundary movement is fully understood and that all movement & disposal follows strict guidelines consistent with the protection of human health/the environment.	Complete listings in the Annex(s) of the waste categories requiring control/those in need of special consideration. The need for any participating party to follow strict guidelines governing the movement of hazardous waste that: ▪ Reduce waste generation to a minimum; ▪ Ensure availability of adequate disposal facilities; ▪ Ensure all necessary steps taken to prevent pollution; ▪ Ensure minimum transboundary movement of waste &, where it occurs, that it is consistent with environmentally sound management of such waste; ▪ Require information on transboundary movement to be provided & any possible effects on human health; ▪ Require relevant packaging, labelling & movement documentation from point of movement to point of disposal; ▪ Prevent the import of hazardous wastes if they are not dealt with in an environmentally sound manner.	http://www.basel.int/text/con-e.pdf

Continued

Table 13.5.1 International Level Waste Management Policy. *Continued*

Title	Venue	Purpose	Key Outcome(s)	Website
Joint Convention on the Safety of Spent Fuel Management & on the Safety of Radioactive Waste Management	Vienna, Austria (1997)	To achieve & maintain a high level of safety world-wide in the management of spent fuel and radioactive waste through the enhancement of national measures and international co-operation and to establish and maintain at all stages of spent fuel and radioactive waste management effective defences against potential hazards so that individuals, society and the environment are protected from harmful effects of ionising radiation.	The Convention imposes obligations on contracting parties in relation to the transboundary movement of spent fuel and radioactive waste based on the concepts contained in the IAEA (International Atomic Energy Agency) Code of Practice on the International Transboundary Movement of Radioactive Waste. Also, contracting parties have the obligation to take appropriate steps to ensure that disused sealed sources are managed safely. The obligations of the Contracting Parties include, in particular, the obligation to establish and maintain a legislative and regulatory framework to govern the safety of spent fuel and radioactive waste management and the obligation to ensure that individuals, society and the environment are adequately protected against radiological and other hazards, *inter alia*, by appropriate siting, design and construction of facilities and by making provisions for ensuring the safety of facilities both during their operation and after their closure.	http://www.iaea.org/Publications/Documents/Infcircs/1997/infcirc546.pdf

Table 13.5.2 European Level Waste Management Policy.

Policy Type	Programme Title	Policy Objective	Community Measure	Document Reference	Website
Council Directives	Disposal of Waste Oil	To promote the safe collection & disposal of waste oils.	Council Directive 75/439/EEC on the disposal of waste oil (16/06/75). Amended by the following: ▪ Council Directive 87/101/EEC (22/12/86) ▪ Council Directive 91/692/EEC (23/12/91).	▪ Off Jnl L194, 25/07/75 ▪ Off Jnl L42, 12/02/87 ▪ Off Jnl L377, 31/12/91	http://europa.eu.int/scadplus/leg/en/lvb/l21206.htm
	Waste Management & Disposal	To set up a system for the co-ordinated management of waste within the Community in order tc limit waste production.	Council Directive 75/442/EEC on waste disposal (15/07/75). Amended by the following measures: ▪ Council Directive 91/156/EEC (18/03/91) ▪ Council Directive 91/692/EEC (23/12/91) ▪ Commission Directive 96/350/EC (24/05/96) ▪ Council Directive 96/59/EC (16/09/96).	▪ Off Jnl L194, 25/07/75 ▪ Off Jnl L78, 26/03/91 ▪ Off Jnl L377, 31/12/91 ▪ Off Jnl L135, 06/06/96 ▪ Off Jnl L243, 24/09/96	http://europa.eu.int/scadplus/leg/en/lvb/l21197.htm
	Disposal of Titanium Dioxide Industrial Waste	To prevent, gradually reduce & ultimately eliminate pollution from titanium dioxide industrial waste.	Council Directive 78/176/EEC on titanium dioxide industrial waste (20/02/78). Amended by the following: ▪ CD 82/883/EEC (03/12/82) ▪ CD 83/29/EEC (24/01/83) ▪ CD 91/692/EEC (23/12/91).	▪ Off Jnl L54, 25/02/78 ▪ Off Jnl L378, 31/12/82 ▪ Off Jnl L32, 31/12/83 ▪ Off Jnl L377, 31/12/91	http://europa.eu.int/scadplus/leg/en/lvb/l21203.htm

Continued

Table 13.5.2 European Level Waste Management Policy. *Continued*

Policy Type	Programme Title	Policy Objective	Community Measure	Document Reference	Website
Council Directives	Surveillance & Monitoring of Titanium Dioxide Waste	To fix common reference methods of measurement for sampling in order to conserve environments concerned by titanium dioxide waste.	Council Directive 82/883/EEC on procedures for the surveillance & monitoring of environments concerned with titanium dioxide waste (03/12/82).	■ Off Jnl L378, 31/12/82	http://europa.eu.int/scadplus/leg/en/lvb/l21204.htm
	Use of Sewage Sludge in Agriculture	Regulating the use of sewage sludge in agriculture in such a way as to prevent harmful effects on soil, vegetation, animals & man.	Council Directive 86/278/EEC on the protection of the environment, & in particular of the soil, when sewage sludge is used in agriculture (12/06/86).	■ COM (97) 23 ■ Off Jnl L181, 04/07/86	http://europa.eu.int/scadplus/leg/en/lvb/l28088.htm
	New Waste Incineration Plants	To reduce emissions of certain pollutants from new municipal waste incineration plants.	Council Directive 89/369/EEC on the prevention of air pollution from new municipal waste incineration plants (08/06/89).	■ Off Jnl L163, 14/06/89	http://europa.eu.int/scadplus/leg/en/lvb/l28030b.htm
	Existing Waste Incineration Plants	To reduce the emissions of certain pollutants from existing municipal waste incineration plants.	Council Directive 89/429/EEC on the reduction of pollution from existing municipal waste incineration plants (21/06/89).	■ CD 89/429/EEC ■ Off Jnl L203, 15/07/89	http://europa.eu.int/scadplus/leg/en/lvb/l28030a.htm

Disposal of Spent Batteries & Accumulators	To introduce measures for the upgrading & controlled disposal of spent batteries & accumulators containing dangerous materials in the Community.	Council Directive 91/157/EEC on batteries & accumulators containing dangerous substances (18/03/91). Amended by the following: ■ Commission Directive 93/86/EEC (04/10/93) ■ Commission Directive 98/101/EC (22/12/98).	■ Off Jnl L78, 26/03/91 ■ Off Jnl L264, 23/10/93 ■ Off Jnl L1, 05/01/99	http://europa.eu.int/scadplus/leg/en/lvb/l21202.htm
Urban Waste Water Treatment	To harmonise measures on urban waste water treatment throughout the Community.	CD 91/271/EEC concerning urban waste water treatment (21/05/91). Amended by Comm Dir 98/15/EC (27/02/98).	■ Off Jnl L135, 30/05/91 ■ Off Jnl L67, 07/03/98	http://europa.eu.int/scadplus/leg/en/lvb/l28008.htm
Controlled Management of Hazardous Waste	The management, recovery & correct disposal of hazardous waste.	Council Directive 91/689/EEC on hazardous waste (12/12/91). Amended by Commission Directive 94/31/EC (27/07/94).	■ Off Jnl L377, 31/12/91 ■ Off Jnl L168, 02/07/94	http://europa.eu.int/scadplus/leg/en/lvb/l21199.htm
Transfer of Radioactive Waste: Supervision & Control	To lay down a system of prior authorisation for all movements of radioactive waste in order to increase protection against the dangers arising from ionising radiation.	Council Directive 92/3/Euratom regarding the supervision & control of transfers of radioactive waste between Member States coming in & out of the Community (03/02/92).	■ Off Jnl L35, 12/02/92	http://europa.eu.int/scadplus/leg/en/lvb/l11020.htm
Reduction of Pollution Caused by Waste from the Titanium Dioxide Industry	To set up, according to CD 78/176/EEC, procedures for the harmonisation of programmes for the reduction of pollution caused by waste from existing industrial establishments, & improve the conditions of completion in the sector of titanium dioxide production.	Council Directive 92/112/EEC on procedures for harmonising the programmes for the reduction & eventual elimination of pollution caused by waste from the titanium dioxide industry (15/12/92).	■ Off Jnl L409, 31/12/92	http://europa.eu.int/scadplus/leg/en/lvb/l21205.htm

Continued

Table 13.5.2 European Level Waste Management Policy. *Continued*

Policy Type	Programme Title	Policy Objective	Community Measure	Document Reference	Website
Council Directives	Basel Convention on the Control of Transboundary Movements of Hazardous Waste	To control, at an international level, Transboundary movements of wastes hazardous to human health, & the environment, & their disposal.	Council Directive 93/98/EEC on the conclusion, on behalf of the Community, of the Convention on the control of Transboundary movements of hazardous wastes & their disposal (Basel Convention) (01/02/93).	▪ Off Jnl L34, 16/02/93	http://europa.eu.int/scadplus/leg/en/lvb/l28043.htm
			Council Decision 97/640/EC on the approval, on behalf of the Community, of the amendment to the convention on the control of Transboundary movements of hazardous wastes & their disposal, as laid down in decision III/I of the conference of parties (22/09/97).	▪ Off Jnl L272, 04/10/97	
	Packaging & Packaging Waste	To harmonise national measures concerning the management of packaging & packaging waste to provide a high level of environmental protection & to ensure the functioning of the internal market.	Council Directive 94/62/EC on packaging & packaging waste (15/12/94).	▪ COM (1999) 596 Final ▪ Off Jnl L365, 31/12/94	http://europa.eu.int/scadplus/leg/en/lvb/l21207.htm
	Incineration of Hazardous Waste	To prevent or reduce the effects of hazardous waste incineration on the environment & the ensuing risks for public health.	Council Directive 94/67/EC on the incineration of hazardous waste (16/12/94).	▪ Off Jnl L365, 31/12/94	http://europa.eu.int/scadplus/leg/en/lvb/l21200.htm
	Disposal of Polychlorinated Biphenyls (PCBs) & Polychlorinated Terphenyls (PCTs)	To approximate the laws of the Member States on the controlled disposal of PCBs, the decontamination or disposal of equipment containing PCBs &/or the disposal of used PCBs in order to eliminate them completely.	Council Directive 96/59/EC on the disposal of PCBs/PCTs (16/09/96).	▪ Off Jnl L243, 24/09/96	http://europa.eu.int/scadplus/leg/en/lvb/l21201.htm

The Landfill of Waste	To prevent or reduce as far as possible negative effects on the environment from the landfilling of waste, by introducing stringent technical requirements for waste & landfills.	Council Directive 99/31/EC on landfill of waste (26/04/99).	Official Jnl L182, 16/07/99	http://europa.eu.int/scadplus/leg/en/lvb/l21208.htm
End-of-Life Vehicles	To prevent waste from end-of-life vehicles & promote the collection, re-use & recycling of their components to protect the environment.	Directive 2000/53/EC of the European Parliament & of the Council on the end of life vehicles (18/09/00).	Off Jnl L269, 21/10/00	http://europa.eu.int/scadplus/leg/en/lvb/l21225.htm
Waste Incineration	To prevent or reduce, as far as possible, air, water & soil pollution caused by the incineration or co-incineration of waste, as well as the resulting risk to human health.	Directive 2000/76/EC on the incineration of waste (04/12/00).	Off Jnl L332, 28/12/00	http://europa.eu.int/scadplus/leg/en/lvb/l28072.htm
The Restriction of the Use of Certain Hazardous Substances in Electrical & Electronic Equipment	The purpose of this Directive is to approximate the laws of the Member States on the restrictions of the use of hazardous substances in electrical & electronic equipment & to contribute to the protection of human health & the environmentally sound recovery & disposal of waste electrical & electronic equipment.	Dir 2002/95/EC of the European Parliament & the Council on the restriction of the use of certain hazardous substances in electrical & electronic equipment (27/01/03).	Off Jnl L037, 13/02/03	http://europa.eu.int/eur-lex/pri/en/oj/dat/2003/l_037/l_03720030213en00190023.pdf
Waste Electrical & Electronic Equipment (WEEE Directive)	The purpose of this Directive is, as a first priority, the prevention of waste electrical and electronic equipment (WEEE), & in addition, the reuse, recycling & other forms of recovery of such wastes so as to reduce the disposal of waste. It also seeks to improve the environmental performance of all operators involved in the life cycle of electrical & electronic equipment, e.g. producers, distributors & consumers & in particular those operators directly involved in the treatment of waste electrical & electronic equipment.	Dir 2002/96/EC of the European Parliament & the Council on waste electrical & electronic equipment (WEEE) (27/01/03).	Off Jnl L037, 13/02/03	http://europa.eu.int/eur-lex/pri/en/oj/dat/2003/l_037/l_03720030213en00240038.pdf
Amendment to the WEEE Directive (Dir 2002/96/EC)	The new Directive amends article 9 of Dir 2002/96/EC due to concerns about the possible financial implications of the wording of article 9 for producers of the equipment concerned.	Directive 2003/108/EC of the European Parliament & the Council amending Directive 2002/96/EC on waste electrical & electronic equipment (WEEE) (08/12/03).	Off Jnl L345/106, 31/12/03	http://europa.eu.int/eur-lex/pri/en/oj/dat/2003/l_345/l_34520031231en01060107.pdf

Continued

Table 13.5.2 European Level Waste Management Policy. *Continued*

Policy Type	Programme Title	Policy Objective	Community Measure	Document Reference	Website
Council Directives	Amendment to Packaging & Packaging Waste Directive (Dir 94/62/EC)	This Directive amends Dir 94/62/EC on packaging & packaging waste notably the definition of packaging contained in Annex 1 has been further defined.	Directive 2004/12/EC of the European Parliament & of the Council amending Directive 94/62/EC on packaging & packaging waste (11/02/04).	■ Off Jnl L 47/26 18/02/04	http://europa.eu.int/eur-lex/pri/en/oj/dat/2004/l_047/l_04720040218en00260031.pdf
Council Decisions	Basel Convention on the Control of Transboundary Movements of Hazardous Waste	To control, at an international level, Transboundary movements of wastes hazardous to human health, & the environment, & their disposal.	Council Decision 97/640/EC on the approval, on behalf of the Community, of the amendment (22/09/97) to the convention on the control of Transboundary movements of hazardous wastes & their disposal (01/02/93), as laid down in decision III/I of the Conference of Parties.	■ Off Jnl L34, 16/02/93 ■ Off Jnl L272, 04/10/97	http://europa.eu.int/scadplus/leg/en/lvb/l28043.htm
	Establishing the Identification System for Packaging Materials Pursuant to European Parliament & Council Directive 94/62/EC on Packaging & Packaging Waste	This Decision, which covers all packaging covered by Directive 94/62/EC aims to establish the numbering & abbreviations on which the identification system is based, indicating the nature of the packaging material(s) used & specifying which materials shall be subject to the identification system.	97/129/EC: Commission Decision establishing the identification system for packaging materials pursuant to European Parliament & Council Directive 94/62/EC on packaging & packaging waste (28/01/97).	■ Off Jnl L050 20/02/97	http://europa.eu.int/smartapi/cgi/sga_doc?smartapi!celexapi!prod!CELEXnumdoc&lg=EN&numdoc=31997D0129&model=guichett
	Establishing a List of Wastes	The present list is a harmonised list of wastes. It will be periodically reviewed on the basis of new knowledge &, in particular, of research results, & if necessary revised in accordance with Article 18 of Directive 75/442/EEC.	Council Decision 2000/532/EC replacing Decision 94/3/EC establishing a list of wastes pursuant to Article 1(a) of Council Directive 75/442/EEC	■ Council Dec 2000/532/EC	http://europa.eu.int/eur-lex/en/consleg/pdf/2000/en_2000D0532_do_001.pdf

			on waste & Council Decision 94/904/EC establishing a list of hazardous waste pursuant to Article 1(4) of Council Directive 91/689/EEC on hazardous waste (03/05/00).		
	Establishing Criteria & Procedures for the Acceptance of Waste at Landfills	This Decision establishes the criteria & procedures for the acceptance of waste at landfills in accordance with the principles set out in Directive 1999/31/EC & in particular Annex II thereto.	Council Dec 2003/33/EC establishing criteria & procedures for the acceptance of waste at landfills (19/12/02).	Off Jnl L11/29, 16/01/03	http://europa.eu.int/eur-lex/pri/en/oj/dat/2003/l_011/l_01120030116en00270049.pdf
	Establishing Component & Material Coding Standards for Vehicles Pursuant to Directive 2000/53/EC of the European Parliament & of the Council on End-of-Life Vehicles	Without prejudice to Article 3(3) of Directive 2000/53/EC, Member States shall take the necessary measures to ensure that producers, in concert with material & equipment manufacturers, use the nomenclature of ISO (International Standards Organization) component & material coding standards referred to in the Annex to this Decision for the labelling and identification of components & materials of vehicles.	Commission Decision establishing component & material coding standards for vehicles pursuant to Directive 2000/53/EC of the European Parliament & of the Council on End-of-Life Vehicles (27/02/03).	Off Jnl L53/58 28/02/03	http://europa.eu.int/eur-lex/pri/en/oj/dat/2003/l_053/l_05320030228en00580059.pdf
Council Communications	Removal & Disposal of Disused Offshore Oil & Gas Installations	To protect the environment by reducing pollution from disused offshore oil & gas installations.	Communication from the Commission to the Council & the European Parliament on the removal & disposal of disused offshore oil & gas installations (18/02/98).	COM (98) 49	http://europa.eu.int/scadplus/leg/en/lvb/l28053.htm
	Competitiveness of the Recycling Industries	To list the major difficulties encountered by recycling businesses in achieving or maintaining viability & to propose a package of measures capable of solving these problems.	Communication from the Commission on the competitiveness of the recycling industries (22/07/98).	COM (98) 463	http://europa.eu.int/scadplus/leg/en/lvb/l28065.htm

Continued

Table 13.5.2 European Level Waste Management Policy. *Continued*

Policy Type	Programme Title	Policy Objective	Community Measure	Document Reference	Website
Council Communications	Present Situation & Prospects for Radioactive Waste Management	To analyse the radioactive waste situation in the EU following the implementation of the Community action plan for 1980–1999.	Communication & Fourth report from the Commission on the present situation & prospects for radioactive waste management in the EU (11/01/99).	▪ COM (98) 799	http://europa.eu.int/scadplus/leg/en/lvb/l28097.htm
	Community Strategy for Dioxins, Furans & Polychlorinated Biphenyls	The scope of this strategy will cover the polychlorinated dibenzodioxins (PCDDs) commonly known as dioxins, polychlorinated dibenzofurans (PCDFs) commonly known as furans & polychlorinated biphenyls (PCBs). Among the PCBs, in terms of toxicity, special attention will be given to a small group of so called dioxin-like PCBs which exhibit dioxin-like toxicity.	Communication from the Commission to the Council, the European Parliament & the Economic & Social Committee – Community strategy for dioxins, furans & polychlorinated biphenyls (17/11/01).	▪ COM (2001) 593 ▪ Off Jnl C 322/2	http://europa.eu.int/comm/environment/waste/pcbs/en.pdf
	Towards a Thematic Strategy on the Prevention & Recycling of Waste	This Communication is a first contribution to the development of a thematic strategy that will cover both waste prevention & recycling. It is the starting point of a consultation process which will continue until the Commission adopts proposals both for the overall framework of the thematic strategy & for the individual measures through which the strategy will be implemented. In relation to waste prevention, the objective of this Communication is to launch, for the first time, a consultation process leading to the development of a comprehensive strategy, including waste prevention targets & the instruments needed to achieve them. For waste recycling, this Communication invites comments on options to promote recycling.	Communication from the Commission – Towards a thematic strategy on the prevention & recycling of waste (27/05/03).	▪ COM (2003) 301	http://europa.eu.int/eur-lex/en/com/cnc/2003/com2003_0301en01.pdf

Miscellaneous	Community Plan of Action in the Field of Radioactive Waste (1980–1999)	To establish a plan of action for the period 1980–1999, in order to allow management of the increasing volume of radioactive waste within the community.	Council Resolution (15/06/92) on the implementation of a Community plan of action (18/02/80) in the field of radioactive waste.	▪ Off Jnl C51, 29/02/80 ▪ Off Jnl C158, 23/06/92	http://europa.eu.int/scadplus/leg/en/lvb/l28096.htm
	Supervision & Control of Transfrontier Shipments of Waste	To establish a system of supervision & control of all movements of waste.	Council Regulation 259/93/EEC on the supervision & control of shipments of waste within, into & out of the European Community (01/02/93). Amended by the following: ▪ Decision 94/721/EC (24/10/94) ▪ Decision 96/660/EC (18/11/96) ▪ Regulation 120/97/EC (27/01/97) ▪ Regulation 2408/98/EC (07/11/98) ▪ Regulation 1420/1999/EC (30/09/99) ▪ Regulation 1547/1999/EC (16/08/99) ▪ Regulation 2557/2001 EC (01/01/02).	▪ Off Jnl L30, 06/02/93	http://europa.eu.int/scadplus/leg/en/lvb/l11022.htm
	Shipments of Radioactive Substances	To set up a Community system for the declaration of shipments of radioactive substances between Member States, following the removal of frontier controls in the Community, so that the competent authorities concerned receive the same level of information as before 1993 as regards radiation protection control.	Council Regulation (Euratom) No. 1493/93 on shipments of radioactive substances between Member States (08/06/93).	▪ Off Jnl L148, 19/06/93	http://europa.eu.int/scadplus/leg/en/lvb/l28099.htm

Continued

Table 13.5.2 European Level Waste Management Policy. *Continued*

Policy Type	Programme Title	Policy Objective	Community Measure	Document Reference	Website
Miscellaneous	End-of-Life Vehicles: Current Basic Data Reflecting the Overall Ecological & Economic Context of the ELV Issue	The aim of this report is to provide up-to-date data reflecting the overall ecological & economic context of the ELV issue in the 15 EU member states. It is intended as a tool to assist DG XI in its further preparatory work on the legislative proposal.	DG Environment Report (July 1996).	▪ DG XI Report	http://europa.eu.int/comm/environment/waste/pdf_comments/elv_study1996.pdf
	Waste Management Statistics	To ensure better monitoring of effective implementation of Community policy on waste disposal.	Draft Council Regulation (EC on waste management statistics (1999).	▪ COM (99) 31 ▪ Off Jnl C87, 29/03/99	http://europa.eu.int/scadplus/leg/en/lvb/l28082.htm
	Construction & Demolition Waste Management Practices, & Their Economic Impacts	The study which is to cover the 15 Member States of the European Union aims at: ▪ identifying the origin & destination of construction & demolition waste; ▪ identifying the quantities of hazardous waste & other waste parts for which a separate collection would be appropriate; ▪ identifying methods & ways which are used within Member States to promote re-utilisation/recovery of construction & demolition waste; ▪ examining the economic implications if construction & demolition waste is sorted in order to separate hazardous & other materials; ▪ examining possibilities & practical measures used to prevent construction & demolition waste going to landfills.	Report to the Directorate-General Environment on construction & demolition waste management (Feb 1999).	▪ Report to the DG XI	http://europa.eu.int/comm/environment/waste/studies/cdw/c&dw_chapter1-6.pdf http://europa.eu.int/comm/environment/waste/studies/cdw/c&dw_chapter7-10.pdf

European Conference on Waste Management Planning	The objectives of this conference were to encourage the drawing up & application of waste management plans, based on obligations in community legislation & more particularly on the framework directive covering waste (which declares the need to draw up plans & defines the minimum conditions for them), as well as the directive relating to hazardous waste & the directive relating to packaging & packaging waste (which provide for specific plans, possibly integrated into the general plan). For some 20 years, this has been translated into very few waste management plans.	Proceedings from the European Conference on waste management planning (June 1999).	Conference proceedings	http://europa.eu.int/comm/environment/waste/plans/conference_en.pdf
EU Focus on Waste Management	This brochure addresses how the EU is seeking to reduce waste mountains. It is intended to inform local & regional players, NGOs, policy-makers at all levels, social partners & consumers, as well as citizens. It is also hoped this information will give readers ideas about making their own particular contribution towards resolving environmental problems.	DG Environment Report (Aug 1999).	DG XI Report	http://europa.eu.int/comm/environment/waste/publications/eufocus_en.pdf
Chemical Recycling of Plastics Waste (PVC & other Resins)	The chemical recycling of PVC, is being carried out by TNO (The Netherlands Organisation for Applied Scientific Research) with DG III as the primary client. The overall aim of the project is to analyse the role that chemical recycling may have in a future European system for PVC waste management. Since both rather pure PVC waste & mixed plastic waste (MPW) containing PVC may in principle be treated by chemical recycling, this study cannot concentrate on PVC alone. Chemical recycling of PVC will therefore be treated within the broader context of chemical recycling of plastic waste.	DG Industry Report (Dec 1999).	DG III Industry Report	http://europa.eu.int/comm/environment/waste/studies/pvc/chem_recycle.pdf

Continued

Table 13.5.2 European Level Waste Management Policy. *Continued*

Policy Type	Programme Title	Policy Objective	Community Measure	Document Reference	Website
Miscellaneous	Implementation of Community Waste Legislation (1995–1997)	This report intends to inform the other Community Institutions, Member States & the interested public of the implementation of waste legislation for the period 1995 to 1997, especially the implementation of: ▪ Directive 75/442/EEC on waste ▪ Directive 91/689/EEC on hazardous waste (replaced Directive 78/319/EEC) ▪ Directive 75/439/EEC on the disposal of waste oils ▪ Directive 86/278/EEC on the protection of the environment, & in particular of the soil, when sewage sludge is used in agriculture.	Report from the Commission to the Council & the European Parliament on implementation of community waste legislation for the period 1995–1997 (10/01/00).	▪ COM (1999) 752	http://europa.eu.int/comm/environment/waste/reporting/en.pdf
	Waste Electrical & Electronic Equipment	To promote re-use, recycling & other forms of recovery of electrical & electronic waste in order to reduce the quantity of such waste to be eliminated & to improve the environmental performance of the economic operators involved in the treatment of such waste.	Proposal for a Directive of the European Parliament & of the Council on waste electrical & electronic equipment and a Proposal for a Directive of the European Parliament & of the Council on the restriction of the use of certain hazardous substances in electrical & electronic equipment (2000).	▪ COM (2000) 347 ▪ Off Jnl C365, 19/12/00	http://europa.eu.int/scadplus/leg/en/lvb/l21210.htm

The Behaviour of PVC in Landfill	The study was carried out with the objective of evaluating the environmental impact of PVC waste after final disposal in landfills. The objective of this study was to: ■ provide a survey of existing analyses & research as regards the behaviour of PVC in landfills ■ evaluate the environmental impact of PVC waste after final disposal in landfills ■ perform practical analysis under simulated landfill conditions ■ estimate the quantities of PVC landfilled as well as costs of landfilling PVC waste, including the environmental costs.	DG Environment Report (Feb 2000).	■ DG XI Report	http://europa.eu.int/comm/environment/waste/studies/pvc/landfill.pdf
Heavy Metals in Vehicles	The objectives of this study include: ■ The generation of a reliable information basis concerning the presence & quantity (including market trends) of heavy metal containing parts & components in vehicle categories M1 & N1; ■ The analysis whether technical alternatives to the use of heavy metals (lead, mercury, cadmium & hexavalent chromium) in vehicles exist; ■ The examination of the technical, environmental & economic implications of the alternatives; ■ The examination of technical options for separation of heavy metal containing parts before shredding, description of technical recycling options, & the analysis of their economic implications.	DG Environment Report (Mar 2000).	■ DG XI Report	http://europa.eu.int/comm/environment/waste/studies/elv/heavymetals1.pdf

Continued

Table 13.5.2 European Level Waste Management Policy. *Continued*

Policy Type	Programme Title	Policy Objective	Community Measure	Document Reference	Website
Miscellaneous	Economic Evaluation of PVC Waste Management	This report is an assessment of the waste management costs of diverting PVC waste away from incineration (& in particular, towards recycling) & the associated environmental costs & benefits. The study covers the EU-15 & the first six Accession Countries over the period 2000–2020. The specific objective of this study is to assess the economic implications of diverting PVC waste away from incineration. The approach adopted evaluates three scenarios for reducing PVC incineration, to be achieved by 2020. The study identifies the major changes in financial costs for PVC waste management & environmental burdens associated with achieving the scenarios, taking account of the impacts of other policy measures adopted at EU or Member State level. The analysis examines the extent to which recycling can consume the PVC diverted from incineration & the waste streams & PVC applications that will contribute most to achieving the targets & it evaluates the costs of support measures to stimulate recycling.	A report produced for European Commission Environment Directorate (June 2000).	■ DG XI Report	http://europa.eu.int/comm/environment/waste/studies/pvc/economic_eval.pdf
	Green Paper on Environmental Issues of PVC	The two objectives of this document are, firstly, to present and assess on a scientific basis, the various environmental issues including related human health aspects that occur during the life cycle of PVC and, secondly, to consider, in view of sustainable development, a number of options to reduce those impacts that need to be addressed. This	Green Paper on the environmental issues of PVC (27/06/00).	■ COM (2000) 469	http://europa.eu.int/comm/environment/waste/pvc/en.pdf

	should serve as a basis for a consultation with stakeholders in order to identify practical solutions to health and environmental issues raised by PVC.			
Economic Evaluation of PVC Waste Management	The study covers the EU-15 & the first 6 Accession Countries over the period 2000–2020. Over this period post-consumer PVC waste (accounting for 88% of all PVC waste) is expected to increase from 3.6 to 6.4 million tonnes per year across this group of countries. Recycling rates for these wastes are very low at only 3% of arisings, reflecting high separation & processing costs. The remaining 12% of the total volume of PVC waste is pre-consumer waste. The analysis presented in this report is based on 3 scenarios. In the first & second of these, recycling rates increase to 15% & 22% respectively, with proportionate decreases in the amount of PVC sent to incineration & landfill. In the third scenario recycling rates are unchanged against the 'business as usual' scenario (BAU), but incineration rates increase to 27% to 30% as a result of the diversion of constructional wastes to landfill, compared with up to 45% incineration forecast under BAU.	Report to the Directorate-General Environment on an economic evaluation of PVC waste management (June 2000).	▪ Report to the DG XI	http://europa.eu.int/comm/environment/waste/studies/pvc/economic_eval.pdf
A Study on the Economic Valuation of Environmental Externalities from Landfill Disposal & Incineration of Waste	This study aims at reviewing & presenting in an easily comprehensible way the parts of the methodological toolkit of cost-benefit analysis where information is not readily available, i.e. environmental externalities from landfill disposal & incineration. More specifically, the study aims to provide 'an overview of the environmental externalities that need to be taken into account when evaluating different waste management policies & how they can be integrated into cost-benefit analysis'.	Report to the Directorate-General Environment on the economic valuation of environmental externalities from landfill disposal & incineration of waste (Oct 2000).	▪ Report to the DG XI	http://europa.eu.int/comm/environment/enveco/waste/cowi_ext_from_landfill.pdf

Continued

Table 13.5.2 European Level Waste Management Policy. *Continued*

Policy Type	Programme Title	Policy Objective	Community Measure	Document Reference	Website
Miscellaneous	Successful Stories on Composting & Separate Collection	This report has been produced as part of a project undertaken for the European Commission, Directorate-General for the Environment, on success stories on composting & separate collection. This project has involved identifying a number of successful centralised & home composting schemes in the six Member States of Spain, France, Ireland, Italy, Portugal & the UK. For each scheme a case study has been completed, highlighting in particular the reasons for the scheme's success. The objectives of the report are to provide information to local authorities across Member States that will assist in introducing successful & cost-effective home composting & biodegradable waste-separation schemes in their localities.	Report to the Directorate-General Environment on composting & separate collection (2000).	▪ Report to the DG XI	http://europa.eu.int/comm/environment/waste/publications/compost_en.pdf
	European Packaging Waste Management Systems – Main Report	The objectives of this study are to provide an overview of the different management systems in operation in each Member State, covering the managerial, technical & economic aspects involved in packaging waste management systems, & to draw up potential scenarios for each Member State for the years 2006 & 2011.	Report to the Directorate-General Environment on European packaging waste management systems (Feb 2001).	▪ Report to the DG XI	http://europa.eu.int/comm/environment/waste/studies/packaging/epwms.pdf
	The Impact of Community Environmental Waste Policies on Economic & Social Cohesion	The aim of this report is to assess the cohesion impacts of European waste management policy. But, since the Cohesion Fund is closely involved in this process, this report will also help to evaluate to what extent environmental projects financed by	Report for the DG Regional Policy (Feb 2001).	▪ DG Regional Policy Report	http://europa.eu.int/comm/regional_policy/sources/docgener/studies/pdf/enviwas.pdf

	the Cohesion Fund of the European Union actually contribute to real social, economic & environmental cohesion towards the richer members of the European Union. Even though the aid from Cohesion Fund is likely to change in the future, especially when the candidate countries enter the Union, its investment efforts will have shown by then successful achievements in the economic & social growth in the Cohesion countries.			
Waste Management Options & Climate Change	This document is the final report of a study undertaken for the European Commission Environment Directorate General to assess the climate change impacts of options for municipal solid waste (MSW) management in the EU. The study covers the 15 member states of the European Union and the time horizon 2000 to 2020. The study assesses climate change impacts in terms of net fluxes of greenhouse gases from various combinations of options used for the management of MSW.	Report to the Directorate-General Environment on waste management options & climate change (July 2001).	▪ Report to the DG XI	http://europa.eu.int/comm/environment/waste/studies/climate_change.pdf
Heavy Metals in Vehicles II – Final Report	The purpose of this study is to provide the Commission with technical information in view of possible amendments of Annex II of the Directive 2000/53/EC of the European Parliament & the Council on End-of-Life Vehicles (ELV). The main results of the study presented herewith show that Article 4(2)(a) of the ELV Directive has indeed highlighted an area where significant improvements with respect to reduced use of hazardous substances can be achieved, thus leading to an improved recyclability of vehicles & a better environmental performance in the recycling sectors, as well as resulting in 'cleaner' waste to be disposed of.	DG Environment Report (July 2001).	▪ DG XI Report	http://europa.eu.int/comm/environment/waste/studies/elv/heavy_metals.pdf

Continued

Table 13.5.2 European Level Waste Management Policy. *Continued*

Policy Type	Programme Title	Policy Objective	Community Measure	Document Reference	Website
Miscellaneous	Applying Compost – Benefits & Needs	This seminar was organised in order to discuss the results of research with experts at the European level. The seminar, with invited international experts, intended to share the outcome of the literature research with the results of the latest findings by scientific research work. Furthermore, the hypothesis was to discuss issues with colleagues not only from the scientific world (plant production, soil science etc.) but also with experts from agricultural organisations & official bodies & authorities at national & EU level of the concerned areas (soil protection, water conservation, agriculture).	Scientific Seminar on applying compost – benefits & needs (Nov 2001).	▪ Seminar Proceedings	http://europa.eu.int/comm/environment/waste/pdf_comments/040119_proceedings.pdf
	Critical Review of Existing Studies & Life Cycle Analysis on the Regeneration & Incineration of Waste Oils	The purpose of this document is to bring a new set of arguments & facts to the debate which should serve to attain an environmentally sound management of WO (Waste Oils) in the revised directive. Its main objective is to undertake a thorough analysis of the literature available in the context of the regeneration of WO & a critical assessment of the comparisons carried out between regeneration & incineration of WO.	DG Environment Report (Dec 2001).	▪ DG XI Report	http://europa.eu.int/comm/environment/waste/studies/oil/waste_oil.pdf

Reuse of Primary Packaging – Final Report	As a first step, this study shows the tendencies in the development of the reuse packaging systems in the EU as a whole & in each individual Member State (see country reports). The second step of the study is to show which are the streams of goods of the EU-internal trade relevant for reuse packaging. The third step is the detailed analysis of the players & their activities as well as the background conditions in which they decide to use or not to use reuse packaging systems. Finally, but not any less important, the different legal activities of the 15 EU-member governments are analysed under the aspect of how they support or hinder reuse packaging & to illustrate which ones are the most efficient and successful instruments to stabilise & to support reuse packaging systems.	Report to the Directorate-General Environment on the reuse of primary packaging (2001).	▪ Report to the DG XI	http://europa.eu.int/comm/environment/waste/studies/packaging/reuse_main.pdf
Heavy Metals in Waste	The overall objective of the present project is to present information concerning sources of heavy metals to waste, harmful effects of heavy metals, the problems posed by the disposal & recycling of heavy metals & heavy metal containing products, & to assess the options for substituting the heavy metals.	Report to the Directorate-General Environment on heavy metals in waste (Feb 2002).	▪ Report to the DG XI	http://europa.eu.int/comm/environment/waste/studies/heavy_metalsreport.pdf
Costs for Municipal Waste Management in the EU	This is an important study given that to date, no comprehensive analysis has been undertaken to give an overview of the costs of different waste management options at local & national levels for all 15 EU member states. There is also a lack of information concerning the variety of financing systems. The study constitutes an attempt on the part of the Commission to generate baseline data for the costs of future policy changes.	Report to the Directorate-General Environment on the costs for municipal waste management in the EU (Mar 2002).	▪ Report to the DG XI	http://europa.eu.int/comm/environment/waste/studies/eucostwaste.pdf

Continued

Table 13.5.2 European Level Waste Management Policy. *Continued*

Policy Type	Programme Title	Policy Objective	Community Measure	Document Reference	Website
Miscellaneous	Financing & Incentive Schemes for Municipal Waste Management – Case Studies	The study was conducted to fulfil part of a wider contract concerning municipal waste management (the companion study concerns the costs of different collection & treatment options for municipal waste). The specific objective of that contract to which this study relates is to carry out 20 case studies on innovative financing systems at local & national level, to be chosen in agreement with the European Commission. These case studies shall in particular focus on systems which were able to give incentives to reduce waste quantities & increase recycling & composting. The experiences & results of such systems are described.	Report to the Directorate-General Environment on financing & incentive schemes for municipal waste management (Mar 2002).	▪ Report to the DG XI	http://europa.eu.int/comm/environment/waste/studies/financingmuncipalwaste_management.pdf
	Study on Hazardous Household Waste (HHW) with a main Emphasis on Hazardous Household Chemicals (HHC)	The Directorate-General Environment has commissioned an evaluation of national experiences of the management of hazardous products likely to become hazardous household waste (HHW), & to make proposals for strategies for the appropriate management of such products within a lifecycle perspective. This is the Final Report on HHW.	Report to the Directorate-General Environment on hazardous waste (July 2002)	▪ Report to the DG XI	http://europa.eu.int/comm/environment/waste/studies/household_report.pdf
	Economic Analysis of Options for Managing Biodegradable Municipal Waste	The main objective of the study is to conduct an economic evaluation (that considers both private & social welfare costs & benefits) of existing options for managing the biodegradable fraction of municipal solid waste (MSW). Although all management options (anaerobic digestion, composting, landfilling, incineration, etc.) are considered in the study, the main emphasis is on the separate collection & recycling of the biodegradable fraction of MSW.	Report to the Directorate-General Environment – economic analysis of options for managing biodegradable municipal waste (Dec 2002).	▪ Report to the DG XI	http://europa.eu.int/comm/environment/waste/compost/econanalysis_finalreport.pdf

Case Studies on Waste Minimisation Practices in Europe	The objective of this report is to support & inspire the work with waste minimisation in EEA member countries by introducing a catalogue of successful examples of waste prevention, recycling & cleaner technology. Thus, the report is supporting EU waste policy manifested in the EU waste strategy & the sixth environmental action programme. The emphasis of the report is placed on case studies on waste prevention & recycling.	European Environment Agency Report (Jan 2003).	EEA Report	http://reports.eea.eu.int/topic_report_2002_2/en/Topic_report_2-2002_web.pdf
Evaluation of Costs & Benefits for the Achievement of Reuse & Recycling Targets for the Different Packaging Materials	Dir 94/62/EC provides that by end 2000 the Council shall fix new targets for the 5 years following June 2001. During the preparation of a proposal, the Commission was called upon to base new targets on an evaluation of costs & benefits of such new targets. This study was commissioned for this purpose. More specifically, the objective of the study was to perform a cost–benefit analysis of packaging recycling & reuse systems.	Report to the Directorate-General Environment on an evaluation of costs & benefits for the achievement of reuse & recycling targets for the different packaging materials in the frame of the Packaging & Packaging Waste Directive 94/62/EC (Mar 2003).	Report to the DG XI	http://europa.eu.int/comm/environment/waste/studies/packaging/costsbenefits.pdf
Preparing a Waste Management Plan – A Methodological Guidance Note	These Guidelines aim to provide a tool for waste management planning & promote the development of more coherent & appropriate planning practices across the EU Member States, in compliance with the requirements of the relevant EU legislation. The Guidelines contain a review of the overall policies & principles applying to planning in the field of waste management in the EU. This includes a review of legislation in force, & in some fields practical methods are indicated for filling in the framework set up by the EU & the various Member States for the contents of management plans. Throughout the Guidelines there are boxes with examples from waste management plans or from other guidelines, as well as links to relevant information when drafting a waste plan.	Report to the Directorate-General Environment on preparing a waste management plan (May 2003).	Report to the DG XI	http://europa.eu.int/comm/environment/waste/plans/wasteguide_final.pdf

Continued

Table 13.5.2 European Level Waste Management Policy. *Continued*

Policy Type	Programme Title	Policy Objective	Community Measure	Document Reference	Website
Miscellaneous	Management of Waste from the Extractive Industries	The Communication sets out three priority actions intended to improve the safety of mining waste management. These are: ▪ An amendment to the Seveso II Directive to bring the processing of mineral ores & mine waste management under the scope of the Directive; ▪ An initiative to promote a new Directive on the management of waste from the extractive industries (this Proposal); ▪ The production of a best available techniques (BAT) document covering the management of waste rock & tailings in mining activities. The approaches of the Communication were endorsed by the international Task Force established to assess the Baia Mare accident. The European Parliament's Resolution in 2001 also welcomed the Communication & strongly supported the need for a Directive on waste from the extractive industries.	Proposal for a Directive of the European Parliament & of the Council on the management of waste from the extractive industries (02/06/03).	▪ COM (2003) 319	http://europa.eu.int/eur-lex/en/com/pdf/2003/com2003_0319en01.pdf
	Shipments of Waste	The proposal has four main objectives: ▪ Implementing the OECD (Organisation for Economic Co-operation and Development) Council Decision C(2001)107 of 14 June 2001 in Community legislation. ▪ Addressing the problems encountered in the application, administration & enforcement of the 1993 Regulation & establishing greater legal clarity.	Proposal for a Regulation of the European Parliament & of the Council on the shipments of waste (30/06/03).	▪ COM (2003) 379	http://europa.eu.int/eur-lex/en/com/pdf/2003/com2003_0379en01.pdf

	▪ Pursuing global harmonisation in the area of transboundary shipments of waste. ▪ Enhancing the structure of the Articles of the Regulation.			
The Implementation of Community Waste Legislation for the Period 1998–2000	This report intends to inform the other Community Institutions, Member States and the interested public of the implementation of waste legislation for the period 1998 to 2000, in particular the implementation of: ▪ Directive 75/442/EEC on waste; ▪ Directive 91/689/EEC on hazardous waste (replaced Directive 78/319/EEC); ▪ Directive 75/439/EEC on the disposal of waste oils; ▪ Directive 86/278/EEC on the protection of the environment, & in particular of the soil, when sewage sludge is used in agriculture; ▪ Directive 94/62/EC on packaging & packaging waste.	Report from the Commission to the Council & the European Parliament on the implementation of community waste legislation for the period 1998–2000 (11/07/03).	▪ COM (2003) 250	http://europa.eu.int/eur-lex/en/com/rpt/2003/com2003_0250en03.pdf
Impact Assessment on Selected Policy Options for Revision of the Battery Directive Final Report	In order to contribute to a proper functioning of the internal market & to establish a high level of environmental protection in the field of waste management of spent batteries & accumulators, the European Commission commissioned a study to perform an analysis of economic, environmental & social impacts of different policy options about batteries & accumulators, in the framework of an extended impact assessment and the positive & negative impacts of different policy options in view of revising the battery directives. The methodology developed is based on the guidelines recently published by the EC about extended impact assessment.	European Commission Directorate General Environment Report (July 2003).	▪ DG XI Report	http://europa.eu.int/comm/environment/waste/batteries/eia_batteries_final.pdf

Continued

Table 13.5.2 European Level Waste Management Policy. *Continued*

Policy Type	Programme Title	Policy Objective	Community Measure	Document Reference	Website
Miscellaneous	Proposal for a Directive on Batteries & Accumulators & Spent Batteries & Accumulators	Current Community legislation on batteries & accumulators is CD 91/157/EEC on batteries & accumulators containing certain dangerous substances, as amended by CD 98/101/EC, which only covers batteries & accumulators containing more than 0.0005% mercury by weight, more than 0.025% cadmium by weight & more than 0.4% lead by weight. This proposed directive now aims to approximate the laws of the Member States on the recycling & controlled disposal of spent batteries & accumulators by: ▪ prohibiting the marketing of batteries & accumulators containing more than 0.0005% mercury from 1 January 2000 onwards; ▪ requiring Member States to ensure the separate collection of the batteries & accumulators covered by the Directive; ▪ requiring Member States to draw up four-yearly programmes designed, among others, to reduce the heavy metal content of batteries & gradually reduce their share in the municipal solid waste stream.	Proposal for a Directive of the European Parliament & the Council on batteries & accumulators & spent batteries & accumulators (21/11/03)	▪ COM (2003) 723	http://europa.eu.int/eur-lex/en/com/pdf/2003/com2003_0723en01.pdf

Proposal for a Directive on Waste	The purpose of this proposal is to undertake a codification of Council Directive 75/442/EEC of 15 July 1975 on waste. The new Directive will supersede the various acts incorporated in it; this proposal fully preserves the content of the acts being codified and hence does no more than bring them together with only such formal amendments as are required by the codification exercise itself.	Proposal for a Directive of the European Parliament & of the Council on waste (27/11/03).	▪ COM (2003) 731	http://europa.eu.int/eur-lex/en/com/pdf/2003/com2003_0731en01.pdf
Proposal for a Council Directive amending Directive 94/62/EC on Packaging & Packaging Waste	Updated proposal taking into account the low level of packaging consumption of the 10 new Member States as the proposal for a revised packaging directive did not contain specific proposals for the acceding States.	Proposal for a European Parliament & Council Directive amending Directive 94/62/EC on packaging & packaging waste (24/02/04).	▪ COM (2004) 127	http://europa.eu.int/eur-lex/en/com/pdf/2004/com2004_0127en01.pdf
Amended Proposal for a Regulation on Shipments of Waste	In order to achieve the objectives in COM (2003) 379, the revision amends various sections & aspects of Council Regulation (EEC) No 259/93. These include: ▪ Changes to its structure; ▪ Changes & clarifications as regards definitions, & clarification of its scope (Title I); ▪ Changes & clarifications as regards the procedures applicable to shipments of waste (Title II–VI); ▪ Changes in other provisions of the Regulation (Title VII).	Amended Proposal for a Regulation of the European Parliament & the Council on shipments of Waste (08/03/04).	▪ COM (2004) 172	http://europa.eu.int/eur-lex/en/com/pdf/2004/com2004_0172en01.pdf

Continued

Table 13.5.2 European Level Waste Management Policy. *Continued*

Policy Type	Programme Title	Policy Objective	Community Measure	Document Reference	Website
Miscellaneous	Heavy Metals & Organic Compounds from Wastes used as Organic Fertilisers	This report is the result of a study carried out on behalf of the Directorate-General for the Environment of the European Commission in the context of the European waste management policy & work on the biological treatment of biodegradable waste. The objectives of the study were as follows: ▪ Identification of the pollutants (heavy metals, organic compounds) likely to be found in foodstuffs for human consumption. ▪ Evaluation of the concentration of most representative pollutants in compost produced from biowaste that is separately collected & from unsorted waste &/or residual municipal waste across the Community. ▪ Assessment of the influence of the different sampling techniques & analytical methods that are used in Member States for determining the concentration of pollutants in compost & stabilised biowaste. ▪ Identification of the applied ISO, CEN or other international or national standards. ▪ Elaboration of a scientific concept for limiting the amount of pollutants in compost & stabilised biowaste with a view to long term soil protection in the Community.	Report prepared for DG XI by Compost Working Group (July 2004).	▪ Working Group Report	http://europa.eu.int/comm/environment/waste/compost/pdf/hm_finalreport.pdf

Study on the Implementation of Directive 94/62/EC on Packaging & Packaging Waste & Options to Strengthen Prevention & Reuse of Packaging	An overall evaluation is given of the impacts of Directive 94/62/EC from an environmental, economic & social perspective. This is done on the basis of available information & appropriately reasoned extrapolations. Critical questions have been verified on the basis of additional work, with the geographical coverage being the EU15.	DG Environment Report (Oct 2004).	▪ DG XI Report	http://europa.eu.int/comm/environment/waste/pdf_comments/041014_final_report.pdf
A Cleaner, Greener Europe: LIFE & the European Union Waste Policy	The LIFE programme plays a key role in contributing to the development of innovative waste prevention & management projects. Between 1996 & 2003, LIFE Environment funded 141 projects on different aspects of waste management & prevention ranging from industrial waste or construction sites, to community composting, to recycling of electronic & electrical waste. This edition of LIFE Focus highlights nine of the many innovative LIFE projects. These projects are proof of the LIFE-Environment programme's commitment to foster & support fully the efforts of the European Community to reach the targets set by the Sixth Environment Action Programme & make the world a healthier place to live today & in the future.	EU LIFE Report (2004).	▪ LIFE Report	http://europa.eu.int/comm/environment/life/infoproducts/lifewaste_en.pdf

Table 13.5.3 UK Level Waste Management Policy.

Policy Type	Document Title	Objective	Contents	Website
Strategies & Reports	Response to the Environment, Transport & Regional Affairs Select Committee Report on Sustainable Waste Management (13/11/98)	This response concentrates on offering initial responses to the Committee's recommendations. The Government's intention was to produce a draft of the new waste management strategy for consultation, & to finalise the strategy, in the light of information from the Environment Agency on waste arisings & the impact of different waste management options, & the responses to that second consultation, later in 1999.	Introduction; Recommendation 1–5; Recommendation 6–10; Recommendation 11–15; Recommendation 16–20; Recommendation 21–25; Recommendation 26–30; Recommendation 31–35; Recommendation 36–40; Recommendation 41–45; Recommendation 46–50; Recommendation 51–56.	http://www.defra.gov.uk/environment/waste/swm/index.htm
	A Way with Waste: A draft waste strategy for England & Wales (Part I & II) (21/07/99 and 03/08/99)	This Strategy gives a blueprint for how waste is to be managed to meet the objectives for the next 20 years. Fundamental to the strategy is the need to curb the growth in the quantity of waste produced.	Part I: Summary; Introduction; Sustainable waste management: shaping the vision; Sustainable waste management into the 21st Century: the challenges ahead; Achieving the vision: a plan of action for the next 5 yrs; The National Assembly of Wales; Conclusions & implementation. Part II: Introduction; Identifying the problem; Waste arisings data; The decision-making framework; Planning & licensing issues for waste management facilities; Waste options; Sectoral interests; International context; Progress with various waste streams; Annex A: Recommendations of the Market Development Group; Annex B: Maps of the facilities; Annex C: Draft regulatory impact assessment: the waste strategy for England & Wales.	http://www.defra.gov.uk/environment/waste/strategy/part1/index.htm http://www.defra.gov.uk/environment/waste/strategy/part2/index2.htm
	Waste Strategy for England & Wales (Part I & II) (25/05/00)	This Strategy describes the vision for managing waste & resources better. It sets out the changes needed to deliver more sustainable development.	Part I: Summary; The need for change; Our vision; Levers for change; Delivering change; Monitoring & review. Part II: Introduction; Identifying the problem; The decision-making framework; Waste stakeholders; Waste management options;	http://www.defra.gov.uk/environment/waste/strategy/cm4693/index.htm

		Handling hazardous waste; Dealing with packaging & packaging waste; Progress with various waste streams; Annex A: Major waste facilities in England & Wales; Annex B: Legal framework for waste legislation; Annex C: Regulatory impact assessment; Annex D: Glossary.	
The Implementation of Council Directive 1999/31/EC on the Landfill of Waste (19/10/00)	This document describes the regulatory requirements set out in the Landfill Directive 1999/31/EC, & considers the likely impact upon both landfill operators & waste producers. It describes how it is intended to implement the Directive, inviting views where there appear to be options for national controls.	Chapter 1: Introduction; Chapter 2: The regulatory framework; Chapter 3: Scope of Landfill Directive; Chapter 4: Landfill classification; Chapter 5: Prohibited wastes; Chapter 6: Treatment of waste; Chapter 7: Landfill permits; Chapter 8: Cost of landfill of waste; Chapter 9: Waste acceptance criteria & procedures; Chapter 10: Control & monitoring procedures; Chapter 11: Closure & after care procedures; Chapter 12: General requirements for all classes of landfill; Chapter 13: Timetable for implementation; Chapter 14: The consultation process; Annex A: Timetable of when Directive provisions take effect; Annex B: Regulatory impact assessment.	http://www.defra.gov.uk/environment/consult/landfill/index.htm
Research Study on International Recycling Experience (April 2001)	The objective of this study was to examine recycling schemes in countries other than the UK & to provide advice on how this experience could be used to increase the rate of recycling of municipal solid waste (MSW) to meet the goals set out in the draft waste strategy *A Way with Waste*. The project considers, where data are available, the breakdown of recycling rates for different materials, recycling targets & rates achieved, regulatory & fiscal tools, the organisation of the recycling scheme, financial costs, public awareness activities & markets for recovered materials. The information is presented in the context of general waste management policy & relevant socio-economic factors. The case studies focus on specific municipalities within a number of countries or states identified by the DETR for investigation.	Executive summary; Introduction; Assessment of Case Studies; Case Study Summaries; Transferability to the UK Context; Annex A – Case Studies; Denmark – Aarhus; France – Niort; Germany – Wiesbaden; Italy – Monza; Netherlands – Arnhem; California – Santa Monica; New York City; Seattle; Ontario – Region of Peel; Australia – Canberra; New Zealand – Christchurch; Annex B – Information Required on Recycling Schemes; Annex C – Contact and References.	http://www.defra.gov.uk/environment/waste/intrecex/index.htm

Continued

Table 13.5.3 UK Level Waste Management Policy. *Continued*

Policy Type	Document Title	Objective	Contents	Website
Strategies & Reports	Distributions of the Aggregates Levy Sustainability Fund – Summary of Responses (11/04/02)	The Department for Environment, Food and Rural Affairs (DEFRA) published a consultation document on proposals for the Distribution of an Aggregates Levy Sustainability Fund in England. The consultation paper built on an earlier Treasury consultation & put forward a number of possibilities. Respondents were asked to provide comments on how these objectives could best be delivered.	Introduction; Consultation issues; Overall response to consultation; General themes from the consultation; Conclusion; Annex A: List of respondents.	http://www.defra.gov.uk/environment/consult/aggregates/response/index.htm
	Model Forecasts of Recycling Participation Rates & Material Capture Rates for Possible Future Recycling Scenarios (July 2002)	The aim of the research was to derive model forecasts of the likely levels of participation & the material captured from the enhancement of recycling provision across England. The specific questions to be addressed were: ■ What levels of participation & material capture could ultimately be achieved; ■ How much improvement could be achieved simply by increasing recycling provisions; & ■ How much improvement would need to be addressed by parallel awareness & educational campaigns.	Introduction; North West England; South West England; Comparison between regions; A survey of behavioural factors influencing household waste recycling participation & capture.	http://www.number-10.gov.uk/su/waste/report/downloads/recycling_participation.pdf
	Digest of Environmental Statistics & Municipal Waste Management Statistics (21/09/02)	This digest covers bulletins containing results from the annual Municipal Waste Management Surveys carried out by the Department for Environment, Food & Rural Affairs (DEFRA). Estimates are based on information from questionnaires sent to all Waste Collection Authorities, Waste Disposal Authorities & Unitary Authorities in England & Wales. Information is collected on the amounts of municipal waste collected & disposed of, & the levels of recycling/recovery of household & municipal waste, on the methods of waste containment, levels of service provision & details of waste collection & disposal contracts.	Municipal Waste Management 1996/97 & 1997/98 Municipal Waste Management 1998/99 Municipal Waste Management Survey 1999/00 Municipal Waste Management Survey 2000/01.	http://www.defra.gov.uk/environment/statistics/wastats/index.htm

The Context for Emissions & Health Impacts Associated with Waste Management (Sept 2002)	The report provides information setting emissions and health effects associated with waste management in context with those in society generally. The aim of this report is to support the Strategy Unit's forthcoming report into waste management. The information is designed to help the Strategy Unit address the following issues: ▪ How do emissions from waste management compare with emissions from other common activities? ▪ How do emissions from the current waste management practices compare with emissions from the main alternatives? ▪ Can any potential health risks be identified which are associated with emissions from waste management, and can these be put in the wider context of other public health/environmental risks?	Executive summary; Introduction; Key issues & substances; Context for emissions to air; Health risks; Factors influencing perception of risk; Appendix 1; Calculation basis.	http://www.number-10.gov.uk/su/waste/report/downloads/enviros.pdf
Public Attitudes Towards Recycling & Waste Management – Quantitative & Qualitative Review (Sept 2002)	The aim of the survey is to provide the Strategy Unit with an overview of public attitudes towards waste and recycling, to feed into the development of the forthcoming review of the UK Waste Strategy. The research covered a wide range of areas, in particular: ▪ Perceptions of environmental issues & 'green' behaviour; ▪ Attitudes to household rubbish; ▪ Attitudes to the options for waste management; ▪ Patterns of behaviour in dealing with household rubbish & recycling; ▪ Motivations to recycle; ▪ Local recycling service provision; ▪ Awareness & information; ▪ Responsibility for waste; ▪ Household management; ▪ Charging for waste.	Introduction; Summary of findings; Waste issues in the public consciousness; Attitudes to waste management options; Reported behaviour; Motivations; Local recycling service provision; Awareness & information; Household waste management; Responsibility for waste; Charging for waste; Appendices.	http://www.number-10.gov.uk/su/waste/report/downloads/mori.pdf

Continued

Table 13.5.3 UK Level Waste Management Policy. *Continued*

Policy Type	Document Title	Objective	Contents	Website
Strategies & Reports	Waste Not, Want Not – A Strategy for Tackling the Waste Problem in England (Nov 2002)	The Strategy Unit was tasked at the end of 2001 with carrying out a review of the Waste Strategy in England. The aim of this review has been: ▪ to analyse the scale of the challenge posed by growing quantities of municipal household waste; ▪ to assess the main causes & drivers behind this growth now & in the future; & ▪ to devise a strategy, with practical & cost-effective measures for addressing the challenge, which will put England on a sustainable path for managing future streams of household waste.	Foreword; Executive summary; Introduction; What is waste & how much is there? Why waste matters – economic & environmental challenge; Barriers to more sustainable waste management; Moving forwards to a new strategy; Economic & regulatory framework; Strategic investment measures; Funding & delivery; Summary of key recommendations & action plan; Annexes.	http://www.number-10.gov.uk/su/waste/report/downloads/wastenot.pdf
	Delivering the Landfill Directive: The Role of New & Emerging Technologies (Nov 2002)	This report sets out to independently advise Phase II of the Strategy Unit's Waste Study, by assessing the potential role of new & emerging technologies in meeting England's obligations under the Landfill Directive. Whilst the focus is on biodegradable municipal waste (BMW) it is envisaged that the report's findings could also help inform waste decision-making surrounding similar commercial & industrial waste arisings.	Introduction & approach to report; Technologies; Environmental releases & health effects in context; Indicative integrated MSW (Municipal Solid Waste) scenarios; Blue sky technologies; Critique of barriers to progress; conclusions & recommendations; References; Appendices.	http://www.number-10.gov.uk/su/waste/report/downloads/technologies.pdf
	Analysis of Household Waste Composition & Factors Driving Waste Increases (Dec 2002)	The first part of this paper focuses on current household waste compositional data with the objective of obtaining the best current estimates. Results of this review have been fed into the Strategy Unit municipal waste modelling exercise. The second part of the	Introduction; Household waste compositional analysis; Factors driving household waste increases; References; Appendix 1: Detailed results of compositional analysis: England 2000/01; Appendix 2: Regulatory model: total household waste.	http://www.number-10.gov.uk/su/waste/report/downloads/composition.pdf

	paper focuses on the factors that influence the quantities of household waste arisings, building on the conclusions from the compositional analysis. Household waste arisings have been increasing in the UK, yet no systematic review of the drivers behind these increases has been conducted.		
Collection Cost Projections (Dec 2002)	This report describes the methodology used to project both the current baseline costs & the costs of the different scenarios described by SU (the Strategy Unit, Cabinet Office). The report summaries the key assumptions & the resulting projected standard costs under a number of different assumptions, for example collection frequency.	Projection of the baseline costs for refuse & recyclables collection in England 2000/2001; Caution; Baseline cost projections for England; Projection costs for 2007, 2014 & 2020 in England; Assumptions for kerbside sorted collections; Co-mingled collections; Compost collections; Bring collections; All collections; Summary of cost projections.	http://www.number-10.gov.uk/su/waste/downloads/collection.pdf
A Study to Estimate the Disamenity Costs of Landfill in Great Britain – Final Report (Feb 2003)	The current study was undertaken to identify & estimate the disamenity costs of landfill in Great Britain, that is those local nuisance costs experienced by households living close to landfill that are associated with it such as odour, dust, litter, noise, vermin, & visual intrusion.	Introduction; Theoretical underpinning & literature review; Landfill data, analysis & assumptions; Housing data, analysis & assumptions; Estimation & results; Conclusions; Appendices.	http://www.defra.gov.uk/environment/waste/landfill/pdf/landfill_disamenity.pdf
Municipal Waste Management Survey Results 2002/03 (Aug 2004)	The 2001/02 survey, covering the period from 1st April 2001 to 31st March 2002 was the seventh survey in the series & was intended to build a more comprehensive picture of municipal waste management in England. This bulletin presents & analyses the results of the 2002/03 survey for England. The key points are shown in summary tables or charts within the text, whilst more detailed results can be found in the tables in Annex A. All figures quoted are based on grossed-up data, meaning that estimates have been made where information was not supplied.	Summary; Introduction; Municipal waste management; Household waste recycling; Waste management in different regions & types of authority; Trends since 1996/97; annexes.	http://www.defra.gov.uk/environment/statistics/wastats/mwb0203/wbtables.htm

Continued

Table 13.5.3 UK Level Waste Management Policy. *Continued*

Policy Type	Document Title	Objective	Contents	Website
Strategies & Reports	Mid-Term Evaluation of the Aggregates Levy Sustainability Fund (Sept 2003)	A commitment was made in the DEFRA consultation on the Aggregates Levy Sustainability Fund (ALSF) to establish research & monitoring arrangements to inform a review of the Fund. Subsequent parliamentary questions committed the department to an evaluation of the operation of the scheme in order to make a decision on its future in readiness for the Pre-Budget Report (PBR) in November 2003. As the ALSF runs until April 2004, the present report, which has been prepared for the PBR, constitutes a mid-term evaluation of the Fund.	Background to the mid-term evaluation; Conclusions & recommendations; Executive summary; Background to the aggregates levy sustainability fund & mid-tern review; The design of the ALSF in response to environmental & economic analysis & the views of stakeholders; Operation & management of the fund by DEFRA; The performance of the distributing bodies; Stakeholders views on the operation, performance & early impact of ALSF; The outcome & impact of the ALSF; Appendices.	http://www.defra.gov.uk/environment/waste/aggregates/pdf/mte.pdf
	'Towards a Thematic Strategy on the Prevention & Recycling of Waste' – United Kingdom Government Response (Nov 2003)	This document constitutes the United Kingdom Government's response to the Commission's express invitation to provide comments on the Communication. It represents a preliminary view on the issues raised by the Communication. The UK anticipates close collaboration with the Commission, Member States & other stakeholders in taking forward the work of developing an EU strategy on waste. The UK will provide the Commission with further considered views as the strategy takes shape over the coming months as part of this ongoing collaboration. Sections 1–5 set the main thrust of the UK's comments. Detailed comments on specific issues in the Communication can be found in the Annex.	Introduction; Developing the thematic strategy; Specific issues arising from the commissioned building blocks; Next steps; Conclusions; Annex: Detailed comments on the text of the Commission Communication.	http://www.defra.gov.uk/environment/waste/thematicstrat/response.pdf

Waste Implementation Programme – Factsheets (Nov 2003)	These factsheets explain the key components of the Waste Implementation Programme. They set out the rationale for the various work streams, how they will be implemented and managed and the outcomes we are trying to achieve. They also provide contact information for those who would like to find out more.	Introduction; Overview; Landfill tax credit scheme; Structure; Monitoring controls; Work stream 1: Local authority support; Work stream 2: Local authority funding; Work stream 3: New technologies; Work stream 4: Data; Work stream 5: Research; Work stream 6: Kerbside good practice & organics; Work stream 7: Waste minimisation; Work stream 8: Waste awareness; Contact details.	http://www.defra.gov.uk/environment/waste/review/factsheet1103.pdf
WEEE & Hazardous Waste (March 2004)	This scoping study was commissioned by the Department for Environment, Food & Rural Affairs (DEFRA) with the objectives of identifying the products & components of present & historic Waste from Electrical & Electronic Equipment (WEEE) that are hazardous as well as WEEE components that display hazardous properties above the threshold criteria as laid out in the Hazardous Waste Directive & the Hazardous Waste List, & determining what renders those products depolluted & non-hazardous.	Introduction; Background; Components in WEEE; Assessment of Hazardousness; Conclusions & recommendations; Appendices.	http://www.defra.gov.uk/environment/waste/topics/electrical/pdf/weee-hazwaste.pdf
Review of Environmental & Health Effects of Waste Management: Municipal Solid Waste (MSW) & Similar Wastes (May 2004)	The report examines the waste management options for treating municipal solid & similar waste. It focuses, as DEFRA requested, on the principal types of facilities that are currently used for dealing with such waste in the UK & in Europe & on what the currently available scientific evidence can tell us about their environmental & health effects. This study reviews the health & environmental effects which can be linked directly to facilities treating MSW & similar wastes.	Introduction; Review of information on emissions; Review of epidemiological research; Quantification of the health consequences of emissions; Quantification of the environmental consequences of emissions; Context for quantified health & environmental risks, & review of public perception issues; Conclusions; Appendices.	http://www.defra.gov.uk/environment/waste/research/health/pdf/health-report.pdf

Continued

Table 13.5.3 UK Level Waste Management Policy. *Continued*

Policy Type	Document Title	Objective	Contents	Website
Strategies & Reports	Government Interpretation of the Landfill (England & Wales) Regulations 2002 (As Amended) (Sept 2004)	The Government considers that there are aspects of the Regulations & the underlying intentions of the Directive for which policy interpretation would be helpful. This document provides that interpretation. It is aimed at waste producers, waste managers, landfill operators & regulators & is intended to provide a clear statement of the Government's intentions in implementing the Directive through the relevant Regulations in England & Wales.	Introduction; Regulation 3: Definition of landfill; Regulation 4: Dredgings; Regulation 5: Planning permission; Regulation 6: Disapplication of the 2000 regulations; Regulation 7: Landfills of different classes in close proximity; Regulation 8: Issues for permit conditions; Regulation 9: Wastes in liquid form; Regulation 10(1): Waste treatment; Regulation 10(2) & (3): Waste acceptance procedures; Regulation 11: Charging the costs of landfill; Regulation 14: Control & monitoring; Regulation 15: Aspects of closure; Regulation 19: Professional technical development & training; Enforcement.	http://www.defra.gov.uk/environment/waste/topics/landfill-dir/pdf/reg-interpret.pdf
	International Waste Prevention & Reduction Practice: Final Report (Oct 2004)	The study set out to identify interesting waste prevention & reduction initiatives from overseas, at the national, regional & local level, & where possible to note the policy & strategy context in which they operate. By interesting initiatives we mean those that: ▪ offer real innovation, i.e. ideas that are new to the UK or not well-known; or ▪ present a different angle on an approach that has been tried in the UK.	Introduction – this report; The waste prevention context – policy & potential; Waste management practices; Packaging eco-design; Container reuse & other deposit-refund schemes; Non-packaging products – eco-design & take back; Eco-labelling, product taxes & retailer initiatives; Product–service systems; Public education & awareness raising projects; Unwanted mail; Community reuse & refurbishment initiatives; Waste reduction in businesses & public organisations; Closing comments; Appendices.	http://www.defra.gov.uk/environment/waste/wip/wastemin/enviros-wasteprevent.pdf
	Clean Neighbourhoods & Environment Bill (Dec 2004)	A bill in relation to: vehicles exposed for sale or being repaired; abandoned vehicles, litter & refuse, graffiti, fly-posting & the display of advertisements; waste; the control of dogs & stray dogs; noise; abandoned shopping &	Part 1 – Crime & disorder; Part 2 – Vehicles; Part 3 – Litter & refuse; Part 4 – Graffiti & other defacement; Part 5 – Waste; transport of waste; deposit & disposal of waste; site waste; Part 6 – Dogs; Part 7 – Noise; Part 8 –	http://www.publications.parliament.uk/pa/cm200405/cmbills/011/2005011.pdf

		luggage trolleys etc.; and to provide for the Commission for Architecture & the Built Environment (CABE) & making grants relating to the quality of the built environment.	Architecture & the built environment; Part 9 – Miscellaneous; Part 10 – General.	
Consultation Papers	Less Waste. More Value. Consultation Paper on the Waste Strategy for England & Wales (09/06/98)	This document describes government action & policy & requests comments on a broad range of waste issues, such as: the waste hierarchy; targets; waste & sustainable transport; waste minimisation; waste reuse & recycling; composting; incineration; landfill/taxes; hazardous waste; householders/consumers; the voluntary/commercial/industrial sectors; the waste management. industry; local authorities; the Environment Agency; Central Government; & the new waste strategy.	Foreword; Summary of issues for consultation; Chapter 1: Introducing the Sustainable Development context; Chapter 2: Principles, instruments & targets; Chapter 3: Taking waste management decisions; Chapter 4: Who should do what; Chapter 5: Conclusions; Appendix A: Coverage of the new waste strategy in England & Wales; Appendix B: Development of the new waste strategy in Wales.	http://www.defra.gov.uk/environment/consult/waste/index.htm
	Limiting Landfill: A Consultation Paper on Limiting Landfill to Meet the EC Directive's Targets for the Landfill of Biodegradable Municipal Waste (05/10/99)	This consultation paper is designed to stimulate a debate on what are the best mechanisms for limiting landfill to meet the Landfill Directive's targets. We are looking to local authorities, the waste management industry, and others who will be most affected by the Directive's targets, to give us a better understanding of what can be achieved practically and cost-effectively.	Foreword; Chapter 1: Introduction; Chapter 2: The challenge; Chapter 3: Principles behind the options; Chapter 4: The options; Chapter 5: Summary of data needs; Chapter 6: Seeking your views – the consultation process; List of all questions contained in the paper; Annex A: Regulatory impact assessment.	http://www.defra.gov.uk/environment/waste/strategy/landfill/index.htm
	Landfill Allowance Trading Scheme Consultation (Aug 2003)	Part 1: Seeks views on the detailed operation of the Landfill Allowance Trading Scheme for England, to be established through regulations under the Waste and Emissions Trading Bill. Part 2: Considers the targets to be set for each country of the UK in target years, & the targets to be set for England in non-target years under the Waste & Emissions Trading Bill. The targets for the Devolved administrations for non-target years will be subject to separate consultation by each administration & subsequent agreement with the Secretary of State.	Scope of this consultation paper; Part 1: Landfill allowance trading scheme regulations; Introduction; Overview of scheme; Allocation of allowances; Banking & borrowing of allowances; Trading allowances; The electronic register; Monitoring the scheme; The penalty system; Part 2: Regulations under section 1 & 2 of Waste & Emissions Trading Bill; Introduction; Division targets in 2010, 2013 & 2020; Non-target year BMW (Biodegradable Municipal Waste) landfill levels in England; Annex A – Mass Balance System.	http://www.defra.gov.uk/corporate/consult/landfill/consultation-doc.pdf

Continued

Table 13.5.3 UK Level Waste Management Policy. *Continued*

Policy Type	Document Title	Objective	Contents	Website
Consultation Papers	Fly-Tipping Strategy – Consultation Document (Feb 2004)	This document concerned government strategy to improve local environmental quality & deal with fly tipping. The document also provided details of the other action being taken to help combat fly tipping. It should be noted that this strategy does not cover the issue of abandoned vehicles, since action is being taken separately to deal with these & they are covered by a different legislative framework.	Introduction; Tackling the illegal disposal of waste; The current situation; Potential future pressures on fly-tipping levels Landfill directive; WEEE directive; Developing the fly-tipping strategy; Beginning to implement the strategy; More flexible & effective enforcement of the duty of care; Ensuring the main fly-tipping offence provides a greater deterrent; More robust powers for clearing waste from land; Consideration of the amount of construction, demolition & excavation waste being fly-tipped; Changes to secondary legislation; Refinements to the way that stop, search & seizure powers work; Additional measures that do not need legislation; The consultation criteria.	http://www.defra.gov.uk/corporate/consult/flytipping-strategy/consultation.pdf
	Review of the Special Waste Regulations 1996 in England – Proposals for replacement Hazardous Waste Regulations & List of Wastes Regulations: A Consultation Paper (July 2004)	This consultation paper seeks comments on proposed new regulations in England to replace the Special Waste Regulations 1996 & to implement the revised European hazardous waste list, now incorporated into the European Waste Catalogue. Such Regulations provide a system of control to ensure that hazardous wastes are soundly managed from the point at which they are produced to the final point of disposal.	Executive Summary; Introduction; An alternative approach to the definition and classification of hazardous waste; Application of the regulations; Mixing hazardous waste; Notification of premises; Movement of hazardous waste; Returns & records; The Agency's functions; Emergencies & grave danger; Enforcement; Charges Transitional provisions; Consequential Amendments; Annexes.	http://www.defra.gov.uk/corporate/consult/wastereg-haz/consult.pdf
	The Landfill (England & Wales) Regulations 2005 – A Consultation Paper (Dec 2004)	The consultation seeks views on the proposed Landfill (England & Wales) Regulations 2005 amending earlier Regulations but also introducing new provisions.	Section 1 – Scope of the consultation; Section 2 – Introduction; Section 3 – Role of the producer in waste characterisation; Section 4 – (a) Limit values for Monolithic Wastes; (b) Bearing Capacity and Physical	http://www.defra.gov.uk/corporate/consult/landfillregs-rev/landfill-consult.pdf

			Stability of Wastes; Section 5 – Limit values for Polycyclic Aromatic Hydrocarbons (PAHs); Section 6 – Application of Article 3(2) exclusions to wet dredgings; Section 7 – Implementation date(s) for bans on landfilling certain non-hazardous wastes and the requirement for pre-treatment of all non-hazardous waste going to landfill; Section 8 – Risk Assessment Option; Section 9 – List of Questions; Annexes.	
	Consultation on Sustainable Waste Management (Dec 2004)	This consultation sets out proposals for a stronger, simpler & more integrated framework to deliver the significant expansion in new waste management facilities needed to meet EU obligations & national policy. The suite of papers are: ▪ Proposed changes to Waste Strategy 2000, which underpin the guidance on Municipal Waste Management Strategies & Planning Policy Statement 10; ▪ Draft guidance on the preparation of Municipal Waste Management Strategies; and ▪ Proposals for the periodic provision of national data & advice on waste arisings, recycling potential & any national capacity needs.	Consultation Paper 1 – Document A – Changes to Waste Strategy 2000; Document B – Partial Regulatory Impact Assessment for Waste Strategy 2000; Document C1 – Environmental report on changes to Waste Strategy 2000 for the Strategic Environmental Assessment; Document C2 – Non technical summary on changes to Waste Strategy 2000 for the Strategic Environmental Assessment; Consultation Paper 2 – Document D – Draft guidance on Municipal Waste Management Strategies; Document E – Draft outline of practice guidance for Municipal Waste Management Strategies; Consultation Paper 3 – Document F – Advice on data & forecasting.	http://www.defra.gov.uk/corporate/consult/wasteman-strat/index.htm
Guidance Notes	Waste Strategy Guidance – Best Value & Waste Management (15/08/00)	'Best Value' gives Local Authorities the framework & the tools to achieve set statutory performance standards for recycling. The goal must be to deliver a fully integrated approach to waste management with Authorities working together with partners in the community. This guidance should help those involved rise to the challenge set down in the Waste Strategy 2000.	Foreword; Part I: Explaining best value; Part II: The application of best value to waste management; Annex A: Useful publications; Annex B: Useful addresses; Annex C: Performance indicator definitions.	http://www.defra.gov.uk/environment/waste/strategy/guidance/bestvalue/index.htm

Continued

Table 13.5.3 UK Level Waste Management Policy. *Continued*

Policy Type	Document Title	Objective	Contents	Website
Guidance Notes	Strategic Planning for Sustainable Waste Management (Oct 2002)	This Guidance seeks to address the limitations of BPEO assessment. It sets out a methodology for appraising strategic waste planning options that takes account of environmental, socio-economic & implementation issues – as well as those normally addressed by BPEO (Best Practicable Environmental Option) assessments. The step-by-step methodology is intended to provide greater transparency in decision-making, as well as a robust & comprehensive approach. The Guidance is aimed primarily at RTABs (Regional Technical Advisory Bodies), to assist their work at the regional level, but may also assist strategic planning authorities engaged in planning for waste management.	Preface; Executive Summary; Introduction; Identifying & Agreeing Appraisal Criteria; Developing Strategic Waste Planning Options; Data Collection; Appraising Strategic Waste Planning Options; Ranking & Valuing Performance; Weighting Indicators; Sensitivity Testing; Appendices.	http://www.odpm.gov.uk/stellent/groups/odpm_planning/documents/pdf/odpm_plan_pdf_606386.pdf
	Guidance on Directive 2000/76/EC on the Incineration of Waste (WID) (2002)	This guide to the Directive, the Regulations & directions describes the scope, regulatory & technical requirements of the WID & how they should be interpreted & applied. It explains the main legal provisions, but the precise legal requirements can only be determined by the Courts. The guide is particularly intended for operators of incineration & co-incineration plants, regulators, waste producers, statutory consultees, waste producers & waste managers, all of whom will be affected by the implementation of the WID.	Executive summary; Introduction; Scope of the directive; The regulatory requirements; Technical requirement of the WID; Annexes.	http://www.defra.gov.uk/environment/ppc/wasteincin/pdf/wid_guidance_pb8005.pdf

The Producer Responsibility Obligations (Packaging Waste) Regulations 1997 (as amended) – User's Guide Second edition (2003)	This Guidance Note fulfils the requirement on government to provide guidance on the Producer Responsibility Obligations (Packaging Waste) Regulations 1997 (as amended). This *User's Guide* is therefore intended to ensure that there is detailed guidance in place on all aspects of the Regulation. The *User's Guide* is therefore in two parts: Part One sets down a summary of the requirements in the Regulations for those that want a succinct overview; Part Two sets down more detailed guidance.	Introduction; User's guide Part 1: Summary of the main requirements in the Regulations; User guide Part 2: Directive targets & business targets – who is obliged & the main requirements in the regulations; Registration & date provision; Recovery & recycling obligations; Calculation of the recovery & recycling obligations; The certifying obligations; The consumer information obligations; Monitoring compliance, offences & penalties; Appendices.	http://www.defra.gov.uk/environment/waste/topics/packaging/pdf/userguide.pdf
Draft Guidance for Waste Collection Authorities on the Household Waste Recycling Act 2003 (July 2004)	The Household Waste Recycling Act 2003 provides that where English waste collection authorities have a general duty to collect waste they shall ensure (except in some circumstances) that by 31 December 2010 they collect at least two types of recyclable waste separate from the rest of the household waste. This Guidance is issued to help waste collection authorities implement the Act, while respecting their freedom to choose the best way of doing so, & to fulfil commitments given by Ministers during the Committee stage of the Bill's progress through Parliament regarding issues that would be addressed in guidance.	Executive Summary; General; Essence of the Act; Aim of this Guidance; Guidance; Materials; Guidance on collection of textiles & shoes; Assisted collections; Dirty MRFs (Material Recycling Facilities); Health & Safety; Best practice; Role of the Waste Collection Authority; Evidence from research; Exceptions; Cost of doing so would be unreasonably high; Comparable alternative arrangements; Delay in implementation; Monitoring; Annexes.	http://www.defra.gov.uk/corporate/consult/waste-collect/consultation.pdf

Table 13.5.4 Northern Ireland Level Waste Management Policy.

Policy Type	Document Title	Objective	Contents	Website
Strategies & Reports	A Guide to the Special Waste Regulations (Northern Ireland), Jan 1998	The Special Waste Regulations (Northern Ireland) 1998 implement the European Hazardous Waste Directive 91/689/EEC. Their purpose was to provide an effective system of control for wastes which are dangerous & difficult to handle, to ensure they are soundly managed from their point of production to their final destination for disposal & recovery.	Part A: The classification of special waste; Part B: Use of the consignment note; Part C: Summary of how to obtain & send consignment notes; Part D: Data sources, other relevant legislation	http://www.ehsni.gov.uk/pubs/publications/guide_special_waste_regsNI_1998_web.pdf
	Waste Management Strategy for NI (March 2000)	The purpose of this document is to set out a Waste Management Strategy for NI in fulfillment of requirements set out in the EC Waste Framework Directive. The Strategy offers a long-term vision for the future development of waste management practices in NI & provides the framework to enable this to become reality. The key aim of this Strategy is to achieve fully sustainable waste management. This means using material resources more efficiently to cut down on the amount of waste produced. Where waste is generated, it means dealing with it in a way which minimises impacts on the environment & contributes positively to economic & social development.	Foreword; Part 1: Introduction; Part 2: Waste management policies; Part 3: The way forward; Annex A: Waste management concepts; Annex B: Waste management data; Annex C: Terms of reference for the advisory board.	http://www.ehsni.gov.uk/pubs/publications/NIWMS.pdf
	Guidelines for Company Reporting on Waste (Jan 2001)	The purpose of these Guidelines is to help businesses to: ▪ Measure the waste they produce; ▪ Identify how their waste management could be improved; ▪ Set targets to reduce their waste; ▪ Achieve savings & improve their performance;	Why business waste is important to us all; The waste reporting guidelines; How to start reporting on waste; Step 1: Identifying a 'waste champion'; Step 2: Identify the types of waste the business produces; Step 3: Measure the wastes; Step 4: Review what is done with the waste; Step 5: Set targets for improvement; Step 6: Identify the benefits; Step 7: Report the outcome; Step 8: Repeat the cycle; Useful	http://www.ehsni.gov.uk/pubs/publications/Waste_Report_Guidelines_PP.pdf

	■ Report on their performance; ■ Meet current & future legislative requirements.	information; Annex 1: Example of proforma for describing waste streams; Annex 2: Most significant waste streams by industrial sector; Annex 3: Typical container weights & waste conversion factors; Annex 4: Top 10 hit list of waste streams for review; Annex 5: Hazardous or special wastes.	
Waste Arisings Survey for NI – 1999–2000 (Aug 2001)	As part of the implementation of the Waste Management Strategy the Environment and Heritage Service (EHS) commissioned the first comprehensive Waste Arisings Survey for NI. This survey was designed to assess & evaluate the quantity of waste produced within Northern Ireland between 1999 & 2000.	Chapter 1: Introduction; Chapter 2: Municipal waste arisings; Chapter 3: Commercial & industrial waste arisings; Chapter 4: Waste industry; Chapter 5: Conclusions.	http://www.ehsni.gov.uk/pubs/publications/Waste_Arisings_Survey_NI_Screen.pdf
Evaluation of the First Phase of the Waste Management Public Awareness & Information Programme (June 2002)	The purpose of this report is to assess the performance of the first stage of the public awareness campaign or waste management issues in order to: ■ Analyse the effectiveness & value for money of the campaign; ■ Inform & direct the development and content of Phase 2. The report addresses Phase 1, covering the launch, & completion & analysis of consultation period on Waste Management Plans to end of May 2002.	Executive summary; Introduction; Components of the Public Awareness & Information Programme & Assessment of Performance; Monitoring Impact of the Public Awareness & Information Programme; Impact of the Public Awareness & Information Programme; Conclusions & Recommendations; Appendices.	http://www.ehsni.gov.uk/pubs/publications/WUTW_eval_Phase1_report_external.pdf
Waste & Contaminated Land (Northern Ireland) Order 1997 – Article 5: Waste Management The Duty of Care – A Code of Practice (2002)	The code of practice consists of the guidance in Sections 1–7 together with their related Annexes. It is issued by the Department of the Environment in accordance with Article 5(9) of the Waste and Contaminated Land (Northern Ireland) Order 1997.	Introduction; The code of practice; Step-by-step guidance; Identify & describe the controlled waste; Keep the waste safely; Transfer to the right person; Receiving waste; Checking up; Expert help & guidance; The duty of care & scrap metal; Summary checklist; Annex A: The law on the duty of care; Annex B: Responsibilities under the duty of care; Annex C: Regulations on keeping records; Annex D: Other legal controls; Annex E: Glossary.	http://www.ehsni.gov.uk/pubs/publications/Duty_of_Care_Code_of_Practice.doc

Continued

Table 13.5.4 Northern Ireland Level Waste Management Policy. *Continued*

Policy Type	Document Title	Objective	Contents	Website
Strategies & Reports	The NW Region Waste Management Group – Waste Management Plan (2002)	The purpose of this plan is to set out the arrangements for the management of wastes arising within the NWR, in fulfillment of the councils' statutory obligations under the EU Waste Framework Directive, The Waste & Contaminated Land (Northern Ireland) Order 1997. It ensures that the waste management practices become more sustainable, whilst ensuring councils meet their statutory obligations & targets. It also ensures that reliable information & data is generated so that the waste management debate within the region takes place within a clearly understood & constructive framework.	Executive summary; Introduction; General description of the area; Legislation & waste planning; Current infrastructure & waste arisings; Waste management methods; Markets for recyclables; Implementation – Municipal waste; Other controlled waste; Monitoring & review; App A: Consultation report; App B: Packing waste; App C: Hazardous waste management; App D: Best Practicable Environmental Option; App E: Detail calculations for recycling & composting; App F: Equality screening for EQIA; App G: Digest of waste management facilities; App H: Costs; App I: Trail schemes.	http://www.northwestwasteplan.org.uk/downloads/5253.00%20NW%20WMP%20Complete.pdf
	Southern Waste Management Partnership (SWaMP) – Waste Management Plan (2002)	The fundamental purpose of the plan is to set out the arrangements for the management of the wastes arising within the SWaMP region, in fulfilment of their statutory obligations under the EU Waste Framework Directive, & The Waste and Contaminated Land (Northern Ireland) Order 1997.	Executive summary; Introduction; General description of the area; Policy, legislation & regional interaction; Current infrastructure & waste arisings; Approaches for the management of waste; Assessment of markets for recyclables; Implementation – municipal waste; Implementation – monitoring & review; App A: Consultation report; App B: Packaging waste; App C: Hazardous waste management; App D: BPEO; App E: Detail calculations for recycling & composting; App F: Equality screening for EQIA; App G: Individual council waste statistics; App H: Digest of waste management facilities; App I: Costs; App J: Trial schemes.	http://www.swampni.org.uk/downloads/5193.00%20SWaMP%20WMP%20-%20Complete.pdf

Interim Needs for Landfill Capacity 2003–2007 (Mar 2003)	This report assesses the need for the development of further landfill in order to ensure that sufficient capacity exists until approximately 2007. The Waste Management Strategy anticipates that decisions on the nature of an integrated infrastructure of waste management facilities will be detailed in Waste Management Plans. Proposals for the next generation of facilities to re-use & recycle waste & markedly reduce our current dependency on landfill should be complete by 2005. However, it is recognised that the development of a complete network of new facilities may require a lead-time beyond this date. Hence the assessment of need until 2007.	Executive summary; Introduction & background purpose; Methodology; Data summary; Discussion; Recommendations; Recommendation summary; Appendices.	http://www.ehsni.gov.uk/pubs/publications/EIC_report_2003.pdf
Draft Biodegradable Waste Strategy for NI (May 2003)	The Biodegradable Waste Strategy for Northern Ireland has been produced in fulfillment of obligations under Articles 5(1) and 5(2) of the Landfill Directive (1999/31/EC). The scope of this Strategy encompasses all waste streams that have a biodegradable content, including the biodegradable fraction of the following waste streams: Municipal Wastes; Commercial & Industrial Wastes; Packaging Wastes; Construction & Demolition Wastes; Sewage Sludge; Agricultural Wastes.	Executive summary; Part 1: Background; Introduction; The policy & legislative context; Part 2: Biodegradable waste production & capacity requirements; Sources & composition of biodegradable waste; Biodegradable municipal waste; Commercial & industrial wastes; Packaging waste; Construction & demolition wastes; Sewage sludge; Agricultural wastes; Part 3: Landfill reduction – an integrated approach; Current measures & initiatives; Infrastructure & waste planning; Actions & commitments; Part 4: Monitoring performance; Implementation, monitoring & review; Annexes.	http://www.ehsni.gov.uk/pubs/publications/draftBiodegradableWasteStrategy.pdf

Continued

Table 13.5.4 Northern Ireland Level Waste Management Policy. *Continued*

Policy Type	Document Title	Objective	Contents	Website
Strategies & Reports	Compliance Monitoring Strategy for 2003 (July 2003)	The Producer Responsibility Obligations (Packaging Waste) Regulations were introduced in Northern Ireland in 1999 as part of the UK implementation of the EC Directive on Packaging & Packaging Waste. This Directive aimed to reduce the impact of packaging on the environment largely by setting mandatory targets for the recovery & recycling of 50%–65% of packaging waste by July 2001. Legal responsibility for contributing towards achieving the Directive targets has been placed on most UK businesses which have an annual turnover exceeding £2 million & who manufacture, convert, pack, sell or import more than 50 tonnes of packaging each year. This document highlights: ▪ the Department's policy in relation to monitoring & enforcement of producer responsibility obligations; & ▪ an indication of the minimum number of persons which it proposes to monitor in the course of that year.	Introduction; Background; The duty to publicise a monitoring strategy; The duty to monitor compliance; The purpose of monitoring; Monitoring activities; Registrants; Certificates of compliance; Compliance scheme operators; Compliance with consumer information obligations; Packaging flows; Accredited reprocessors & exporters; Monitoring strategy; 2003 compliance monitoring targets; Target groups for compliance monitoring; Operators of compliance schemes; Accredited reprocessors & exporters; Provision of advice & guidance; Funding; Enforcement.	http://www.ehsni.gov.uk/pubs/publications/CompMonStrat4.pdf
	Wake up to Waste II Evaluation Report (Sept 2003)	The purpose of this report is to the assess the performance of Phase II & provide recommendations for the way forward in Phase III of the campaign, specifically: ▪ Review the context of the campaign as a whole ▪ Analyse the effectiveness & value for money of the campaign ▪ Compare its impact with that of Phase I ▪ Inform & direct the development of Phase III.	Introduction; Wake up to Waste Programme: Phase II; Impact on attitudes & behaviour; Conclusions & Future Directions; Appendices.	http://www.ehsni.gov.uk/pubs/publications/phase2final.pdf

Factors Affecting the Future of Hazardous Waste in Northern Ireland (Nov 2003)	This paper stated that the way in which hazardous waste is managed would change significantly over the following years, driven by a series of European Directives affecting both resource use & waste management. These Directives would direct hazardous waste away from landfill, impose more stringent requirements for waste treatment & incineration & increase the number & types of waste defined as hazardous. This document highlights the main Directives and their impact on hazardous waste management in NI.	Classification of hazardous waste; Hazardous waste producers; Legislative drivers; The EC landfill Directive; The proposed hazardous waste regulations & waste classification changes; Pollution prevention & control (PPC); Waste electrical & electronic equipment Directive (WEEE); Restriction of hazardous substances Directives (RoHS); End-of-life (ELV) vehicles Directive; Batteries Directive; Waste incineration Directive (WID); Solvent emissions Directive (SED).	http://www.ehsni.gov.uk/pubs/publications/Hazardwaste.pdf
Northern Ireland Waste Management Strategy Review Report (June 2004)	In developing its recommendations the Waste Management Advisory Board for Northern Ireland has reviewed the progress against the key aims & targets identified within the text of the NIWMS. The progress review is intended to be complementary to that prepared by the external consultants for the Department, & to reflect the Board's independent & considered analysis.	Statement from the Chairman; Executive Summary; Summary of Recommendations; The Waste Management Advisory Board for Northern Ireland; Key Issues & Recommendations; Context & Background of the NIWMS; Reduction, Recycling & Recovery; Strategy Leadership; Regulations & Guidance; Planning & Infrastructure; Improving our Understanding; Marketing the strategy; Implementation of Waste Strategy; The Role of the Waste Management Advisory Board; Appendices.	http://www.ehsni.gov.uk/pubs/publications/WasteManagementStrategyReview.pdf
Hazardous Waste in Northern Ireland: An Action Plan for its Environmentally Sound Management (June 2004)	This Action Plan makes a number of recommendations for action & is the beginning of a process to change the way in which hazardous waste is managed. The remainder of this Action Plan sets out the specific objectives & recommended actions by stakeholders, together with the specific issues & relevant background information in relation to each of the recommended actions.	Executive summary; Introduction; Northern Ireland Hazardous Waste Forum – Members; Northern Ireland Hazardous Waste Forum – Terms of Reference; Action Plan; Glossary – List of terms/abbreviations; Annex 1 – Factors affecting future Hazardous Waste Management in NI; Annex 2 – Quantifying Hazardous Waste Capacity Needs; Appendix 1 – Assessment of Potential Changes.	http://www.ehsni.gov.uk/pubs/publications/Hazardous_Waste.pdf

Continued

Table 13.5.4 Northern Ireland Level Waste Management Policy. *Continued*

Policy Type	Document Title	Objective	Contents	Website
Strategies & Reports	BPEO (Best Practicable Environmental Option) for the Management of Waste Asbestos (Oct 2004)	This report describes various options for the management of asbestos waste in Northern Ireland based on a study carried out on behalf of the Environment and Heritage Service (EHS) in March 2004. The guidance is intended for use by EHS staff in assessing applications for treatment & disposal of asbestos waste & to inform operators on the approach to asbestos waste management in Northern Ireland.	Introduction; Treatment options; BPEO; Sensitivity analysis; Conclusions; Annex – Legislative requirements.	http://www.ehsni.gov.uk/pubs/publications/BPEOAsbestos.pdf
	Baseline Waste Audit (Nov 2004)	The aim of the audit was to provide clear & accurate information relating to resource use & waste management practices within the Department of the Environment & its four agencies, as listed below: ▪ Environment & Heritage Service (EHS) ▪ Planning Service ▪ Driver & Vehicle Testing Agency (DVTA) ▪ Driver & Vehicle Licensing Northern Ireland (DVLNI). A representative sample of 20 sites, covering the Department & its four agencies, were chosen for detailed analysis. The audit was designed to include those sites where the greatest proportion of DoE staff worked & account for diversity of function. Internal staff will audit the rest of the DOE sites but the waste will not be compositionally analysed. A tool for predicting estimates of site waste arisings has been developed based on the robust data generated from the audit. The baseline data provides a basis on which to develop appropriate & effective methods of improvement in waste management practices across the Department.	Introduction; Legislation review; Data collection methodology; Results; Analysis of results; Recommendations; Appendix 1 – Individual site reports; Appendix 2 – European waste catalogue; Appendix 3 – Site proforma; Appendix 4 – Action plan; Appendix 5 – Recyclates flow chart.	http://www.ehsni.gov.uk/pubs/publications/WasteAuditJRAFinalRep.pdf

Consultation Papers	ARC21: Eastern Region Waste Mangt Group – Consultation Waste Management Plan (Jan 2002)	The purpose of this Consultation Waste Management Plan is to set out waste management arrangements for the Eastern Region (ER) to fulfil the requirements of the EC Waste Framework Directive, The Waste & Contaminated Land (Northern Ireland) Order 1997 & the NI Waste Management Strategy 2000. The Plan provides a framework for waste management provision & a regional network of facilities for all controlled wastes produced within the ER for the next 20 yrs. It establishes the need for waste management capacity & details the arrangements to deal with the wastes produced. These include collection, recovery, recycling, treatment & disposal services.	Executive summary; Part 1: Background & issues; Chapter 1: Introduction; Chapter 2: Environmental context; Chapter 3: Legislative & policy context; Chapter 4: Political & socio-economic context; Chapter 5: Baseline waste management review; Chapter 6: Review of waste management options; Part 2: Policies & proposals; Chapter 7: Future waste management requirements; Chapter 8: Waste policies; Chapter 9: Review of potential development sites; Part 3: Implementation; Chapter 10: Future waste management requirements; References; Glossary; Part 4: Figures; Part 5: Appendices.	http://cobweb.businesscollaborator.com/arc21/cwmpframe.htm
	Consultation Paper on Recovery & Recycling Targets in NI for Packaging Waste in 2002 (March 2002)	This NI consultation paper sets down the proposed business recovery & recycling targets for 2002 under the Producer Responsibility Obligations (Packaging Waste) Regulations (Northern Ireland) 1999.	Introduction; Recovery & recycling targets for 2002; Annex 1: Regulatory impact assessment; Annex 2: Draft regulations; Annex 3: Equality screening document.	http://www.ehsni.gov.uk/pubs/publications/Recovery_and_Recycling_Targets_for_Packagin_Waste_in_2002.pdf
	Consultation Paper on a Proposal of the European Parliament & Council to Amend the Directive on Packaging & Packaging Waste (May 2002)	This consultation paper set out the content of the proposed amendment to Directive 94/62/EC on Packaging & Packaging Waste, & sought views on the recovery & recycling targets proposed for 2006. It also sought views on the suggested extension to the proposed deadline to 2008 & on the proposed new definitions. It commented that there was no doubt the proposed new targets would be extremely challenging for UK business & NI business in particular and that a proposal for a more comprehensive amendment of the Directive was expected during 2002.	Introduction; Part 1: The issues; Part 2: The targets; Annex 1: Partial regulatory impact assessment; Annex 2: UK recycling achievements for packaging waste by material in 2000; Annex 3: Advice from the advisory committee on packaging; Annex 4: Possible timetable for the passage of the proposed amendments to Directive 94/62/EC; Annex 5: Proposed amendment to Directive 94/62/EC on packaging & packaging waste; Annex I: Detailed calculations of total costs & benefits of the proposal; Annex II: Bibliography; Annex III: Impact of proposal on business; Annex IV: Draft of amended Directive; Annex V: New 'Annex 1'; Annex VI: Summary of questions contained in consultation document.	http://www.ehsni.gov.uk/pubs/publications/CP_Packaging_Waste.pdf

Continued

Table 13.5.4 Northern Ireland Level Waste Management Policy. *Continued*

Policy Type	Document Title	Objective	Contents	Website
Consultation Papers	A Consultation Paper on Limiting Landfill to meet the EC Landfill Directive's Targets for Reducing the Landfill of Biodegradable Municipal Waste (Oct 2002)	This consultation paper discussed options for implementation of part of the EC Landfill Directive (1999/31/EC) in NI. It described the three legally binding targets for reductions in the landfill of biodegradable municipal waste (BMW) set by the Directive & the challenge faced in meeting them over the next two decades. The paper set out a number of possible options for achieving the three BMW landfill targets in NI. It explored some of the merits & disadvantages of each of these options, & asked questions relating to their design & effect. The approaches described have been written in the context of the proposed new waste management licensing system & proposed IPPC (Integrated Pollution Prevention and Control) controls.	Chapter 1: Introduction; Chapter 2: The challenge; Chapter 3: Principles behind the options; Chapter 4: The options; Chapter 5: Summary of data needs; Chapter 6: Seeking your views: the consultation process; Annex A: List of questions contained in consultation paper; Annex B: Response form; Annex C: Screening for EQI; Annex D: Human Rights Act 1998; Annex E: Glossary; Annex F: Partial regulatory impact assessment.	http://www.ehsni.gov.uk/pubs/publications/LANDFILL_DIRECTIVE-ARTICLE5_BMW_TARGETS_CONPAPER.pdf
	Consultation Paper on the Implementation of Council Directive 2000/76/EC on the Incineration of Waste (April 2003)	This consultation paper addressed how the Department of the Environment intended to implement EC Directive 2000/76/EC on the Incineration of Waste (WID) & invited views where there appeared to be options for implementation. It was aimed at operators of incineration & co-incineration plant, regulators, waste producers & waste managers, all of whom would be affected by the implementation of WID. This document was also aimed at waste & environmental interest groups who have an interest in environmentally sound regulation.	Introduction; The regulatory framework; The scope of the Directive; Issues arising from the Directive; Explanatory notes; Draft regulations & directions; Annexes.	http://www.ehsni.gov.uk/pubs/publications/ConsPaperIncinerWaste.pdf
	Consultation Paper on Implementation of EC Directive	This document described the regulatory requirements set out in the Landfill Directive 1999/31/EC & considered the likely impact	Introduction; The regulatory framework; Scope of the landfill directive; Landfill classification; Prohibited wastes; Treatment of waste; Landfill	http://www.ehsni.gov.uk/pubs/publications/LFD_Con_Paper.pdf

	1999/31/EC on the Landfill of Waste (June 2003)	upon both landfill operators & waste producers. It described how the Department intended to implement the Landfill Directive, inviting views where there appeared to be options for implementation.	permits; Construction of the landfill of waste; Waste acceptance criteria & procedures; Control & monitoring procedures; Closure & after-care procedures; General requirements for all classes of landfills; Annexes.	
	Consultation on the Transposition of Article 6 & Annex I of the End-of-Life Vehicles Directive (2000/53/EC) – Proposals for Regulations (June 2003)	This Northern Ireland consultation papers sought views on proposed implementation options for transposing Article 6 and Annex I (storage & treatment) of the End-of-Life Vehicles Directive. There would be a separate consultation issued in due course on draft Regulations in respect of the remaining provisions of the Directive regarding 'free take-back' & recycling & recovery targets.	Introduction; Purpose of consultation; App 1: Partial regulatory impact assessment; App 2: Draft transposing regulations; App 3: Definition of 'vehicle'; App 4: Minimum technical requirements; App 5: General requirements of waste management licensing; App 6: Equality screening; App 7: Human rights screening.	http://www.ehsni.gov.uk/pubs/publications/EPDCondocELV.pdf
	Waste Management & Contaminated Land – Proposed Fees & Charging Scheme (Consultation Paper) (Dec 2003)	This document described the Department's charging scheme for various activities. Section 3 of the document contains the Waste Management Charging (Northern Ireland) Scheme 2003/2004. The Scheme applies to Northern Ireland only.	Section 1: Introduction; Section 2: Guidance on the 2003/04 charging scheme; Section 3: The waste management charging (NI) scheme 2003/04; Section 4: Charges, registration of exemptions, carriers & brokers; Annexes.	http://www.ehsni.gov.uk/pubs/publications/WML_Fees_cons.pdf
Guidance Notes	Tackling Fly Tipping – A Guidance Note & Response Protocol (Mar 1999)	The purpose of this document is to provide operational guidelines, which may be referred to as a response protocol or best practice guidance, in respect of dealing with incidents of fly-tipping in Northern Ireland.	Introduction; Duties, powers & responsibilities in respect of fly-tipping; Duties/responsibilities & powers with regard to litter; A response strategy; An information strategy; Upon discovery fly-tipping; Appendices.	http://www.ehsni.gov.uk/pubs/publications/FlyTippingGuidance.pdf
	Best Practicable Environmental Option (BPEO) – Decision-Makers Guide (Sept 2000)	A guidance document on decision making for Best Practicable Environmental Option (BPEO) was published by SNIFFER (Scotland and Northern Ireland Forum for Environmental Research) in September 2000. This guidance document provides a stand-alone methodology to district councils in NI for determining BPEO, but may also be used as a complement to the SNIFFER guidance. Furthermore, there may be other approaches to identifying BPEO that the councils might choose to adopt.	Executive summary; Purpose; Introduction to BPEO; This guidance document; The BPEO process; Step 1: Define study objectives; Step 2: Identify decision criteria; Step 3: Develop options; Step 4: Apply constraints; Step 5: Evaluate options; Step 6: Weight decision criteria; Step 7: Generate final option scores; Step 8: Sensitivity analysis; Step 9: Create shortlist/iterate; Step 10: Identify the BPEO; Annex A: NI BPEO decision criteria; Annex B: Waste management methods; Annex C: BPEO practitioners' guide.	http://www.ehsni.gov.uk/pubs/publications/NI_BPEO_Guidance.pdf

Continued

Table 13.5.4 Northern Ireland Level Waste Management Policy. *Continued*

Policy Type	Document Title	Objective	Contents	Website
Guidance Notes	Municipal Waste Data Monitoring & Reporting – Interim Guidelines (Mar 2003)	The Environment & Heritage Service of the Department of the Environment has prepared guidance on municipal waste data reporting for use by District Councils & the sub-regional waste management planning groups in their regular monitoring & annual performance reviews. The objectives of the monitoring & reporting guidelines are: ▪ To identify appropriate key performance indicators that need to be measured to assess progress towards achieving the Waste Strategy Targets; ▪ To enable compliance with the reporting requirements of EU Directives, the Department & other UK legislation; ▪ To ensure uniform & consistent measurement & presentation of the key performance indicators across the whole of Northern Ireland.	Introduction; The need for waste data reporting; Interim guidelines; Data quality requirements; Conclusion; Annexes.	http://www.ehsni.gov.uk/pubs/publications/MunicipalWasteDataGuidelines1.pdf
	Hazardous Waste – Interpretation of the Definition & Classification of Hazardous Waste (Nov 2003)	This Technical Guidance document has been developed to provide guidance on the assessment & classification of hazardous waste based on the Hazardous Waste Directive definition of hazardous waste. It is intended to provide guidance to all involved in the production, management & control of hazardous waste & to be a reference document for all legislation related to hazardous waste & its management.	Introduction; Regulatory framework; Hazardous waste assessment framework; App A – Consolidated European waste catalogue; App B – Waste & potential hazards for absolute & mirror entries in the European waste catalogue; App C – Hazardous property assessment; App D – Data sources.	http://www.ehsni.gov.uk/pubs/publications/HazWasteTGN.pdf
	De-polluting End-of-Life Vehicles Guidance for Authorised Treatment Facilities (Jan 2004)	This is guidance on how to depollute an ELV to achieve a depolluted ELV classed as non-hazardous. The particular guidance methods are not mandatory (provided the same levels of depollution are demonstrated). Non-destructive methods of removing hazardous components may be carried out, provided that component integrity is not compromised.	Introduction; Example of the depollution process; Preliminary activities; Removal of fluids & other items; Removal or deployment of air bags; End of depollution procedure.	http://www.ehsni.gov.uk/pubs/publications/DepollutingELVs.pdf

14 Water

14.1 Water and the Built Environment

Water, air and green cells are essential for life in all communities across the world. The Planet Earth is considered to be the 'Blue Planet' as three-fourths of the surface is covered by water. Over 97% of the earth's water inventory is contained in seas and oceans which are saline, while a large proportion of the remaining water is locked up in glaciers and icecaps. This leaves a limited amount, estimated at 0.65%, in the rivers, lakes, surface water and groundwater as potentially available as potable water for human and animal consumption.

Human habitation and the built environment demand water, with global population growth and more sophisticated lifestyles increasing this demand. However as water remains a finite resource these increased water demands will need to be curtailed. This pattern of growing water poverty and scarcity is sporadic across different regions, is heavily influenced by periods of drought caused by climate change and remains a source of international and political conflict. The impact of water is also being increasingly felt through increased levels and frequency of flooding, occurrences of small and large tsunamis and anticipated rising sea levels. The production of high quality drinking water in developed nations has required substantial financial investment, but stressed water regions, with frequently inadequate water supply and sanitation, has lacked this investment. Thus in developing nations high infant mortality rates are prevalent due to the effects of water-borne diseases and there is a related impact on the economy, culture and societal stability.

Water plays an essential role in the built environment as it provides for and facilitates basic public health, and this quest for a wholesome water supply meets the needs of domestic, industrial, commercial and public service demands. However, water is also a source for and means of supporting landscapes, bathing areas, aquatic life, water recreation, biodiversity, wetlands, all manner of animals and mammals and for cultural pursuits. Scientists have analysed the physical, chemical and biological processes involved in both the natural environment and in monitoring man-made water systems. Likewise, engineers have been instrumental in developing water and waste water treatment and distribution systems, and are sometimes referred to as Public Health Engineers, following their engineering ingenuity in identifying cholera and other water-related diseases in sewage flows in the nineteenth century.

Significant international attention has been focusing on the need to preserve, protect and enhance water supplies and sanitation provision. The United Nations addressed this global issue by designating 1981–1990 as the International Decade of Water Supply and Sanitation, in an effort to provide all communities with adequate water supply and appropriate sanitation, and this was supported by the World Bank and the World Health Organisation. Water was cited in the Rio de Janeiro Earth Summit in 1992 as a key area for urgent attention, and this drive for global water management was highlighted again at Earth Summit II in Johannesburg with an agreement to 'halve world water needs by 2015'. At the UN Assembly in 2003, it was declared that 'water is critical for sustainable development, including environmental integrity, alleviation of poverty and hunger and is indispensable for human health and well-being'. This resulted in a further International Decade for Action entitled 'Water for Life 2005–2015' and commenced on World Water Day, 22 March 2005.

The European Community (EC) has recognised the complex task of legislating for water issues, and implemented several EC Directives related to water through quality objectives, emission standards and specific activities such as pollution impacts. The EC developed a Water Framework Directive (EC/2000/60) in order to secure 'good water status' for each of its regions, employing catchment management planning and the use of a river basin model on a European-wide approach, in order to establish a balance between environmental issues, costs and water yield, and eliminating a piece-meal approach to specific directives by separate national governments.

14.2 Key Water Elements

The provision of water and wastewater systems is central to most national governments and normally demands significant funding due to their key role in maintaining public health, supporting industry and enhancing townscapes and landscapes. While there are varieties of approach in legislating for and procuring these roles and services, the elements and stages of feasibility, design, construction, monitoring, maintenance and final removal or replacement of water services are similar. The following paragraphs highlight some of the key water elements:

- *Reliable water yields* are predicted by hydrologists who study the patterns and quantities of naturally sourced water from upland catchments, lowland catchments and groundwater locations. In turn, *water demands* are estimated by referring to land plans, population growth patterns, increased usage of water and specific industrial, commercial and public utility requirements, as well as water usage at water treatment plants and 'unaccounted for water', arising from leakages and illegal usage.

- *Water supply* is met by the design and construction of dams and reservoirs at upland catchments, by lowland river and lake abstractions and by groundwater borewells, with appropriate water treatment and quality monitoring. The protection of riparian rights of adjacent landowners and the sensitive attention to the protection of the environment and aquatic life is important to ensure the longevity of supply and its catchment. Development of desalination plants has allowed sea or coastal water to be converted to potable water, but this is considered to be only economic when incorporated with a coastal power station, in which excess steam and hot water are employed in the desalination process.

- *Wastewater services* include the provision of sewers and stormwater drains with appropriate ancillary works and sewage treatment plants. Due to the increasing demand for water supply, the treatment of sewage and industrial effluents should be considered in a catchment and river basin management model. Hence, the development of low-cost, environmentally friendly treatment processes is important.

- *Infrastructure* is required to deliver both water supply and wastewater systems, and this involves the provision, maintenance and rehabilitation of watermains, trunk mains, sewers, stormwater overflows, stormwater drains, pumping stations and all related fittings. This activity requires an asset management planning approach, necessitating the need for adequate records of all infrastructure as well as the use of modelling tools to predict the performance and replacement policies across the entire networks.

- *Flood management* is needed to predict risks due to an increase in frequency and severity of floodwaters, caused by global warming. Engineers and planners are required to strategically plan flood defences to protect vulnerable coastal zones and the built environment infrastructure in those areas, and to control construction in flood plains. Flood warning systems are needed at local and oceanic locations to minimise the impact of the movement of large bodies of water due to exceptional precipitation, high tides and geological movements. Coastal erosion is also occurring on a long-term basis, requiring engineering interventions and monitoring programmes.

14.3 Sustainability Challenges for Water

Water has parallel high level challenges in both providing life to all societies as a means of nutrition and hydration and in having a finite lifespan, as indicated by water droughts and an increase in areas of water stress. These high level challenges indicate the critical need to fully understand and build robust strategies and legislation to ensure integrity of water and wastewater services, and these are synonymous with and contribute to all four objectives of sustainable development. The links are as follows:

- *Effective protection of the environment* through water abstraction systems which are responsive to and sensitive to adjacent land, waters and biodiversity. Wastewater systems need to minimise air and noise pollution, ensure that watercourse status is not damaged and prevent all infrastructural elements from polluting groundwater. The impacts of irrigation systems and agricultural waste should be understood and measured to avoid environmental damage, while the maintenance and enhancement of wetlands is important to all aspects of biodiversity.
- *Social progress which recognises the needs of everyone* through providing a quality of life which is relevant to and fitting for various communities across the globe. Water provides nutrition, essential food and disease control to those who live in poverty, while it is also a means of recreation and pleasure to communities in the developed world. Access to water and sanitation for poor people can allow development of culture pursuits, alter the role of women who have been traditionally used as water-carriers, and facilitate industrial development.
- *Prudent use of natural resources* can be achieved by reducing dependency on new water resources and greater emphasis on water recycling through use of grey water or by collecting rain water for harvesting. The recent development of Sustainable Urban Drainage Systems (SUDS) to collect surface runoff is healthier for the natural environment and reduces pollution. Likewise recycling treated sewage via river courses can alleviate the need for potable water being used for flushing, gardening and cleaning activities.
- *Maintenance of high and stable levels of economic growth and employment* through the efficient and economic use and supply of water and wastewater systems. Proper systems monitoring and maintenance will avoid excessive wastage and leakage, thereby increasing efficiency, reducing actual costs and allowing limited funding to be used to provide rehabilitation or replacement of systems. The water industry is an essential element of a stable society thereby providing sustained employment, while areas which are recovering from wars or disasters will require key professional and manual workers to reconstitute water systems. All these activities contribute to national economies.

Several water issues and challenges exist for environmentalists, economists, decision-makers and strategists in delivering sustainable water management. The critical water challenges which contribute to this broader sustainability agenda include:

- *Eradication of water poverty* and areas of water stress across the globe by international strategies, allocation of adequate funding, provision of appropriate technologies, existence of stable national governments, sensitive use of non-governmental organisations and recognition of water status in each catchment.
- *Development of catchment management planning* in developed nations to ensure full integration of water and wastewater systems, consistency of legislation, a proper balance of issues, allocation of adequate funding and greater attention to and solutions for 'unaccounted for water' usage.
- *Greater public appreciation* of the value and significance of a wholesome water supply, and the role of water in meeting the needs of society, providing attractive landscapes, ensuring adequate fire fighting water supply, facilitating water recreation and sports and allowing all aspects of biodiversity to survive.
- *Reduction in water-borne diseases* through improved water supply and distribution systems, to reduce infant mortality rates, allow societies to have a better quality of life and improved nutrition levels, help increase life expectancy, reduce the allocation of hospital beds to water disease-related illnesses and increase employment and productivity.

- *Increase international water research and development* into the nature of and potential for exploiting water in polar ice-caps and icebergs, thereby conserving water resources.
- *Protection of river and lake lowland waters* from eutrophication, by controlling use of phosphate-ridden detergents which discharge to sewage treatment plants and surface runoff containing nitrates from farmland intensification, both of which discharge to watercourses and lakes.
- *Use of water efficiency measures in buildings* at design, construction and operational phases. Integration with Building Environmental Management Systems (BEMS) will encourage awareness and efficient use of water-saving devices as well as knowledge of actual water costs.
- *Development of water charging systems* which ensure equity to all consumers, encourage water conservation and complement central funding for enhancing the water infrastructure.
- *Application of water energy systems* to allow greater use of renewable energy through hydro-electric schemes. Use of appropriate electricity and other energy tariffs for operation of pumps and other elements of the infrastructure can reduce energy demands and increase economic efficiency.
- *Greater use of groundwater sources of water* by full investigation of all aquifers to reduce dependency on upland reservoirs and potentially polluted lowland sources.

14.4 Water Theme Index

Jurisdiction	Tier	Title	Date	Page No.
EUROPEAN	Council Directives	Surface Freshwater: Methods of Measurement & Analysis *(CD 79/869/EEC Amended by CD 81/855/EEC; CD 90/656/EEC; CD 91/692/EEC)*	1979 1981 1990 1991	607
		Shellfish Waters *(CD 79/923/EEC Amended by CD 91/692/EEC)*	1979 1991	607
		Other Dangerous Substances: Protection of Groundwater *(CD 80/68/EEC Amended by CD 90/565/EEC; CD 91/692/EEC)*	1980 1990 1991	607
		General Provisions on the Quality of Drinking Water *(CD 80/778/EEC Amended by CD 81/858/EEC; CD 90/656/EEC; CD 91/692/EEC; CD 98/83/EC)*	1980 1981 1990 1991 1998	607
		Discharges of Dangerous Substances – Mercury *(CD 82/176/EEC Amended by CD 90/656/EEC; CD 91/692/EEC)*	1982 1990 1991	607
		Urban Waste Water Treatment *(CD 91/271/EEC Amended by Comm Dir 98/15/EC)*	1991 1998	608
		Pollution Caused by Nitrates from Agricultural Sources *(CD 91/676/EEC)*	1991	608
		Integrated Pollution Prevention & Control *(CD 96/61/EC)*	1996	608
		Quality of Drinking Water: New Requirements *(CD 98/83/EC)*	1998	608
		Framework Directive in the Field of Water Policy *(Dir 2000/60/EC Amended by Dec No. 2455/2001/EC)*	2000 2001	609
	Council Decisions	The Convention for the Protection of the Mediterranean Sea against Pollution & the Protocol for the Prevention of Pollution of the Mediterranean Sea by Dumping from Ships & Aircraft *(C Dec 77/585/EEC)*	1977	609
		The Protocol concerning co-operation in Combating Pollution of the Mediterranean Sea by Oil & other Harmful Substances in Cases of Emergency *(C Dec 81/420/EEC)*	1981	609
		The Protocol for the Protection of the Mediterranean Sea against Pollution from Land-based Sources *(C Dec 83/101/EEC)*	1983	609
		The Protocol concerning the Mediterranean's Specially Protected Areas *(C Dec 84/132/EEC)*	1984	609
		The Convention on the protection of the marine environment of the Baltic Sea (Helsinki Convention) *(C Dec 94/156/EEC C Dec 94/157/EEC)*	1994	610
		Helsinki Convention: Transboundary Watercourses & International Lakes *(C Dec 95/308/EC)*	1995	610
		The Protection of the Marine Environment of the North East Atlantic (Paris Convention) *(C Dec 98/249/EC)*	1998	610
		Convention for the Protection of the Rhine *(C Dec 2000/706/EC)*	2000	611
		Setting up a Community Framework for Co-operation in the Field of Accidental or Deliberate Marine Pollution *(Dec No. 2850/2000/EC)*	2000	611

Jurisdiction	Tier	Title	Date	Page No.
EUROPEAN	Council Decisions	Priority Substances in the Field of Water Policy *(Dec No. 2455/2001/EC)*	2001	611
		Co-operation in Preventing Pollution from Ships &, in Cases of Emergency, Combating Pollution of the Mediterranean Sea *(C Dec 2004/575/EC)*	2004	611
	Council Communications	Community Water Protection & Management Policy *(COM (96) 59)*	1996	612
		Pricing & Long-term Management of Water *(COM (2000) 477)*	2000	612
		Danube – Black Sea Region *(COM (2001) 615)*	2001	612
		Water Management in Developing Countries Policy & Priorities for EU Development Co-operation *(COM (2002) 132)*	2002	612
		Towards a Strategy to Protect & Conserve the Marine Environment *(COM (2002) 539)*	2002	613
		Flood Risk Management: Flood Prevention, Protection & Mitigation *(COM (2004) 472)*	2004	613
	Miscellaneous	Towards Sustainable Water Resources Management – A Strategic Approach	1998	614
		Towards a European Integrated Coastal Zone Management (ICZM) Strategy – General Principles & Policy Options	1999	615
		Lessons from the European Community Demonstration Programme on Integrated Coastal Zone Management (ICZM)	1999	615
		EU Focus on Clean Water	1999	615
		Sustainable Water Use in Europe Part 1: Sectoral Use of Water	1999	616
		Sustainable Use of Europe's Water? State, Prospects & Issues	2000	616
		Maritime Safety: Compensation Fund for Oil Pollution Damage in European Waters *(COM (2000) 802)*	2000	617
		Study on the Impact of Community Environment – Water Policies on Economic & Social Cohesion	2001	617
		Sustainable Water Use in Europe Part 2: Demand Management	2001	618
		Common Implementation Strategy for the Water Framework Directive	2001	618
		EU Focus on Coastal Zones – Turning the Tide for Europe's Coastal Zones	2001	619
		Sustainable Water Use in Europe Part 3: Extreme Hydrological Events – Floods & Droughts	2001	619
		Extensive Wastewater Treatment Processes: Adapted to Small & Medium Sized Communities	2002	619
		Implementation of Council Directive 91/271/EEC: Urban Wastewater Treatment, as amended by Commission Directive 98/15/EC – Summary Report	2002	620
		Phosphates & Alternative Detergent Builders – Final Report	2002	620
		Proposal for a Directive on the Quality of Bathing Water *(COM (2002) 581)*	2002	621

Jurisdiction	Tier	Title	Date	Page No.
EUROPEAN	Miscellaneous	Implementation of Council Directive 91/676/EEC: Protection of Waters Against Pollution Caused by Nitrates from Agricultural Sources – Synthesis from year 2000 Member States Reports	2002	621
		Water an Essential Resource: LIFE & the New European Water Policy	2002	621
		Eutrophication & Health	2002	622
		The Water Framework Directive: Tap into it!	2002	622
		Economic Assessment of Groundwater Protection: A Survey of the Literature Final Report	2003	622
		Carrying forward the Common Implementation Strategy for the Water Framework Directive – Progress & Work Programme for 2003–04	2003	623
		Ship-Source Pollution & the Introduction of Sanctions, Including Criminal Sanctions, for Pollution Offences *(COM (2003) 92)*	2003	623
		Common Implementation Strategy for the Water Framework Directive (2000/60/EC) – Economics & the Environment: The Implementation Challenge of the Water Framework Directive	2003	623
		Water for Life EU Water Initiative: International Co-operation – From Knowledge to Action	2003	624
		Proposal for a Directive on the Protection of Groundwater Against Pollution *(COM (2003) 550)*	2003	624
		Reporting for Water – Concept Report Towards a Shared Information System for Water	2003	624
		Europe's Water: An Indicator Based Assessment	2003	625
		Proposal for a Directive on Pollution Caused by Certain Dangerous Substances Discharged into the Aquatic Environment of the Community *(COM (2003) 847)*	2003	625
		Proposal for a Directive on the Quality of Fresh Waters Needing Protection or Improvement in order to Support Fish Life *(COM (2004) 19)*	2004	625
		Implementation of Council Directive 91/271/EEC of 21 May 1991 concerning Urban Waste Water Treatment, as amended by Commission Directive 98/15/EC	2004	626
UK	Strategies & Reports	Public Health & Drinking Water – Preventing *Cryptosporidium* Getting into Public Drinking Water Supplies	1998	627
		Potential Costs & Benefits of Implementing the Proposed Water Resources Framework Directive	1999	627
		Water Prices & the Environment – Government Response to the Environmental Audit Commission Seventh Report	2001	628
		Addressing the Water Crisis – Healthier & More Productive Lives for Poor People	2001	628
		To What Degree Can the October/November 2000 Flood Events be Attributed to Climate Change?	2001	628
		National Appraisal of Assets at Risk from Flooding and Coastal Erosion, including the potential impact of climate change	2001	629

Jurisdiction	Tier	Title	Date	Page No.
UK	Strategies & Reports	*Cryptosporidium* in Water Supplies	2001	629
		Learning to Live with Rivers – ICE Report	2001	630
		Best Practice Principles in the Economic Level of Leakage Calculation	2002	630
		Leakage Performance Indicators	2002	630
		Sewage Treatment in the UK: UK Implementation of the EU Urban Waste Water Treatment Directive	2002	631
		Safeguarding our Seas: A Strategy for the Conservation & Sustainable Development of our Marine Environment	2002	631
		The 2004 Periodic Review: Research into Customers' Views	2002	631
		Directing the Flow: Priorities for the Future of Water Policy	2002	632
		Strategic Review of Diffuse Water Pollution from Agriculture – Discussion Document	2003	632
		Water Framework Directive – Indicative Costs of Agricultural Measures	2003	632
		Water Bill – Regulatory Impact Assessment, Environmental & Equal Treatment Appraisals – 2nd Edition	2003	633
		Structure of the Water Industry in England: Does it Remain Fit for Purpose	2003	634
		National Assessment of Defence Needs & Costs for Flood & Coastal Erosion Management (NADNAC) – Summary Report	2004	634
		Drinking Water 2003 – A report by the Chief Inspector Drinking Water Inspectorate	2004	634
		Review of Marine Nature Conservation – Working Group Report to Government	2004	634
		Cross-Government Review of Water Affordability	2004	635
	Consultation Papers	Outcomes of the Online Consultation – www.waterquality-debate.org	2002	636
		Second Consultation Paper on the Implementation of the EC Water Framework Directive (2000/60/EC)	2002	636
		Seas of Change – The Government's Consultation Paper to Help Deliver Our Vision for the Marine Environment	2002	637
		Reductions for Vulnerable Groups: A Consultation Paper	2003	637
		Third Consultation Paper on the Implementation of the EC Water Framework Directive (2000/60/EC)	2003	637
		Developing Measures to Promote Catchment-Sensitive Farming: A joint DEFRA-HM Treasury Consultation	2004	637
		Making Space for Water: Developing a new Government Strategy for Flood & Coastal Erosion Risk Management in England – A Consultation Exercise	2004	638

Jurisdiction	Tier	Title	Date	Page No.
UK	Guidance Notes	Water Quality – A Guide to Water Protection in England	2000	638
		Environmental Reporting Guidelines for Company Reporting on Water	2000	639
		Shoreline Management Plans – A Guide for Coastal Defence Authorities	2001	639
		Planning a Future for the Inland Waterways – A Good Practice Guide	2001	640
		Preparing for Floods – Interim Guidance for Improving the Flood Resistance of Domestic & Small Business Properties	2003	640
		Principal Guidance from the Secretary of State to the Director General of Water Services – 2004 Periodic Review of Water Price Limits	2004	641
		Managing Coastal Activities: A Guide for Local Authorities	2004	641
NORTHERN IRELAND	Strategies & Reports	A Groundwater Monitoring Strategy for NI	2000	642
		Eutrophication in NI – Proposals for a Strategy to Control Nutrient Enrichment	2000	642
		Sustainable Urban Drainage Systems – An Introduction	2001	642
		A River Water Quality Monitoring Strategy for NI	2001	642
		Policy & Practice for the Protection of Groundwater in NI	2001	643
		Policy for Setting & Delivering Water Quality Targets	2001	643
		NI Regional Report of the National Monitoring Programme	2002	643
		Report on the Environmental Aspects of the Nitrates Directive in NI	2002	643
		Managing the Water Environment in NI – 2000	2002	644
		Delineation of Nitrate Vulnerable Zones for Four Locations in Northern Ireland	2002	644
		NI Drinking Water Quality Report 2001	2002	645
		Options for Management of Northern Ireland's Coastal Zone: Scoping Study Examining the Potential Establishment of a Northern Ireland Coastal Forum	2003	645
		Water Pollution Incident & Prosecution Statistics 2001	2003	645
		Regulation of Water Service Discharges – Report for the Year 2001	2003	646
		Northern Ireland Drinking Water Quality 2002	2003	646
		An Evaluation of Nitrogen Sources & Inputs to Tidal Waters in Northern Ireland	2004	647
		Northern Ireland Drinking Water Quality 2003	2004	647
		Regulation of Water Service Discharges 2002–2003	2004	647
	Consultation Papers	Northern Ireland River Conservation Strategy – A Draft for Consultation	2000	648
		A Consultation Paper on the Water Quality (Water Supply) Regulations (Northern Ireland) 2002	2002	648

Jurisdiction	Tier	Title	Date	Page No.
NORTHERN IRELAND	Consultation Papers	Managing our Shared Waters – A Joint North/South Consultation Paper on International River Basin Districts & Administrative Arrangements for Implementation of the EC Water Framework Directive (2000/60/EC)	2003	648
		A Consultation Paper on Nitrates & the Protection of Groundwater in Northern Ireland	2003	649
		Nitrates Directive 2nd Consultation Paper – Proposal for the Protection of Northern Ireland's Surface & Groundwater	2004	649
	Guidance Notes	Pollution Prevention Guideline (PPG1): General Advice	2001	649
		PPG5: Works in, Near or Liable to Affect Watercourses	2001	649
		PPG6: Working at Demolition & Construction Sites	2001	649
		PPG11: Industrial Sites	2001	650
		PPG15: Retail Premises	2001	650
		Guidance for the Monitoring of *Cryptosporidium* in Treated Water Supplies in Northern Ireland	2003	650

14.5 Water Policy Tables

Table 14.5.1 International Level Water Policy.

Title	Venue	Purpose	Key Outcome(s)	Website
International Conference on Freshwater	Bonn, Germany (2001)	The international conference on Freshwater reviewed the role of water in sustainable development, taking stock on progress in the implementation of Agenda 21 whilst identifying how this implementation can improve. It built on many previous efforts & conferences which defined the challenges & developed principles & policies related to water & sustainable development. This conference helped to narrow the gap between Sustainable Development policies & putting these into practice, through focusing on practical ideas.	The Ministerial Declaration agreed that governments, the international community, the private sector, NGOs, & all other stakeholders need to base their actions on governance, the funding gap, the role of the international community, capacity building & technology transfer, gender, & possible next steps. An Action list of 27 recommendations was drawn up under the above headings. The Bonn Keys developed highlighting five ways of managing water to help achieve Sustainable Development. ▪ Meet the water security needs of the poor. ▪ Local level action where national policy meets community needs. ▪ Partnerships the key to better water outreach. ▪ Long-term harmony with nature & neighbours through co-operative arrangements at water basin level. ▪ Stronger, better performing governance arrangements.	http://www.water–2001.de/ConferenceReport.pdf

Continued

Table 14.5.1 International Level Water Policy. *Continued*

Title	Venue	Purpose	Key Outcome(s)	Website
2nd IWA Leading-Edge Conference on Water and Wastewater Treatment Technologies – LET2004	Prague, Czech Republic (June 2004)	The conference has two parallel streams, Drinking Water Treatment Technologies and Wastewater Treatment Technologies.	The sessions in this conference were. Leading-edge Drinking Water Treatment Technologies: ▪ Membranes in drinking water ▪ UV disinfection ▪ Natural organic matter removal ▪ Distribution system water quality ▪ Advances in biological filtration. Leading-edge Wastewater Treatment Technologies: ▪ Modelling, sensors & control ▪ Treatment technologies for new trace pollutants ▪ Novel Reactors & Technologies ▪ Sludge management ▪ Rethinking the wastewater treatment concept.	http://www.let2004.com/templates/Conferences/LET/let.aspx?ObjectId=10499
1st IWA Leading-Edge Conference on Strategic Asset Management – LEA2004	San Francisco, USA (July 2004)	A rough estimate suggests that 70% of a water utility's operating costs are those related to infrastructure development & maintenance, & 70% of infrastructure costs are related to the pipeline network. In many cities, water & sewerage infrastructure expanded rapidly after World War II. This infrastructure is now 50–60 years old & its gradual deterioration suggests that investment requirements for maintenance & replacement will rise significantly in coming years. These issues will be discussed at the first international Leading-Edge Asset Management Conference. A new approach to asset investment will require buy-in from stakeholders, including regulators, politicians & customers. Understanding how to achieve this buy-in will therefore also be a particular feature of the event.	This International Water Association conference focused on water infrastructure, specifically: ▪ Infrastructure development & maintenance; ▪ Pipeline costs; ▪ Asset investment; & ▪ Stakeholder buy-in.	http://www.lea2004.iwa-conferences.org/templates/Conferences/LEAM/home.aspx?ObjectId=28598

2nd IWA Leading-Edge Conference on Sustainability: Sustainability in Water Limited Environments – LES2004	Sydney, Australia (Nov 2004)	For this Conference the question was posed: 'How do we ensure that the systems we design meet the needs of the environments in which they are located?' Water limited environments demand efficiency in water provision – but is this achieved through demand management, recycling, economic incentives or stormwater reuse (to give but a few examples)? How are the demands of the community & investors taken into account & what risk factors should be considered? The problems of making the best decisions, recognising that probably no one decision is 'right' when measured against all criteria.	This International Water Association conference will focus on the theme of Sustainability in Water Limited Environments and will be revisited every second year, alternating with Sustainability in Water Rich Environments. However, there will be many themes in common between the two conferences, and which will be relevant to all practitioners interested in issues of sustainability, whether from water limited or water rich areas.	http://www.les2004.iwa-conferences.org/templates/Conferences/LES/les_std.aspx?ObjectId=27914
Convention on the Prevention of Marine Pollution by Dumping Waste & Other Matter (London Dumping Convention)	London, UK (1972)	To improve protection of the marine environment by encouraging States with a common interest in particular geographical areas to enter into appropriate agreements supplementary to this Convention. (Superseded by 1996 Protocol)	An agreement that Contracting Parties shall individually and collectively promote the effective control of all sources of pollution of the marine environment, and pledge themselves especially to take all practicable steps to prevent the pollution of the sea by the dumping of waste and other matter that is liable to create hazards to human health, to harm living resources and marine life, to damage amenities or to interfere with other legitimate uses of the sea. ■ A comprehensive listing of substances which are prohibited from ocean disposal. ■ Regulations for the control of incineration of wastes and other matter at sea.	http://www.londonconvention.org/documents/lc72/LC1972.pdf

Continued

Table 14.5.1 International Level Water Policy. *Continued*

Title	Venue	Purpose	Key Outcome(s)	Website
International Convention for the Prevention of Pollution from Ships (MARPOL 73/78)	London, UK (1973)	The MARPOL Convention is the main international convention covering prevention of pollution of the marine environment by ships from operational or accidental causes. It is a combination of two treaties adopted in 1973 and 1978 respectively and has been updated by amendments through the years. The Convention covered pollution by oil, chemicals, harmful substances in packaged form, sewage and garbage. (Modified by the Protocol of 1978 relating thereto (MARPOL 73/78).)	The Convention includes regulations aimed at preventing and minimising pollution from ships – both accidental pollution and that from routine operations – and currently includes six technical Annexes: ■ Annex I – Regulations for the Prevention of Pollution by Oil ■ Annex II – Regulations for the Control of Pollution by Noxious Liquid Substances in Bulk ■ Annex III – Prevention of Pollution by Harmful Substances Carried by Sea in Packaged Form ■ Annex IV – Prevention of Pollution by Sewage from Ships (not yet in force) ■ Annex V – Prevention of Pollution by Garbage from Ships ■ Annex VI – Prevention of Air Pollution from Ships (adopted September 1997 – not yet in force).	http://www.imo.org/Conventions/mainframe.asp?topic_id=255
UN Convention on the Law of the Sea (UNCLOS)	Montego Bay, Jamaica (1982)	Lays down a comprehensive regime of law and order in the world's oceans and seas establishing rules governing all uses of the oceans and their resources. It enshrines the notion that all problems of ocean space are closely interrelated and need to be addressed as a whole.	The Convention embodied in one instrument traditional rules for the uses of the oceans and at the same time introduced new legal concepts and regimes and addressed new concerns. The Convention also provided the framework for further development of specific areas of the law of the sea.	http://www.un.org/Depts/los/convention_agreements/convention_overview_convention.htm

			The Convention comprises 320 articles and nine annexes, governing all aspects of ocean space, such as delimitation, environmental control, marine scientific research, economic and commercial activities, transfer of technology and the settlement of disputes relating to ocean matters.	
Convention for the Protection of the Marine Environment of the North East Atlantic (OSPAR)	Paris, France (1992)	It replaces the Oslo and Paris Conventions, but Decisions, Recommendations and all other agreements adopted under those Conventions will continue to be applicable, unaltered in their legal nature, unless they are terminated by new measures adopted under the 1992 OSPAR Convention.	Convention consists of a series of provisions notably the application of the precautionary principle, the polluter pays principle, best available techniques (BAT) & best environmental practice (BEP), including clean technology. It contains a series of annexes dealing with specific areas such as: - I, II, III – prevention & elimination of pollution from land-based resources; by dumping or incineration; and from offshore sources - IV – assessment of the quality of the marine environment - Establishment of OSPAR commission to administer the convention & develop policy/inter. agreements.	http://www.ospar.org/eng/html/welcome.html
Convention on the Protection & Use of Transboundary Watercourses & International Lakes	Helsinki, Finland (1992)	To prevent and control pollution of transboundary watercourses and international lakes by developing international co-operation.	It establishes a framework for co-operation between the member countries of the United Nations Economic Commission for Europe on the prevention and control of pollution of transboundary watercourses by ensuring rational use of water resources with a view to sustainable development.	http://www.unece.org/env/water/

Table 14.5.2 European Level Water Policy.

Policy Type	Programme Title	Policy Objective	Community Measure	Document Reference	Website
Council Directives	Surface Freshwater: Quality & Control Requirements	To reduce & prevent pollution of surface water intended for the abstraction of drinking water.	CD 75/440/EEC concerning the quality required of surface water intended for the abstraction of drinking water in the Member States (16/06/75) Amended by CD 79/869/EEC (09/10/79), CD 90/656/EEC (04/12/90) and CD 91/692/EEC (23/12/91).	▪ Off Jnl L271, 29/10/79 ▪ Off Jnl L353, 17/12/90 ▪ Off Jnl L377, 31/12/91	http://europa.eu.int/scadplus/leg/en/lvb/l28006a.htm
	Quality of Bathing Water	To reduce & prevent the pollution of bathing water.	CD 76/160/EEC concerning the quality of bathing water (08/12/75) Amended by CD 90/656/EEC (04/12/90) and CD 91/692/EEC (23/12/91).	▪ Off Jnl L31, 05/02/76 ▪ Off Jnl L353, 17/12/90 ▪ Off Jnl L377, 31/12/91	http://europa.eu.int/scadplus/leg/en/lvb/l28007.htm
	Other Substances: Protection of the Aquatic Environment	To harmonise the legislation of Member States on discharges of certain dangerous substances into the aquatic environment & to take preventative action at source.	CD 76/464/EEC on pollution caused by certain dangerous substances discharged into the aquatic environment of the Community (04/05/76) Amended by CD 90/656/EEC (04/12/90) and CD 91/692/EEC (23/12/91).	▪ Off Jnl L129, 18/05/76 ▪ Off Jnl L353, 17/12/90 ▪ Off Jnl L377, 31/12/91	http://europa.eu.int/scadplus/leg/en/lvb/l28017a.htm
	Water Suitable for Fish-Breeding	To protect fresh waters in order to safeguard fish species.	CD 78/659/EEC on the quality of fresh waters needing protection or improvement in order to support fish life (18/07/78) Amended by CD 90/656/EEC (04/12/90) and CD 91/692/EEC (23/12/91).	▪ Off Jnl L222, 14/08/78 ▪ Off Jnl L353, 17/12/90 ▪ Off Jnl L377, 31/12/91	http://europa.eu.int/scadplus/leg/en/lvb/l28010.htm

Surface Freshwater: Methods of Measurement & Analysis	To harmonise the national regulations on monitoring the quality of surface water.	CD 79/869/EEC concerning the methods of measurement & frequencies of sampling & analysis of surface water intended for the abstraction of drinking water in the Member States (09/10/79) Amended by CD 81/855/EEC (19/10/81), CD 90/656/EEC (04/12/90) and CD 91/692/EEC (23/12/91).	• Off Jnl L271, 29/10/79 • Off Jnl L319, 07/11/81 • Off Jnl L353, 17/12/90 • Off Jnl L377, 31/12/91	http://europa.eu.int/scadplus/leg/en/lvb/l28006b.htm
Shellfish Waters	To improve & to protect the quality of shellfish waters.	CD 79/923/EEC on the quality required of shellfish waters (30/10/79) Amended by CD 91/692/EEC (23/12/91).	• Off Jnl L281, 10/11/79 • Off Jnl L377, 31/12/91	http://europa.eu.int/scadplus/leg/en/lvb/l28009.htm
Other Dangerous Substances: Protection of Groundwater	To combat pollution by harmonising the laws of Member States on the discharge of certain dangerous substances into groundwater & by establishing systematic monitoring of the quality of such water.	CD 80/68/EEC on the protection of groundwater against pollution caused by certain dangerous substances (17/12/79) Amended by CD 90/565/EEC (04/12/90) and CD 91/692/EEC (23/12/91).	• Off Jnl L20, 26/01/80 • Off Jnl L353, 17/12/90 • Off Jnl L377, 31/12/91	http://europa.eu.int/scadplus/leg/en/lvb/l28017b.htm
General Provisions on the Quality of Drinking Water	To lay down at Community level minimum quality & control standards for water intended for human consumption.	CD 80/778/EEC relating to the quality of water intended for human consumption (15/07/80) Amended by CD 81/858/EEC (19/10/81), CD 90/656/EEC (04/12/90), CD 91/692/EEC (23/12/91) and CD 98/83/EC (03/11/98).	• Off Jnl L229, 30/08/80 • Off Jnl L319, 07/11/81 • Off Jnl L353, 17/12/90 • Off Jnl L377, 31/12/91 • Off Jnl L330, 05/12/98	http://europa.eu.int/eur-lex/en/consleg/pdf/1980/en_1980L0778_do_001.pdf
Discharges of Dangerous Substances – Mercury	Limit discharges of mercury into the aquatic environment of the Community.	CD 82/176/EEC on limit values & quality objectives for mercury discharges by the chlor-alkali electrolysis industry (22/03/82) Amended by CD 90/656/EEC (04/12/90) and CD 91/692/EEC (23/12/91).	• Off Jnl L81, 27/03/82 • Off Jnl L353, 17/12/90 • Off Jnl L377, 31/12/91	http://europa.eu.int/scadplus/leg/en/lvb/l28014.htm

Continued

Table 14.5.2 European Level Water Policy. *Continued*

Policy Type	Programme Title	Policy Objective	Community Measure	Document Reference	Website
Council Directives	Urban Waste Water Treatment	To harmonise measures on urban waste water treatment throughout the Community.	CD 91/271/EEC concerning urban waste water treatment (21/05/91) Amended by Comm Dir 98/15/EC (27/02/98).	■ Off Jnl L135, 30/05/91 ■ Off Jnl L67, 07/03/98	http://europa.eu.int/scadplus/leg/en/lvb/l28008.htm
	Pollution Caused by Nitrates from Agricultural Sources	To reduce or prevent water pollution caused or induced by nitrates from agricultural sources.	CD 91/676/EEC concerning the protection of waters against pollution caused by nitrates from agricultural sources (12/12/91).	■ Off Jnl L375, 31/12/91	http://europa.eu.int/scadplus/leg/en/lvb/l28013.htm
	Integrated Pollution Prevention & Control	The purpose of this Directive is to achieve integrated prevention & control of pollution arising from the activities listed in Annex I. It lays down measures designed to prevent or, where that is not practicable, to reduce emissions in the air, water & land from such activities, including measures concerning waste, in order to achieve a high level of protection of the environment taken as a whole, without prejudice to Directive 85/337/EEC & other relevant Community provisions.	Council Directive 96/61/EC concerning integrated pollution prevention & control (24/09/96).	■ Off Jnl L257, 10/10/96	http://europa.eu.int/smartapi/cgi/sga_doc?smartapi!celexplus!prod!CELEXnumdoc&lg=en&numdoc=31996L0061
	Quality of Drinking Water: New Requirements	To define the essential quality standards which water intended for human consumption must meet.	CD 98/83/EC on the quality of water intended for human consumption (03/11/98).	■ Off Jnl L330, 05/12/98	http://europa.eu.int/scadplus/leg/en/lvb/l28079.htm

	Framework Directive in the Field of Water Policy	To establish a Community Framework for the protection of inland surface waters, coastal waters & groundwater, in order to prevent & reduce pollution, promote sustainable water use, protect the aquatic environment, improve the status of aquatic ecosystems & mitigate the effects of floods & droughts.	Directive 2000/60/EC of the European Parliament & the Council, establishing a framework for Community action in the field of water policy (23/10/00) Amended by Decision No. 2455/2001/EC of the European Parliament & the Council (20/11/01).	▪ Off Jnl L327, 22/12/01 ▪ Off Jnl L331, 15/12/01	http://europa.eu.int/scadplus/leg/en/lvb/l28002b.htm
Council Decisions	Barcelona Convention: Protecting the Mediterranean Sea	Reducing pollution in the region of the Mediterranean Sea.	Council Decision 77/585/EEC concluding the Convention for the protection of the Mediterranean Sea against pollution & the protocol for the prevention of the pollution of the Mediterranean Sea by dumping from ships & aircraft (25/07/77).	▪ Off Jnl L240, 19/09/77	http://europa.eu.int/scadplus/leg/en/lvb/l28084.htm
			Council Decision 81/420/EEC on the conclusion of the protocol concerning co-operation in combating pollution of the Mediterranean Sea by oil & other harmful substances in cases of emergency (19/05/81).	▪ Off Jnl L162, 19/06/81	
			Council Decision 83/101/EEC concluding the protocol for the protection of the Mediterranean Sea against pollution from land-based sources (28/02/83).	▪ Off Jnl L67, 12/03/83	
			Council Decision 84/132/EEC on the conclusion of the protocol concerning the Mediterranean's specially protected areas (01/03/84).	▪ Off Jnl L68, 10/03/84	

Continued

Table 14.5.2 European Level Water Policy. *Continued*

Policy Type	Programme Title	Policy Objective	Community Measure	Document Reference	Website
Council Decisions	Helsinki Convention on the Protection of the Baltic Sea	To help reduce pollution in the Baltic Sea area.	Council Decision 94/156/EC on the accession of the Community to the Convention on the protection of the marine environment of the Baltic Sea (Helsinki Convention) (21/02/94). Council Decision 94/157/EC on the conclusion, on behalf of the Community, of the convention on the protection of the marine environment of the Baltic Sea area (Helsinki Convention as revised in 1992).	▪ Off Jnl L73, 16/03/94	http://europa.eu.int/scadplus/leg/en/lvb/l28089.htm
	Helsinki Convention: Transboundary Watercourses & International Lakes	To prevent & control pollution of transboundary watercourses & international lakes by developing international co-operation.	Council Decision 95/308/EC on the conclusion, on behalf of the Community, of the Convention on the protection & use of transboundary watercourses & international lakes (Helsinki Convention) (27/07/95).	▪ Off Jnl L186, 05/08/95	http://europa.eu.int/scadplus/leg/en/lvb/l28059.htm
	Paris Convention	To prevent & eliminate pollution & to protect the maritime area against the adverse effects of human activities.	Council Decision 98/249/EC on the conclusion of the Convention for the protection of the marine environment of the North East Atlantic (Paris Convention) (07/10/97).	▪ Off Jnl L104, 03/04/98	http://europa.eu.int/scadplus/leg/en/lvb/l28061.htm

Convention for the Protection of the Rhine	To strengthen co-operation between the Community & the Rhine riparian states in order to preserve & improve the ecosystem of the river.	Council Decision 2000/706/EC concerning the conclusion, on behalf of the Community, of the Convention for the protection of the Rhine (07/11/00).	■ Off Jnl L289, 16/11/00	http://europa.eu.int/scadplus/leg/en/lvb/l28115.htm
Accidental Marine Pollution	To establish a Community framework for co-operation between Member States in the field of accidental or deliberate marine pollution.	Decision No. 2850/2000/EC of the European Parliament & the Council, on setting up a Community framework for co-operation in the field of accidental or deliberate marine pollution (20/12/00).	■ Off Jnl L332, 28/12/00	http://europa.eu.int/scadplus/leg/en/lvb/l28085.htm
Priority Substances in the Field of Water Policy	To establish a list of priority substances in the field of water policy, for which quality standards & measurements for the reduction of emission controls will be set at Community level.	Decision No. 2455/2001/EC of the European Parliament & the Council, establishing the list of priority substances in the field of water policy & amending Directive 2000/60/EC (20/11/01).	■ Off Jnl L331, 15/12/01	http://europa.eu.int/scadplus/leg/en/lvb/l28108.htm
Cooperation in Preventing Pollution from Ships &, in Cases of Emergency, Combating Pollution of the Mediterranean Sea	The Protocol updates the legal instruments of the Barcelona Convention, to include cooperation on the prevention of pollution from ships, to make more effective co-operation in response to pollution incidents & to promote the implementation of the applicable international regulations.	Council Decision on the conclusion, on behalf of the European Community, of the Protocol to the Barcelona Convention for the Protection of the Mediterranean Sea against Pollution, concerning co-operation in preventing pollution from ships and, in cases of emergency, combating pollution of the Mediterranean Sea (29/04/04).	■ C Dec 2004/575/EC ■ Off Jnl L261/40, 06/08/04	http://europa.eu.int/eur-lex/pri/en/oj/dat/2004/l_261/l_26120040806en00400040.pdf

Continued

Table 14.5.2 European Level Water Policy. *Continued*

Policy Type	Programme Title	Policy Objective	Community Measure	Document Reference	Website
Council Communications	Community Water Protection & Management Policy	To establish an integrated water management policy.	Commission Communication on Community water policy (1996).	▪ COM (96) 59	http://europa.eu.int/scadplus/leg/en/lvb/l28002a.htm
	Pricing & Long-term Management of Water	To describe the questions & options in connection with defining water pricing policies enabling the sustainability of water resources to be boosted.	Communication from the Commission to the Council, European Parliament & the Economic & Social Committee, on pricing & sustainable management of water resources (27/7/00).	▪ COM (2000) 477	http://europa.eu.int/scadplus/leg/en/lvb/l28112.htm
	Danube–Black Sea Region	Highlight actions to be taken to improve environmental quality in the Danube–Black Sea region & the outline of a strategy aimed at protecting the environment of the region.	Communication from the Commission – Environmental co-operation in the Danube–Black Sea region (2001).	▪ COM (2001) 615	http://europa.eu.int/scadplus/leg/en/lvb/l28016.htm
	Water Management in Developing Countries Policy & Priorities for EU Development Co-operation	This Communication guides the European Community's support to water resources management in developing countries for achieving the main development goals laid down in the Treaty establishing the European Community – poverty reduction, sustainable economic & social development, integration of developing countries into the global economy.	Communication from the Commission to the Council & the European Parliament on water management in developing countries policy & priorities for EU development co-operation (12/03/02).	▪ COM (2002) 132	http://europa.eu.int/eur-lex/en/com/cnc/2002/com2002_0132en01.pdf

Towards a Strategy to Protect & Conserve the Marine Environment	In particular the present document will: ■ review the current information concerning the environmental status of the seas & oceans & identify the main threats; ■ review the present situation with regard to the development & implementation of policies to control these threats, both within the EU & at regional & international level; ■ identify gaps in knowledge & review the present situation with regard to monitoring assessment & research; ■ draw operational conclusions as to what needs to be done to improve the current situation; ■ identify the appropriate operational & institutional objectives for the EU; ■ set out an action plan & a work programme for the Commission, the Member States, Candidate Countries & all relevant stakeholders to work together between now & 2004 in order to define & develop a thematic strategy for the protection & sustainable use of the marine environment.	Communication from the Commission to the Council & the European Parliament – Towards a Strategy to Protect & Conserve the Marine Environment (2002).	■ COM (2002) 539	http://europa.eu.int/scadplus/leg/en/lvb/l28129.htm http://europa.eu.int/smartapi/cgi/sga_doc?smartapi!celexplus!prod!DocNumber&lg=en&type_doc=COMfinal&an_doc=2002&nu_doc=539
Flood Risk Management: Flood Prevention, Protection & Mitigation	Flood events have the potential to undermine the EU's drive towards sustainable development. Moreover, the frequency & severity of severe flood events is likely to increase significantly in the future due to global warming. The problem can only be tackled comprehensively through concerted action in each affected river basin & coastal area. This communication highlights the EU approach to minimising flood risk, through protection & mitigation.	Communication from the Commission to the Council, the European Parliament, the European Economic & Social Committee & the Committee of the Regions – Flood Risk Management: Flood Prevention, Protection & Mitigation (12/07/04).	■ COM (2004) 472	http://europa.eu.int/eur-lex/en/com/cnc/2004/com2004_0472en01.pdf

Continued

Table 14.5.2 European Level Water Policy. *Continued*

Policy Type	Programme Title	Policy Objective	Community Measure	Document Reference	Website
Miscellaneous	Towards Sustainable Water Resources Management – A Strategic Approach	The centrepiece of the Guidelines is a 'strategic approach for the equitable, efficient & sustainable management of water resources'. The approach is based on internationally agreed core principles concerning the need to protect the eco-system, & to extend the health-giving & productive properties of freshwater resources equitably, efficiently & sustainably among humankind, with special emphasis on poorer & underserved people. It provides a comprehensive framework for all activities relating to water resources development; its application involves a radical change in traditional attitudes towards water management, & the introduction of good practice consistent with the internationally agreed core principles. The strategy covers the full cycle of activity, from national policy-making through implementation of programmes & projects & the subsequent operation & maintenance of services.	DG Environment Report (1998).	■ DG XI Report	http://europa.eu.int/comm/development/body/publications/water/en/tableofcontents_en.htm

Towards a European Integrated Coastal Zone Management (ICZM) Strategy – General Principles & Policy Options	This present document has two purposes: ▪ To compile and diffuse the principal policy lessons emerging from the experiences of the Demonstration Programme to date; ▪ To stimulate debate and develop consensus on a European ICZM Strategy, designed to reverse the trend towards unsustainability that is presently pervasive in coastal zones across Europe.	A European Commission Report to the Environment Directorate-General (April 1999).	▪ DG XI Report	http://europa.eu.int/comm/environment/iczm/vol1.pdf
Lessons from the European Community Demonstration Programme on Integrated Coastal Zone Management (ICZM)	The Demonstration Programme has yielded a wealth of technical information about ICZM mechanisms & solutions to specific problems faced in the coastal zone. However, as the purpose of the present document is to generate discussion concerning a possible strategy for promoting ICZM in Europe, only those lessons with policy implications are discussed, leaving the more technical details to the final reports of the thematic studies. Furthermore, this document is focused on the cross-cutting, non-sectoral issues of common interest to the range of stakeholders in the coastal zone, although some particularly important issues of a more specific sectoral, geographic or technical nature are introduced in Annex I of this document.	A European Commission Report to the Environment Directorate-General (April 1999).	▪ DG XI Report	http://europa.eu.int/comm/environment/iczm/vol2.pdf
EU Focus on Clean Water	The European Union is conducting an extensive range of essentially legislative measures in order to ensure a sustainable management of water quantity & quality in Europe. This policy is addressed in this brochure. It is intended to inform local & regional players, NGOs, policymakers at all levels, social partners & consumers, as well as citizens.	Report to the Directorate-General Environment, Nuclear Safety & Civil Protection (1999).	▪ DG XI Report	http://europa.eu.int/comm/environment/eufocus/clean_water.pdf

Continued

Table 14.5.2 European Level Water Policy. *Continued*

Policy Type	Programme Title	Policy Objective	Community Measure	Document Reference	Website
Miscellaneous	Sustainable Water Use in Europe Part 1: Sectoral Use of Water	This report is the first in a series on 'Sustainable Water Use in Europe', & as such examines & assesses the available information on the Sectoral Use of Water. The report aims to inform & provide information for policy & decision-makers at both the national & European level. It will also be of value to Non-Governmental Organisations & of general interest to members of the public. The report highlights the need to improve existing information by establishing reliable definitions, a common understanding, & quantitative & consistent records on a European scale. Reliable & comparable information is required to ensure that existing policies are effective & that any new policies are targeted correctly so that there can be a sustainable use of water in Europe.	European Environment Agency Report (1999).	▪ EEA Report	http://reports.eea.eu.int/binaryeenviasses01pdf/en/enviasses01.pdf
	Sustainable Use of Europe's Water? State, Prospects & Issues	This report is intended to give ministers, senior civil servants, other policy-makers & others with an interest to protect waters a broad overview of the major water issues in Europe. The report provides, for each issue, a summary of scientific & technical knowledge of the problem, an analysis of its causes, an indication of the actions	European Environment Agency Report (Oct 2000).	▪ EEA Report	http://reports.eea.eu.int/water_assmnt07/en/water_assmnt07_en.pdf

	taken & their effects, & an assessment of what further needs to be done. It is written for the non-scientist &, to help readers gain the maximum of relevant information in the most efficient way, much of the content is provided in colour-coded text boxes.			
Maritime Safety: Compensation Fund for Oil Pollution Damage in European Waters	To ensure fair compensation for oil pollution damage occurring in European waters by complementing the existing international liability & compensation regime and to provide for financial penalties to be imposed on any person liable for causing oil pollution.	Proposal for a Regulation of the European Parliament & the Council, on the establishment of a fund for the compensation for oil pollution damage in European waters & related measures (06/12/00).	■ COM (2000) 802	http://europa.eu.int/scadplus/leg/en/lvb/l24238.htm
Study on the Impact of Community Environment – Water Policies on Economic & Social Cohesion	The study aims at providing material relevant to the production of a *Second Report on Economic & Social Cohesion*, in relation to the impact of environmental policies in the area of water policy. Section 2 provides relevant background material & a general introduction to the issues. Section 3 presents the results of the work carried out in identifying the linkages between water policies & cohesion units. Sections 4, 5, 6 & 7 present the quantification of first order & final effects of the environment–water policies on economic & social cohesion for the selected linkages.	Report for the DG Regional Policy (Feb 2001).	■ DG Regional Policy Report	http://europa.eu.int/comm/regional_policy/sources/docgener/studies/pdf/enviwat.pdf

Continued

Table 14.5.2 European Level Water Policy. *Continued*

Policy Type	Programme Title	Policy Objective	Community Measure	Document Reference	Website
Miscellaneous	Sustainable Water Use in Europe Part 2: Demand Management	This is the second report from the European Environment Agency on sustainable water use in Europe & focuses on how the demand side of water management is being approached across Europe. The report is concerned mainly with measures which aim to achieve increases in the efficiency of use of water over the medium to long term. A distinction is made between urban, industrial & agricultural uses since these vary considerably & water demand management programmes need to be designed specifically for each sector. In addition to sectoral differences, there are considerable differences between & within countries depending on socio-economic, geographical & climatological factors.	European Environment Agency Report (April 2002).	▪ EEA Report	http://reports.eea.eu.int/Environmental_Issues_No_19/en/Environmental_Issues_No_19.pdf
	Common Implementation Strategy for the Water Framework Directive	The aim of this Strategic Document is to allow, as far as possible, a coherent & harmonious implementation of the framework directive. Most of the challenges & difficulties arising will inevitably be common to all Member States and many of the European river basins are shared, crossing administrative & territorial borders, where a common understanding & approach is crucial to successful & effective implementation. A Common Strategy could limit the risks of bad application of the Directive & subsequent dispute.	DG Environment Report (May 2001).	▪ DG XI Report	http://europa.eu.int/comm/environment/water/water-framework/strategy.pdf

EU Focus on Coastal Zones – Turning the Tide for Europe's Coastal Zones	The increasing demand for coastal resources is leading to their degradation, reduced water quality & quantity, accelerated erosion, accumulation of pollution, loss of fisheries resources, etc. Moreover, this degradation has negative social & economic consequences. All of these factors suggest that Europe's coastal regions are entitled to special attention from the Union's policy-makers. This is why the European Union is working to introduce a co-ordinated policy for the Union's coastal regions. As well as taking steps to improve the EU policies that influence coastal zones, the European Commission is calling on Member States to put in place national strategies for what is known as integrated coastal zone management (ICZM).	A European Commission Report to the Environment Directorate-General (July 2001).	DG XI Report	http://europa.eu.int/comm/environment/iczm/2000brochure_en.pdf
Sustainable Water Use in Europe Part 3: Extreme Hydrological Events – Floods & Droughts	This report presents an overview of the main natural & artificial causes & impacts of extreme hydrological events, such as floods & droughts in European countries. It also gives an overview on policy responses to prevent such events & reduce their impact.	European Environment Agency Report (Aug 2001).	EEA Report	http://reports.eea.eu.int/Environmental_Issues_No_21/en/enviissue21.pdf
Extensive Wastewater Treatment Processes: Adapted to Small & Medium Sized Communities	The purpose of this tool is to facilitate the implementation of the Directive by the development of demonstration efforts & innovative technologies adapted to environmental problems that need to be resolved. Furthermore, the Environment Directorate-General supports the distribution of these techniques, via the development of technical exchanges & advisories. This document & the development of aids such as structural funds & cohesion funds are examples of this.	A European Commission Report to the Environment Directorate-General (April 2002).	DG XI Report	http://europa.eu.int/comm/environment/water/water-urbanwaste/waterguide_en.pdf

Continued

Table 14.5.2 European Level Water Policy. *Continued*

Policy Type	Programme Title	Policy Objective	Community Measure	Document Reference	Website
Miscellaneous	Implementation of Council Directive 91/271/EEC: Urban Wastewater Treatment, as amended by Commission Directive 98/15/EC – Summary Report	The directive is one of the cornerstones of Community water policy and its aim is to protect the environment from the adverse effects of urban wastewater discharges. This report updates implementation progress at the critical enforcement date of Dec 1998.	A European Commission Report to the Environment Directorate-General (May 2002).	■ DG XI Report	http://europa.eu.int/comm/environment/water/water-urbanwaste/report2/2001_1669_en.pdf
	Phosphates & Alternative Detergent Builders – Final Report	The European Commission (EC) has implemented this study to address the current use of phosphates in detergents throughout the European Union (EU) & recommend appropriate measures to improve the current situation. The study covers the fifteen Member States of the EU & the three accession countries Poland, Hungary & the Czech Republic. The aim of the study is to investigate the costs & benefits of substituting phosphorus in detergents with other appropriate builders & to provide recommendations on the most appropriate method of reducing phosphorus concentrations in surface waters, through either improving wastewater treatment, banning the use of phosphates as detergent builders, or a combination of the two approaches.	DG Environment Report (June 2002).	■ DG XI Report	http://europa.eu.int/comm/environment/water/phosphates.pdf

Proposal for a Directive on the Quality of Bathing Water	Initiatives to revise the Bathing Water Directive actually started in 1994, with the Commission presenting a Proposal for a revision. This Proposal saw first reading in the European Parliament, but was never further negotiated by the Council. It was preferred that a new Directive was developed, based on new scientific evidence and on a broad consultation. However, this Proposal gave the incentive for further studies & developments on bathing water quality, as regards parameters as well as the managerial approach.	Proposal for a Directive of the European Parliament & of the Council concerning the quality of bathing water (24/10/02).	■ COM (2002) 581	http://europa.eu.int/eur-lex/en/com/pdf/2002/com2002_0581en01.pdf
Implementation of Council Directive 91/676/EEC: Protection of Waters Against Pollution Caused by Nitrates from Agricultural Sources – Synthesis from year 2000 Member States Reports	This report was prepared in order to provide an overview of the current situation with regard to the directive together with possible pathways for the future. It illustrates, with some case studies, the positive effects of some farm practices on the quality of water. However, it is emphasised that there is a considerable time lag between improvements at farm and soil level and a response in water-body quality.	A Report to the DG Environment on the implementation of CD 91/676/EEC (Nov 2002).	■ DG Environment Report	http://europa.eu.int/comm/environment/water/water-nitrates/91_676_eec_en.pdf
Water an Essential Resource: LIFE & the New European Water Policy	This first *LIFE Focus* publication presents numerous examples of projects dealing with water, carried out under LIFE-Environment & LIFE-Nature. The presentation of projects, first-hand descriptions or reports from the field are intended to show the reader that the protection of water is a matter for us all. This brochure reflects the efforts of many players in the field, from all over Europe, to find solutions that will provide future generations with the water resources they need.	A European Commission Report to the Environment Directorate-General (Nov 2002).	■ DG XI Report	http://europa.eu.int/comm/environment/life/result/water_en.pdf

Continued

Table 14.5.2 European Level Water Policy. *Continued*

Policy Type	Programme Title	Policy Objective	Community Measure	Document Reference	Website
Miscellaneous	Eutrophication & Health	The purpose of this booklet is to describe in a simple way the causes of eutrophication, the environmental effects, the associated nuisances & health risk as well as the preventive & mitigating measures.	World Health Organisation Report for the European Commission (2002).	■ WHO Report	http://europa.eu.int/comm/environment/water/water-nitrates/eutrophication.pdf
	The Water Framework Directive: Tap into it!	Water is essential to life on the planet. Human existence as well as economic activities are totally dependent upon this precious resource & yet, at a global level, water is often a limited resource. This brochure is intended to provide an overview on the goals & provisions of the European Water Framework Directive.	EU Environment Brochure (2002).	■ EU Env Brochure	http://europa.eu.int/comm/environment/water/water-framework/pdf/brochure_en.pdf
	Economic Assessment of Groundwater Protection: A Survey of the Literature Final Report	The main objective of this literature review has been to report results from case studies that have assessed the costs & the benefits of groundwater protection & remediation, either as a qualitative description or, where possible, in monetary terms. In assessing the costs & benefits of groundwater protection, this paper takes account of the focus of the Water Framework Directive, which is primarily concerned with the qualitative aspects of groundwater protection. Quantitative aspects will only be considered in so far as they have a connection to qualitative problems.	DG Environment Report (April 2003).	■ DG XI Report	http://europa.eu.int/comm/environment/water/water-framework/pdf/grounwater_protection.pdf

Carrying forward the Common Implementation Strategy for the Water Framework Directive – Progress & Work Programme for 2003–04	The European Community and the Norwegian Water Directors identified a number of elements for a Common Implementation Strategy of the Water Framework Directive (WFD) such as the need to integrate activities on different horizontal issues for the effective development of river basin management plans & implementation of the WFD & the need to establish working groups and develop informal guiding & supporting documents on key aspects of the WFD.	A European Commission Report to the Environment Directorate-General (June 2003).	DG XI Report	http://europa.eu.int/comm/environment/water/water-framework/strategy2.pdf
Ship-Source Pollution & the Introduction of Sanctions, Including Criminal Sanctions, for Pollution Offences	This proposal consists of two distinct, but equally important measures. Firstly, it incorporates the applicable international discharge rules for ship-source pollution into Community law & regulates the enforcement of these rules in detail. This part of the proposal includes certain important new features, notably the inclusion of violations that have taken place in the high seas (sea areas beyond the jurisdiction of any State). Secondly, the proposed Directive establishes that violations of the discharge rules shall be criminal offences & provides guidance on the nature of the penalties to be awarded.	Proposal for a Directive of the European Parliament and of the Council on ship-source pollution and on the introduction of sanctions, including criminal sanctions, for pollution offences (2003).	COM (2003) 92	http://europa.eu.int/smartapi/cgi/sga_doc?smartapi!celexplus!prod!DocNumber&lg=en&type_doc=COMfinal&an_doc=2003&nu_doc=92
Common Implementation Strategy for the Water Framework Directive (2000/60/EC) – Economics & the Environment: The Implementation Challenge of the Water Framework Directive	This document aims at guiding experts & stakeholders in the implementation of the Directive 2000/60/EC establishing a framework for Community action in the field of water policy (the Water Framework Directive). It focuses on the implementation of its economic elements in the broader context of the development of integrated river basin management plans as required by the Directive.	Report produced by Working Group 2.6 – WATECO (2003).	Working Group Report	http://www.wrrl-info.de/docs/Guidance_doc_6_intercalibration_klein.pdf

Continued

Table 14.5.2 European Level Water Policy. *Continued*

Policy Type	Programme Title	Policy Objective	Community Measure	Document Reference	Website
Miscellaneous	Water for Life EU Water Initiative: International Co-operation – From Knowledge to Action	This brochure shows the European Commission's work on issues addressed by the Water Initiative & its commitment to pursue its goals by partnering with others.	EU Environment Brochure (2003).	▪ EU Env Brochure	http://europa.eu.int/comm/research/water-initiative/pdf/water-for-life_en.pdf
	Proposal for a Directive on the Protection of Groundwater Against Pollution	In addition to the rules in Directive 80/68/EEC on the protection of groundwater against pollution caused by certain dangerous substances, groundwater protection is also a feature of the Water Framework Directive 2000/60/EC (WFD), which is the basic legislation for the protection of Europe's aquatic environment. Article 17 of the WFD requires that on the basis of a proposal from the Commission, the European Parliament & the Council shall adopt specific measures to prevent & control groundwater pollution by defining common criteria on good chemical status & on quality trends. This proposal for a Groundwater Daughter Directive meets that requirement.	Proposal for a Directive of the European Parliament & of the Council on the protection of groundwater against pollution (19/09/03).	▪ COM (2003) 550	http://europa.eu.int/eur-lex/en/com/pdf/2003/com2003_0550en01.pdf
	Reporting for Water – Concept Report Towards a Shared Information System for Water	This document introduces the reporting requirements of the WFD & summarises existing reporting requirements. It goes on to discuss the purposes of reporting & the uses to which information supplied can be put. Opportunities for the sharing of data & information are identified & the principles of a shared & accessible data & information management system are outlined & the potential for coherence with other reporting mechanisms is discussed.	DG Environment Report (Nov 2003).	▪ DG XI Report	http://europa.eu.int/comm/environment/water/pdf/concept_report.pdf

Europe's Water: An Indicator Based Assessment	This report assesses the quality & quantity of Europe's water. Its geographical scope is the European Union, EFTA (European Free Trade Association) & EU accession & candidate countries. Four water issues are assessed: ecological quality, eutrophication & organic pollution of water, hazardous substances & water quantity. This is done on the basis of 57 indicators selected for their representativeness & relevance. Using these indicators, the report seeks to answer a number of questions that have been formulated to assess whether the broad objectives & targets of EU water policy are being achieved & to indicate where policy gaps may be occurring.	European Environment Agency Report (Nov 2003).	EEA Report	http://reports.eea.eu.int/topic_report_2003_1/en/Topic_1_2003_web.pdf
Proposal for a Directive on Pollution Caused by Certain Dangerous Substances Discharged into the Aquatic Environment of the Community	The purpose of this proposal is to undertake a codification of Council Directive 76/464/EEC on pollution caused by certain dangerous substances discharged into the aquatic environment of the Community. The new Directive will supersede the various acts incorporated in it; this proposal fully preserves the content of the acts being codified & hence does no more than bring them together with only such formal amendments as are required by the codification exercise itself.	Proposal for a Directive of the European Parliament & of the Council on pollution caused by certain dangerous substances discharged into the aquatic environment of the Community (07/01/04).	COM (2003) 847	http://europa.eu.int/eur-lex/en/com/pdf/2003/com2003_0847en01.pdf
Proposal for a Directive on the Quality of Fresh Waters Needing Protection or Improvement in order to Support Fish Life	The purpose of this proposal is to undertake a codification of Council Directive 78/659/EEC on the quality of fresh waters needing protection or improvement in order to support fish life. The new Directive will supersede the various acts incorporated in it; this proposal fully preserves the content of the acts being codified & hence does no more than bring them together with only such formal amendments as are required by the codification exercise itself.	Proposal for a Directive of the European Parliament & of the Council on the quality of fresh waters needing protection or improvement in order to support fish life (19/01/04).	COM (2004) 19	http://europa.eu.int/eur-lex/en/com/pdf/2004/com2004_0019en01.pdf

Continued

Table 14.5.2 European Level Water Policy. *Continued*

Policy Type	Programme Title	Policy Objective	Community Measure	Document Reference	Website
Miscellaneous	Implementation of Council Directive 91/271/EEC concerning Urban Waste Water Treatment, as amended by Commission Directive 98/15/EC	The report presents the measures implemented by the Member States in relation to the deadline of 31 December 2000. In addition, it presents the improvements having taken place by 2002 concerning agglomerations discharging into sensitive areas & subject to the deadline of 31 December 1998. Beyond the updated situation of waste water treatment infrastructure, the report also includes information on the treatment performance of waste water treatment plants discharging their effluents into sensitive areas in the year after the deadline had expired. Furthermore the report gives an overview of waste water treatment in major cities, as well as a general review of the progress of Member States towards the implementation of the directive.	Report from the Commission to the Council, the European Parliament, the European Economic & Social Committee & the Committee of the Regions Implementation of Council Directive 91/271/EEC concerning urban waste water treatment, as amended by Commission Directive 98/15/EC (23/04/04).	■ COM (2004) 248	http://europa.eu.int/eur-lex/en/com/rpt/2004/com2004_0248en01.pdf

Table 14.5.3 UK Level Water Policy.

Policy Type	Document Title	Objective	Contents	Website
Strategies & Reports	Public Health & Drinking Water – Preventing *Cryptosporidium* Getting into Public Drinking Water Supplies (June 1998)	The Secretary of State for the Environment, Transport & the Regions, & the Secretary of State for Wales, are minded to make new Regulations. They believe that these Regulations are needed to provide additional protection to the public by reducing the risk of *Cryptosporidium* being found in drinking water supplies. But before reaching a decision on whether to make the Regulations they wish to know the views of interested parties on the broad approach that it is proposed to adopt. This consultation paper explains the proposed new Regulations, & seeks views on a number of specific points.	Introduction; Background; The nature of the problem; Current legislation; The policy solution proposed; Other aspects of the new regulations; Regulatory appraisal; Issues for consultation; Annexes.	http://www.defra.gov.uk/environment/water/industry/consult/crypto/index.htm
	Potential Costs & Benefits of Implementing the Proposed Water Resources Framework Directive (Jan 1999)	In September 1997, the Department of the Environment, Transport & the Regions (DETR) commissioned a study designed to examine the costs & associated benefits of implementing the WFD in the UK. Although many of the detailed technical specifications required to implement the directive were not in place at this time, the study was intended to provide the DETR with an initial assessment of the potential impact of the WFD in the UK. The scope & nature of the study was designed to be flexible so that it could reflect changes made to the WFD as & when they became apparent. Due to the uncertainty, during the study, about the way in which the Directive would finally be framed & implemented, a number of assumptions had to be made during the course of the project.	Executive summary; Introduction; Objectives & overall approach; Administrative arrangements; Planning processes & information; Monitoring; Assessment of good water status; Water status & gap assessment in England & Wales; Water status & gap assessment in Northern Ireland; Water status & gap assessment in Scotland; Cost assessment for water status improvements; Benefit assessment methodology; Overall appraisal.	http://www.defra.gov.uk/environment/water/wfd/pdf/wrc99.pdf

Continued

Table 14.5.3 UK Level Water Policy. *Continued*

Policy Type	Document Title	Objective	Contents	Website
Strategies & Reports	Water Prices & the Environment – Government Response to the Environmental Audit Commission Seventh Report (06/02/01)	The Government welcomes the Environmental Audit Committee's report on water prices and the environment, published on 14 November 2000. The report has made an important contribution to the understanding & further improvement of the process for setting water industry price limits in England and Wales.	Introduction; Response to specific recommendations; The outcome; Wider issues.	http://www.defra.gov.uk/environment/response/waterprice/index.htm
	Addressing the Water Crisis – Healthier & More Productive Lives for Poor People (March 2001)	This paper is about water & its links to poverty elimination & to the achievement of the international development targets (IDTs). Sustainable access to safe water for drinking, and water for agriculture & the environment, plays an important part in achieving a number of key UN targets. This paper looks at the water supply & sanitation & water for food in the context of broader water resource issues.	Foreword; Executive summary; The challenges; Target statements; Experience to date; Meeting the challenge; Priorities for DfID (Department for International Development); Monitoring progress; Annex 1: Global & regional international development indicators; Annex 2: Water & sanitation coverage by country; Annex 3: Water resource & water abstraction data by country.	http://www.dfid.gov.uk/Pubs/files/tspwater.pdf
	To What Degree Can the October/November 2000 Flood Events be Attributed to Climate Change? (March 2001)	Following the severe floods of Oct/Nov 2000, there is a general public perception that both the occurrence & magnitude of flooding & associated extreme rainfall in the UK is increasing. There has also been speculation that the autumn 2000 floods were themselves 'caused by' climate change. There is therefore an urgent need to determine how unusual the flooding episode of late 2000 was, & whether or not it can be linked to climate change.	Executive summary; Background & purpose of the study; Research methods; Key scientific findings; Assessment of the evidence; Conclusions; Recommendations for future research; References.	http://www.defra.gov.uk/environ/fcd/floodincidents/fd2304fr.pdf

National Appraisal of Assets at Risk from Flooding and Coastal Erosion, including the potential impact of climate change (July 2001)	The Department of Environment, Food & Rural Affairs has commissioned this report to update the findings of the initial National Assets at Risk from Flooding & Coastal Erosion report produced in June 2000. This report extends the original to include: ■ Damage estimates for Wales. ■ The incorporation of additional data on Flood Defences. ■ Assessment of agricultural losses. ■ Consideration of the impact of road transportation disruption. ■ Provision of preliminary estimates of the impacts of future climate change.	Introduction; Approach; Results; Extent of risk; Do Nothing; Maintain current standards; Meeting indicative standards; Maintaining present investment levels; Influence of climate change on flooding; Conclusions & recommendations; Appendix A: Mapped results.	http://www.defra.gov.uk/environ/fcd/policy/NAAR1101textonly.pdf
Cryptosporidium in Water Supplies (July 2001)	The Report is written in the context of the administrative arrangements for water supply, health care & local government in England & Wales & in Scotland. The Water Supply Regulations are administered in England & Wales by the DWI (Drinking Water Inspectorate) & the Scottish Office Environment Department & the Northern Ireland Environment & Heritage Department DWI have broadly similar functions in Scotland & Northern Ireland respectively. This Report reviews knowledge gathered & research carried out since the 1995 report, the lessons learnt from the analysis of outbreaks of water-borne cryptosporidiosis, the functioning of Incident Management Teams (IMTs) & Outbreak Control Teams (OCTs), the limitations of epidemiology & analysis, & the potential risks to groundwater. Where appropriate, recommendations for improvement are made. Advice is given to water utilities, IMTs & OCTs, immunocompromised people, the food industry & users of private water supplies.	Summary & recommendations; Introduction; Lessons learnt from outbreaks of water-borne cryptosporidiosis; Assessment of contamination risk from groundwater resources; Advice to water utilities; Advice on management of water-borne outbreaks of cryptosporidiosis; Guidance on the epidemiological investigation of outbreaks of infection; Advice to the immunocompromised individual; Current therapeutic approaches to cryptosporidiosis; Advice on private water supplies; Review of research; Appendices.	http://www.dwi.gov.uk/pubs/bouchier/index.htm

Continued

Table 14.5.3 UK Level Water Policy. *Continued*

Policy Type	Document Title	Objective	Contents	Website
Strategies & Reports	Learning to Live with Rivers – ICE Report (Nov 2001)	In the light of experience from the autumn 2000 floods the Government invited the President of the Institution of Civil Engineers to establish a Commission to consider what approaches to managing fluvial flood risk are appropriate to the twenty-first century, in particular to review: ▪ current methods of estimating & reducing flood risk ▪ whether a more strategic catchment-based approach to fluvial flood alleviation is appropriate ▪ the impact of flood defences on the natural & built environment ▪ whether flood risk management can make more use of natural processes ▪ possible impact of climate change ▪ experience in other countries.	Introduction; Flood risk; Catchment & river basin management; Urban drainage; Engineered solutions; Flood management tools; Impact of climate change & catchment dynamics; Skills; International; Supply & delivery; Conclusions & recommendations; Annexes.	http://www.ice.org.uk/rtfpdf/iceflooding.pdf
	Best Practice Principles in the Economic Level of Leakage Calculation (Mar 2002)	This report presents a best practice approach to leakage target setting for water companies in England & Wales. This report builds on the earlier report, *Key principles in the Economic Level of Leakage Calculation*, (January 2001). A parallel report considers the development of Leakage Key Performance Indicators.	Summary; Introduction & overview; Leakage target setting – initial considerations; Costs & benefits of leakage reduction; Calculating the economic level of leakage; Alterative methods of target setting; References; Appendices.	http://www.ofwat.gov.uk/aptrix/ofwat/publish.nsf/AttachmentsByTitle/tripartite_best_practice_principles.pdf/$FILE/tripartite_best_practice_principles.pdf
	Leakage Performance Indicators (Mar 2002)	This report presents details of the development of a set of Leakage Performance Indicators (LPIs) that can be used to assess leakage management performance in support of an economic level of leakage analysis.	Summary; Introduction; Initial set of LPIs & explanatory factors; Testing of LPIs & explanatory factors using real company data; Final set of LPIs; Appendices.	http://www.ofwat.gov.uk/aptrix/ofwat/publish.nsf/AttachmentsByTitle/tripartite_leakage_performance_indicators.pdf/$FILE/tripartite_leakage_performance_indicators.pdf

Sewage Treatment in the UK: UK Implementation of the EU Urban Waste Water Treatment Directive (25/04/02)	This booklet explains how sewage is treated in the UK & what action is being taken to improve this. It fulfils the UK's obligations under Article 16 of the EU Waste Water Treatment Directive, to produce periodically national situation reports on the collection & treatment of waste water, & the reuse or disposal of the by-product of this treatment, sewage sludge.	Introduction; Background on urban waste water; The urban waste water treatment directive; Industrial discharges (not to sewer); Improvements on the collection of waste water; Improvements in the treatment of waste water; Further improvements to come; Monitoring to ensure protection of the environment; Re-use & disposal of sewage sludge; Investment; What can you do?; Glossary; Further information sources: web sites & documents.	http://www.defra.gov.uk/environment/water/quality/uwwtd/report02/pdf/uwwtreport2.pdf
Safeguarding our Seas: A Strategy for the Conservation & Sustainable Development of our Marine Environment (01/05/02)	This report sets out how we will work to achieve our vision by developing new ideas & initiatives, particularly at the cross-cutting level. New initiatives include: ▪ Protecting important habitats ▪ Improving marine conservation in the UK ▪ Pressing for sustainability ▪ Becoming more integrated ▪ Working more effectively ▪ Improving co-ordination in government ▪ Assessing & monitoring progress.	Foreword; Executive summary; Chapter 1: Our vision & its delivery; Chapter 2: Protecting marine biodiversity; Chapter 3: Integrated coastal management; Chapter 4: Tackling pollution from land-based sources & dumping at sea; Chapter 5: The importance of shipping & ports; Chapter 6: The contribution of offshore activities; Chapter 7: Sustainable fisheries; Chapter 8: Facing up to climate change; Chapter 9: Making the best of marine science; Annex A: Summary of targets & time-scales for delivering our vision; Annex B: Glossary of terms.	http://www.defra.gov.uk/environment/marine/stewardship/pdf/marine_stewardship.pdf
The 2004 Periodic Review: Research into Customers' Views (Aug 2002)	The aim of this customer research project is to inform decision makers on customers' priorities for the water industry by providing a better understanding of customers' views on the scope and pace of potential improvement programmes during 2005 to 2010.	Executive summary; Introduction; Background & context; Environmental priorities; Perceptions of the current levels of water & sewerage services & the need for improvement; The views of small businesses; Appendices.	http://www.defra.gov.uk/environment/water/industry/research/pdf/customer_view.pdf

Continued

Table 14.5.3 UK Level Water Policy. *Continued*

Policy Type	Document Title	Objective	Contents	Website
Strategies & Reports	Directing the Flow: Priorities for the Future of Water Policy (Nov 2002)	This document takes the first step towards setting strategic aims & summarising proposed future priorities & direction for the water environment, water resources, water management & sewerage systems. It takes account of a range of discussions held with key stakeholders & it aims to take a long-term view, by looking at the next 20 years & beyond, & setting the context within which specific actions will need to be taken.	Executive summary; Introduction; Water policy objectives & our vision for the future; General principles – sustainable development & policy integration; Delivery – the story so far; Challenges we now face & responses to them; Summary of action points; Annex A: Existing EU directives with major impacts on water; Annex B: Timetable for implementation & negotiation of EU directives; Annex C: Planning policy guidance documents that refer to water; Annex D: Existing water management plans.	http://www.defra.gov.uk/environment/water/strategy/pdf/directing_the_flow.pdf
	Strategic Review of Diffuse Water Pollution from Agriculture – Discussion Document (April 2003)	Since the strategic review of diffuse water pollution from agriculture started at the beginning of 2002, much work has been done on understanding the problem & developing potential solutions. This document is intended to show the results of the progress that has been made, to set out the current position & to identify the way forward.	Part I – Aims & objectives; Introduction; Aims, objectives & timescales; Part II – The extent of the problem; Impacts of diffuse water pollution; Sources of diffuse water pollution; Extent of problem; Part III – Action needed; Improving land management; Developing our understanding; Policy measures; Monitoring; Next steps; annexes.	http://www.defra.gov.uk/environment/water/quality/diffuse/agri/reports/pdf/dwpa07-a.pdf
	Water Framework Directive – Indicative Costs of Agricultural Measures (July 2003)	In the context of this study examining agricultural impacts these are principally associated with the definition of 'good ecological status' of waters. The initial concerns and areas for work in this study have been highlighted by DEFRA as follows: ▪ To verify a suitable proxy for 'good ecological status';	Executive summary; Introduction; Background; Issues for the calculation of costs & benefits; Summary of agricultural impacts & possible responses; Developing cost estimates; Conclusions & recommendations; Annexes.	http://www.defra.gov.uk/corporate/consult/waterframe3/agricria-full.pdf

	■ To review the WRC (Water Research Centre) (1999) study & RIA (Regulatory Impact Assessment) sections on costs for agriculture; ■ To consider, given the extent to which it is known what agricultural practices contribute to the pollution of water, what measures to mitigate pollution could be adopted, & the costs/loss of income involved; ■ To make a distinction & separate out what will be needed to meet the requirements of the WFD where these are additional to the existing Bathing Waters Directive, Freshwater Fish Directive, and Nitrates Directive; ■ To establish a rough characterisation of the variation in regional impacts – comparing environmental sensitivity with agricultural activities.		
Water Bill – Regulatory Impact Assessment, Environmental & Equal Treatment Appraisals – 2nd Edition (July 2003)	This Regulatory Impact Assessment (RIA) assesses the impact of measures to be included in the Water Bill. It was originally published alongside the *Water Bill: Consultation on Draft Legislation*, in November 2000, and takes into account policy changes outlined in the Government Response published in May 2002.	Regulatory impact assessment; Overview; Abstraction & impounding, drought, information conservation; New regulatory arrangements; Competition; The drinking water inspectorate; Fluoridation of water; Water resale; Flood defence; Reservoirs; Fire hydrants; Coal mine water pollution; Contaminated land; Sewerage; Transfer of discharge consents; Changes to trade effluent regime; Laterals; Requisitioning & mains self-lay; Environmental appraisal; Abstraction & impounding, drought, information, conservation; New regulatory arrangements; Competition; Fluoridation; Coal mine water pollution; Contaminated land; Sewerage; Changes to trade effluent regime; Equal treatment appraisal; Abstraction & impounding, drought, information conservation; New regulatory arrangements; Competition; Water resale.	http://www.defra.gov.uk/environment/water/legislation/pdf/riaupdate_030722.pdf

Continued

Table 14.5.3 UK Level Water Policy. *Continued*

Policy Type	Document Title	Objective	Contents	Website
Strategies & Reports	Structure of the Water Industry in England: Does it Remain Fit for Purpose? (Nov 2003)	This report, which was commissioned in July 2003 by DEFRA and Ofwat, in consultation with the Treasury and DTI, seeks to provide further advice on the implications of recent trends in the pattern of ownership & financing of water companies for the industry's ability to meet Government objectives for the sector.	Executive summary; Purpose, scope & method of project; Government objectives for the water sector; Concerns about recent developments in the sector; Description of the main changes; The Regulatory response to developments; Assessment of public policy issues; Implications for government & regulatory policy; Annexes.	http://www.defra.gov.uk/environment/water/industry/structure-report/finalreport.pdf
	National Assessment of Defence Needs & Costs for Flood & Coastal Erosion Management (NADNAC) – Summary Report (June 2004)	This report summarises analysis undertaken by a team comprising staff from the Private Sector, working with officials from DEFRA's Flood Management Division to appraise flood & coastal defence needs & costs. The study aims to provide a rigorous estimate of the costs & benefits associated with investment in flood & coastal defence infrastructure in England.	Introduction; Approach; Results; Discussion of results; Conclusions; References; Appendix A: Detailed methodology.	http://www.defra.gov.uk/environ/fcd/policy/nadnac0604.pdf
	Drinking Water 2003 – A report by the Chief Inspector Drinking Water Inspectorate (July 2004)	This Report is intended for the reader wanting general information about the work of the Inspectorate & the quality of drinking water supplies. Whilst briefly covering the full range of the Inspectorate's activities, this Report focuses on the extent to which water companies comply with the legal requirements concerning drinking water quality, & the actions taken by the water companies & the Inspectorate where requirements are not being met.	Chief Inspector's statement; Introduction; Regulatory framework; Part 1: Overview of water quality in England & Wales; Part 2: Individual water company performance 2003; Part 3: Inspectorate Objectives & Key Results; Annexes.	http://www.dwi.gov.uk/pubs/annrep03/contents.shtm
	Review of Marine Nature Conservation – Working Group	The Review of Marine Nature Conservation was established in 1999 to examine the effectiveness of the system for protecting nature conservation in the marine environment & develop practical &	Preface; Background to the Review; The wider policy context; Human activity & the marine environment; A marine nature conservation framework;	http://www.defra.gov.uk/wildlife-countryside/ewd/rmnc/pdf/rmnc-report-0704.pdf

	Report to Government (July 2004)	proportionate proposals for its improvement. This report to Government represents the Working Group's conclusions, based upon completing the recommendations from the Interim Report, a pilot study in the Irish Sea, & from the reports of sub-groups created to consider issues of information provision & legislative mechanisms.	Conservation objectives; Implementing the marine nature conservation framework through the appropriate management of human activities; Assessing the impact of human activities; Monitoring; Effective communication & information provision; Enforcement; Governance; Prioritising action; Summary of recommendations; References; Appendices.	
	Cross-Government Review of Water Affordability (Dec 2004)	This report is the result of a cross-government review of water affordability. A steering group, led by DEFRA, & comprising Ofwat, HM Treasury, Department for Work & Pensions, Department of Trade & Industry & Welsh Assembly Government, looked at the way in which lower income households are helped with their water & sewerage charges. A wide range of interested parties from outside government were also invited to participate in the group's work through a series of seminars & bilateral meetings. This report contains the evidence gathered on the scale & nature of water affordability in England & Wales & the Government's decisions on how it, along with Ofwat & the water industry, should progress in tackling the issue.	Executive Summary; Introduction; Section 1: The Evidence Base; 1(a): Is there a water affordability problem in England and Wales? 1(b): Existing measures relating to affordability; 1(c): Conclusions on the extent of the affordability problem; Section 2: Government's decisions on ways forward; Conclusions; Annexes.	http://www.defra.gov.uk/environment/water/industry/affordability/pdf/wateraffordability.pdf
Consultation Papers	Outcomes of the Online Water Quality Consultation (March 2002)	This report provides the main findings of an on-line pre-consultation on the Government's forthcoming water quality strategy. The consultation, commissioned by DEFRA, consisted of an interactive members-only electronic forum giving 250 selected stakeholders & experts the opportunity to contribute their views & concerns about the strategy. The forum, developed & managed by the UK Centre for Economic & Environmental Development (UK CEED), was designed to provide DEFRA with a steer on important issues that may be addressed in future water related consultations.	Executive summary; Methodology; Questions; Open forum; Appendix 1: Response to the forum; Appendix 2: Summary of all responses.	http://www.defra.gov.uk/environment/water/strategy/pdf/water_stakeholder_debate.pdf

Continued

Table 14.5.3 UK Level Water Policy. *Continued*

Policy Type	Document Title	Objective	Contents	Website
Consultation Papers	Second Consultation Paper on the Implementation of the EC Water Framework Directive (2000/60/EC) (Oct 2002)	At the World Summit on Sustainable Development in Johannesburg, the sustainable management of water was recognised as one of the keys to sustainable development, with the Water Framework Directive held up as a model of how to manage water quality in a way that promotes sustainable water use based on the long-term protection of available water resources. This is the second consultation paper on implementation of the Directive in England & Wales. It invites views on the key issues arising from transposition & implementation of the Directive. It will be followed up with a further consultation paper, containing draft regulations, which must be in force by December 2003. Further key deadlines follow, beginning with the need to have completed major preparatory analyses in relation to all waters by the end of 2004.	Executive summary; Chapter 1: Introduction; Chapter 2: Making links; Chapter 3: Summary of the 1st consultation; Chapter 4: Carrying forward the regulatory impact assessment; Chapter 5: Meeting environmental objectives; Chapter 6: River basin districts: identification; Chapter 7: River basin districts: management plans; Chapter 8: River basin districts: characterisation, review of human activity & economic analysis of water use; Chapter 9: Register of protected areas; Chapter 10: Monitoring; Chapter 11: Programme of measures; Chapter 12: Penalties; Chapter 13: Wetlands; Chapter 14: Public information & consultation; Chapter 15: Future consultation plans; Chapter 16: Summary of consultation questions; Appendices.	http://www.defra.gov.uk/environment/consult/waterframe2/pdf/waterframe2.pdf
	Seas of Change – The Government's Consultation Paper to Help Deliver Our Vision for the Marine Environment (Nov 2002)	*Safeguarding our Seas* contained a commitment to consult on whether current arrangements for stakeholder involvement are adequate & on the scope & content of future Marine Stewardship Reports. This document fulfils that commitment. In addition, the Government is taking the opportunity to invite initial views on how the principles underpinning the ecosystem-based approach can be fleshed out & applied in practice; & on strategic goals for the marine environment.	Making progress; Developing the concept of marine stewardship; Future stakeholder involvement; The scope & content of future reports; Consultation procedure; Annex A – list of consultees.	http://www.defra.gov.uk/environment/consult/seas/pdf/seas_of_change.pdf

Reductions for Vulnerable Groups: A Consultation Paper (Feb 2003)	This consultation paper invites views from individuals & organisations on the arrangements for protecting vulnerable water consumers in England. Under the current arrangements certain low-income households with water meters who need to use a great deal of water can apply to have their bills capped at the average for their water company area. This paper discusses a range of options for amending the regulations to deliver more assistance to vulnerable households.	Section 1 Summary; Section 2 Introduction; Section 3 Qualifying benefits/Tax Credits – essential changes; Section 4 Help for families; Section 5 Medical conditions; Section 6 Promoting good practice; Section 7 Responding to this consultation paper; Annexes.	http://www.defra.gov.uk/corporate/consult/vulnerable/waterchrgvulgps.pdf
Third Consultation Paper on the Implementation of the EC Water Framework Directive (2000/60/EC) (Aug 2003)	This third consultation paper invites views on transposition of the EC Water Framework Directive (WFD) in England & Wales, which Member States are required to achieve before the transposition deadline of December 2003 specified in the Directive. The paper takes forward issues raised in the second consultation which was published in October 2002 & explains how the Government intends to meet the transposition requirements of the Directive through Regulations. It also contains a revised partial Regulatory Impact Assessment to update the original estimates of the Directive's benefits and costs.	Executive summary; Chapter 1: Introduction; Chapter 2: Summary of second consultation paper responses & government response; Chapter 3: Commentary on draft regulations; Chapter 4: Next steps; Appendices.	http://www.defra.gov.uk/corporate/consult/waterframe3/consultdoc.pdf
Developing Measures to Promote Catchment-Sensitive Farming: A joint Defra-HM Treasury Consultation (June 2004)	Our aim in this consultation is to take forward the development of measures that will reduce diffuse pollution from agriculture without adding unnecessary regulatory burdens. This will increase the sustainability of English agriculture & allow farming to play its part in making water in England as clean & as healthy as possible. This consultation invites views on the approach & possible measures to improve water quality through catchment-sensitive farming. Catchment sensitive farming aims to improve water quality within a river catchment by managing land in a way that keeps emissions to water to levels consistent with the ecological requirements of receiving waterbodies.	Executive Summary; Summary of Questions; Introduction; Section 1 The Need for Action; Section 2 Catchment-Sensitive Farming; Section 3 The Policy Context; Section 4 Policy Instruments; Section 5 Better Regulation & Costs of Options; Section 6 Outcomes, Targets & Indicators; Annex A Information & Skills; Annex B Voluntary Approaches; Annex C Regulation; Annex D Grant Aid; Annex E Economic Instruments.	http://www.defra.gov.uk/corporate/consult/csf-june04/csf-consultdoc-screen.pdf

Continued

Table 14.5.3 UK Level Water Policy. *Continued*

Policy Type	Document Title	Objective	Contents	Website
Consultation Papers	Making Space for Water: Developing a new Government Strategy for Flood & Coastal Erosion Risk Management in England – A Consultation Exercise (July 2004)	The report highlights the need to allow space for water so that we can manage the adverse consequences for people & the economy that can result from flooding & coastal erosion while achieving environmental & social benefits in line with wider Government objectives. The key issues examined in this report include risk management, strengthening the sustainable approach, planning & building awareness, coastal issues & funding.	Foreword; Summary; How to respond to this consultation exercise; Consultation questions; Introduction: A new strategy; The pressures we currently face; Our vision & aim; Assessing & managing the risk of flooding from rivers & the sea, & of coastal erosion; Strengthening the sustainable approach: rural land use & managed realignment of floodplains & the coast; The role of rural land management; Measures to reduce flood risk through land-use planning; Integration of drainage management in urban areas; Flooding from sewers; Flooding from groundwater; Flooding of and from the transport network; Managing the consequences of flooding through flood resistance & resilience measures; Raising awareness; Flood warning systems & emergency responses; Coastal issues; Funding issues; Next steps; Further information; Glossary; Annexes.	http://www.defra.gov.uk/corporate/consult/waterspace/consultation.pdf
Guidance Notes	Water Quality – A Guide to Water Protection in England (June 2000)	This document outlines what is being done to manage, protect & enhance the quality of our natural water resources such as rivers & lakes in a sustainable way, safeguarding them for future generations. It states who is responsible for this work, what they do & with what results.	Water – A Vital Resource; Water Policy in England: Sustainable use – balancing supply & demand; Water Quality Principles: Preserving, protecting & improving the quality of water & of the environment; A Fragile Resource: Water quality & how it can be affected; Water Quality & the Law: Important legislation affecting water policy; Monitoring Water Quality: Some ways in which water quality is measured; Water Quality in England: The present situation; Guarding the Resource: How do we protect water	http://www.defra.gov.uk/environment/water/quality/guide/water.htm

		quality? Water Quality – The Challenges: Important inputs to the aquatic environment; Who protects the Water of England? Bodies responsible for our water quality.	
Environmental Reporting Guidelines for Company Reporting on Water (Dec 2000)	These guidelines are designed to help businesses to: ▪ measure their water usage; ▪ review water costs, use & discharges & detect possible leaks; ▪ set targets to reduce their water & wastewater usage; ▪ achieve savings & improve performance & competitiveness; ▪ report on their performance to demonstrate to shareholders, consumers & other stakeholders that theirs is a well-run, environmentally responsible business; ▪ build commitment within the business; ▪ meet current & future legislative requirements.	Why reporting is important; The purpose & the scope of these guidelines; How these guidelines work; Step 1: Identify a water management champion; Step 2: Review water costs, use & discharge; Step 3: Set targets for improvement; Step 4: Achieve your targets; Step 5: Report the outcome; Step 6: Repeat the cycle; Annexes.	http://www.defra.gov.uk/environment/envrp/water/pdf/water.pdf
Shoreline Management Plans – A Guide for Coastal Defence Authorities (June 2001)	A Shoreline Management Plan (SMP) provides a large-scale assessment of the risks associated with coastal processes & presents a policy framework to reduce these risks to people & the developed, historic & natural environment in a sustainable manner. In doing so, an SMP is a high-level document that forms an important element of the Strategy for Flood & Coastal Defence of the Department for Environment, Food & Rural Affairs (DEFRA) & the National Assembly for Wales (NAW). The strategy aims to reduce risks by: ▪ encouraging the provision of adequate & cost-effective flood warning systems; ▪ encouraging the provision of adequate, technically, environmentally & economically sound & sustainable flood & coastal defence measures; ▪ discouraging inappropriate development in areas at risk from flooding or coastal erosion.	Introduction; Aims, objectives & general principles; Policy appraisal: Key issues; Procedure for the production of a plan; DEFRA grant-aid, submissions & approvals; Annex 1 Nature conservation designations; Annex 2: Implications of shoreline management policies on habitats & species protected under European legislation; Annex 3: Suggested model brief for revising a Plan; Annex 4: Recommended format & content of a Plan.	http://www.defra.gov.uk/environ/fcd/pubs/smp/revisedsmpguidancefinal.pdf

Continued

Table 14.5.3 UK Level Water Policy. *Continued*

Policy Type	Document Title	Objective	Contents	Website
Guidance Notes	Planning a Future for the Inland Waterways – A Good Practice Guide (Dec 2001)	The inland waterways of England & Wales are playing an increasingly central role in many aspects of urban & rural life. The Government is determined to increase the contribution they make by ensuring that their social, economic & environmental potential is realised. To achieve this, it is vital that the potential of the waterways is clearly understood, & that effective use is made of the planning process to achieve improvement & change. The Guide draws together advice & experience from across the waterways sector, gathered through an extensive literature review & consultation on all aspects of planning & regeneration associated with the waterways. It uses examples & case studies to demonstrate what can be achieved on the waterways & provides practical advice on how successful waterway projects have been promoted & implemented.	Executive summary; Introduction to the guide; Understanding the waterways; Developing the vision for the waterways; Waterways & the planning system; Project delivery; Appendix 1 – case studies; Appendix 2 – quality of life capital; Appendix 3 – bibliography.	http://www.defra.gov.uk/environment/bw/iwaac/pdf/iwaac_screen.pdf
	Preparing for Floods – Interim Guidance for Improving the Flood Resistance of Domestic & Small Business Properties (Oct 2003)	The aim of this guide is to provide guidance to property owners on how they can improve the flood resistance of their properties. The guide is also intended for use by developers, local planning authorities & others involved in construction of new buildings, & renovation of existing buildings, at risk of flooding. If adopted, the principles set out within this guide should help reduce the stress & disruption of flooding & provide a more sustainable approach to flood risk.	Foreword; Introduction; The impact of flooding; Existing property owners; New development; Parliament measures to reduce flood damage – new & existing properties; Summary; Further reading; Environment Agency Guidance.	http://www.odpm.gov.uk/stellent/groups/odpm_buildreg/documents/page/odpm_breg_600451.pdf

Principal Guidance from the Secretary of State to the Director General of Water Services – 2004 Periodic Review of Water Price Limits (Mar 2004)	This guidance conveys the main decisions & policies affecting the 2004 periodic review of water price limits. This guidance from the Secretary of State for Environment, Food & Rural Affairs (DEFRA) to the Director-General of Water Services (Ofwat) is the government's second & principal public contribution to the next periodic review of water price limits. This principal guidance sets out the main decisions on those policies and programmes that are for Ministers to determine, taking account of the information and advice so far to hand, including the proposals & costings put forward by companies in their draft business plans in August 2003, views on these cost estimates from Ofwat & further policy advice from Ofwat & from the other regulators.	Foreword; Key messages; Introduction; Approach; Setting policies for water; The supply-demand balance & water resources; Drinking water; The water environment.	http://www.defra.gov.uk/environment/water/industry/review/pdf/min-guidance04.pdf
Managing Coastal Activities: A Guide for Local Authorities (2004)	This guide has been developed through consultation with local authorities, recreational representatives & users, harbour authorities, the Maritime & Coastguard Agency & others. Whilst focusing on local authorities in England, the guide should also be useful for harbour authorities & other bodies with coastal responsibilities. This guide will help them to: ▪ identify current issues about coastal activities; ▪ promote management solutions through voluntary action or self regulation; ▪ encourage partnership arrangements in managing recreational activities; ▪ encourage a strategic & co-ordinated approach to managing recreational activities; & ▪ promote & disseminate good practice in the planning & management of recreational activities at the coast.	Ministerial foreword; Introduction; Organisations with a role in the coast; Voluntary approaches; Byelaws; Designing a management scheme; Enforcement; Disseminating information; App 1 – Coastal recreational activities; App 2 – Government departments, agencies, LAs & other organisations; App 3 – Information & guidance; App 4 – Other legislation.	http://www.defra.gov.uk/wildlife-countryside/issues/coastal/coastal-guidance.pdf

Table 14.5.4 Northern Ireland Level Water Policy.

Policy Type	Document Title	Objective	Contents	Website
Strategies & Reports	A Groundwater Monitoring Strategy for NI (Sept 2000)	This document outlines the EHS commitment to monitoring the groundwater resources of NI in line with national & international commitments, through a programme of sampling from an identified representative network of boreholes & springs.	Introduction; Legislative framework; Monitoring of groundwaters to date; Monitoring programme 2000; Review; App 1: Location of index boreholes; App 2: Groundwater monitoring – quarterly; App 3: Groundwater monitoring – annually.	http://www.ehsni.gov.uk/pubs/publications/Groundwater_Monitoring_StrategyA_CM.pdf
	Eutrophication in NI – Proposals for a Strategy to Control Nutrient Enrichment (Sept 2000)	This document presents a current assessment of eutrophication in NI summaries the action currently being taken to minimise its impact on these waters & proposes possible future management activities. A key element of these proposals is a partnership approach to the management of eutrophication & the purpose of this document is therefore to seek the views & opinions of all stakeholders whose activities influence eutrophication & those who are affected by its impacts.	Introduction; Impacts of eutrophication; Current information on trophic status of waters in NI; Sources of nutrients & trends in eutrophication; Current management measures; Targets for trophic status & proposed management options; Next steps; Glossary.	http://www.ehsni.gov.uk/pubs/publications/EutroStrat_VC_A3_PDF.pdf
	Sustainable Urban Drainage Systems – An Introduction (Mar 2001)	This booklet highlights problems caused by conventional urban drainage systems and identifies alternative approaches, referred to as Sustainable Urban Drainage Systems (SUDS).	Introduction; Development, drainage & the legal basis for control; The environmental effects of urban drainage; Sustainable Urban Drainage Systems; Permeable conveyance systems; Passive treatment systems; SUDS as amenities; Conclusion; References & Bibliogrpahy.	http://www.ehsni.gov.uk/pubs/publications/SUDS_introduction_1999pages1–11.pdf http://www.ehsni.gov.uk/pubs/publications/SUDS_introduction_1999pages12–17.pdf http://www.ehsni.gov.uk/pubs/publications/SUDS_introduction_1999pages18–23.pdf
	A River Water Quality Monitoring Strategy for NI (May 2001)	This document sets out the history of the development of the monitoring programme for rivers in NI since 1972. It looks at the reasons why EHS carries out the monitoring currently undertaken & outlines the results of a review of the network undertaken & outlines the results of a review of the network undertaken in 1999 & implemented in 2000.	Introduction; Legislative framework; EC legislation; Local legislation; Review of EHS river water quality monitoring network 1972–1999; River water quality monitoring by other agencies in NI; River water quality monitoring from 2000; Appendices.	http://www.ehsni.gov.uk/pubs/publications/River_Monitoring.pdf

Policy & Practice for the Protection of Groundwater in NI (July 2001)	This document outlines the approach Environment and Heritage Service (EHS) will take towards the protection of both groundwater resources & groundwater quality in NI. It represents a major step towards addressing, in a comprehensive & consistent manner, those activities which have the potential to have an impact on this valuable resource.	Summary; Introduction; Key principles; Powers & responsibilities for groundwater protection; Vulnerability & risk; Description of the vulnerability of groundwater sources; Definition of source protection zones; Groundwater protection policy statements; Appendices.	http://www.ehsni.gov.uk/pubs/publications/Policy_and_Practice_for_the_Protection_of_Groundwater_in_Northern_Ireland.pdf
Policy for Setting & Delivering Water Quality Targets (Aug 2001)	This document summarises the development, to date, of water quality targets for NI's waters & outlines the role of the inter-agency Water Quality Management Committee in helping to achieve those targets. It also sets out the current EHS position on setting water quality targets, on measuring achievement of targets & on the development of partnerships to co-ordinate environmental protection & improvement.	Summary; Introduction; Water quality targets setting in NI; Water quality management planning in NI; Target catchment initiative & water quality management committee; EC water framework directive; Water quality monitoring strategy in NI; Rivers; Lakes & groundwaters; Transitional (Estuarine) & coastal waters; Delivery of water quality targets; Appendices.	http://www.ehsni.gov.uk/pubs/publications/Water_Quality_Targets_(Web).pdf
NI Regional Report of the National Monitoring Programme (NMP) (April 2002)	In addition to the UK NMP Report on the Spatial Survey 1993–95, MPMMG (Marine Pollution Monitoring Management Group) decided that regional reports for NI, Scotland, England & Wales would be prepared. This report fulfils that obligation covering the NI Spatial Survey.	Introduction; Methods; Sample plan; Sample design; Biological effects; Quality control; Chemical data; Biological data; Benthos; Biological effects; Bio-accumulation; Water Sediments; Conclusions; Recommendations; References; Appendices.	http://www.ehsni.gov.uk/pubs/publications/NMP_Report.pdf
Report on the Environmental Aspects of the Nitrates Directive in NI (Aug 2002)	The purpose of this Report is to inform the Department of the Environment (DoE) & the Department of Agriculture & Rural Development (DARD) to enable them to fulfil their obligations under the EC Nitrates Directive (91/676/EEC). In particular, the Report summarises the scientific knowledge concerning the extent to which groundwaters & surface waters in NI contain elevated nitrate levels within the meaning of the Directive (i.e., in excess of 50 mg NO_3 / l), & the status of waters in NI with respect to eutrophication.	Executive summary; Overview; Introduction; Compliance of groundwaters & surface waters in NI with the Nitrates Directive; Eutrophication & the Nitrates directive in NI; Agricultural nitrogen & nitrate losses; Nitrate sources in NI; Implications of options for possible NVZ designations in NI; Denitrifications; References.	http://www.ehsni.gov.uk/pubs/publications/DARDDOE23OCT02.pdf

Continued

Table 14.5.4 Northern Ireland Level Water Policy. *Continued*

Policy Type	Document Title	Objective	Contents	Website
Strategies & Reports	Managing the Water Environment in NI – 2000 (Sept 2002)	This is the latest in a series of five-yearly reports on the water environment. Previous reports have focused mainly on river water quality. This report, however, provides information on the quality of all waters during 2000, including rivers, lakes, groundwaters & marine waters. It also summarises all of the key activities of the WMU (Water Management Unit) during 2000, including monitoring of water quality, preparing action plans to improve water quality, controlling effluent discharges & combating, or minimising the effects of pollution. The report describes the status of NI waters & highlights the main challenges that lie ahead in managing this invaluable resource.	Foreword; Summary; Introduction; Water quality policy; River monitoring; Lake monitoring; Groundwater monitoring; Estuarine/coastal water monitoring; Other environmental monitoring; Regulation; Pollution incidents & prosecutions; Appendices; Glossary.	http://www.ehsni.gov.uk/pubs/publications/Water_Report_Web.pdf
	Delineation of Nitrate Vulnerable Zones for Four Locations in Northern Ireland (Nov 2002)	A review of 2000–2001 monitoring data from the EHS regional groundwater monitoring network identified a number of sites where nitrate concentrations were of concern. Further investigation and analysis identified four areas that were considered by EHS to warrant designation as Nitrate Vulnerable Zones as required by the EC Nitrates Directive (91/676/EEC). Geological Survey of Northern Ireland (GSNI) were commissioned by EHS to review available hydrogeological and monitoring data and propose 'soft boundary zones' for the four areas. This report details the background to the work and the methodology used for determining the soft boundaries for each of the four areas.	Summary; 1 – Introduction & scope of report; 2 – Nitrates in Groundwater; 3 – General Hydrogeological Settings; 4 – Methodology; 5 – Area Reports; 6 – Area 1; 7 – Area 2; 8 – Area 3; 9 – Area 4; References.	http://www.ehsni.gov.uk/pubs/publications/DelineationReport4April2003.pdf

NI Drinking Water Quality Report 2001 (Dec 2002)	The Drinking Water Inspectorate is a unit within the Environment & Heritage Service, an agency of the Department of the Environment in NI. The DWI has a responsibility to regulate drinking water quality in NI under the Water Quality Regulations (Northern Ireland) 1994 & the Private Water Supplies Regulations (Northern Ireland) 1994. This report details the work of the DWI & reviews drinking water quality in NI for 2001. This is the sixth annual report prepared by the DWI, &, as with the previous reports, it will act as a criterion against which water quality in NI will be measured in future years.	Summary; Introduction; The regulatory framework; The role & activities of the drinking water inspectorate; the technical audit; Overview of drinking water quality in NI 2001; Drinking water quality incidents & complaints; Private water supplies; Drinking water research; Definitions & glossary.	http://www.ehsni.gov.uk/pubs/publications/Drinking_Water_2001.pdf
Options for Management of Northern Ireland's Coastal Zone: Scoping Study Examining the Potential Establishment of a Northern Ireland Coastal Forum (Jan 2003)	This report examines the potential for the formation of a Coastal Forum for Northern Ireland in the context of recent national, international & European developments in the field of Integrated Coastal Zone Management (ICZM). The report indicates that while several UK initiatives have precipitated changes in coastal management, the European Commission's demonstration programme on Integrated Coastal Zone Management & its resultant recommendation regarding the implementation of ICZM provide the impetus for integration of management activities at the coastline.	Executive summary; Terms of reference; Background; UK coastal fora review; Questionnaire survey & consultation; Recommendations; References; Appendices.	http://www.ehsni.gov.uk/pubs/publications/final_scoping_study_Jan23.pdf
Water Pollution Incident & Prosecution Statistics 2001 (Feb 2003)	This report presents statistics of water pollution incidents throughout Northern Ireland investigated by, or on behalf of, the Environment & Heritage Service (EHS) in 2001. Comparisons are made with data from earlier years to allow the description of historical trends. The data has been analysed by source, severity, pollutant type (category) & incident cause. This report also includes details of prosecution statistics for 1999/2000, brought under the Water Act (Northern Ireland) 1972.	Summary; Introduction; Analysis of incidents; Pollution prevention; Enforcement action; Annexes.	http://www.ehsni.gov.uk/pubs/publications/Water_Pollution_Incidents_2001.pdf

Continued

Table 14.5.4 Northern Ireland Level Water Policy. *Continued*

Policy Type	Document Title	Objective	Contents	Website
Strategies & Reports	Regulation of Water Service Discharges – Report for the Year 2001 (Dec 2003)	This is the first report on the performance of effluent discharges made by Water Service, an agency of the Department for Regional Development. It provides a summary of the compliance of Water Service discharges with Registered Standards set by Environment & Heritage Service, an agency of the Department of the Environment, to control the quality of such discharges. Discharges included are the effluents from 268 Waste Water Treatment Works (WWTW) serving a population equivalent greater than 250 & the effluents from 25 Water Treatment Works. Details of compliance with the requirements of the EU Urban Waste Water Treatment Regulations (Northern Ireland) 1995 are also presented for the 37 WWTWs that currently qualify for inclusion under the conditions of the Regulations.	Summary; Introduction; Regulatory system; Registered standards; Waste water treatment works performance; Compliance with the Urban Waste Water Treatment Regulations in 2001; Sewerage system assessment; Water treatment works performance; Water service regulation audits; Pollution incident investigation; Comparisons; Appendices.	http://www.ehsni.gov.uk/pubs/publications/Reg_WSD.pdf
	Northern Ireland Drinking Water Quality 2002 (Dec 2003)	This report is the seventh prepared by the Drinking Water Inspectorate Unit of Environment & Heritage Service, acting in its regulatory role in matters of drinking water quality. It describes the work of the DWI & provides an overview of drinking water quality in Northern Ireland for 2002.	Summary; Introduction; The Regulatory framework; The role & activities of the drinking water inspectorate; The technical audit; Assessment of water quality; The 2002 inspection programme; Overview of drinking water quality in NI in 2002; Physical/chemical quality in water supply zones; Drinking water quality incidents & complaints; Private water supplies; Drinking water research; Definitions & glossary of terms.	http://www.ehsni.gov.uk/pubs/publications/DWI_Annual_Report_2002.pdf

An Evaluation of Nitrogen Sources & Inputs to Tidal Waters in Northern Ireland (March 2004)	This report was completed for Environment & Heritage Service, draws on chemical monitoring data of rivers, waste water treatment works & industrial discharges in Northern Ireland that were supplied by Environment and Heritage Service. The primary focus of the study was on nitrate loadings, which were partitioned between the following sources: agricultural land, uplands, forest land, other land, waste water treatment works & industrial inputs.	Introduction; Methodology; River loadings; Calculations of point source loadings; Diffuse inputs; Direct drainage areas; Nutrient sources & inputs to tidal waters & sea Loughs Introduction; Catchment land use; Dissolved inorganic nitrogen loads; Nitrate sources; Ammonium sources; Summer & winter loadings; Dissolved reactive phosphorus sources; Magnitude of nutrient loss rates from agriculture; Modelled loss rates of nutrients from agriculture; Background loadings from agriculture; Trophic status estimates of nutrient loadings; References; Appendices.	http://www.ehsni.gov.uk/pubs/publications/Ni_sources_inputs.pdf
Northern Ireland Drinking Water Quality 2003 (Nov 2004)	This report is the eighth prepared by the Drinking Water Inspectorate Unit of Environment and Heritage Service, acting in its regulatory role in matters of drinking water quality. It describes the work of the DWI and provides an overview of drinking water quality in Northern Ireland for 2003. New Regulations came into operation in December 2003 and future reports will be made against the new regulatory framework.	Summary; Introduction; The Regulatory Framework; The Role & Activities of the Drinking Water Inspectorate; The Technical Audit; Overview of Drinking Water Quality in Northern Ireland in 2003; New Regulations & Associated Programmes of Work Drinking Water Quality Incidents & Complaints; Private Water Supplies; Drinking Water Research; Definitions & Glossary of Terms.	http://www.ehsni.gov.uk/pubs/publications/Drinking_Water_Quality_Report_2003.pdf
Regulation of Water Service Discharges 2002–2003 (Dec 2004)	Environment and Heritage Service published the first public Report on the Regulation of Water Service Discharges for the year 2001, in December 2003. This document updates that Report for 2002 and 2003. It demonstrates changes in compliance over that period and provides updated information on capital works schemes that are in progress or have been completed.	Executive Summary; Section 1 – Waste Water Treatment Works; Section 2 – Compliance with the Urban Waste Water Treatment Directive in 2002 and 2003; Section 3 – Collection Systems; Section 4 – Water Treatment Works; Section 5 – Water Service Audits; Section 6 – Pollution Incident Investigation; Section 7 – Current & Future Performance; Appendices.	http://www.ehsni.gov.uk/pubs/publications/WaterServiceReport0203.pdf

Continued

Table 14.5.4 Northern Ireland Level Water Policy. *Continued*

Policy Type	Document Title	Objective	Contents	Website
Consultation Papers	Northern Ireland River Conservation Strategy – A Draft for Consultation (Aug 2000)	In the strategy, the rich natural & built heritage associated with rivers in NI is outlined & the impacts of greatest significance are discussed. Seven main objectives towards conserving NI's rivers are set out & for each one the actions that EHS will take to achieve them are described.	Introduction; Rivers in NI – their conservation value & interest; Factors impacting on the conservation value of rivers in NI; The role & responsibilities of EHS in river conservation; Objectives of the river conservation strategy; Delivering the river conservation strategy; Action plan; Appendix: glossary & abbreviations.	http://www.ehsni.gov.uk/pubs/publications/River_Conservation_Strategy_JL(2).pdf
	A Consultation Paper on the Water Quality (Water Supply) Regulations (Northern Ireland) 2002 (March 2002)	This document sets out how the Department proposes to implement the Directive on Water Intended for Human Consumption (1998) in respect of public water supplies.	Introduction; Background; The new directive; Content of the draft regulations; Regulatory impact assessment; Issues for consultation; Equality/human rights; Annex 1: The draft regulations; Annex 2: Regulatory Impact assessment; Annex 3: Screening for Equality Impact Assessment.	http://www.ehsni.gov.uk/pubs/publications/Water_Supply_Regulations_2002.pdf
	Managing our Shared Waters – A Joint North/South Consultation Paper on International River Basin Districts & Administrative Arrangements for Implementation of the EC Water Framework Directive (2000/60/EC) (Mar 2003)	The publication of this Joint Consultation Paper marks an important step in translating the Water Framework Directive into practical action on the island of Ireland. It identifies the three international river basin districts proposed by our Departments, & invites views on how best to protect and conserve these shared river catchments. The Paper is being sent to a wide range of interested parties including agricultural, commercial, industrial, recreational & community interests, environmental organisations & individual citizens.	Part 1: Introduction; Background; Further information; Part 2: Requirements of the Directive; Water framework Directive; Co-ordination & common approaches; River basins: internal; River basins: cross-border; Subsidiarity; Part 3: Implementation proposals; North/South co-ordination; Co-ordination group of officials; International river basin districts (IRBDs); Participation by interested parties; Part 4: Questions; Annexes.	http://www.ehsni.gov.uk/pubs/publications/Waters_INSIDE.pdf

	A Consultation Paper on Nitrates & the Protection of Groundwaters in Northern Ireland (Mar 2003)	This Consultation Paper seeks views on the proposals by the Department of Environment (DOE) & the Department of Agriculture & Rural Development (DARD) to extend the designation of Nitrate Vulnerable Zones (NVZs) in Northern Ireland. It is the first of a series of consultations on the subject of the implementation of the Nitrates Directive. The further consultations will consider the issue of eutrophication & how best to address it, either by the establishment of additional NVZs or Action Programmes for the whole of Northern Ireland.	Summary; Requirements of the EC Nitrates Directives 91/676/EEC; Impact of Nitrate loss from agriculture; Historic implementation of Nitrates Directive in NI; Recent progress in the rest of UK & EU; Proposals to bring NI into compliance with the Nitrates Directive; Impact on farmers within Nitrate Vulnerable Zones (NVZs); Regulatory Impact Assessment; Equality Impact Assessment; Human Rights Act 1998; What we would like to know; Responding to the proposals; Annexes.	http://www.ehsni.gov.uk/pubs/publications/ConsultationPaper4April2003.pdf
	Nitrates Directive 2nd Consultation Paper – Proposal for the Protection of Northern Ireland's Surface & Groundwater (April 2004)	This consultation paper invites views on the proposal by DOE & DARD to declare 'total territory' under the Nitrates Directive. In effect, this will mean the establishment of mandatory measures, which will apply to all farmers, to control nutrient losses to water from agriculture.	Summary; Introduction; The need to reduce water pollution; The legal context; Benefits of total territory; Proposed action programme measures; Farm waste management scheme; Mandatory consultation requirements; Annexes.	http://www.ehsni.gov.uk/pubs/publications/nitrates_condoc.pdf
Guidance Notes	Pollution Prevention Guideline (PPG1): General Advice (April 2001)	These guidelines are an introduction to both pollution prevention & the guidance notes on this subject produced by the Environment Agency for England & Wales, the Scottish Environment Protection Agency & the Environment & Heritage Service in Northern Ireland.	Introduction; Availability; Legal framework; Site drainage; Waste storage & disposal; Deliveries & security; Oil storage & pipelines; Contingency plans; Construction & demolition; Agriculture; Groundwater pollution; References.	http://www.ehsni.gov.uk/pubs/publications/PPG01.pdf
	PPG5: Works in, Near or Liable to Affect Watercourses (April 2001)	These guidelines have been drawn up to assist all those who may have cause to work in or near watercourses. Compliance with this guidance should minimise the risk of pollution occurring.	Legal framework; Introduction; General precautions; Silt; Concrete & cement; Oil & chemicals; Bridge cleaning & repainting; Herbicide use; Emergencies; References.	http://www.ehsni.gov.uk/pubs/publications/PPG05.pdf
	PPG6: Working at Demolition & Construction Sites (April 2001)	These guidelines are intended to assist those in the construction & demolition industry with responsibility for managing the environmental impact of their activities. Compliance with these should minimise the effect of the work on the environment.	Legal framework; Introduction; Planning & preparation; Site drainage; Deliveries; Storage; Waste management; Silt; Refuelling; Concrete; Emergencies; References.	http://www.ehsni.gov.uk/pubs/publications/PPG06.pdf

Continued

Table 14.5.4 Northern Ireland Level Water Policy. *Continued*

Policy Type	Document Title	Objective	Contents	Website
Guidance Notes	PPG11: Industrial Sites (April 2001)	These guidelines are intended to provide some information to industrial users on the potential impact of their practices on the surrounding environment.	Legal framework; Introduction; Site drainage; Sewage & waste water disposal; Surface water drainage; Delivery & handling of materials; Storage facilities; Waste management; Routine inspection & maintenance; Cleaning & degreasing; Demolition & construction; Staff training; Emergency plans; Security; References.	http://www.ehsni.gov.uk/pubs/publications/PPG11.pdf
	PPG15: Retail Premises (April 2001)	This document has been drawn up to assist those who design & operate major retail outlets & similar sites, such as distribution warehouses. Compliance with these guidelines should minimise the risk of pollution occurring.	Introduction; Site drainage; Sewage & waste water disposal; Surface water drainage; Waste management; Oil storage; References.	http://www.ehsni.gov.uk/pubs/publications/PPG15.pdf
	Guidance for the Monitoring of *Cryptosporidium* in Treated Water Supplies in Northern Ireland (Feb 2003)	This document provides guidance from the Drinking Water Inspectorate in Northern Ireland, on behalf of the Department of the Environment, on the water treatment requirements associated with the Water Supply (Water Quality) Regulations (Northern Ireland) 2002.	Introduction; General regulatory background; *Cryptosporidium* risk assessment; Monitoring of *cryptosporidium* oocysts in treated water supplies – sampling, transportation & analysis; Reporting procedures; Records & information; References; Glossary; Appendices.	http://www.ehsni.gov.uk/pubs/publications/Crypto_Guidance_feb03.pdf

Index